D1685733

Preharvest and Postharvest Food Safety

Contemporary Issues and Future Directions

The *IFT Press* series reflects the mission of the Institute of Food Technologists—advancing the science and technology of food through the exchange of knowledge. Developed in partnership with Blackwell Publishing, *IFT Press* books serve as essential textbooks for academic programs and as leading edge handbooks for industrial application and reference. Crafted through rigorous peer review and meticulous research, *IFT Press* publications represent the latest, most significant resources available to food scientists and related agriculture professionals worldwide.

- *Microbiology and Technology of Fermented Foods* (Robert W. Hutkins)
- *Regulation of Functional Foods and Nutraceuticals: A Global Perspective* (Clare M. Hasler)
- *Water Activity in Foods: Fundamentals and Applications* (Gustavo V. Barbosa-Canovas, Anthony J. Fontana Jr., Shelly J. Schmidt, and Theodore P. Labuza)
- *Sensory and Consumer Research in Food Product Development* (Howard R. Moskowitz, Jacqueline H. Beckley, and Edgar Chambers IV)

Preharvest and Postharvest Food Safety

Contemporary Issues and Future Directions

Ross C. Beier, Suresh D. Pillai, Timothy D. Phillips EDITORS
Richard L. Ziprin ASSOCIATE EDITOR

Foreword by the Honorable Elsa A. Murano
Afterword by Neville P. Clarke

Blackwell Publishing Professional
2121 State Avenue, Ames, Iowa 50014, USA

Orders: 1-800-862-6657
Office: 1-515-292-0140
Fax: 1-515-292-3348
Web site: www.blackwellprofessional.com

Blackwell Publishing Ltd
9600 Garsington Road, Oxford OX4 2DQ, UK
Tel.: +44 (0)1865 776868

Blackwell Publishing Asia
550 Swanston Street, Carlton, Victoria 3053, Australia
Tel.: +61 (0)3 8359 1011

Authorization to photocopy items for internal or personal use, or the internal or personal use of specific clients, is granted by Blackwell Publishing, provided that the base fee of $.10 per copy is paid directly to the Copyright Clearance Center, 222 Rosewood Drive, Danvers, MA 01923. For those organizations that have been granted a photocopy license by CCC, a separate system of payments has been arranged. The fee code for users of the Transactional Reporting Service is 0-8138-0884-7/2004 $.10.

Printed on acid-free paper in the United States of America

First edition, 2004

The swine picture on the front cover was provided courtesy of the National Pork Board, 1776 N.W. 114th Street, Clive, IA.

The microarray picture on the front cover was provided courtesy of Robert S. Chapkin Intercollegiate Faculty of Nutrition, Texas A&M University, College Station, TX.

The electron micrograph of the bacteria shown on the cover was provided courtesy of Robert E. Droleskey, USDA, ARS, Southern Plains Agricultural Research Center, College Station, TX 77845.

Library of Congress Cataloging-in-Publication Data

Preharvest and postharvest food safety: contemporary issues and future directions / Ross C. Beier ... [et al.] editors; foreword written by Elsa A. Murano. — 1st ed.
    p. cm.
Includes bibliographical references and index.
    ISBN 0-8138-0884-7 (alk. paper)

1. Food—Microbiology. 2. Food—Safety measures. 3. Agricultural microbiology.
I. Beier, Ross C., 1946-   QR115.P735 2004
    664   dc22

2003023172

The last digit is the print number: 9 8 7 6 5 4 3 2 1

The concept for this book originated in a meeting of the Center for Food Safety, Institute of Food Science and Engineering (IFSE), Texas A&M University, College Station, Texas. We began this book by looking at many of the U.S. government's high priorities in food safety and matching these priorities to Center for Food Safety scientists engaged in those areas of research. Later, other government and academic scientists were added to broaden the scope of the book. This book is the result of collaboration of researchers engaged in the area of food safety and sponsored by IFSE.

Center for Food Safety

Institute of Food Science and Engineering
Texas A&M University
1500 Research Parkway, Suite A220
College Station, TX 77845

Contents

Contributors

Adams, L. Garry
Department of Veterinary Pathobiology
College of Veterinary Medicine
Texas A&M University
College Station, TX 77843-4467
Tel: 979-845-9816
Fax: 979-862-1088
e-mail: [gadams@cvm.tamu.edu].

Anderson, Robin C.
Food and Feed Safety Research Unit
U.S. Department of Agriculture
Agricultural Research Service,
Southern Plains Agricultural Research
  Center
2881 F&B Road
College Station, TX 77845-4988
Tel: 979-260-9317
Fax: 979-260-9332
e-mail: [anderson@ffsru.tamu.edu].

Balog, Janice M.
U.S. Department of Agriculture
Agricultural Research Service
Poultry Science Center
University of Arkansas
Fayetteville, AR 72701-1201
Tel: 479-575-6299
Fax: 479-575-4202
e-mail: [jbalog@uark.edu].

Bäumler, Andreas J.
Department of Medical Microbiology &
  Immunology
Texas A&M University
College Station, TX 77843-1114
Tel: 979-862-7756
Fax: 979-845-3479
e-mail: [abaumler@medicine.tamu.edu].

Beier, Ross C.
Food and Feed Safety Research Unit
U.S. Department of Agriculture
Agricultural Research Service
Southern Plains Agricultural Research
  Center
2881 F&B Road
College Station, TX 77845-4988
Tel: 979-260-9411
Fax: 979-260-9332
e-mail: [rcbeier@ffsru.tamu.edu].

Bischoff, Kenneth M.
Fermentation Biotechnology Research Unit
U.S. Department of Agriculture
Agricultural Research Service
National Center for Agricultural
  Utilization Research
1815 N. University St.
Peoria, IL 61604-3999
Tel: 309-681-6067
Fax: 309-681-6427
e-mail: [bischoffk@ncaur.usda.gov].

Brandl, Maria T.
Produce Safety and Microbiology
    Research Unit
U.S. Department of Agriculture
Agricultural Research Service
Western Regional Research Center
800 Buchanan Street, W308
Albany, CA 94710-1105
Tel: 510-559-5885
Fax: 510-559-6162
e-mail: [mbrandl@pw.usda.gov].

Brashears, Mindy
Animal and Food Science
Texas Tech University
Lubbock, TX 79409-2141
Tel: 806-742-2469
Fax: 806-742-2427
e-mail: [mindy.brashears@ttu.edu].

Byrd II, J. Allen
Food and Feed Safety Research Unit
U.S. Department of Agriculture
Agricultural Research Service
Southern Plains Agricultural Research
    Center
2881 F&B Road
College Station, TX 77845-4988
Tel: 979-260-9331
Fax: 979-260-9332
e-mail: [byrd@ffsru.tamu.edu].

Callaway, Todd R.
Food and Feed Safety Research Unit
U.S. Department of Agriculture
Agricultural Research Service
Southern Plains Agricultural Research
    Center
2881 F&B Road
College Station, TX 77845-4988
Tel: 979-260-9374
Fax: 979-260-9332
e-mail: [callaway@ffsru.tamu.edu].

Castell-Perez, M. Elena
Department of Biological and Agricultural
    Engineering
Texas A&M University
College Station, TX 77843-2117
Tel: 979-862-7645
Fax: 979-847-8828
e-mail: [ecastell@tamu.edu].

Castillo, Alejandro
Animal Science Department
Faculty of Food Science
Texas A&M University
College Station, TX 77843-2471
Tel: 979-845-3565
Fax: 979-862-3475
e-mail: [a-castillo@tamu.edu].

Clarke, Neville P.
Director
Institute for Countermeasures against
    Agricultural Bioterrorism
Texas A&M University System
College Station, TX 77843-2129
Tel: 979-845-2855
Fax: 979-845-6574
e-mail: [n-clarke@tamu.edu].

Donoghue, Annie M.
U.S. Department of Agriculture
Agricultural Research Service
    Poultry Science Center
University of Arkansas
Fayetteville, AR 72701-1201
Tel: 479-575-2413
Fax: 479-575-4202
e-mail: [donoghue@uark.edu].

Dowd, Scot E.
Livestock Issues Research Unit
U.S. Department of Agriculture
Agricultural Research Service
Route 3
Box 215
FM Road 1294
Lubbock, TX 79403-9750
Tel: 806-746-5356, Ext. 241
Fax: 806-744-4402
e-mail: [sdowd@lbk.ars.usda.gov].

Eddy, Sarah M.
Department of Animal Science
Texas A&M University
College Station, TX 77843-2471
Tel: 979-845-3957
Fax: 979-845-9454
e-mail: [seddy@neo.tamu.edu].

Edrington, Thomas S.
Food and Feed Safety Research Unit
U.S. Department of Agriculture
Agricultural Research Service
Southern Plains Agricultural Research
    Center
2881 F&B Road
College Station, TX 77845-4988
Tel: 979-260-3757
Fax: 979-260-9332
e-mail: [edrington@ffsru.tamu.edu].

Ficht, Thomas A.
Department of Veterinary Pathobiology
    R&G Base
Texas A&M University
College Station, TX 77843-4467
Tel: 979-845-4118
Fax: 979-862-1088
e-mail: [tficht@cvm.tamu.edu].

Foley, Steven L.
Department of Biology
University of Central Arkansas
180 Lewis Science Center
Conway, AR 72035-0001
Tel: 501-450-5826
Fax: 501-450-5914
e-mail: [sfoley@mail.uca.edu].

Genovese, Kenneth J.
Food and Feed Safety Research Unit
U.S. Department of Agriculture
Agricultural Research Service
Southern Plains Agricultural Research
    Center
2881 F&B Road
College Station, TX 77845-4988
Tel: 979-260-3756
Fax: 979-260-9332
e-mail: [genovese@ffsru.tamu.edu].

Gerba, Charles P.
Department of Soil Water and
    Environmental Science
University of Arizona
Tucson, AR 85721-0038
Tel: 520-621-6906
Fax: 520-621-6163
e-mail: [gerba@ag.arizona.edu].

Goyal, Sagar M.
Department of Veterinary Diagnostic
    Medicine
College of Veterinary Medicine
1333 Gortner Ave.
University of Minnesota
St. Paul, MN 55108-1098
Tel: 612-625-2714
Fax: 612-624-8707
e-mail: [goyal001@tc.umn.edu].

Harris, Kerri B.
International HACCP Alliance
College Station, TX 77843-2471
Tel: 979-862-3643
Fax: 979-862-3075
e-mail: [kharris@tamu.edu].

Harvey, Roger B.
Food and Feed Safety Research Unit
U.S. Department of Agriculture
Agricultural Research Service
Southern Plains Agricultural Research
    Center
2881 F&B Road
College Station, TX 77845-4988
Tel: 979-260-9259
Fax: 979-260-9332
e-mail: [harvey@ffsru.tamu.edu].

Herrera, Paul
Center for Food Safety and Intercollegiate
    Faculty of Toxicology
Veterinary Anatomy & Public Health
College of Veterinary Medicine
Texas A&M University
College Station, TX 77843-4458
Tel: 979-862-4976
Fax: 979-862-4929
e-mail: [pherrera@cvm.tamu.edu].

Huebner, Henry J.
Center for Food Safety and Intercollegiate
    Faculty of Toxicology
Veterinary Anatomy & Public Health
College of Veterinary Medicine
Texas A&M University
College Station, TX 77843-4458
Tel: 979-862-4976
Fax: 979-862-4929
e-mail: [hhuebner@cvm.tamu.edu].

Huff, Gerry R.
U.S. Department of Agriculture
Agricultural Research Service
Poultry Science Center
University of Arkansas
Fayetteville, AR 72701-1201
Tel: 479-575-7966
Fax: 479-575-4202
e-mail: [grhuff@uark.edu].

Huff, William E.
U.S. Department of Agriculture
Agricultural Research Service
Poultry Science Center
University of Arkansas
Fayetteville, AR 72701-1201
Tel: 479-575-2104
Fax: 479-575-4202
e-mail: [huff@uark.edu].

Joseph, Sam W.
Department of Cell Biology and Molecular
    Genetics
University of Maryland
Microbiology Bldg. 231
College Park, MD 20742-4451
Tel: 301-405-5452
Fax: 301-314-9489
e-mail: [sj13@umail.umd.edu].

Keeton, Jimmy T.
Department of Animal Science
Texas A&M University
College Station, TX 77843-2471
Tel: 979-845-3975
Fax: 979-845-9454
e-mail: [jkeeton@tamu.edu].

Khare, Sangeeta
Veterinary Pathobiology
Texas A&M University
College Station, TX 77843-4467
Tel: 979-845-9814
Fax: 979-862-1088
e-mail: [skhare@cvm.tamu.edu].

Kubena, Leon F.
Food and Feed Safety Research Unit
U.S. Department of Agriculture
Agricultural Research Service
Southern Plains Agricultural Research
    Center
2881 F&B Road
College Station, TX 77845-4988
Tel: 979-260-9249
Fax: 979-260-9332
e-mail: [kubena@ffsru.tamu.edu].

Kwon, Young Min
Center of Excellence for Poultry Science
University of Arkansas
Fayetteville, AR 72701-3100
Tel: 479-575-4935
Fax: 479-575-8775
e-mail: [ykwon@uark.edu].

Mandrell, Robert E.
Food Safety and Health Research
U.S. Department of Agriculture
Agricultural Research Service
Western Regional Research Center
Albany, CA 94710-1105
Tel: 510-559-5829
Fax: 510-559-6162
e-mail: [mandrell@pw.usda.gov].

Marshall, Douglas L.
Department of Food Science &
    Technology
Mississippi Agricultural and Forestry
    Experiment Station
Mississippi State University
P.O. Box 9805
Mississippi State, MS 39762-9805
Tel: 662-325-8722
Fax: 662-325-8728
e-mail: [microman@ra.msstate.edu].

McDermott, Patrick F.
Division of Animal and Food Microbiology
U.S. Food and Drug Administration
Center for Veterinary Medicine
8401 Muirkirk Rd.
Laurel, MD 20708-2482
Tel: 301-827-8024
Fax: 301-827-8127
e-mail: [pmcdermo@cvm.fda.gov].

McIntosh, Wm. Alex
Department of Rural Sociology
Texas A&M University
College Station, TX 77843-2125
Tel: 979-845-5332
Fax: 979-845-8525
e-mail: [w-mcintosh@tamu.edu].

Mena, Kristina D.
University of Texas Health Science Center
    at Houston
School of Public Health
El Paso Campus
El Paso, TX 79902-4129
Tel: 915-747-8514
Fax: 915-747-8512
e-mail: [kristina.d.mena@uth.tmc.edu].

Meng, Jianghong
Department of Nutrition and Food Science
University of Maryland
College Park, MD 20742-0001
Tel: 301-405-1399
Fax: 301-314-3313
e-mail: [jm332@umail.umd.edu].

Moreira, Rosana G.
Department of Biological and Agricultural
    Engineering
Texas A&M University
College Station, TX 77843-2117
Tel: 979-847-8794
Fax: 979-845-3932
e-mail: [rmoreira@tamu.edu].

Murano, Elsa A.
Undersecretary for Food Safety at USDA
1400 Independence Ave., S.W.
Washington, DC 20250-3700
Tel: 202-720-0350
Fax: 202-690-0820
e-mail: [elsa.murano@usda.gov].

Nisbet, David J.
Food and Feed Safety Research Unit
U.S. Department of Agriculture
Agricultural Research Service
Southern Plains Agricultural Research
    Center
2881 F&B Road
College Station, TX 77845-4988
Tel: 979-260-9484
Fax: 979-260-9332
e-mail: [nisbet@ffsru.tamu.edu].

O'Shea, Brian
Veterinary Pathobiology
Texas A&M University
College Station, TX 77843-4467
Tel: 979-845-4185
Fax: 979-862-1088
e-mail: [boshea@cvm.tamu.edu].

Phillips, Timothy D.
Center for Food Safety and Intercollegiate
    Faculty of Toxicology
Veterinary Anatomy & Public Health
College of Veterinary Medicine
Texas A&M University
College Station, TX 77843-4458
Tel: 979-845-6414
Fax: 979-862-4929
e-mail: [tphillips@cvm.tamu.edu].

Pillai, Suresh D.
Department of Poultry Science
418 Kleberg Center
Texas A&M University
College Station, TX 77843-2472
Tel: 979-845-2994
Fax: 979-845-1921
e-mail: [spillai@poultry.tamu.edu].

Poole, Toni L.
Food and Feed Safety Research Unit
U.S. Department of Agriculture
Agricultural Research Service
Southern Plains Agricultural Research
    Center
2881 F&B Road
College Station, TX 77845-4988
Tel: 979-845-2855
Fax: 979-260-9332
e-mail: [poole@ffsru.tamu.edu].

Rath, Narayan C.
U.S. Department of Agriculture
Agricultural Research Service
Poultry Science Center
University of Arkansas
Fayetteville, AK 72701-1201
Tel: 479-575-6189
Fax: 479-575-4202
e-mail: [nrath@uark.edu].

Rice-Ficht, Allison C.
Medical Biochemistry and Genetics
Texas A&M Health Science Center
College Station, TX 77843-1114
Tel: 979-845-2728
Fax: 979-847-9481
e-mail: [a-ficht@tamu.edu].

Ricke, Steven C.
Poultry Science Department
101 Kleberg Center
Texas A&M University
College Station, TX 77843-2472
Tel: 979-862-1528
Fax: 979-845-2377
e-mail: [sricke@poultry.tamu.edu].

Rodríguez-García, M. Ofelia
Department of Pharmacy and Biology
University of Guadalajara
Guadalajara, Jal. 44430, Mexico
Tel: +52(333)650-0374
Fax: +52(333)619-4028
e-mail: [rodriguezofelia@hotmail.com].

Rose, Joan B.
Homer Nowlin Chair in Water Research
Department of Fisheries and Wildlife
13 Natural Resources Building
Michigan State University
East Lansing, MI 48824-1222
Tel: 517-432-4412
Fax: 517-432-1699
e-mail: [rosejo@msu.edu].

Schmerr, Mary Jo
Ames Laboratory
Iowa State University
Ames, IA 50011-3110
Tel: 515-294-0949
Fax: 515-294-0266
e-mail: [mschmerr@ameslab.gov].

Scott, H. Morgan
Epidemiology, College of Veterinary
    Anatomy and Public Health
College of Veterinary Medicine
Texas A&M University
College Station, TX 77843-4458
Tel: 979-458-3501
Fax: 979-847-8981
e-mail: [hmscott@cvm.tamu.edu].

Simjee, Shabbir
Division of Animal and Food Microbiology
U.S. Food and Drug Administration
Center for Veterinary Medicine
8401 Muirkirk Rd.
Laurel, MD 20708-2482
Tel: 301-827-8024
Fax: 301-827-8250
e-mail: [simjesas@excite.com].

Tesh, Vernon L.
Department of Microbiology and
    Immunology
Texas A&M University
College Station, TX 77843-1114
Tel: 979-862-4113
Fax: 979-845-3479
e-mail: [tesh@medicine.tamu.edu].

Thurston-Enriquez, Jeanette A.
Soil, Water, and Conservation Research
    Unit
U.S. Department of Agriculture
Agricultural Research Service
Lincoln, NE 68583-0934
Tel: 402-472-8935
Fax: 402-472-0516
e-mail: [jthurston2@unl.edu].

Todd, Ewen C. D.
National Food Safety & Toxicology Center
Michigan State University
East Lansing, MI 48824-1314
Tel: 517-432-3100
Fax: 517-432-2310
e-mail: [toddewen@cvm.msu.edu].

Waghela, Suryakant D.
Department of Veterinary Pathobiology
College of Veterinary Medicine
Texas A&M University
College Station, TX 77843-4467
Tel: 979-845-5176
Fax: 979-862-1147
e-mail: [swaghela@cvm.tamu.edu].

Walker, Robert D.
Division of Animal and Food Microbiology
U.S. Food and Drug Administration
Center for Veterinary Medicine
8401 Muirkirk Rd.
Laurel, MD 20708-2482
Tel: 301-827-8019
Fax: 301-827-8250
e-mail: [rwalker@cvm.fda.gov].

White, David G.
Division of Animal and Food Microbiology
U.S. Food and Drug Administration
Center for Veterinary Medicine
8401 Muirkirk Rd.
Laurel, MD 20708-2482
Tel: 301-827-8037
Fax: 301-827-8127
e-mail: [dwhite@cvm.fda.gov].

Woodward, Casendra L.
Poultry Science Department
101 Kleberg Center
Texas A&M University
College Station, TX 77843-2472
Tel: 979-845-4818
Fax: 979-845-2377
e-mail: [woodward@poultry.tamu.edu].

Zhao, Shaohua
Division of Animal and Food
    Microbiology
U.S. Food and Drug Administration
Center for Veterinary Medicine
8401 Muirkirk Rd.
Laurel, MD 20708-2482
Tel: 301-827-8139
Fax: 301-827-8250
e-mail: [szhao@cvm.fda.gov].

Ziprin, Richard L.
Food and Feed Safety Research Unit
U.S. Department of Agriculture
Agricultural Research Service
Southern Plains Agricultural Research
    Center
2881 F&B Road
College Station, TX 77845-4988
Tel: 979-260-9302
Fax: 979-260-9332
e-mail: [ziprin@ffsru.tamu.edu].

Foreword

The Honorable Elsa A. Murano[1]

Over the last decade, food safety has emerged as a tremendously significant issue, not only for consumers, regulatory agencies, and industry, but also for the research community. It is the latter on which the others depend to identify and understand foodborne hazards and to develop strategies to mitigate the risks associated with them. As such, there are many questions that have already been answered through the efforts of leading researchers in food safety. For one, the last decade has seen great progress made in determining the prevalence of certain microbial pathogens on foods. These data have been used by regulatory agencies such as the Food Safety and Inspection Service to establish criteria to determine whether food safety programs used in meat and poultry processing plants are sound and effective. Second, the world of diagnostic microbiology has also benefited from the focus paid to research in this area over the last few years. This is evident in the fact that health agencies such as the Centers for Disease Control and Prevention have completely moved toward the use of advanced methods such as pulsed field gel electrophoresis in identifying and typing isolates during epidemiological investigations. Finally, we have seen an explosion in the variety and sophistication of decontamination methods used by the food industry to reduce, and even eliminate pathogenic bacteria. From organic acid rinses on beef carcasses, to vacuum-steam-vacuum application on ready-to-eat deli products, to ionizing radiation, significant effort and resources have been invested in research on intervention strategies to improve the safety of our food supply.

In spite of these accomplishments, a simple perusal through the scientific literature suggests that there are many questions yet to be answered. For example, although knowing the prevalence of hazards in food is important, we have come to realize that elucidating how various hazards interact with their environment, and with each other, can provide us with even more essential information on how these factors can be manipulated to minimize food contamination. Similarly, we know that even the sophisticated diagnostic methods of today have limitations, lacking the ability to provide real-time results. In addition, the lack of nondestructive sampling protocols, requiring that sample collection be based on statistical probability of finding contaminants, limits its usefulness for ensuring the safety of individual production lots. Last, there remains a need for better intervention strategies capable of achieving a greater degree of hazard reduction without adversely affecting product quality. Such strategies are available to some extent at present, but they certainly are not applicable to all products, nor are they available to be used at all stages of the farm-to-table continuum.

The authors of this book, all experts in their respective fields, have attempted to provide a thoughtful treatise on the specific research questions that remain to be answered in food safety. They have indeed succeeded in their task, having resisted the temptation to list all possible areas of study, focusing instead on those key questions that are fundamental to advancing our knowledge and, therefore, that are the most important to answer. Chapters dealing with specific pathogens are very useful, providing sound reviews of the state of the knowledge regarding their prevalence and survival in the environment. Research needs related to virulence of *Escherichia coli* O157:H7, for example, are discussed in various chapters, including the need to characterize colonization and adherence factors using protein profiling and other molecular dissection methods. The need to determine the nature and origin of antimicrobial resistance is also included, with suggestions for examining the factors that inhibit dissemination of plasmids, and that bring about point mutations that confer resistance. In addition, the topic of resistance to biocides by pathogens, and how this may engender cross-resistance to antibiotics, is also presented as a fundamental question in need of further exploration.

The book is indeed thorough, also covering questions on topics that are of practical importance, and through which great immediate benefits could be derived. For example, there is a need to further study the effect of diet on carriage rate of human pathogens by food animals. Similarly, examining how rearing practices, feed withdrawal, and transport stress can be successfully manipulated by producers to effect a reduction in the presence of salmonellae and other pathogens at the farm and feedlot, is presented as a topic deserving of significant research focus. Development of better detection and identification methods for pathogens such as *Campylobacter* is also identified as a topic worthy of attention, one that can undoubtedly improve our knowledge of the distribution of this and other pathogens in the environment, in foods, and in clinical samples. Excellent discussions on novel decontamination systems are also included, with the development and standardization of processing parameters with respect to the kinetics of microbial inactivation of various pathogens and in various food matrices being identified as necessary for progress to continue in this area.

There are also chapters that tackle the all-important topics of risk management and risk communication. Food safety objectives and acceptable level of protection are discussed, offering suggestions on how these can be used by regulatory agencies as public health tools. However, the authors point out that before these can be applied, research is needed to determine the maximum correlation of microbial hazards that could be considered as acceptable in achieving consumer protection. A presentation of the Health Belief Model is also included, with suggestions on how research that provides information on the susceptibility, severity, and cost versus benefit of behavioral changes could help predict how consumers may respond to food safety messages.

The authors of the book make the point that these, and many other important questions, need to be addressed if we are to protect the public from foodborne hazards. Obviously, this is no easy task. It will require our keenest scientific minds and most dedicated efforts. As British essayist Samuel Johnson once said, "To expect that the intricacies of science will be pierced by a careless glance, or the eminences of fame ascended without labor, is to expect a peculiar privilege, a power denied to the rest of mankind; but to suppose that the maze is inscrutable to diligence, or the heights inaccessible to perseverance, is to submit tamely to the tyranny of fancy, and enchain the mind in voluntary shackles."

This book is indeed a call to arms for the researcher in the field of food safety. It provides a type of roadmap that can help us engage in the most rewarding work one could endeavor, that of advancing knowledge that will ultimately enhance the health of our citizenry. As a scientist, I can attest to the fact that a more rewarding challenge would be difficult to find. I commend each and every author for their labor in compiling the necessary information in reviews of their individual topics and for the many hours of critical thinking that undoubtedly went into their excellent analysis of research questions yet to be answered. We all look forward to the great scientific work that will emerge in the future as a result of their efforts.

Notes

[1]Dr. Elsa A. Murano currently serves as Undersecretary for Food Safety at the United States Department of Agriculture, Washington, D.C.

Preface

The purpose of this book is to present material on preharvest and postharvest food safety issues, and it covers a wide range of food safety–related research. Food safety is so vast a subject area that one certainly could compile an entire library on this subject alone. The uniqueness of this book stems from an effort made by each author to not only review their area of research but also to provide insights into important unanswered questions, newly advanced theories, and future directions in their area of research.

The concept of the book originated from a brainstorming session that was chaired by Dr. Elsa Murano when she was Director of the Center for Food Safety, a unit of the Institute for Food Science and Engineering, at Texas A&M University. We were searching for ways to bring together Texas A&M University faculty members and Agricultural Research Service/U.S. Department of Agriculture research scientists located in College Station, Texas, who were members of the Institute for Food Science and Engineering. An idea was put forth that one way to catalyze interactions might be by writing short chapters that could later be expanded into a marketable book. Each participant was asked not only to indicate the nature of their personal research but to be creative and forward looking and to attempt to discuss the future research needs and directions in each of their areas.

At the outset, we felt that this exercise would bring some cohesion and unity to the wide array of scientists in College Station and also result in a book with some unique features. Most food safety books are written as textbooks or subject matter reviews. We aimed for something different, a book that points to the future; one that not only reviews the known but that examines the unknown in each of the areas of research.

The six sections in this book were selected on the basis of research priorities identified by an American Academy of Microbiology report (Doores 1999), and the specific areas are the following: (1) Pathogen/Host Interactions; (2) Ecology, Distribution, and Spread of Foodborne Hazards; (3) Antimicrobial Resistance; (4) Verification Tests; (5) Decontamination and Prevention Strategies; and (6) Risk Analysis.

We are positive that you will find that each of the chapters is well conceived and directly relates to the scope of the book. We would also like to point out that some chapters are very innovative for a book of this type. For instance, Chapter 7 discusses the potential for the organism that may be responsible for chronic inflammatory bowel disease of ruminants (Johne's) to be transferred through the food supply to humans, potentially causing inflammatory bowel disease in humans (i.e., Crohn's disease). We felt that though this is a highly controversial issue, it is a contemporary one that was worthy of inclusion. Chapter 14 is also quite contemporary, for it discusses transmissible spongiform encephalopathies, or prion diseases, including BSE, and the risks to animal and human health from them. Chapter 20 nicely spans the breadth of the whole book with a superb discussion of "The Hazard Analysis and Critical Control Point System and Importance of Verification Procedures." This chapter presents the nuts and bolts of hazard analysis and critical control point systems and would be useful to anyone involved in hazard analysis and critical control point systems, from the technician to the quality control supervisor, or even perhaps the plant manager. Chapter 29 is interesting because it discusses a new approach to controlling chemical and microbial hazards in food by sorption of these contaminants on clays. Another example of the uniqueness of this book is found in Chapter 30, which discusses both risk communication and consumer food-handling behavior. The author suggests that high schools may be an important place for instructing young people about proper food-handling procedures and food safety philosophies.

Though we could continue mentioning brief anecdotes about the other chapters, we will leave further discoveries of the important concepts presented by each author to the reader. We are confident that the broad range of material covered in this book, and the quality of the material presented, will appeal to the reader.

As we began the task of finding authors for chapters, we realized that this book could not be limited to only the Institute of Food Science and Engineering faculty. We reached out to additional Agricultural Research Service scientists, to other government scientists, and to faculty at other universities to bring quality, balance, and the broadest scope possible to this book.

When we approached various prospective publishers, some felt that a contemporary food safety book should contain a considerable amount of material on the subjects of food biosecurity and bioterrorism. The editors discussed whether or not to include such material and decided against including a section on food biosecurity or bioterrorism. Although we were well aware of the topical nature and market enhancing value of such chapters, we feared that publishing such material might somehow serve as a roadmap, guideline, or manual for those who would misuse the information to harm us.

However, Neville Clarke, DVM, Ph.D., former Dean of the Texas Agriculture Experiment Station, presently serving as the Director of the Texas A&M University Institute for Counter Measures Against Agricultural Bioterrorism, who also serves in a key role in Homeland Security for the state of Texas, has written some brief comments on the issues of agricultural biosecurity and bioterrorism and these comments are found in the Afterword.

References

Doores, S. 1999. Food Safety: Current Status and Future Needs, pp. 1-29. A report of the American Academy of Microbiology, Avaialble at : http://www.asmusa.org/acasrc/aca1.htm (click on "Food Safety: Current Status and Future Needs").

Ross C. Beier
Suresh D. Pillai
Timothy D. Phillips
Richard L. Ziprin

Part I

Pathogen/Host Interactions

1 Foodborne *Salmonella* Infections

Andreas J. Bäumler

Introduction

Nontyphoidal *Salmonella* serotypes continue to be among the most prominent food safety problems in the United States, as illustrated by the large annual numbers of human cases and outbreaks with which these pathogens are associated. Between 1993 and 1997, *Salmonella* serotypes were the leading cause of foodborne disease outbreaks in the United States, accounting for 41% of outbreaks with known etiology (Olsen et al. 2000). A recent estimate from the Centers of Disease Control and Prevention (CDC) suggests that *Salmonella* serotypes cause approximately 1.4 million illnesses annually. This makes salmonellosis the second most common cause of bacterial foodborne disease of known etiology in the United States (Mead et al. 1999). In addition, nontyphoidal *Salmonella* serotypes are the single most common cause of death from foodborne illness associated with viruses, parasites, or bacteria in the United States, causing an estimated 550 fatal cases each year (Mead et al. 1999). The high morbidity and mortality resulting from foodborne infections with *Salmonella* serotypes in the United States is associated with significant economic losses. The CDC's Foodborne Diseases Active Surveillance Network (Food Net) estimates the annual cost of medical care and lost productivity caused by foodborne salmonellosis in the United States to range from $0.5 billion to $2.3 billion (Frenzen et al. 1999).

Salmonella serotypes implicated in foodborne infections are commonly associated with animal reservoirs, suggesting that infections in man result from animal-to-human transmission (St. Louis et al. 1988, Mishu et al. 1994). In outbreaks within the United States in which the sources were identified, the food vehicles most commonly implicated were chicken, beef, turkey, and eggs (Tauxe 1991, Olsen et al. 2000). Meat, meat products, eggs, or egg products may contain *Salmonella* serotypes either because animals are infected or because fecal contamination occurs during processing (Galbraith 1961). There are currently 2,449 known *Salmonella* serotypes (Brenner et al. 2000). Approximately 70% of the *Salmonella* serotypes have been isolated from human cases of disease (Kelterborn 1967), but only five *Salmonella* serotypes accounted for 61% of the human cases reported to the CDC between 1987 and 1997 (Olsen et al. 2000). These five serotypes include *S. enterica* serotype Typhimurium (23% of isolations between 1987 and 1997), *S. enterica* serotype Enteritidis (21%), *S. enterica* serotype Heidelberg (8%), *S. enterica* serotype Newport (5%), and *S. enterica* serotype Hadar (4%). Collectively, these surveillance data illustrate that the persistence of *Salmonella* serotypes in livestock and domestic fowl is directly responsible for their subsequent introduction into the derived food products and that some *Salmonella* serotypes are epidemiologically more successful than others. Both the preharvest occurrence of *Salmonella* in food animals and the epidemiological success of certain *Salmonella* serotypes are of importance for food safety. However, the molecular mechanisms responsible for these phenomena are poorly understood. For the rational design of innovative prevention strategies, research is needed on the basic mechanisms that enable certain *Salmonella* serotypes to persist successfully in food animal reservoirs.

Preharvest Occurrence of *Salmonella* Serotypes

Salmonella serotypes pose a threat to food safety in the United States by persisting in apparently healthy animals, thereby leading to introduction of these pathogens into animal-derived food products. Transmission between animals before harvest is therefore of prime importance for food safety.

Horizontal Transmission

Between 1% and 6% of farm animals in the United States test positive for intestinal carriage of *Salmonella* serotypes (Ebel et al. 1992, Fedorka-Cray et al. 1998, Byrd et al. 1999, Corrier et al. 1999, Dargatz et al. 2000, Fedorka-Cray et al. 2000, McKean et al. 2001, Wells et al. 2001). *Salmonella* serotypes shed with the feces of livestock and domestic fowl are largely responsible for persistence in these reservoirs. That is, *Salmonella* serotypes spread horizontally between animals via the fecal–oral route through fecal contamination of the environment (Wray et al. 1987, McLaren and Wray 1991, Wray et al. 1991, Hurd et al. 2001a). In addition, fecal contamination by rodents has been proposed to play a role in transmission of *Salmonella* serotypes on chicken farms (Sato et al. 1970, Krabisch and Dorn 1980, Henzler and Opitz 1992, Kinde et al. 1996, Guard-Petter et al. 1997). In contrast, airborne transmission (Proux et al. 2001) and contamination of feed (Funk et al. 2001) do not appear to play important roles in the spread of *Salmonella* serotypes among farm animals. Before slaughter, stress during transport and long periods with intermittent feeding accelerate the spread of *Salmonella* serotypes by the fecal–oral route (Brownlie and Grau 1967, Grau et al. 1968, Samuel et al. 1979, Moo et al. 1980, Samuel et al. 1981, Corrier et al. 1999, Isaacson et al. 1999, Hurd et al. 2001b, McKean et al. 2001). As a result, the prevalence of intestinal carriage among livestock and domestic fowl increases above 10% before slaughter (Corrier et al. 1999, McKean et al. 2001, Wells et al. 2001). Intestinal carriage may result in fecal contamination of hide and carcass surfaces at slaughter (Gronstol et al. 1974), and in addition it may result in colonization of the draining lymph nodes, contaminating food products containing this material (Samuel et al. 1979, Moo et al. 1980, Samuel et al. 1981).

Given the importance of fecal–oral transmission, it is not surprising that control measures aimed at reducing fecal contamination of the environment have been shown to be effective in lowering the preharvest occurrence of *Salmonella* serotypes (Dahl et al. 1997, Davies et al. 1997). However, proper sanitation my not suffice to reduce fecal contamination to a level that would eliminate *Salmonella* serotypes from livestock reservoirs (Twiddy et al. 1988).

Vertical Transmission

Vertical transmission (from an adult to its offspring) is important for the spread of *Salmonella* serotypes in two food animal reservoirs, namely, chickens and cattle. In cattle, *Salmonella* serotypes can persist in the udders of cows for up to 2.5 years, resulting in vertical transmission of the organism through milk (Giles et al. 1989), which may contribute to persistence of *Salmonella* serotypes in dairy herds (Richardson 1973, Giles et al. 1989). However, the widespread pasteurization of milk and dairy products has reduced the importance, for human health, of this mode of transmission in the United States (Tauxe 1991).

In chickens, eggs can become contaminated by transovarian contamination with *Salmonella* serotypes (Snoeyenbos et al. 1969). The importance of transovarian contamination for transmission of *Salmonella* serotypes among chickens has not been investigated. However, transovarian transmission has been proposed to be a source of human infection, primarily because of contamination of eggs with serotype *S. enterica* Enteritidis (St. Louis et al. 1988). This represents a significant problem, as *S. enterica* Enteritidis accounted for the largest number of outbreaks caused by foodborne pathogens in the United States during 1993-1997, and most of these outbreaks were attributed to eating eggs (Olsen et al. 2000).

Mechanisms of Persistence

Mechanisms important for the acute phase of interaction between *Salmonella* serotypes and their host are under intense investigation, and this work has significantly improved our understanding of how these pathogens cause disease (for recent review articles, see Fierer and Guiney 2001, Galán 2001, Groisman 2001, Brumell et al. 2002, Kingsley and Bäumler 2002, Zhang et al. 2003). In contrast, mechanisms important for persistent carriage in apparently healthy animals, which may develop during subclinical infections or after signs of disease have subsided, are poorly understood.

Salmonella serotypes encode a type III secretion system (TTSS-1) encoded by *Salmonella* pathogenicity island 1, which allows these pathogens to alter host cell physiology (Galán 2001). TTSS-1 is required by *S. enterica* Typhimurium for colonizing Peyer's patches of mice (Galán and Curtiss 1989) and for causing diarrhea and lethal morbidity in calves (Watson et al. 1998, Tsolis et al. 1999). However, although the TTSS-1 is clearly a major virulence factor required for the intestinal phase of infection, it is not required for persistent fecal shedding of *S. enterica* Typhimurium from mice (Kingsley et al. 2000).

Work on *Escherichia coli*, a close relative of *Salmonella* serotypes, has implicated type 1 fimbriae in the bacteria's ability to be transmitted from infected rats to their littermates (Bloch et al. 1992). Type 1 fimbriae may promote communicability by mediating colonization of the oropharyngeal mucosa, thereby enabling the organism to multiply before passage through the stomach, which increases the chances that a small inoculum is able to complete the fecal–oral cycle despite the acid sensitivity of the organism (Bloch et al. 1992). However, type 1 fimbriae are not required for intestinal persistence of *E. coli* (McCormick et al. 1989, Bloch and Orndorff 1990). Type 1 fimbriae of *S. enterica* Enteritidis have been shown to be required for intestinal colonization of the rat at 6 days (early) but not 21 days (late) postinfection, suggesting that this adhesin is not required for long-term intestinal persistence of *Salmonella* serotypes (Naughton et al. 2001).

The *S. enterica* Enteritidis wild type is recovered at higher numbers than a mutant unable to elaborate flagella from the ceca of chickens 35 days post oral infection, indicating that motility is required for intestinal persistence (Allen-Vercoe and Woodward 1999). *S. enterica* Typhimurium lipopolysaccharide-deficient mutants are deficient in colonizing the murine large intestine (Nevola et al. 1985), presumably because they possess a reduced ability to penetrate the intestinal mucus layer (Licht et al. 1996). These studies suggest that for *Salmonella* serotypes to persist in the intestine, they require the ability to reach their preferred site of colonization.

Recently, a nonfimbrial adhesin of *S. enterica* Typhimurium, termed ShdA, has been shown to be required for chronic intestinal carriage in mice (Kingsley et al. 2000). The reduced recovery of the ShdA mutant from the feces of mice appears to be the result of its reduced ability to colonize the murine cecum (Kingsley et al. 2003). A ShdA fusion protein binds to areas of the cecum that contain fibronectin, and binding of this extracellular matrix protein to the bacterial surface depends on expression of this protein (Kingsley et al. 2002). Furthermore, immunohistochemical analysis shows that *S. enterica* Typhimurium colonizes the cecal mucosa at areas of epithelial erosion where the extracellular matrix is exposed to the intestinal lumen (Kingsley et al. 2002). These data suggest that ShdA-mediated binding of fibronectin in lesions of the murine cecal mucosa is a mechanism for intestinal carriage of *S. enterica* Typhimurium. Further research on this and other molecular mechanisms of intestinal persistence may identify targets for novel strategies to reduce the preharvest occurrence of *Salmonella* serotypes in livestock.

Epidemiological Success of *Salmonella* Serotypes

S. enterica serotypes Typhimurium, Enteritidis, Heidelberg, Newport, and Hadar are isolated more frequently than other *Salmonella* serotypes from cases of human disease in the United States (Olsen et al. 2000). A better understanding of the reasons for the epidemiological success of these serotypes is thus highly relevant for food safety.

Reservoirs for Serotypes Frequently Associated with Human Disease

S. enterica Typhimurium is the serotype isolated most frequently from patients in the United States (Olsen et al. 2001). It is associated with a large number of animal reservoirs (Rabsch et al. 2002), among which chickens, cattle, and pigs appear to be the most important for food safety. Persistence in these reservoirs commonly leads to the introduction of *S. enterica* Typhimurium into carcasses and meat products. According to Food Safety and Inspection Service's (FSIS's) nationwide microbial baseline studies, *S. enterica* Typhimurium represents 14% of total *Salmonella* isolations from beef carcasses, 10% of isolations from chicken carcasses, 10% of isolations from swine carcasses, 18% of isolations from raw ground pork, and 15% of isolations from raw ground beef in the United States (Schlosser et al. 2000).

    S. enterica Enteritidis is the second most frequently isolated serotype from patients in the United States (Olsen et al. 2001), but this serotype is not a frequent isolate from meat or meat products in the United States (Schlosser et al. 2000). Instead, the majority of *S. enterica* Enteritidis outbreaks in Europe and the United States can be traced back to foods containing undercooked chicken eggs (Coyle et al. 1988, St. Louis et al. 1988, Cowden et al. 1989, Henzler et al. 1994). During experimental infection, *S. enterica* Enteritidis colonizes the reproductive organs of chickens better than other *Salmonella* serotypes, which may be one of the reasons why this serotype is associated with this food source (Okamura et al. 2001). The prevalence of *S. enterica* Enteritidis in eggs produced in the United States is estimated to be 1 in every 20,000 (Ebel and Schlosser 2000).

    S. enterica Heidelberg is a serotype commonly associated with poultry (Hird et al. 1993, Riemann et al. 1998, Byrd et al. 1999). According to FSIS's nationwide microbial baseline studies, the organism represents 26% of total *Salmonella* isolations from chicken carcasses, 14% from turkey carcasses, and 30% from ground raw chicken in the United States

(Schlosser et al. 2000), and *S. enterica* Heidelberg represents 6% of total *Salmonella* isolations from raw ground pork in the United States (Schlosser et al. 2000).

The majority of human infections with *S. enterica* Newport in which a source is known have been traced back to consumption of beef. These isolates commonly originated from dairy cattle, thereby suggesting a bovine animal reservoir for this serotype (Spika et al. 1987, Zansky et al. 2002). According to FSIS's nationwide microbial baseline studies, *S. enterica* Newport represents 2% of the total *Salmonella* isolations from raw ground beef in the United States (Schlosser et al. 2000).

S. enterica Hadar is a serotype associated with turkey in the United States. According to FSIS's nationwide microbial baseline studies, the serotype represents 15% of the total *Salmonella* isolations from turkey carcasses and 24% of isolations from ground raw turkey (Schlosser et al. 2000).

These data illustrate that a high incidence of human illness caused by a *Salmonella* serotype commonly correlates with the serotype's ability to persist in one or several animal reservoirs and its frequent isolation from the derived food products. A better insight into the factors that may allow one particular serotype to dominate within an animal reservoir is key to understanding its epidemiological success revealed by human surveillance.

Emerging Problems

Although the animal reservoirs from which human infections with *S. enterica* serotypes Typhimurium, Enteritidis, Heidelberg, Newport, and Hadar originate are well established, factors contributing to the preponderance of these serotypes in the United States are poorly understood. Our limited grasp of factors responsible for the epidemiological success of these *Salmonella* serotypes represents a key gap in our knowledge of pathogens that are important for food safety.

One serotype, *S. enterica* Enteritidis, has emerged only recently as an important food safety problem in the United States (Angulo and Swerdlow 1998). The incidence of human infections with *S. enterica* Enteritidis steadily increased from 1963 until 1990, when this serotype became the one reported most frequently in the United States. (Aserkoff et al. 1970, Mishu et al. 1994). The epidemic peaked in 1995, and in 1997 *S. enterica* Typhimurium replaced *S. enterica* Enteritidis again as the serovar most frequently isolated in the United States. (CDC 1999). The emergence of *S. enterica* Enteritidis in the second half of the twentieth century provides a unique opportunity to identify factors important for its epidemiological success by determining what triggered the current human epidemic. The emergence of *S. enterica* Enteritidis as an egg-associated pathogen in the United States and Europe coincided with eradication of *S. enterica* serotype Gallinarum from poultry (Bäumler et al. 2000, Rabsch et al. 2001). *S. enterica* Gallinarum is a chicken-adapted serotype expressing an immuno-dominant antigen (lipopolysaccharide) that is identical to that of *S. enterica* Enteritidis (Kingsley and Bäumler 2000). As a result, birds previously exposed to *S. enterica* Gallinarum are immune to organ colonization with *S. enterica* Enteritidis (Sterner and Hein 1998, Witvliet et al. 1998). Mathematical models combining epidemiology with population biology suggest that the presence of *S. enterica* Gallinarum in chicken flocks provided population-wide immunity against *S. enterica* Enteritidis infection at the beginning of the twentieth century (Rabsch et al. 2000). Eradication of *S. enterica* Gallinarum resulted in the loss of population-wide immunity against its lipopolysaccharide, thereby opening an ecological niche that was subsequently filled by *S. enterica* Enteritidis (Bäumler et al. 2000). According to this theory, the human *S. enterica* Enteritidis epidemic was triggered by eliminating a competitor (i.e., *S. enterica* Gallinarum) from chicken flocks, thereby allowing *S. enterica* Enteritidis to enter the human food supply through chicken eggs. This example illustrates that future strategies for improving the safety of our food supply should be based on a better understanding of the selecting forces acting on pathogen populations.

Conclusions

Salmonella is the second most common cause of bacterial foodborne disease of known etiology and the single most common cause of death from foodborne illnesses associated with viruses, parasites, or bacteria in the United States. Although the majority of the 2,449 different *Salmonella* serotypes have been associated with human enterocolitis, only five serotypes account for approximately 60% of the cases. Factors important for the epidemiological success of these serotypes are largely unknown, which represents a key gap in our knowledge of microbes that pose a threat to food safety. Research is needed to elucidate the molecular mechanisms that allow *Salmonella* serotypes to persist successfully in food animals. These mechanisms would be logical targets for the development of new and innovative approaches to improve food safety. Furthermore, research on the origins of the current *S. enterica* Enteritidis epidemic suggests that future intervention strategies should be based on a better understanding of selective forces acting on pathogen populations.

Acknowledgements

Work in Dr. Bäumler's laboratory is supported by USDA/NRICGP grant 2002-35204-12247 and Public Health Service grants AI40124 and AI44170.

References

Allen-Vercoe, E. and Woodward, M. J. 1999. Colonisation of the chicken caecum by afimbriate and aflagellate derivatives of *Salmonella enterica* serotype Enteritidis. *Veterinary Microbiology* **69**:265–275.

Angulo, F. J. and Swerdlow, D. L. 1998. *Salmonella enteritidis* infections in the United States. *Journal of the American Veterinary Medical Association* **213**:1729–1731.

Aserkoff, B., Schroeder, S. A. and Brachman, P. S. 1970. Salmonellosis in the United States—a five-year review. *American Journal of Epidemiology* **92**:13–24.

Bäumler, A. J., Hargis, B. M. and Tsolis, R. M. 2000. Tracing the origins of *Salmonella* outbreaks. *Science* **287**:50–52.

Bloch, C. A. and Orndorff, P. E. 1990. Impaired colonization by and full invasiveness of *Escherichia coli* K1 bearing a site-directed mutation in the type 1 pilin gene. *Infection and Immunity* **58**:275–278.

Bloch, C. A., Stocker, B. A. and Orndorff, P. E. 1992. A key role for type 1 pili in enterobacterial communicability. *Molecular Microbiology* **6**:697–701.

Brenner, F. W., Villar, R. G., Angulo, F. J., Tauxe, R. and Swaminathan, B. 2000. *Salmonella* nomenclature. *Journal of Clinical Microbiology* **38**:2465–2467.

Brownlie, L. E. and Grau, F. H. 1967. Effect of food intake on growth and survival of salmonellas and *Escherichia coli* in the bovine rumen. *Journal of General Microbiology* **46**:125–134.

Brumell, J. H., Perrin, A. J., Goosney, D. L. and Finlay, B. B. 2002. Microbial pathogenesis: new niches for *Salmonella*. *Current Biology* **12**:R15–R17.

Byrd, J. A., DeLoach, J. R., Corrier, D. E., Nisbet, D. J. and Stanker, L. H. 1999. Evaluation of *Salmonella* serotype distributions from commercial broiler hatcheries and grower houses. *Avian Diseases* **43**:39–47.

Centers for Disease Control. 1999. *Salmonella* surveillance: annual tabulation summary, 1998. Atlanta, Ga.: U.S. Department of Health and Human Services.

Corrier, D. E., Byrd, J. A., Hargis, B. M., Hume, M. E., Bailey, R. H. and Stanker, L. H. 1999. Presence of Salmonella in the crop and ceca of broiler chickens before and after preslaughter feed withdrawal. *Poultry Science* **78**:45–49.

Cowden, J. M., Chisholm, D., O'Mahony, M., Lynch, D., Mawer, S. L., Spain, G. E., Ward, L. and Rowe, B. 1989. Two outbreaks of *Salmonella enteritidis* phage type 4 infection associated with the consumption of fresh shell-egg products. *Epidemiology and Infection* **103**:47–52.

Coyle, E. F., Palmer, S. R., Ribeiro, C. D., Jones, H. I., Howard, A. J., Ward, L. and Rowe, B. 1988. *Salmonella enteritidis* phage type 4 infection: association with hen's eggs. *Lancet* **2**:1295–1297.

Dahl, J., Wingstrand, A., Nielsen, B. and Baggesen, D. L. 1997. Elimination of *Salmonella typhimurium* infection by the strategic movement of pigs. *Veterinary Record* **140**:679–681.

Dargatz, D. A., Fedorka-Cray, P. J., Ladely, S. R. and Ferris, K. E. 2000. Survey of *Salmonella* serotypes shed in feces of beef cows and their antimicrobial susceptibility patterns. *Journal of Food Protection* **63**:1648–1653.

Davies, P. R., Morrow, W. E., Jones, F. T., Deen, J., Fedorka-Cray, P. J. and Gray, J. T. 1997. Risk of shedding *Salmonella* organisms by market-age hogs in a barn with open-flush gutters. *Journal of the American Veterinary Medical Association* **210**:386–389.

Ebel, E. and Schlosser, W. 2000. Estimating the annual fraction of eggs contaminated with *Salmonella enteritidis* in the United States. *International Journal of Food Microbiology* **61**:51–62.

Ebel, E. D., David, M. J. and Mason, J. 1992. Occurrence of *Salmonella enteritidis* in the U.S. commercial egg industry: report on a national spent hen survey. *Avian Diseases* **36**:646–654.

Fedorka-Cray, P. J., Dargatz, D. A., Thomas, L. A. and Gray, J. T. 1998. Survey of *Salmonella* serotypes in feedlot cattle. *Journal of Food Protection* **61**:525–530.

Fedorka-Cray, P. J., Gray, J. T. and Wray, C. 2000. *Salmonella* infections in pigs, pp. 191–207. *In Salmonella* in Domestic Animals, C. Wray, A. Wray (Eds.); New York: CABI Publishing.

Fierer, J. and Guiney, D. G. 2001. Diverse virulence traits underlying different clinical outcomes of *Salmonella* infection. *Journal of Clinical Investigation* **107**:775–780.

Frenzen, P., Riggs, T., Buzby, J., Breuer, T., Roberts, T., Voetsch, D., Reddy, S. and Group, T. F. W. 1999. *Salmonella* cost estimate update using FoodNet data. *Food Review* **22**:10–15.

Funk, J. A., Davies, P. R. and Nichols, M. A. 2001. Longitudinal study of *Salmonella enterica* in growing pigs reared in multiple-site swine production systems. *Veterinary Microbiology* **83**:45–60.

Galán, J. E. 2001. *Salmonella* interactions with host cells: type III secretion at work. *Annual Reviews of Cellular and Developmental Biology* **17**:53–86.

Galán, J. E. and Curtiss III, R. 1989. Cloning and molecular characterization of genes whose products allow *Salmonella typhimurium* to penetrate tissue culture cells. *Proceedings of the National Academy of Sciences* **86**:6383–6387.

Galbraith, N. S. 1961. Studies of human salmonellosis in relation to infection in animals. *Veterinary Record* **73**:1296–1303.

Giles, N., Hopper, S. A. and Wray, C. 1989. Persistence of *S. typhimurium* in a large dairy herd. *Epidemiology and Infection* **103**:235–241.

Grau, F. H., Brownlie, L. E. and Roberts, E. A. 1968. Effect of some preslaughter treatments on the *Salmonella* population in the bovine rumen and faeces. *Journal of Applied Bacteriology* **31**:157–163.

Groisman, E. A. 2001. The pleiotropic two-component regulatory system PhoP-PhoQ. *Journal of Bacteriology* **183**:1835–1842.

Gronstol, H., Osborne, A. D. and Pethiyagoda, S. 1974. Experimental *Salmonella* infection in calves. 1. The effect of stress factors on the carrier state. *Journal of Hygene (London)* **72**:155–162.

Guard-Petter, J., Henzler, D. J., Rahman, M. M. and Carlson, R. W. 1997. On-farm monitoring of mouse-invasive *Salmonella enterica* serovar Enteritidis and a model for its association with the production of contaminated eggs. *Applied and Environmental Microbiology* **63**:1588–1593.

Henzler, D. J., Ebel, E., Sanders, J., Kradel, D. and Mason, J. 1994. *Salmonella enteritidis* in eggs from commercial chicken layer flocks implicated in human outbreaks. *Avian Diseases* **38**:37–43.

Henzler, D. J. and Opitz, H. M. 1992. The role of mice in the epizootiology of *Salmonella enteritidis* infection on chicken layer farms. *Avian Diseases* **36**:625–631.

Hird, D. W., Kinde, H., Case, J. T., Charlton, B. R., Chin, R. P. and Walker, R. L. 1993. Serotypes of *Salmonella* isolated from California turkey flocks and their environment in 1984-89 and comparison with human isolates. *Avian Diseases* **37**:715–719.

Hurd, H. S., Gailey, J. K., McKean, J. D. and Rostagno, M. H. 2001a. Rapid infection in market-weight swine following exposure to a *Salmonella typhimurium*–contaminated environment. *American Journal of Veterinary Research* **62**:1194–1197.

Hurd, H. S., McKean, J. D., Wesley, I. V. and Karriker, L. A. 2001b. The effect of lairage on *Salmonella* isolation from market swine. *Journal of Food Protection* **64**:939–944.

Isaacson, R. E., Firkins, L. D., Weigel, R. M., Zuckermann, F. A. and DiPietro, J. A. 1999. Effect of transportation and feed withdrawal on shedding of *Salmonella typhimurium* among experimentally infected pigs. *American Journal of Veterinary Research* **60**:1155–1158.

Kelterborn, E. 1967. *Salmonella*-species. First isolations, names and occurrence, pp. 23–33. Karl-Marx-Stadt: S. Hirzel.

Kinde, H., Read, D. H., Ardans, A., Breitmeyer, R. E., Willoughby, D., Little, H. E., Kerr, D., Gireesh, R. and Nagaraja, K. V. 1996. Sewage effluent: likely source of *Salmonella enteritidis*, phage type 4 infection in a commercial chicken layer flock in southern California. *Avian Diseases* **40**:672–676.

Kingsley, R. A. and Bäumler, A. J. 2000. Host adaptation and the emergence of infectious disease: the *Salmonella* paradigm. *Molecular Microbiology* **36**:1006–1014.

Kingsley, R. A. and Bäumler, A. J. 2002. Pathogenicity islands and host adaptation of *Salmonella* serovars. *Current Topics in Microbiology and Immunology* **264**:67–87.

Kingsley, R. A., Humphries, A. D., Weening, E., de Zoete, M., Winter, S., Papaconstantinopoulou, A., Dougan, G. and Bäumler, A. J. 2003. Molecular and phenotypic analysis of the CS54 island of *Salmonella enterica* serotype Typhimurium. *Infection and Immunity* **71**:629–640.

Kingsley, R. A., Santos, R. L., Keestra, A. M., Adams, L. G. and Bäumler, A. J. 2002. *Salmonella enterica* serotype Typhimurium ShdA is an outer membrane fibronectin-binding protein that is expressed in the intestine. *Molecular Microbiology* **43**:895–905.

Kingsley, R. A., van Amsterdam, K., Kramer, N. and Bäumler, A. J. 2000. The *shdA* gene is restricted to serotypes of *Salmonella enterica* subspecies I and contributes to efficient and prolonged fecal shedding. *Infection and Immunity* **68**:2720–2727.

Krabisch, P. and Dorn, P. 1980. Epidemiologic significance of live vectors in the transmission of *Salmonella* infections in broiler flocks. *Berliner und Münchener Tierärztliche Wochenschrift* **93**:232–235.

Licht, T. R., Krogfelt, K. A., Cohen, P. S., Poulsen, L. K., Urbance, J. and Molin, S. 1996. Role of lipopolysaccharide in colonization of the mouse intestine by *Salmonella typhimurium* studied by in situ hybridization. *Infection and Immunity* **64**:3811–3817.

McCormick, B. A., Franklin, D. P., Laux, D. C. and Cohen, P. S. 1989. Type 1 pili are not necessary for colonization of the streptomycin-treated mouse large intestine by type 1-piliated *Escherichia coli* F-18 and *E. coli* K-12. *Infection and Immunity* **57**:3022–3029.

McKean, J. D., Hurd, H. S., Larsen, S., Rostagno, M., Griffith, R. and Wesley, I. 2001. Impact of commercial preharvest processes on the prevalence of *Salmonella enterica* in cull sows. *Berliner und Münchener Tierärztliche Wochenschrift* **114**:353–355.

McLaren, I. M. and Wray, C. 1991. Epidemiology of *Salmonella typhimurium* infection in calves: persistence of salmonellae on calf units. *Veterinary Record* **129**:461–462.

Mead, P. S., Slutsker, L., Dietz, V., McCaig, L. F., Bresee, J. S., Shapiro, C., Griffin, P. M. and Tauxe, R. V. 1999. Food-related illness and death in the United States. *Emerging Infectious Diseases* **5**:607–625.

Mishu, B., Koehler, J., Lee, L. A., Rodrigue, D., Brenner, F. H., Blake, P. and Tauxe, R. V. 1994. Outbreaks of *Salmonella enteritidis* infections in the United States, 1985–1991. *Journal of Infectious Diseases* **169**:547–552.

Moo, D., O'Boyle, D., Mathers, W. and Frost, A. J. 1980. The isolation of *Salmonella* from jejunal and caecal lymph nodes of slaughtered animals. *Australian Veterinary Journal* **56**:181–183.

Naughton, P. J., Grant, G., Bardocz, S., Allen-Vercoe, E., Woodward, M. J. and Pusztai, A. 2001. Expression of type 1 fimbriae (SEF 21) of *Salmonella enterica* serotype Enteritidis in the early colonisation of the rat intestine. *Journal of Medical Microbiology* **50**:191–197.

Nevola, J. J., Stocker, B. A., Laux, D. C. and Cohen, P. S. 1985. Colonization of the mouse intestine by an avirulent *Salmonella typhimurium* strain and its lipopolysaccharide-defective mutants. *Infection and Immunity* **50**:152–159.

Okamura, M., Kamijima, Y., Miyamoto, T., Tani, H., Sasai, K. and Baba, E. 2001. Differences among six *Salmonella* serovars in abilities to colonize reproductive organs and to contaminate eggs in laying hens. *Avian Diseases* 45:61–69.

Olsen, S. J., Bishop, R., Brenner, F. W., Roels, T. H., Bean, N., Tauxe, R. V. and Slutsker, L. 2001. The changing epidemiology of *Salmonella*: trends in serotypes isolated from humans in the United States, 1987–1997. *Journal of Infectious Diseases* **183**:753–761.

Olsen, S. J., MacKinnon, L. C., Goulding, J. S., Bean, N. H. and Slutsker, L. 2000. Surveillance for foodborne-disease outbreaks—United States, 1993–1997. *Morbidity and Mortality Weekly Report CDC Surveillance Summary* **49**:1–62.

Proux, K., Cariolet, R., Fravalo, P., Houdayer, C., Keranflech, A. and Madec, F. 2001. Contamination of pigs by nose-to-nose contact or airborne transmission of *Salmonella typhimurium*. *Veterinary Research* **32**:591–600.

Rabsch, W., Andrews-Polymenis, H. L., Kingsley, R. A., Prager, R., Tschape, H., Adams, L. G. and Bäumler, A. J. 2002. *Salmonella enterica* serotype Typhimurium and its host-adapted variants. *Infection and Immunity* **70**:2249–2255.

Rabsch, W., Hargis, B. M., Tsolis, R. M., Kingsley, R. A., Hinz, K. H., Tschäpe, H. and Bäumler, A. J. 2000. Competitive exclusion of *Salmonella enteritidis* by *Salmonella gallinarum* from poultry. *Emerging Infectious Diseases* **6**:443–448.

Rabsch, W., Tschäpe, H. and Bäumler, A. J. 2001. Non-typhoidal salmonellosis: emerging problems. *Microbes and Infection* **3**:237–247.

Richardson, A. 1973. The transmission of *Salmonella dublin* to calves from adult carrier cows. *Veterinary Record* **92**:112–115.

Riemann, H., Himathongkham, S., Willoughby, D., Tarbell, R. and Breitmeyer, R. 1998. A survey for *Salmonella* by drag swabbing manure piles in California egg ranches. *Avian Diseases* **42**:67–71.

Samuel, J. L., Eccles, J. A. and Francis, J. 1981. *Salmonella* in the intestinal tract and associated lymph nodes of sheep and cattle. *Journal of Hygiene* **87**:225–232.

Samuel, J. L., O'Boyle, D. A., Mathers, W. J. and Frost, A. J. 1979. Isolation of *Salmonella* from mesenteric lymph nodes of healthy cattle at slaughter. *Research in Veterinary Science* **28**:238–241.

Sato, G., Miyamae, T. and Miura, S. 1970. A long term epizootiological study of chicken salmonellosis on a farm with reference to elimination of paratyphoid infection by cloacal swab culture. *Japanese Journal of Veterinary Research* **18**:47–62.

Schlosser, W., Hogue, A., Ebel, E., Rose, B., Umholtz, R., Ferris, K. and James, W. 2000. Analysis of *Salmonella* serotypes from selected carcasses and raw ground products sampled prior to implementation of the pathogen reduction; hazard analysis and critical control point final rule in the U.S. *International Journal of Food Microbiology* **58**:107–111.

Snoeyenbos, G. H., Smyser, C. F. and Van Roekel, H. 1969. *Salmonella* infections of the ovary and peritoneum of chickens. *Avian Diseases* **13**:668–670.

Spika, J. S., Waterman, S. H., Hoo, G. W., St Louis, M. E., Pacer, R. E., James, S. M., Bissett, M. L., Mayer, L. W., Chiu, J. Y. and Hall, B. 1987. Chloramphenicol-resistant *Salmonella newport* traced through hamburger to dairy farms. A major persisting source of human salmonellosis in California. *New England Journal of Medicine* **316**:565–570.

St. Louis, M. E., Morse, D. L., Potter, M. E., DeMelfi, T. M., Guzewich, J. J., Tauxe, R. V. and Blake, P. A. 1988. The emergence of grade A eggs as a major source of *Salmonella enteritidis* infections. New implications for the control of salmonellosis. *Journal of the American Medical Association* **259**:2103-2107.

Sterner, F. and Hein, R. 1998. An attenuated *Salmonella gallinarum* live vaccine induces long-term protection against *Salmonella enteritidis*. International Symposium on Food-Borne *Salmonella* in Poultry, July 25–26, Baltimore, MA.

Tauxe, V. T. 1991. *Salmonella*: A postmodern pathogen. *Journal of Food Protection* **54**:563–568.

Tsolis, R. M., Adams, L. G., Ficht, T. A. and Bäumler, A. J. 1999. Contribution of *Salmonella typhimurium* virulence factors to diarrheal disease in calves. *Infection and Immunity* **67**:4879–4885.

Twiddy, N., Hopper, D. W., Wray, C. and McLaren, I. 1988. Persistence of *S. typhimurium* in calf rearing premises. *Veterinary Record* **122**:399.

Watson, P. R., Galyov, E. E., Paulin, S. M., Jones, P. W. and Wallis, T. S. 1998. Mutation of invH, but not stn, reduces *Salmonella*-induced enteritis in cattle. *Infection and Immunity* **66**:1432–1438.

Wells, S. J., Fedorka-Cray, P. J., Dargatz, D. A., Ferris, K. and Green, A. 2001. Fecal shedding of *Salmonella* spp. by dairy cows on farm and at cull cow markets. *Journal of Food Protection* **64**:3–11.

Witvliet, M., Vostermans, T., van den Bosch, J., de Vries, T. S. and Pennings, A. 1998. Induction of cross-protection against *S. enteritidis* by the *S. gallinarum* 9R vaccine. International Symposium on Food-Borne *Salmonella* in Poultry, July 25–26, Baltimore, MA.

Wray, C., Todd, J. N. and Hinton, M. 1987. Epidemiology of *Salmonella typhimurium* infection in calves: excretion of *S. typhimurium* in the faeces of calves in different management systems. *Veterinary Record* **121**:293–296.

Wray, C., Todd, N., McLaren, I. M. and Beedell, Y. E. 1991. The epidemiology of *Salmonella* in calves: the role of markets and vehicles. *Epidemiology and Infection* **107**:521–525.

Zansky, S., Wallace, B., Schoonmaker-Bopp, D., Smith, P., Ramsey, F., Painter, J., Gupta, A., Kalluri, P. and Noviello, S. 2002. From the Centers for Disease Control and Prevention. Outbreak of multi-drug resistant *Salmonella Newport*—United States, January–April 2002. *Journal of the American Medical Association* **288**:951–953.

Zhang, S., Kingsley, R. A., Santos, R. L., Andrews-Polymenis, H., Raffatellu, M., Figueiredo, J., Nunes, J., Tsolis, R. M., Adams, L. G. and Bäumler, A. J. 2003. Molecular Pathogenesis of *Salmonella enterica* serotype Typhimurium-induced diarrhea. *Infection and Immunity* **71**:1–12.

2   Pathogenic *Escherichia coli*

Suryakant D. Waghela

Introduction

Escherichia coli was first reported as *Bacterium coli commune* by the German pediatrician and bacteriologist Theodor Escherich, who observed a rod-shaped bacterium to be more abundant in the fecal flora of infant babies that had suckled colostrum compared to the flora in the meconium (Escherich 1885, Bettlheim 1988). For some 40 years this bacterium was considered to be a ubiquitous commensal organism until it was suspected to be the cause of an outbreak of neonatal diarrhea (Dulaney and Michelson 1935). Although most *E. coli* are still considered nonpathogenic, acquisition of certain virulence genes has made increasing numbers of isolates capable of causing a variety of intestinal and systemic infections (Robins-Browne 1987, Clarke et al. 2002).

    E. coli is the type species in the family *Enterobacteriaceae*—a group of facultative anaerobic, nonsporing, Gram-negative rods that inhabit the gastrointestinal tract of humans and animals. This family of about 30 genera, with approximately 100 species, includes some natural pathogens such as *Salmonella* spp., *Yersinia* spp., *Shigella* spp., and others like *Escherichia*, *Klebsiella*, *Citrobacter*, *Proteus*, and *Klebsiella*, which are usually commensals. However, some strains within each of the latter species may be associated with the pathogenesis of a disease syndrome. Colonization of the newborns' gut by the fecal–oral route occurs within 4–12 hours, as their birth environment is contaminated with *E. coli*. However, only a few of the several hundred strains of *E. coli* that form nearly 1% of the total enterobacterial population of the gut may be capable of causing a disease. The pathogenic *E. coli* are different from the nonpathogenic strains in that they have virulence genes located either extrachromosomally or chromosomally. The circular chromosomal DNA of *E. coli* is about 0.15 cm in length and consists of 4×10^6 base pairs. The virulence genes, which encode proteins for toxin activity, cellular adherence, and invasion, were most likely transferred to the pathogenic strains from related enterobacteria. The strains can be differentiated with antisera specific to the somatic lipopolysaccharide (O) antigens, capsular (K) antigens, and flagellar (H) antigens (Kauffmann 1947). There are nearly 700 different antigenic types based on the 100 different K groups, 180 O groups, and more than 60 H antigens (Robins-Browne and Hartland 2002). In general, *E. coli* causes three types of infections: enteric infections, urinary tract infections, and septicemic infections. The urinary tract infections can be asymptomatic bacteriuria or urosepsis, whereas the septicemic infections lead to meningitis, pneumonia, cholecystitis, and wound infection. Many serologically typed enteric *E. coli* isolates have been broadly grouped into six different pathotypes: enterotoxigenic *E. coli* (ETEC), enteroaggregative *E. coli* (EAEC), diffuse-adherence *E. coli* (DAEC), enteropathogenic *E. coli* (EPEC), enterohemorrhagic *E. coli* (EHEC), and enteroinvasive *E. coli* (EIEC). However, certain isolates have been placed in a seventh group: necrotoxigenic *E. coli* (NTEC). In this chapter, we will consider the above pathotypes (See Chapter 3 for EHEC) that cause enteric infections.

ETEC

De et al. (1956) established the pathogenicity of *Bacterium coli* (*E. coli*) from human cases of gastroenteritis, using the ligated loop of rabbit intestine. Later, enterpathogenicity of several *E. coli* isolates causing diarrhea and death in newborn cattle and pigs was demonstrated as being caused by secreted toxins (Smith and Halls 1967). Similar ETEC strains expressing plasmid-encoded enterotoxins were associated with infantile diarrhea and traveler's diarrhea in humans (Sack 1975).

ETEC attach to the surface of the intestinal epithelial cells by colonizing factors (CFs), multiply, and produce either a heat-labile toxin (LT) or a heat-stable toxin (ST) or both the enterotoxins. The CFs are rigid 5–7-nm rodlike fimbriae, flexible 2–4-nm thin fibrillae, or a very fine mesh of nonfimbrial fibrils. Biogenesis of CFs requires a cluster of genes, either on a plasmid or the chromosome, that includes genes for the pilin subunit, chaperones for transporting the subunits across the periplasmic space, ushers for the assembly of the fimbria on the outer membrane, and regulatory proteins. The adhesive moiety, which binds specifically to a carbohydrate-bearing receptor on the epithelial cells, may be on the tip of the fimbriae or spread out on the entire length of the fimbriae (Mol and Oudega 1996). Because the receptor oligosaccharide will vary with the tissue as well as the host species, ETEC colonization is host specific. ETEC strains isolated from piglets and calves possess any of the following fimbriae: F1 (type 1), F4 (K88), F5 (K99), F6 (987P), F17, F18, F41, F42, and F165. F4, for example, is encoded by an eight-gene operon (*faeA-J*) on a plasmid, with the pilus primarily composed of the *faeG* gene product, and is the subunit responsible for binding to the host receptor containing Gal α(1-3)Gal on porcine enterocytes. Expression of the F4 operon genes is tightly regulated by specific genes, including cyclic AMP receptor protein (*Crp*) and leucine-responsive proteins (*Lrp*; Mol and Oudega 1996). Animal ETEC isolates express different CF from those of human ETEC and are not considered zoonotic pathogens. About 20 different CF structures, designated colonization factor antigens (CFAs), which can be either fimbriae such as CFA/I or fibrillae composed of two fibrils formed in a double helix such as CFA/II or CFA/IV, have been isolated from human strains (Gaastra and Svennerholm 1996).

The structure and the mode of action of the LT enterotoxins are similar to the cholera toxin. The 84–86-kDa holotoxin, encoded by *elt* genes on a plasmid, is a multimer protein made up of an enzymatically active (A) subunit surrounded by five identical binding (B) subunits. The holotoxin is assembled for transport through the outer membrane by the type II secretory pathway after the subunits containing a signal sequence are translocated across the cytoplasmic membrane into the periplasmic space (Tauschek et al. 2002). LTs have been classified into two subclasses, LT-I and LT-II, because of differences in the B subunits. LT-II is expressed primarily by ETEC isolates from animals, and LT-I is expressed by both human (LTh-I) and animal (LTp-I) isolates (Sears and Kaper 1996).

Endocytosis of LT occurs when B units bind to surface ganglioside receptors on enterocytes, LT-1 to GM1 and LT-II to GD1 (Fukuta et al. 1988). Proteolytic cleavage of the A unit following endocytosis yields disulfide bond conjoined A1 and A2 subunits. A1 catalyzes ADP–ribosylation of the α subunit of GTP-binding-regulatory-protein Gs, causing an irreversible activation of the adenylate cyclase. The resulting cAMP (adenosine $3',5'$-monophosphate) buildup activates protein kinase A to excessively phosphorylate the Cl^- channel, CFTR, thereby increasing the Cl^- secretion across the crypt cells and also leading to malabsorption of sodium chloride at the villus tips. Diarrhea results from water passively drawn into the intestinal lumen by the increased ion content (Sears and Kaper 1996).

The ST enterotoxins, classified as ST-I and ST-II, are a family of toxins whose 72–amino acid preprotein, coded by *estA* gene on a plasmid, is processed to produce an extracellular biologically active component of about 2 kDa that is rich in cysteine residues (Betley et al. 1986, Rasheed et al. 1990). ST-I causes an increase in cyclic GMP in host cell cytoplasm by binding to an extracellular domain of the 120-kDa guanylate cyclase C (GC), found on the apical membranes of host cells, leading to effects similar to those seen with the increase in cAMP. ST-I binds competitively with guanylin, a natural hormonal peptide, that activates GC and stimulates Cl^- secretion (Sears and Kaper 1996) and with the inhibition of NaCl co-transport in the villi cells resulting in diarrhea (Moon 1978). ST-II is mostly found in ETEC isolates from pigs, although it is secreted by some human strains. The mature ST-II is about 5 kDa, processed from a 71-kDa preprotein encoded by *estB* gene on a plasmid, and causes villi atrophy with loss of epithelial cells (Whipp et al. 1986). The mechanism of action of ST-II is still unknown as there is no stimulation of either adenylate or guanylate cyclase; however, activation of a GTP–binding protein–regulated calcium channel has been suggested as a target of ST-II (Dreyfus et al. 1993). The mechanism may involve transport of bicarbonate rather than Cl^- ions in the increased intestinal secretion (Peterson and Whipp 1995).

ETEC cause a cholera-like syndrome with acute diarrhea in humans as well as animals, especially in developing countries, where morbidity and mortality in the young can be high. In developed countries, ETEC may be isolated from people who have recently visited countries where sanitation and hygiene are poor, and infections are acquired by ingestion of contaminated food and water.

EAEC

Recent epidemiological evidence from developing countries and studies in volunteers confirm the importance of some EAEC strains that cause diarrhea (Okeke and Nataro 2001, Jiang et al. 2002). These *E. coli* strains, initially classified as EPEC, attach to epithelial cells in a diverse aggregative pattern resembling stacked bricks. This attachment is accompanied by the presence of thick mucus on the epithelium, which may have a role in the persistence of infection (Tzipori et al. 1992). EAEC may adhere to the intestinal epithelium initially through adhesins such as aggregative adherence fimbriae (AAF) I, II, or III, which are expressed by some isolates and are coded by a gene cluster on an approximately 110-kb EAEC virulence plasmid (pAA; Okeke and Nataro 2001, Bernier et al. 2002).

Most of these strains produce three toxins, which may likely stimulate intestinal secretion. *E. coli* heat-stable enterotoxin (EAST), encoded by the gene *astA*, is a 38–amino acid protein with 50% homology to ST-I (Sears and Kaper 1996). Although less potent than ST-I, it increases cyclic GMP in the intestinal mucosa by activating GC-C, but does not produce any microscopically visible changes (Uzzau and Fasano 2000). A few isolates other than EAEC may either carry the gene for EAST as a transposon or express the toxin, but EAST's role in the pathogenesis of disease following infection with these *E. coli* is still unclear (Savarino et al. 1996, Zhou et al. 2002). A 120-kDa heat-labile toxin related to the hemolysin produced by uropathogenic *E. coli* strains has been identified in EAEC, but its role in pathogenesis is unknown (Sears and Kaper 1996). Another protein, Pet, has been identified as a putative cause of intestinal secretion resulting from acute inflammation. This 108-kDa protein, encoded by a gene on the plasmid pAA and a member of serine protease autotransporters of *Enterobacteriaceae* (SPATE), causes rounding and detachment of HEp-2 cell monolayers probably because of the breakdown of fodrin, an isoform of spectrin—a protein that maintains red blood cell shape and flexibility by linking cytoskeleton to the plasma membrane. Pet also cleaves Factor V and pepsin (Esalva et al. 1998, Villasenca et al. 2000, Henderson and Nataro 2001, Dutta et al. 2002). Another SPATE (Pic) present in some EAEC strains and involved in intestinal colonization does not cleave spectrin but can cleave mucin and Factor V (Henderson et al. 1999). Both Pet and Pic are related to elastases based on specific activity on peptides and by phylogenetic analysis (Dutta et al. 2002).

Aggregated EAEC penetrate into mucus for dispersion in the intestine by releasing a 10.2-kDa protein, dispersin, that probably decreases the binding affinity of AAF by attaching to the fimbriae at or near the site of attachment to host cell receptors (Sheikh et al. 2002). Dispersin is encoded by the *aap* gene upstream of the AraC transcriptional activator (*AggR*) necessary for AAF expression.

EAEC are associated with long-term watery diarrhea in young children, especially in developing countries (Okeke and Nataro 2001), but unlike ETEC infections, the diarrhea with concurrent increase in IL-8 can be inflammatory and sometimes bloody. However, growth retardation in infants associated with EAEC infections may be more important than diarrhea (Steiner et al. 1998, Nataro and Kaper 1998).

DAEC

Different from the localized (EPEC) and aggregative (EAEC) pattern of adhesion to cultured HEp-2 cells, some isolates of *E. coli* exhibit a more diffuse attachment pattern, in which the bacteria uniformly cover the entire cell surface, and hence are termed "diffuse-adherence *E. coli*" (Scaletsky et al. 1984). Implicated in the diffuse adherence of DAEC strains is a family of very fine fibrillar afimbrial adhesins and invasins (Afa), whose genes are within a cluster either in an 11.6-kb chromosomal region or on a plasmid (Keller et al. 2002). The product of *afa* genes shows homology to the adhesins that bind to a Dr–blood group antigen on decay-accelerating factor (Nowicki et al. 2001). *E. coli* strains expressing this group of adhesions cause intestinal infections especially in children older than 2 years (Nataro and Kaper 1998). *E. coli* strains possessing gene clusters related to the *afa* family have been isolated from calves with diarrhea and septicemia (Lalioui et al. 1999). However, the Afa of the bovine *E. coli* isolates bind to different receptors from those of the human strains, indicating that such *afa* carrying isolates from animals may not be important in the cause of disease in humans.

E. coli secretory protein homologues have been identified in the DAEC isolates, but other genes for proteins such as type three secretion (TTS) system or effector proteins, which subvert host cellular signaling for producing, attaching, and effacing lesions of EPEC infections, may be missing (Beinke et al. 1998). However, a DAEC isolate from a diarrheagenic child had part of a pathogenicity island nucleotide sequence that is usually found in uropathogenic *E. coli* (Blanc-Potard et al. 2002). It is suggested that Afa/Dr adhesin binding to the host cell surface receptor including decay-accelerating factor may be the main mechanism by which enteropathogenicity of DAEC is initiated by induction of a proinflammatory response. Contact between leukocytes and the epithelial cell basolateral membrane may increase permeability by enhanced phosphorylation of myosin L chain at the tight junctions of the epithelial cells (Edens et al. 2002).

Opinions on the virulence of DAEC as the cause of diarrheal disease tend to be contentious (Le Bouguénec 1999). DAEC did not induce diarrhea in healthy adult volunteers, suggesting that this heterogeneous group of *E. coli* may cause disease in immunosuppressed individuals and malnourished children and may be an important cause of a persistent diarrhea rather than an acute diarrhea in developing countries (Nataro and Kaper 1998).

EPEC

Several *E. coli* with certain O antigens first identified by Bray in 1945 are the most common causes of infantile diarrhea in the developing countries of the world (Nataro and Kaper 1998). Such non-enterotoxigenic and non-invasive strains (Levine et al. 1978) are locally adherent *in vitro* to epithelial cells in a distinctive pattern (Scaletsky et al. 1984). The severe, sometimes fatal, watery diarrhea is caused by a characteristic "attaching and effacing" (A/E) lesion in the intestinal epithelium (Moon et al. 1983). Microvilli disappear as EPEC come into close—but loose—contact with the enterocyte surface to form microcolonies in localized areas of the intestinal epithelium. This prerequisite for the formation of the A/E lesion is associated with the expression of about 14 genes on the approximately 80-kb virulence EPEC adherence factor (EAF) plasmid required for the biogenesis of the bundle-forming pilus (BFP)—a type IV fimbria (Nataro and Kaper 1998). BfpA is the basic subunit required to form BFP in association with a global regulatory protein Per and a periplasmic enzyme encoded by *dsbA* (DeVinney et al. 1999). Tobe and Sasakawa (2001) showed that BFP-expressing EPEC adhered directly to the cell surface instead of to preformed microcolonies on the cell surface. However, structures other than BFP, such as flagella, may be involved in the initial loose attachment, and BFP solely contributes to the bacterial aggregation (Giron et al. 2002). BFP undergoes a structural transformation during the late log and stationary phases of the bacterial growth, as the aggregates disperse, and it is associated with the expression of a nucleotide binding protein, bfpF (Frankel et al. 1998, Knutton et al. 1999).

Once attached, EPEC start translocating onto or into the host cell several of its own proteins essential in the induction of signal transduction for the formation of A/E lesions via a specialized TTS system (Jarvis et al. 1995, Frankel et al. 1998). These proteins are expressed by a gene cluster (locus of enterocyte effacement, LEE) within a 35-kb pathogenicity island in the chromosome (Kenny 2002). This LEE of about 41 genes containing three functional domains—the region for the *E. coli* secreted proteins (Esps) and their chaperones (Ces), the region containing translocated intimin receptor (Tir) and intimin for intimate adherence, and the region of TTS pathway proteins—is presumed to have been transferred horizontally (Deng et al. 2001).

On bacterial attachment, EspA proteins form a tubelike conduit from the EPEC surface to the host cell surface for translocating several correlated proteins, including EspB and EspD chaperoned by CesD to create a pore (translocon) in the enterocyte membrane (DeVinney et al. 1999, Daniell et al. 2001). Then the 78-kDa protein, Tir (EspE or Hp90), is transported by CesT into host cells, where Tir may undergo some processing by a codelivered bacterial protein. Tir, which is phosphorylated at one or more of its tyrosines before insertion into the host cell membrane (DeVinney et al. 1999), has two transmembrane domains with an extracellular central fragment that acts as a docking unit for intimin, a 94-kDa outer-membrane protein of EPEC encoded by the *eae* gene in the LEE locus (Jerse et al. 1990, de Grado et al. 1999). The intimin family of proteins binds to Tir with a disulfide bond–bridged 76–amino acid loop of their C-terminus 280 amino acids (Kelly et al. 1998). Part of this C-terminus of intimin may bind to other host cell–encoded intimin receptors such as β1 integrins and nucleolin (DeVinney et al. 1999, Sinclair and O'Brien 2002). As the bacterium intimately adheres, an active cytoskeletal rearrangement is initiated within the host cell for the pedestal formation and phospholipase Cγ activation in the cell signaling cycle (Kenny 2002). The cytoskeletal arrangement is initiated with Tir binding to a host cell protein, Nck, that starts a process of actin polymerization just below where the EPEC is attached, which also involves other host cytoskeletal proteins such as α-actinin, ezrin, talin, and villin that bind to actin, either directly or indirectly. This polymerization requires neural Wiscott-Aldrich syndrome protein and Arp 2/3 complex mobilization (Thrasher 2002) that is dependent on Tir phosphorylation, a distinct process separate from the one involved in EHEC A/E lesions (DeVinney 2001, Kenny 2002). Tir associates directly with host α-actinin through its *N*-terminus that aids anchoring of EPEC to the pedestals. Other proteins like BipA may play an enhancing role in this actin remodeling (Farris et al. 1998).

The bacterial adhesion triggers an increase in Ca^{2+} within the enterocytes, Ca induced by the release from inositol phosphatase–sensitive intracellular stores because of phospholipase Cγ1 activation. This leads to decreased absorption of NaCl and an increase in the chloride ion secretion, leading to diarrhea (DeVinney et al. 1999). The elevated Ca^{2+} levels activate calmodulin-dependent protein kinases by phosphorylation of myosin light chain, which may increase the permeability of enterocyte tight junctions. This permeability is related to a progressive shift of occludin—a major integral membrane protein component localized primarily at tight junctions to an intracellular compartment (Frankel et al. 1998, Simonovic et al. 2000).

EPEC strains initiate an inflammatory response during the formation of the A/E lesions that is associated with increased levels of IL-8 because of NF-κB activation, which leads to paracellular permeability and may also stimulate chloride ion secretion. There is interference in the signal transduction within the damaged infected cells that also plays a role in the diarrhea and other symptoms of EPEC infections (DeVinney et al. 1999).

Cultured epithelial cells show necrosis as well as apoptosis following attachment with EPEC. This induction of cell death has been directly correlated with the expression of BFP (Abul-Milh et al. 2001) and binding of phosphatidylethanolamine to BFP, which triggers downstream events involving some of the key mediators of apoptosis. However, cell death is considered multifactorial; for example, a gene upstream of *tir* encodes an additional effector protein, Orf19 or mitochondrial-associated protein, whose delivery is dependent on EspB, but whose target is the host mitochondria. This protein appears to interfere with the mitochondrial membrane potential, which may impair either energy production or control of cell death. An alternate function of this molecule may involve cytoskeletal arrangement (Kenny 2002). Orf30 in the LEE encodes a protein EspF, which has three almost identical repeats of a 19–amino acid proline-rich sequence, a motif present in other eukaryotic signaling proteins. EspF could contribute to the disruption of tight junctions of the epithelial cells (Crane et al. 2001, Kenny 2002) and not be involved in cell detachment (Shifrin et al. 2002). EPEC also secrete a 110-kDa SPATE, which digests spectrin, Factor V, and pepsin similarly to Pet of EAEC; however, unlike Pet, it does not cause any cytopathic effects (Dutta et al. 2002). Recently, it was suggested that EPEC translocate proteins to effect detachment of infected host cells by dephosphorylating a kinase in vinculin—a cytoskeletal protein associated with actin binding proteins contained in the focal adhesions by which the cells attach to the extracellular matrix (Shifrin et al. 2002).

Typical EPEC isolates, found only in humans, do not elaborate LT, ST, Shiga toxin, Shiga-like toxin, or verocytotoxins, but they possess the EAF plasmid, whereas atypical EPEC do not possess this plasmid but may produce a *Shigella*-like toxin. The majority of typical EPEC strains group into certain O:H serotypes and can be classified into two clonotypes based on sequence variation in the housekeeping genes (Frankel et al. 1998). Both clonotypes may be of various O serotypes, but clone 1 mostly has flagellar antigen H6, whereas clone 2 has H2 or is nonmotile. Atypical EPECs lacking the EAF plasmid have been increasingly reported to occur both in humans and animals in developing countries and may be emerging pathogens of significance (Trabulsi et al. 2002).

EIEC

EIEC closely resemble *Shigella* in their invasive nature and pathogenesis in producing dysentery-like illness, which is sometimes fatal in young children. The first association of EIEC with human disease was mistaken as shigellosis because of the similarities in the clinical symptoms and the biochemical properties. It was only when a few isolates from patients did not match biochemically to *Shigella* that they were termed Shigella-like (Ewing and Gravatti 1947). Later, a similar isolate was incriminated as the cause of an acute gastroenteritis that was eventually typed as *E. coli* (Ewing et al. 1958). Conclusive evidence came from studies by DuPont et al. (1971), when volunteers fed EIEC developed diarrhea. EIEC invade the intestinal cells, especially those of the colon, where they multiply and cause severe inflammation, with extensive damage. Membranous epithelial cells (M cells) in the intestinal epithelium are the most likely targets for attachment and invasion by EIEC because of their resemblance to *Shigella* (Sansonetti and Phalipon 1999). The invasion ability of EIEC is associated with the presence of a 140-MDa plasmid that carries a cluster of virulence genes in a 20-kb region. This plasmid (pINV) has sequence homology to plasmids found in certain strains of *Shigella flexneri*, *Shigella boydii*, and *Shigella dysenteriae* (Lan et al. 2001). With *Shigella*, bacterial invasion of epithelial cells is prevented if mutations are created in four of these plasmid encoded genes (*ipaA, ipaB, ipaC,* and *ipaD*; Baudry et al. 1987), whose products bind to $\alpha_5\beta1$ integrin on the enterocytes, inducing membrane ruffling and leading to bacterial internalization (Watarai et al. 1996). Because EIEC do not possess any fimbriae, a nonfimbrial outer membrane protein is considered likely to be an adhesin for the intestinal cells. These invasive *E. coli* do not produce LT, ST, EAST, and unlike *Shigella*, the verocytotoxins (Shiga toxins). However, in rabbit ligated loops, supernatants from EIEC cultures stimulate some degree of secretion, which may be a result of an enterotoxin (EIEC enterotoxin or *Shigella* enterotoxin 2) of about 62 kDa, encoded by gene *sen* on the pInv plasmid (Nataro et al. 1995). The cause of diarrhea could be the action of this enterotoxin, which alters expression or up-regulates the inducible NO synthase and cyclooxygenase 2, leading to the modulation of Cl^- secretion and barrier function in the intestinal epithelial cells (Resta-Lenert and Barrett 2002). In addition, a small, 30-kDa protein that demonstrates a mild cytotoxicity for Vero cells has been identified, but its activity is unknown (Sears and Kaper 1996).

Although EIEC infections still occur in developed countries, they are more common in developing countries. The clinical syndrome usually includes vomiting, fever, and an initial watery diarrhea, which can become mucoid and bloody in some cases.

NTEC

Several isolates of *E. coli* from cases of diarrhea in pigs, calves, and humans, and sometimes from healthy individuals, produce cytotoxic necrotizing factors (CNFs; De Rycke et al. 1999). Two antigenic types of CNF—CNF-1 of about 115 kDa and CNF-2 of about

110 kDa—cause the formation of giant cells in HeLa and Vero cell monolayers (Pohl et al. 1993). CNF-1 is encoded by genes located on the chromosome, whereas CNF-2 is encoded by a plasmid designated Vir (Oswald et al. 1989, Falbo et al. 1992). Experimental studies indicate that these toxins are associated with the pathogenesis of NTEC infections (Wray et al. 1993). CNF-2-producing *E. coli* (NTEC-2) are isolated mainly from animals, and NTEC-1-producing CNF-1 have been isolated from humans and various pet and food animals (De Rycke et al. 1999, Van Bost et al. 2001). Aside from diarrhea, other syndromes associated with septicemia and urinary tract infections can occur. CNF-2 increases the activity of a GTP-binding-Rho protein (Oswald et al. 1994). Other toxins such as LT-II, EAST1, and cytolethal-distending toxin type III (Oswald et al. 1991, Peres et al. 1997, Bertin et al. 1998) may be produced by some isolates of NTEC. Cytolethal-distending toxin type III prevents the activation of cdc2 protein kinase by the blocking of host cells in G2/M phase (Comayras et al. 1997, Peres et al. 1997).

Conclusions

The current research trend in trying to understand the pathogenesis of the various *E. coli* isolates and to group them by a common factor relative to virulence is providing valuable information on the results of the *in vitro* interaction between the host cell and enteric *E. coli*. However, the elucidation of molecular mechanisms in the process of attachment of certain pathotypes or downstream events in the host cells is incomplete. Characterization of molecular interactions is necessary in the development of intervention strategies to prevent or treat infections. This need can be addressed by the use of *in vivo* expression technology supplemented by DNA array use to study differential transcription following attachment of various *E. coli* pathotypes to the host cells (Merrel and Camilli 2000). For example, such studies can provide information on the co-regulation of toxin production *in vivo*. However, as there can be a weak correlation between the total mRNA transcribed and the levels of proteins, a broad proteome analysis may identify relevant molecules that generate a detrimental response from the enterocytes (Washburn and Yates 2000). Protein profiling under conditions conducive for infections adds an advantage in the study of bacterial protein–host cell protein interactions either in yeast two-hybrid or in protein microarray screening systems (Kodadek 2001, Uetz 2002). These techniques can allow finding target proteins for modulation, for example, inhibition of binding, by drugs or other proteins and for immunological studies. The assembly of virulence-associated pili, PapG, is blocked when an octamer peptide representing the C-terminus of the pilus protein binds to the usher protein PapD (Flemmer et al. 1995). Similar proteins including fimbria can be targeted for competitive inhibition by identifying peptides using peptide phage libraries.

Supplements of specific antibodies to either whole *E. coli* or one of its components can prevent or reduce infections by preventing the early but critical step of attachment of the bacteria to the enterocytes (Morter 1984, Korhonen et al. 2000). However, this method is not commonly used for prevention, most likely because of problems associated with the means of production and storage of the antibodies. Antibody engineering techniques allow identification and development of antibody binding sites (Fv) for an epitope (or its variants) on fimbria, which may be the adhesive sites for the host cell receptor. The relevant Fv or a cocktail of Fvs can be produced cheaply in several expression systems including plants (Gavilondo and Larrick 2000). Similarly, other proteins can be targeted to reduce lesions; for example, dispersin of EAEC (Sheikh et al. 2002). The ability of EAEC-Pet to degrade spectrin and fodrin is inhibited by monospecific polyclonal antibodies (Villasenca et al. 2000), indicating that an Fv directed against certain epitopes of Pet may prevent or reduce cell death and associated pathogenesis.

Thus, molecular dissection with the state-of-the-art techniques will provide virulence-related indicators that will be useful in identifying drugs or inhibitory molecules and immunological relevant molecules for use as vaccines. The signature molecules may also provide a basis for the development of diagnostic tests for relevant strains intended for epidemiologic analysis. These tests can be useful in reducing the contamination of food for prevention of enteric *E. coli* infections.

References

Abul-Milh, M., Wu, Y., Lau, B., Lingwood, C. A. and Barnett Foster, D. 2001. Induction of epithelial cell death including apoptosis by enteropathogenic *Escherichia coli* expressing bundle-forming pili. *Infection and Immunity* **69**:7356–7364.

Baudry, B., Maurelli, A. T., Clerc, P., Sadoff, J. C. and Sansonetti, P. J. 1987. Localization of plasmid loci necessary for the entry of *Shigella flexneri* into HeLa cells, and characterization of one locus encoding four immunogenic polypeptides. *Journal of General Microbiology* **133**:3403–3413.

Beinke, C., Laarmann, S., Wachter, C., Karch, H., Greune, L. and Schmidt, M. A. 1998. Diffusely adhering *Escherichia coli* strains induce attaching and effacing phenotypes and secrete homologs of Esp proteins. *Infection and Immunity* **66**:528–539.

Bernier, C., Gounon, P. and Le Bouguenec, C. 2002. Identification of an aggregative adhesion fimbria (AAF) type III-encoding operon in enteroaggregative *Escherichia coli* as a sensitive probe for detecting the AAF-encoding operon family. *Infection and Immunity* **70**:4302–4311.

Bertin, Y., Martin, C., Girardeau, J. P., Pohl, P. and Contrepois, M. 1998. Association of genes encoding P fimbriae, CS31A antigen and EAST 1 toxin among CNF1-producing *Escherichia coli* strains from cattle with septicemia and diarrhea. *FEMS Microbiology Letters* **162**:235–239.

Betley, M. J., Miller, V. L. and Mekalanos, J. J. 1986. Genetics of bacterial enterotoxins. *Annual Review of Microbiology* **40**:577–605.

Bettlheim, K. S. 1988. The intestinal bacteria of the neonate and the breast-fed infant. *Review of Infectious Diseases* **10**:1220–1225.

Blanc-Potard, A.-B., Tinsley, C., Scaletsky, I., Le Bouguenec, C., Guignot, J., Servin, A. L., Nassif, X. and Bernet-Camard, M.-F. 2002. Representational difference analysis between Afa/Dr diffusely adhering *Escherichia coli* and non-pathogenic *E. coli* K-12. *Infection and Immunity* **70**:5503–5511.

Clarke, S. C., Haigh, R. D., Freestone, F. F. E. and Williams, P. H. 2002. Enteropathogenic *Escherichia coli* infection: history and clinical aspects. *British Journal of Biomedical Science* **59**:123–127.

Comayras, C., Tascas, C., Peres, S. Y., Ducommun, B., Oswald, E. and De Rycke, J. 1997. *Escherichia coli* cytolethal distending toxin blocks the HeLa cell cycle at G2/M transition by preventing cdc2 protein kinase dephosphorylation and activation. *Infection and Immunity* **65**:5088–5095.

Crane, J. K., McNamara, B. P. and Donnenberg, M. S. 2001. Role of EspF in host cell death induced by enteropathogenic *Escherichia coli*. *Cellular Microbiology* **3**:197–211.

Daniell, S. J., Takahashi, N., Wilson, R., Friedberg, D., Rosenshine, I., Booy, F. P., Shaw, R. K., Knutton, S., Frankel, G. and Aizawa, S-I. 2001. The filamentous type III secretion translocon of enteropathogenic *Escherichia coli*. *Cellular Microbiology* **3**:865–871.

De, S. N., Bhattacharya, K. and Sarkar, J. K. 1956. A study of the pathogenicity of *Bacterium coli* from acute and chronic enteritis. *Journal of Pathology and Bacteriology* **71**:201–209.

de Grado, M., Abe, A., Gauthier, A., Steele-Mortimer, O., DeVinney, R. and Finlay, B. B. 1999. Identification of the intimin-binding domain of Tir of enteropathogenic *Escherichia coli*. *Cellular Microbiology* **1**:7–17.

De Rycke, J., Milon, A. and Oswald, E. 1999. Necrotoxigenic *Escherichia coli* (NTEC): two emerging categories of human and animal pathogens. *Veterinary Research* **30**:221–233.

Deng, W., Li, Y., Vallance, B. A. and Finlay, B. B. 2001. Locus of enterocyte effacement from *Citrobacter rodentium*: sequence analysis and evidence for horizontal transfer among attaching and effacing pathogens. *Infection and Immunity* **69**:6323–6335.

DeVinney, R., Knoechel, D. G. and Finlay, B. B. 1999. Enteropathogenic *Escherichia coli*: cellular harassment. *Current Opinion in Microbiology* **2**:83–88.

DeVinney, R., Puente, J. L., Gauthier, A., Goosney, D. and Finlay, B. B. 2001. Enterohaemorrhagic and enteropathogenic *Escherichia coli* use a different Tir-based mechanism for pedestal formation. *Molecular Microbiology* **41**:1445–1458.

Dreyfus, L. A., Harville, B., Howard, D. E., Shaban, R., Beatty, D. M. and Morris, S. J. 1993. Calcium influx mediated by the *Escherichia coli* heat-stable enterotoxin B (ST$_b$). *Proceedings of the National Academy of Sciences (USA)* **90**:3202–3206.

Dulaney, A. D. and Michelson, I. D. 1935. A study of *B. coli* mutabile from an outbreak of diarrhea in the newborn. *American Journal of Public Health* **25**:1241–1251.

DuPont, H. L., Formal, S. B., Hornick, R. B., Snyder, M. J., Labonati, J. P., Sheanan, D. G., La Brec, E. H. and Kalas, J. P. 1971. Pathogenesis of *Escherichia coli* diarrhea. *New England Journal of Medicine* **285**:1–9.

Dutta, P. R., Cappello, R., Navarro-Garcia, F. and Nataro, J. P. 2002. Functional comparison of serine protease autotransporters of *Enterobacteriaceae*. *Infection and Immunity* **70**:7105–7113.

Edens, H. A., Levi, B. P., Jaye, D. L., Walsh, S., Reaves, T. A., Turner, J. R., Nusrat, A. and Parkos, C. A. 2002. Neutrophil transepithelial migration: evidence for sequential, contact-dependent signaling events and enhanced paracellular permeability independent of transjunctional migration. *Journal of Immunology* **169**:476–486.

Esalva, C., Navarro-Garcia, F., Czeczulin, J. R., Henderson, I. R., Cravioto, A. and Nataro, J. P. 1998. Pet, an autotransporter enterotoxin from enteroaggregative *Escherichia coli*. *Infection and Immunity* **66**:3155–3163.

Escherich, T. 1885. Bakteriologische untersuchngen über frauenmilch. *Forteschritte der Medizin* **3**:231–236.

Ewing, W. H. and Gravatti, J. L. 1947. *Shigella* types encountered in the area. *Journal of Bacteriology* **53**:191–195.

Ewing, W. H., Reavis, R. W. and Davis, B. R. 1958. Provisional *Shigella* serotypes. *Canadian Journal of Microbiology* **4**:89–107.

Falbo, V., Famiglietti, M. and Caprioli, A. 1992. Gene block encoding production of cytotoxic necrotizing factor 1 and hemolysin in *Escherichia coli* isolates from extra-intestinal infections. *Infection and Immunity* **60**:2182–2187.

Farris, M., Grant, A., Richardson, T. B. and O'Connor, C. D. 1998. BipA: a tyrosine-phosphorylated GTPase that mediates interactions between enteropathogenic *Escherichia coli* (EPEC) and epithelial cells. *Molecular Microbiology* **28**:265–279.

Flemmer, K., Xu, Z., Pinkner, J. S., Hultgren, S. J. and Kihlberg, J. 1995. Peptides inhibit complexation of the bacterial chaperone PapD and reveal potential to block assembly of virulence associated pili. *Bioorganic and Medicinal Chemistry Letters* **5**:927–932.

Frankel, G., Phillips, A. D., Rosenshine, I., Dougan, G., Kaper, J. B. and Knutton, S. 1998. Enteropathogenic and enterohaemorrhagic *Escherichia coli*: more subversive elements. *Molecular Microbiology* **30**:911–921.

Fukuta, S., Magnani, J. L., Twiddy, E. M., Holmes, R. K. and Ginsburg, V. 1988. Comparison of the carbohydrate-binding specificities of cholera toxin and *Escherichia coli* heat-labile enterotoxins LTh-I; LT-1a, and LT-1b. *Infection and Immunity* **56**:1748–1753.

Gaastra, W. and Svennerholm, A.-M. 1996. Colonization factors of human enterotoxigenic *Escherichia coli* (ETEC). *Trends in Microbiology* **4**:444–452.

Gavilondo, J. V. and Larrick, J. W. 2000. Antibody engineering at the millennium. *BioTechniques* **29**:128–149.

Giron, J. A., Torres, A. G., Freer, E. and Kaper, J. B. 2002. The flagella of enteropathogenic *Escherichia coli* mediate adherence to epithelial cells. *Molecular Microbiology* **44**:361–379.

Henderson, I. R., Czeczulin, J., Esalva, C., Noreiga, F. and Nataro, J. P. 1999. Characterization of Pic, a secreted protease of *Shigella flexneri* and enteroaggregative *Escherichia coli*. *Infection and Immunity* **67**:5587–5596.

Henderson, I. R. and Nataro, J. P. 2001. Virulence functions for auotransporter proteins. *Infection and Immunity* **69**:1231–1243.

Jarvis, K. G., Giron, J. A., Jerse, A. E., McDaniel, T. K., Donnenberg, M. S. and Kaper, J. B. 1995. Enteropathogenic *Escherichia coli* contains a putative type III secretion system necessary for the export of proteins involved in attaching and effacing lesion formation. *Proceedings of the National Academy of Sciences (USA)* **92**:7994–8000.

Jerse, A. E., Yu, J., Tall, B. D. and Kaper, J. B. 1990. A genetic locus of enteropathogenic *Escherichia coli* necessary for the production of attaching and effacing lesions on tissue culture cells. *Proceedings of the National Academy of Sciences (USA)* **84**:7839–7843.

Jiang, Z. D., Greenberg, D., Nataro, J. P., Steffen, R. and DuPont, H. L. 2002. Rate of occurrence and pathogenic effect of enteroaggregative *Escherichia coli* virulence factors in international travelers. *Journal of Clinical Microbiology* **40**:4185–4190.

Kauffmann, F. 1947. The serology of the coli group. *Journal of Immunology* **57**:71–100.

Keller, R., Ordonez, J. G., De Oliveira, R. R., Trabulsi, L. R., Baldwin, T. J. and Knutton, S. 2002. Afa, a diffuse adherence fibrillar adhesin associated with enteropathogenic *Escherichia coli*. *Infection and Immunity* **70**:2681–2689.

Kelly, G., Prasannan, S., Daniel, S., Frankel, G., Dougan, G. and Connerton, I. 1998. Sequential assignment of the triple labeled 30.1 kDa cell adhesion domain of intimin from enteropathogenic *E. coli*. *Journal of Biomolecular NMR* **12**:189–191.

Kenny, B. 2002. Enteropathogenic *Escherichia coli* (EPEC)—a crafty subversive little bug. *Microbiology* **148**:1967–1978.

Knutton, S., Shaw, R. K., Anantha, R. V., Donnenberg, M. S. and Zorgani, A. A. 1999. The type IV bundle-forming pilus of enteropathogenic *Escherichia coli* undergoes dramatic alterations in structure associated with bacterial adherence, aggregation and dispersal. *Molecular Microbiology* **33**:499–509.

Kodadek, T. 2001. Protein microarrays: prospect and problems. *Chemistry and Biology* **8**:105–115.

Korhonen, H., Marnila, P. and Gill, H. S. 2000. Bovine milk antibodies for health. *British Journal of Nutrition* **84**:S135–S146.

Lalioui, L., Jouve, M., Gounon, P. and Le Bouguénec, C. 1999. Molecular cloning and characterization of the *afa-7* and *afa-8* gene clusters encoding afimbrial adhesins in *Escherichia coli* strains associated with diarrhea or septicemia in calves. *Infection and Immunity* **67**:5048–5059.

Lan, R., Lumb, B., Ryan, D. and Reeves, P. R. 2001. Molecular evolution of large virulence plasmid in *Shigella* clones and enteroinvasive *Escherichia coli*. *Infection and Immunity* **69**:6303–6309.

Le Bouguénec, C. 1999. Diarrhea-associated diffusely adherent *Escherichia coli*. *Clinical Microbiology Reviews* **12**:180–181.

Levine, M. M., Bergquist, E. J., Nalin, D. R., Waterman, D. H., Hornick, R. B., Young, C. R., Sotman, S. and Rowe, B. 1978. *Escherichia coli* strains that cause diarrhea but do not produce heat-labile or heat-stable enterotoxins and are non-invasive. *Lancet* **1**:1119–1122.

Merrel, S. and Camilli, A. 2000. Detection and analysis of gene expression during infection by in vivo expression technology. *Philosophical Transactions: Biological Sciences* **355**:587–599.

Mol, O. and Oudega, B. 1996. Molecular and structural aspects of fimbriae biosynthesis and assembly in *Escherichia coli*. *FEMS Microbiology Reviews* **19**:25–52.

Moon, H. W. 1978. Mechanisms in the pathogenesis of diarrhea: a review. *Journal of American Veterinary Medical Association* **172**:443–448.

Moon, H. W., Whipp, S. C., Argenzio, R. A., Levine, M. M. and Gianella, R. A. 1983. Attaching and effacing activities of rabbit and human enteropathogenic *Escherichia coli* in pig and rabbit intestines. *Infection and Immunity* **53**:1340–1351.

Morter, R. L. 1984. Genetically engineered monoclonal antibody for *E. coli* diarrhea in calves. *Modern Veterinary Practice* **65**:427–428.

Nataro, J. P., Seriwatana, J., Fasano, A., Maneval, D. R., Guers, L. D., Noreiga, F., Dubovsky, F., Levine, M. M. and Morris Jr., J. G. 1995. Identification and cloning of a novel plasmid-encoded enterotoxin of enteroinvasive *E. coli* and *Shigella* strains. *Infection and Immunity* **63**:4721–4728.

Nataro, J. P. and Kaper, J. B. 1998. Diarrheagenic *Escherichia coli*. *Clinical Microbiology Reviews* **11**:142–201.

Nowicki, B., Selvarangan, R. and Nowicki, S. 2001. Family of *Escherichia coli* Dr adhesins: decay-accelerating factor receptor recognition and invasiveness. *Journal of Infectious Diseases* **183**:S24–S27.

Okeke I. N. and Nataro, J. P. 2001. Enteroaggregative *Escherichia coli*. *Lancet Infectious Diseases* **1**:304–313.

Oswald, E., De Rycke, J., Guillot, J. F. and Boivin, R. 1989. Cytotoxic effect of multinucleation in HeLa cell cultures associated with the presence of Vir plasmid in *Escherichia coli* strains. *FEMS Microbiology Letters* **49**:95–99.

Oswald, E., De Rycke, J., Lintermans, P., Van Muylen, K., Mainil, J., Daube, G. and Pohl, P. 1991. Virulence factors associated with cytotoxic necrotizing factor type-2 in bovine diarrheic and septicemic strains of *Escherichia coli*. *Journal of Clinical Microbiology* **29**:2522–2527.

Oswald, E., Sugai, M., Labigne, A., Wu, H. C., Fiorenti, C., Boquet, P. and O'Brien, A. D. 1994. Cytotoxic necrotizing factor type-2 produced by virulent *Escherichia coli* modifies the small GTP-binding proteins Rho involved in assembly of actin stress fibers. *Proceedings of the National Academy of Sciences (USA)* **91**:3814–3818.

Peres, S. Y., Marches, O., Daigle, F., Nougareyde, J.-P., Herault, F., Tasca, C., De Rycke, J. and Oswald, E. 1997. A new cytolethal distending toxin (CDT) from *Escherichia coli* producing CNF2 blocks HeLa cell division in G2/M phase. *Molecular Microbiology* **24**:1095–1107.

Peterson, J. and Whipp, S. C. 1995. Comparison of the mechanisms of action of cholera toxin and the heat-stable enterotoxins of *Escherichia coli*. *Infection and Immunity* **63**:1452–1461.

Pohl, P., Oswald, E., Van Muylen, K., Jacquemin, E., Lintermans, P. and Mainil, J. 1993. *Escherichia coli* producing CNF1 and CNF2 cytotoxins in animals with different disorders. *Veterinary Research* **24**:311–315.

Rasheed, J. K., Guzman-Verduzco, L.-M. and Kupersztoch, Y. M. 1990. Two precursors of the heat-stable enterotoxin of *Escherichia coli*: evidence of extracellular processing. *Molecular Microbiology* **4**:265–273.

Resta-Lenert, S. and Barrett, K. E. 2002. Enteroinvasive bacteria alter barrier and transport properties of human intestinal epithelium: Role of iNOS and COX. *Gastroenterology* **122**:1070–1087.

Robins-Browne, R. M. 1987. Traditional enteropathogenic *Escherichia coli* of infantile diarrhea. *Reviews of Infectious Diseases* **9**:28–53.

Robins-Browne, R. M. and Hartland, E. L. 2002. *Escherichia coli* as a cause of diarrhea. *Journal of Gastroenterology and Hepatology* **17**:467–475.

Sack, B. R. 1975. Human diarrheal diseases caused by enterotoxigenic *Escherichia coli*. *Annual Review of Microbiology* **29**:333–353.

Sansonetti, P. J. and Phalipon, A. 1999. M cells as ports of entry for enteroinvasive pathogens: mechanisms of interaction, consequences for the disease process. *Seminars in Immunology* **11**:193–203.

Savarino, S. J., McVeigh, A., Watson, J., Cravioto, A., Molina, J., Echeverria, P., Bhan, M. K., Levine, M. M. and Fasano, A. 1996. Enteroaggregative *Escherichia coli* heat-stable enterotoxin is not restricted to enteroaggregative *E. coli*. *Journal of Infectious Diseases* **173**:1019–1022.

Scaletsky, I. C. A., Silva, M. L. M. and Trabulsi, L. R. 1984. Distinctive patterns of adherence of enteropathogenic *Escherichia coli* to HeLa cells. *Infection and Immunity* **45**:534–536.

Sears, C. L. and Kaper, J. B. 1996. Enteric bacterial toxins: mechanisms of action and linkage to intestinal secretion. *Microbiology Reviews* **60**:167–215.

Sheikh, J., Czeczulin, J. R., Harrington, S., Hicks, S., Henderson, I. R., Le Bouguenec, C., Gounon, P., Phillips, A., and Nataro, J. P. 2002. A novel dispersin protein in enteroaggregative *Escherichia coli*. *Journal of Clinical Investigation* **110**:1329–1337.

Shifrin, Y., Kirschner, J., Gelger, B. and Rosenshine, I. 2002. Enteropathogenic *Escherichia coli* induces modification of the focal adhesions of infected host cells. *Cellular Microbiology* **4**:235–243.

Simonovic, I., Rosenberg, J., Koutsouris, A. and Hecht, G. 2000. Enteropathogenic *Escherichia coli* dephosphorylates and dissociates occludin from intestinal epithelial tight junctions. *Cellular Microbiology* **2**:305–315.

Sinclair, J. F. and O'Brien, A. D. 2002. Cell-surface localized nucleolin is a eukaryotic receptor for the adhesin intimin-gamma of enterohemorrhagic *Escherichia coli* O157:H7. *Journal of Biological Chemistry* **277**:2876–2885.

Smith, H. W. and Halls, S. 1967. Studies on *Escherichia coli* enterotoxin. *Journal of Pathology and Bacteriology* **93**:531–543.

Steiner, T. S., Lima, A. A., Nataro, J. P. and Guerranto, R. I. 1998. Enteroaggregative *Escherichia coli* produce intestinal inflammation and growth impairment and cause interleukin-8 release from intestinal epithelial cells. *Journal of Infectious Diseases* **177**:88–96.

Tauschek, M., Gorrell, R. J., Strugnell, R. A. and Robins-Browne, R. M. 2002. Identification of a protein secretory pathway for the secretion of heat-labile enterotoxin by an enterotoxigenic strain of *Escherichia coli*. *Proceedings of the National Academy of Sciences (USA)* **99**:7066–7071.

Thrasher, A. J. 2002. WASp in immune-system organization and function. *Nature Reviews Immunology* **2**:635–646.

Tobe, T. and Sasakawa, C. 2001. Role of bundle-forming pilus of enteropathogenic *Escherichia coli* in host cell adherence and in microcolony development. *Cellular Microbiology* **3**:579–585.

Trabulsi, L. R., Keller, R. and Gomes, A. T. 2002. Typical and atypical enteropathogenic *Escherichia coli*. *Emerging Infectious Diseases* **8**:508–513.

Tzipori, S., Montanaro, J., Robins-Browne, R. M., Vial, P., Gibson, R. and Levine, M. M. 1992. Studies with enteroaggregative *Escherichia coli* in the gnotobiotic piglet gastroenteritis model. *Infection and Immunity* **60**:5302–5306.

Uetz, P. 2002. Two-hybrid arrays. *Current Opinion in Chemical Biology* **6**:57–62.

Uzzau, S. and Fasano, A. 2000. Cross-talk between enteric pathogens and the intestine. *Cellular Microbiology* **2**:83–89.

Van Bost, S., Roels, S. and Mainil, J. 2001. Necrotoxigenic *Escherichia coli* type-2 invade and cause diarrhoea during experimental infection in colostrum-restricted newborn calves. *Veterinary Microbiology* **81**:315–329.

Villasenca, J. M., Navarro-Garcia, F., Mendonza-Hernandez, G., Nataro, J. P., Cravioto, A. and Esalva, C. 2000. Pet toxin from enteroaggregative *Escherichia coli* produces cellular damage associated with fodrin disruption. *Infection and Immunity* **68**:5920–5927.

Washburn, M. P. and Yates III, J. R. 2000. Analysis of the microbial proteome. *Current Opinion in Microbiology* **3**:292–297.

Watarai, M., Funato, S. and Sasakawa, C. 1996. Interaction of Ipa proteins of *Shigella flexneri* with $\alpha_5\beta_1$ integrin promotes entry of the bacteria into mammalian cells. *Journal of Experimental Medicine* **185**:281–292.

Whipp, S. C., Moseley, S. L. and Moon, H. W. 1986. Microscopic alterations in jejunal epithelium of 3-week-old pigs induced by pig-specific, mouse-negative, heat-stable *Escherichia coli* enterotoxin. *American Journal of Veterinary Research* **47**:615–618.

Wray, C., Piercy, D. W. T., Carrol, P. J. and Cooley, W. A. 1993. Experimental infection of neonatal pigs with CNF toxin-producing strains of *Escherichia coli*. *Research in Veterinary Science* **54**:290–298.

Zhou, Z., Ogasawara, J., Nishikawa, Y., Seto, Y., Helander, A., Hase, A., Iritani, N., Nakamura, H., Arikawa, K., Kai, A., Kamata, Y., Hoshi, H. and Haruki, K. 2002. An outbreak of gastroenteritis in Osaka, Japan due to *Escherichia coli* serogroup O166:H15 that had a coding gene for enteroaggregative *E. coli* heat-stable enterotoxin 1 (EAST 1). *Epidemiology and Infection* **128**:363–371.

3 Foodborne Enterohemorrhagic *Escherichia coli* Infections

Vernon L. Tesh

Introduction

The vast majority of the Gram-negative bacteria classified as *Escherichia coli* are harmless commensals found in the intestinal tracts of mammals and birds. Along with a number of obligate anaerobic bacteria, *E. coli* are major components of the normal intestinal flora of humans, commonly found in densities of approximately 10^6 *E. coli* per gram of fecal contents. It has become apparent, however, that some strains of *E. coli* have acquired genes encoding virulence determinants which confer to the bacteria the ability to cause disease (Selander et al. 1987). Pathogenic *E. coli* may be characterized based on O:H serotypes, the expression of specific virulence determinants, and the clinical presentation and epidemiology of disease. *E. coli* capable of causing gastroenteritis are categorized as enterotoxigenic *E. coli* (ETEC), enteropathogenic *E. coli* (EPEC), enterohemorrhagic *E. coli* (EHEC), enteroinvasive *E. coli* (EIEC), and enteroaggregative *E. coli* (EAEC)(Levine 1987). This chapter will focus on one subset of pathogenic *E. coli*, the EHEC. EHEC are the causative agents of diarrhea, bloody diarrhea, or hemorrhagic colitis that may progress to deadly complications such as acute renal failure and central nervous system damage. EHEC comprise a select group of *E. coli* serotypes that express virulence factors that allow the organisms to attach to and alter the cellular morphology of intestinal epithelial cells. The bacteria also produce potent cytotoxins. The remarkable potential of these bacteria to cause widespread outbreaks of disease will be reviewed. Precisely how EHEC cause disease remains a subject of intense research scrutiny, but both what is known about EHEC virulence determinants and pathogenesis and future directions in research will be discussed.

The Sakai City Outbreak: A Case in Point

On the morning of Saturday, July 13, 1996, the Public Health Department of Sakai City, Japan, received an unusual report from Sakai City Hospital. That morning doctors had examined 10 children who had developed abdominal cramps and diarrhea on Friday night. By the end of the day, there were 255 children receiving medical attention for the same symptoms, and by Sunday morning, over 2,000 people had been treated. Within the next 24 hours, all hospital beds in the Sakai City area were filled with patients, mostly schoolchildren, with the same chief complaints: abdominal cramping and diarrhea with blood frequently noted in the stool. Stool samples from the patients tested positive for *E. coli* O157:H7, an EHEC serotype. Pulsed field gel electrophoresis analysis (PFGE) showed that the patient isolates were identical, suggesting that the organisms were clonal in nature and may have been acquired from a common source (Takatorige et al. 1997, Izumiya et al. 1997).

Because most of the patients were school children (first through sixth grade) and their teachers, local health officials immediately suspected school lunches as a possible source of the outbreak. Over 1,700 food and environmental samples were tested from 615 facilities. *E. coli* O157:H7 was not detected in any sample. Water and milk were quickly ruled out as possible sources. Most of the ill children attended a school in two of the six Sakai City school districts. The only uncooked food items that were served in common in the two school districts contained white radish sprouts (kaiware-daikon). The radish sprouts were supplied to the school by a single producer. *E. coli* O157:H7 was not isolated from radish sprouts, well water, or drainage water samples taken from the suspected farm. However, while the *E. coli* O157:H7 outbreak at Sakai City schools was going on, another, smaller, outbreak in a nursing home outside Sakai City was reported. Ninety-eight patients reported symptoms including abdominal cramping and bloody diarrhea. The only meal shared among the 33 nursing home residents with laboratory-confirmed *E. coli* O157:H7–positive stool cultures was beef curry, pickled scallions, and a salad containing white radish sprouts. The radish sprouts were supplied from the same farm implicated in the Sakai City school outbreak. PFGE analysis showed that the isolates from the nursing home patients had the same DNA pattern as the Sakai City strain (Michino et al. 1999). Thus, even though it was not confirmed by culture of the organism, it is highly probable that radish sprouts served as the vehicle for these outbreaks. In December 1997 the Task Force on the Mass Outbreak of Diarrhea in Schoolchildren of Sakai City reported that the total number of confirmed and suspected cases in this outbreak was 16,111 (Takatorige et al. 1997).

The Sakai City outbreak highlights the potential for EHEC to cause widespread outbreaks of diarrheal disease. Although Sakai City was a particularly large outbreak, smaller outbreaks of diarrhea caused by EHEC are consistently reported year after year in many developed countries. In the United States, for example, between 1982 and 1996, there were 139 reported outbreaks of *E. coli* O157:H7. Approximately 22% of patients in these outbreaks were hospitalized, 6% developed severe complications, and 0.6% died (Griffin 1998). EHEC also cause sporadic or individual cases of diarrhea or bloody diarrhea. Between 1990 and 1992, 10 U.S. hospitals collaborated in a study to calculate the frequency of enteric pathogens isolated from patients with diarrhea (Slutsker et al. 1997). From the 30,463 stool specimens evaluated in the study, *E. coli* O157:H7 was isolated from only 0.4% of the specimens, compared with *Campylobacter* (2.3%), *Salmonella* (1.8%), and *Shigella* (1.1%). Even though *E. coli* O157:H7 was rarely isolated, the clinical disease was more severe, resulting in a 47% hospitalization rate compared with hospitalization rates for *Campylobacter* (21%), *Salmonella* (38%), and *Shigella* (21%) infections. Mead et al. (1999) compiled data from multiple surveillance programs to estimate the annual numbers of pathogen-related foodborne illnesses in the United States. In this study the investigators factored in an estimated 20-fold underreporting of disease caused by EHEC. On the basis of their calculations, Mead et al. (1999) estimate that there are 73,480 cases of bloody diarrhea per year caused by *E. coli* O157:H7 and an additional 36,740 cases caused by EHEC other than serotype O157:H7. These cases are estimated to result in about 4.3% of the total numbers of deaths per year caused by foodborne bacterial, viral, and parasitic pathogens.

The Sakai City outbreak emphasized two disturbing aspects of foodborne EHEC infections. First was the high incidence of secondary cases; that is, family members and acquaintances of schoolchildren infected by person-to-person transmission. Secondary transmission of the bacteria was associated with inadequate hand washing, sharing hand towels and bath towels, and sharing bath water or pools. The second disturbing fact was that the causative organism was not isolated from the implicated food source or from other samples taken at the suspected farm. This indicates that the infective dose of EHEC causing widespread outbreaks may be very low. In 1993 an outbreak of bloody diarrhea associated with the ingestion of undercooked *E. coli* O157:H7–contaminated hamburgers served by fast food restaurants occurred in the Pacific Northwest of the United States (Bell et al. 1994). Frozen hamburger patties from this outbreak contained less than 700 *E. coli* O157:H7 per patty on thawing (Griffin 1998). If one considers that the numbers of bacteria would be reduced by cooking (or partial cooking) and that patients were infected after only partial consumption of hamburger patties, then the infectious dose that led to this outbreak (involving over 500 cases) may have been on the order of 100 bacteria. Indeed, in an outbreak of bloody diarrhea in Australia linked to ingestion of contaminated mettwurst sausage prepared from raw pork, beef, and lamb, less than one EHEC organism per 10 grams of meat was detected (Paton et al. 1996). The low infective dose makes epidemiologic "trace-back" investigations difficult. Detection of EHEC contamination in foods during processing is also problematic because the sensitivities of detection assays must be sufficient to detect small numbers of organisms in large quantities of foods.

The Organisms

How is it that EHEC are capable of causing such widespread outbreaks of diarrheal disease that may progress to lethal complications? In 2001 Dr. Frederick Blattner and colleagues published the genomic sequence of an *E. coli* O157:H7 strain and compared it with the genome of a non-pathogenic *E. coli* strain (Perna et al. 2001). Although both strains possessed a 4.1-Mb "backbone" of homologous DNA, the EHEC genome was larger, encoding an additional 1,387 EHEC-specific genes. Thus, as is often the case with pathogenic bacteria, the EHEC have acquired a collection of additional genes, many of which encode virulence factors that elevate the organisms from the obscurity of commensalism to the spotlight of the pathogen.

Attaching and Effacing Lesions

EHEC strains have acquired large (ca. 35 kb) DNA inserts called the locus of enterocyte effacement (LEE) pathogenicity-associated island (McDaniel et al. 1995). The LEE pathogenicity-associated island usually integrates in the *E. coli* chromosome within a gene encoding a selenocysteine tRNA, although exceptions to this rule have been documented (Wieler et al. 1997). The LEE pathogenicity-associated island contains a cluster of genes that confers on EHEC the ability to adhere to intestinal epithelial cells and to reorganize the normal cellular architecture of the brush border, a histopathologic change called attaching and effacing (A/E) lesions. Following the relatively superficial attachment of EHEC to intestinal epithelial cells, filamentous actin within microvilli begins to depolymerize. Actin provides the structural support for microvilli, and its breakdown causes the effacement or loss of microvilli. There subsequently appears, immediately adjacent to adherent EHEC, a single large structure called a pedestal (Figure 3.1). The pedestal contains accumulated filamentous actin (Knutton et al. 1989). Recent studies suggest that *E. coli* O157:H7 may initially bind and cause A/E lesions on human epithelial cells associated with Peyer's patches (Phillips et al. 2000). Thus, EHEC are capable of mediating alterations in human intestinal epithelial cell structure by the disassembly and reassembly of host cell cytoskeletal elements. The transfer and expression of the *E. coli* O157:H7 LEE pathogenicity-associated island into a nonpathogenic *E. coli* laboratory strain failed to confer to the recipient the ability to cause A/E lesions,

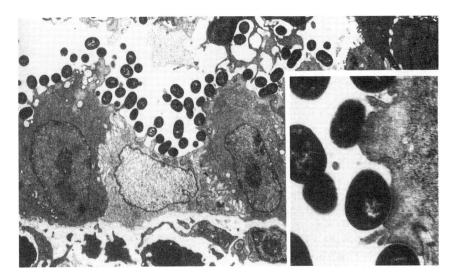

Figure 3.1 Electron micrograph (original magnification = 3,450×) of A/E lesions in the intestinal epithelium of a piglet fed *Escherichia coli* O157:H7. Note the loss of microvilli. The higher magnification (15,870×) inset shows the characteristic pedestal formation. Reprinted from Donnenberg et al. 1993, with permission.

suggesting that additional structural or regulatory genes may be important for expression of the A/E lesion phenotype (Elliott et al. 1999).

The gene encoding the EHEC adherence factor intimin is found within the LEE pathogenicity-associated island, and the expression of intimin is essential for A/E lesion formation. Also found within the LEE pathogenicity island are a number of genes encoding a type III secretion apparatus. The secretion system is essential for transporting bacterial proteins across bacterial membranes and into the host target cell (Jarvis and Kaper 1996). Many bacterial pathogens use host cell proteins as receptors, and intimin has been shown to bind to β1-integrins and cell surface–associated nucleolin. However, an additional receptor for intimin on the surface of intestinal epithelial cells is, in fact, a LEE-encoded protein referred to as Tir (translocated intimin receptor) that is transported by the type III secretion system. Why EHEC translocate their own receptor into the target cell membrane is unknown. It may be that host-derived receptors target EHEC for internalization, a deleterious event for an extracellular pathogen. The advantages of injecting a specific receptor into the target cell membrane may include blocking the internalization process, targeting adherence to a single cell type and the triggering of cytoskeletal rearrangements which stabilize adherence (Frankel et al. 1998). The structural changes orchestrated by EHEC are more complex than the disassembly and reassembly of actin filaments, and it is becoming increasingly clear that a large number of actin-associated proteins are recruited into the pedestal following EHEC adherence (reviewed in Goosney et al. 2001). In addition, evidence suggests that many of the LEE-encoded translocated proteins may alter host cell intracellular signaling cascades (Kresse et al. 2001). In the future, the characterization of the molecular events occurring at the interface of the host cell cytoskeleton and EHEC Tir-intimin complex will be the focus of intense interest, as effective strategies to prevent colonization will require an increased understanding of EHEC adherence mechanisms and the mucosal immune response to adherent bacteria.

Shiga Toxins

In addition to the acquisition of the LEE gene cluster and the ability to cause A/E lesions, EHEC have been lysogenized with one or more lambdoid bacteriophage that encode the genes for a family of genetically and structurally related cytotoxins (Schmidt 2001). These cytotoxins are called Shiga toxins (Stxs). The cytotoxic activities of Stxs are measured in the laboratory using cultured Vero cells (African green monkey renal epithelial cells), and the alternative designation of Stxs is verotoxins. The Stxs produced by EHEC can be categorized into two types based on their antigenic similarity to the prototypical Shiga toxin produced by *Shigella dysenteriae* serotype 1. Shiga toxin type 1 (Stx1) differs by a single amino acid from *S. dysenteriae* Shiga toxin, whereas Shiga toxin type 2 (Stx2) is approximately 56% homologous to Shiga toxin/Stx1 at the deduced amino acid sequence level (Jackson et al. 1987). There are a number of Stx2 variant toxins with amino acid sequence homologies ranging from 84% to 99% compared with Stx2 (Melton-Celsa and O'Brien 1998). EHEC may be lysogenized with multiple toxin-converting bacteriophage, and therefore, the bacteria may express Stx1, Stx2, Stx2 variants, or any combination of the toxins. Epidemiologic studies of outbreaks have shown that *E. coli* producing Stx2, either alone or in combination with other Stxs, are more likely to cause serious disease in humans.

Despite variability in genetic sequence, members of the Shiga toxin family share functional and structural features. Stxs are AB_5 toxins; that is, holotoxins that consist of a single "A" subunit of approximately 32 kDa in noncovalent association with five identical "B" subunits (Figure 3.2; Fraser et al. 1994). Each B subunit is about 7.7 kDa in size. The Stx B subunits are responsible for binding the toxins to cells expressing the Stx receptor. The toxin receptor is a neutral glycolipid called globotriaosylceramide (Gb_3; galactoseα(1 → 4) galactoseβ(1 → 4) glucoseβ(1 → 1)ceramide), which may also be referred to as CD77 or the P^k blood antigen (Figure 3.3; Lingwood 1996). One Stx2 variant toxin, called Stx2e, is the causative agent of edema disease of swine. In contrast to the other members of the Shiga toxin family, Stx2e appears to preferentially bind the glycolipid globotetraosylceramide (Gb_4). Once bound to the target cell via Gb_3, Stxs are internalized in endosomes via clathrin-coated pits and transported to the trans-Golgi network through the stacks of the Golgi apparatus to the membranes of the endoplasmic reticulum and nucleus (Figure 3.4; Sandvig and van Deurs 1996). This intracellular trafficking pattern is referred to as retrograde transport, as the toxins appear to be transported against the normal trafficking pattern used to secrete correctly folded host-derived proteins. During the intracellular translocation process, the A subunit is proteolytically cleaved by furin or a furin-like protease, but the two fragments of the A subunit remain linked via a disulfide bond. Following reduction of the disulfide bond, the A_1 fragment (ca. 27 kDa) is thought to translocate across the endoplasmic reticulum membrane, whereas the smaller (ca. 4 kDa) A_2 fragment remains associated with the B pentamer. The A_1 fragment possesses *N*-glycosidic activity, which removes a single adenine residue found in a prominent loop structure near the 3'-end of the 28S rRNA component of eukaryotic ribosomes (Endo et al. 1987, Saxena et al. 1989). Following the depurination reaction, elongation factor-1-dependent binding of aminoacyl-tRNAs is blocked and protein synthesis is inhibited. The mechanism of action of Stxs is identical to that of the plant toxin ricin, although the Stxs and ricin do not share extensive genetic or structural similarities. Stxs are extremely effective protein synthesis inhibitors, being cytotoxic for cultured epithelial cell lines in quantities of picograms per milliliter.

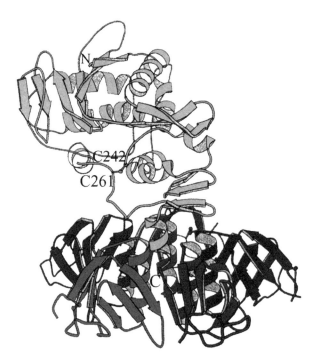

Figure 3.2 Ribbon diagram of the AB_5 structure of the Shiga toxin molecule. The enzymatic A subunit (gray) noncovalently associates with a pentameric ring (black) of receptor-binding B subunits. Following proteolysis by furin or a furin-like protease, a disulfide bond between cysteine residues C242 and C261 links the A_1 and A_2 fragments. Reprinted from Fraser et al. 1994, with permission from publisher and authors.

Crystallographic analysis of purified Stx1 B subunits showed that the pentamer forms a symmetrical ringed structure with a central pore lined with α helices composed of neutral and nonpolar residues (Stein et al. 1992). X-ray crystallography of the purified Shiga toxin holotoxin molecule (A subunit plus five B subunits) revealed that the carboxy-terminus of the A subunit interacts with the central pore of the B pentamers (Fraser et al. 1994). Interestingly, a number of toxins produced by enteric pathogens have been shown to possess AB_5 structures. Despite having very little amino acid sequence homology, heat-labile enterotoxin produced by ETEC, cholera toxin produced by *Vibrio cholerae,* and the Stxs all have AB_5 structures (Merritt and Hol 1995). Why the AB_5 molecular organization should be favored by organisms that deliver toxins across intestinal epithelial barriers remains to be fully elucidated. The structures and enzymatic activities of the A subunits are extensively varied, yet similar pentameric binding structures are used by the toxins to bind polysaccharide components of membrane glycolipids. The structural similarity among the binding components of the AB_5 toxins would seem to suggest that they evolved from a common primordial protein capable of binding carbohydrate residues. However, the lack of amino acid sequence homology among the toxin B subunits does not support a divergent evolutionary relationship. Alternatively, it has been suggested that the AB_5 toxins arose by convergent evolution (Sixma et al. 1993). Thus, EHEC, ETEC, and *V. cholerae* may have independently evolved the capacity to express AB_5 toxins to fulfill a number of requirements of enteric pathogens;

Figure 3.3 Structure of the Stx-binding glycolipid receptor, Gb_3. Reprinted from Lingwood 1996, with permission.

namely, a means of delivery of an enzyme (A subunit) into intestinal epithelial cells using glycolipids or glycoproteins as receptors, the formation of a stable "docking site" (the central pore) for the enzyme, and a means for the dissociation of the activated catalytic subunit from the cell-binding B subunits by limited proteolysis of the A subunit. Recent structural studies of Stxs bound to Gb_3 have revealed that there may be 12–15 receptor binding sites per B pentamer, an observation that explains in part the high affinity (ca. 10^{-9} M) of Stxs for Gb_3 in cell membranes (Nyholm et al. 1996, Ling et al. 1998). Advances in our understanding of the physical interactions of Stxs with their membrane glycolipid receptor have led to the development of a number of interventional strategies to prevent the development of systemic disease after the ingestion of EHEC (see Interventional Strategies, below).

Other EHEC Virulence Determinants

Although intimin is essential for the formation of A/E lesions, EHEC strains lacking intimin are still capable of colonizing some portions of the intestinal tract and causing

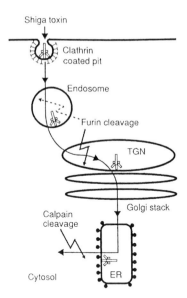

Figure 3.4 Retrograde transport of Shiga toxins from the cell membrane to the endoplasmic reticulum. Reprinted from Sandvig and van Deurs 1996, with permission.

bloody diarrhea, suggesting that bacterial factors other than intimin may be involved in EHEC adherence (Dytoc et al. 1994). Furthermore, it is unlikely that all EHEC serotypes will necessarily use comparable adherence mechanisms.

Dr. Philip Tarr and his colleagues have identified a chromosomally encoded EHEC outer membrane protein that may serve as an adhesin (Tarr et al. 2000). This adherence factor was named the IrgA homologue adhesin (Iha) because of its homology to *V. cholerae* iron-regulated gene A (IrgA). Although *V. cholerae* iron-regulated gene A is involved in iron acquisition by the pathogen, it is known that *V. cholerae irgA* mutants are reduced in their ability to colonize infant mice. Expression of the *iha* gene in a nonadherent *E. coli* laboratory strain conferred on the recipient the ability to adhere to HeLa cells *in vitro*. An additional chromosomally encoded adherence factor called EHEC factor for adherence (Efa1) expressed by *E. coli* O111:H⁻ has been characterized. Efa1 facilitates the attachment of the bacteria to CHO cells *in vitro* and is required for the efficient colonization of 4- and 11-day-old conventional calves (Nicholls et al. 2000, Stevens et al. 2002). Examination of the *E. coli* O157:H7 genome reveals that this particular EHEC serotype lacks the full-length *efa1* gene. However, transposon mutagenesis studies suggest that truncated versions of Efa1 may still contribute to EHEC adherence and intestinal colonization. The *efa1* gene is 97.4% identical to the EPEC gene encoding lymphostatin (*lifA*). Lymphostatin inhibits interleukin-2 (IL-2), IL-4, IL-5, and interferon-γ production by human lymphocytes, suggesting that EPEC may alter the mucosal immune response *in vivo*. Future experiments will determine whether Efa1 and truncated Efa1 products also possess immunomodulatory properties.

Many *E. coli* O157:H7 strains possess a large (ca. 93 kb) plasmid designated pO157. Genes encoding a catalase-peroxidase, a secreted serine protease, four genes that make up the hemolysin operon, and 13 genes encoding a type II (general secretory pathway) secretion system have been mapped to pO157. The catalase-peroxidase is a periplasmic enzyme that presumably protects the bacteria from oxidative stress. The serine protease cleaves human coagulation factor V and may contribute to pathogenesis by exacerbating vascular damage and hemorrhage. The EHEC hemolysin is a pore-forming toxin that damages mammalian cells by disruption of cell membrane integrity. The pO157 plasmids from two *E. coli* O157:H7 isolates were independently sequenced and shown to encode 100 open reading frames (possible genes; Burland et al. 1998, Makino et al. 1998). An exciting discovery from the analysis of the pO157 sequence was an open reading frame designated *toxB* with homology to the genes encoding glucosyltransferase toxins of the enteric pathogen *Clostridium difficile*. Recently, Tatsuno et al. (2001) showed that the expression of *toxB* by EHEC may be necessary for the secretion of LEE-encoded proteins and for full bacterial adherence to cultured intestinal epithelial cells. Interestingly, the EHEC ToxB and Efa1 proteins share approximately 40% amino acid sequence homology over 480 amino acids found at the amino termini, suggesting that the mechanisms of adherence mediated by ToxB and Efa1 may be similar. A cluster of six genes responsible for the expression of fimbriae has been mapped to the large plasmid of sorbitol-fermenting *E. coli* O157:H⁻ strains (Brunder et al. 2001). The gene encoding the major pilus subunit of *E. coli* O157:H⁻ fimbriae showed homology to P-fimbriae expressed by uropathogenic *E. coli* and Pmp fimbriae of *Proteus mirabilis*. The *E. coli* O157:H⁻ fimbriae mediate mannose-resistant hemagglutination, but their role in pathogenesis remains to be fully characterized. Finally, Dr. James Kaper and his colleagues have identified a six-gene operon in the *E. coli* O157:H7 chromosome with operon organization and sequence similarity to the long polar fimbriae (*lpf*) operon of *Salmonella enterica* serovar Typhimurium (Torres et al. 2002). Transfer of the *E. coli* O157:H7 *lpf* operon into a non-fimbriated *E. coli* laboratory strain conferred on the recipient the ability to express fimbriae, as detected by electron microscopy, and the ability to adhere to HeLa cells. *E. coli* O157:H7 *lpf* mutants displayed reduced adherence and microcolony formation *in vitro*. Given the well-characterized role of *Salmonella* long polar fimbriae in targeting the organisms to Peyer's patches (Bäumler et al. 1996), and the recent work indicating that EHEC may initially target Peyer's patches for colonization (Phillips et al. 2000), a comparable role for EHEC long polar fimbriae might be expected.

Pathogenesis

Following ingestion of contaminated foods or water containing small numbers of EHEC, the bacteria travel to the stomach. The acidic environment of the stomach is known to be an important and effective bactericidal barrier against infection with ingested microorganisms (Peterson et al. 1989). However, EHEC have been shown to survive in acidic media (pH = 2.5–3.0) at 37°C for up to 5 hours, conditions that reproduce the gastric environment (Benjamin and Datta 1995). The organisms pass through the stomach and colonize the distal small intestine and large intestine, causing A/E lesions. The formation of A/E lesions may result in the initial onset of watery diarrhea associated with increased intracellular permeability, increased chloride secretion, and malabsorption caused by microvillus effacement. The adherent bacteria also synthesize Stxs. Increased intestinal permeability caused by A/E lesions and by inflammatory cells extravasating into the bowel may allow the toxins to gain access to the lamina propria and colonic microvasculature. There are studies, however, suggesting that the toxins may be actively transported across the gut epithelial cell barrier without disruption of the cells (Hurley et al. 1999). In the submucosa, the toxins bind to cells expressing the glycolipid toxin receptor. Vascular endothelial cells express Gb_3, so that Stxs may damage capillaries serving the distal small bowel and colon, resulting in hemorrhage and blood in the stool. The toxins may cause necrotic cell death by protein synthesis inhibition, but studies have also shown that Stxs and purified Stx B subunits induce apoptotic cell death *in vitro* (Mangeney et al. 1993). Stx-mediated damage to blood vessels serving the gut creates an entry site for Stxs and other bacterial products into the bloodstream. Once in the bloodstream, the toxins may damage endothelial cells lining small blood vessels in many organs, although the kidneys and the central nervous system (CNS) are the principal targets of destruction. It has proven difficult to detect free Stxs in the circulation of patients, and it was recently shown that the toxins bind to white blood cells and may be disseminated throughout the body while "piggybacking" on these cells (te Loo et al. 2000). Finally, several studies have demonstrated that Stxs induce the expression of proinflammatory cytokines from human monocytes and macrophage *in vitro*, indicating that the host innate immune response may contribute to inflammation and tissue destruction (Ramegowda and Tesh 1996, van Setten et al. 1996).

The cytotoxic action of Stxs on glomerular capillaries leads to the development of the hemolytic uremic syndrome (HUS). HUS is a disease that is clinically defined by the development of hemolytic anemia, thrombocytopenia, and acute renal failure associated with the deposition of fibrin microthrombi within glomeruli (Siegler 1995). Children and the elderly are at an increased risk for developing HUS following diarrhea or bloody diarrhea caused by EHEC. The median time interval between the onset of diarrhea and the diagnosis of HUS is 5–7 days. HUS is now recognized as the major cause of pediatric acute kidney failure in many developed countries. Stxs also damage blood vessels serving the CNS. Inflammation that occurs in response to vascular damage may lead to increased cerebral edema and result in strokes, seizures, coma, and death (Siegler 1994). With the increased awareness of the association between EHEC infection and acute renal failure, physicians are instituting dialysis and other measures of renal support early in the course of the disease. As a result, the renal crises are being averted, only to find that neurologic complications following bloody diarrhea are now a common cause of death in children. In a recent prospective study of HUS in the United States, 25% of the patients developed neurologic complications (Banatvala et al. 2001).

Interventional Strategies

Advances in our understanding of the interaction of Stxs with Gb_3 led to the development of synthetic toxin receptor analogues for use as potential therapeutic agents. Dr. Glen Armstrong and his colleagues covalently coupled the trisaccharide component of Gb_3 to diatomaceous silicon dioxide. This toxin receptor analogue, called Synsorb-Pk, was shown to reduce cytotoxicity when Vero cells were co-treated with Synsorb-Pk and Stxs *in vitro* (Armstrong et al. 1991). It was proposed that the indigestible Synsorb-Pk could be fed to patients with diarrhea or bloody diarrhea who are at risk for progression to systemic complications associated with EHEC infections. Unfortunately, although Synsorb-Pk was readily tolerated by human volunteers, Phase II clinical trials have not demonstrated an efficacious trend in preventing progression to HUS (Armstrong et al. 1998). In light of recent evidence indicating that Stxs may be actively transported across intestinal epithelial cells and, therefore, may be unavailable to bind orally ingested toxin receptor analogues, the focus of efforts to intervene in disease has shifted to the development of injectable toxin receptor analogues. A major advancement that came from the structural studies on the interaction of Stxs with Gb_3 was the realization that effective toxin receptor analogues must be "tailored" to engage multiple binding sites on the B pentamer (Kitov et al. 2000). Nishikawa et al. (2002) developed carbosilane carrier molecules to which were coupled multiple Gb_3 trisaccharides. Optimal binding to Stxs was detected with a molecule called SuperTwig 1(6), which contained six Gb_3 trisaccharide units. Most importantly, it was shown that the intravenous administration of SuperTwig 1(6) protected mice against lethality when the animals were challenged with a lethal dose of purified Stx2 or fed *E. coli* O157:H7. The mechanism of protection may involve the selective routing of toxin-SuperTwig complexes to phagocytic cells in murine spleen and liver. Additional experiments will be required to confirm that multivalent toxin receptor analogues will react with Stxs in association with leukocyte or "carrier cell" membranes.

The passive administration of anti-toxin antibodies, and the induction of active humoral immunity using toxoids or subunit vaccines, has been used to protect humans from numerous toxin-mediated diseases. Both anti-Stx antibodies and subunit toxin vaccines are being developed to treat patients with, or protect patients from, EHEC infections. Recently, Mukherjee et al. (2002) developed a series of anti-Stx2 humanized monoclonal antibodies derived from Stx2 toxoid-treated transgenic mice expressing human immunoglobulin loci. Following oral feeding with a Stx2-expressing *E. coli* O157:H7 strain, anti-Stx2 humanized monoclonal antibodies protected gnotobiotic piglets from the development of CNS signs and pathology, although trends in prolonged survival in the animals were often not significant when compared with controls. The development of Stx subunit vaccines for the induction of active acquired immunity against the toxins has been complicated by the finding that purified B subunits induce apoptosis in some cell types *in vitro*. Nevertheless, Marcato et al. (2001) purified Stx2 B subunits and showed that rabbits developed an anti-Stx2 B-subunit humoral immune response following immunization and were protected from the Stx2-mediated pathologic changes occurring in non-immunized rabbits. Future clinical trials to assess the efficacy of passively administered anti-toxin antibodies and toxin subunit vaccines are clearly warranted.

EHEC and Foods

Cattle have emerged as the major reservoir of EHEC. Adult cattle and weaned calves are asymptomatic carriers of EHEC; that is, cattle presented for slaughter do not appear ill. Initial attempts to calculate the prevalence of *E. coli* O157:H7 in cattle produced estimates in the range of 0.2%–1.8%. More recently, Elder et al. (2000) employed sensitive bacterial isolation techniques including immunomagnetic separation of *E. coli* O157 from bovine intestinal flora and culturing the bacteria in enriched media containing antibiotics to screen for EHEC carriage rates. These improved detection techniques for *E. coli* O157 were used in sampling the feces, hides, and carcasses of cattle presented for slaughter at four Midwestern meat-processing plants. Surprisingly, 27.8% of the animals had *E. coli* O157 in their feces, 10.7% had the organisms on hides, and the bacteria were detected on 43.4% of the pre-evisceration carcasses. Even though this study was carried out during the summer, and summer is known to be the time of peak carriage and shedding of EHEC in herds, the prevalence of *E. coli* O157 carriage was much higher than had previously been reported. On a more positive note, sampling of carcasses "postprocessing" (following steam pasteurization, hot water washes, or organic acid washes) detected *E. coli* O157 on only 1.8% of the carcasses, suggesting that in-plant antimicrobial intervention strategies are effective. The study does highlight, however, the need for improved techniques to lower EHEC carriage rates in cattle before slaughter.

Although *E. coli* O157:H7 is the predominant EHEC serotype isolated in the United States, over 200 different serotypes of Shiga toxin–producing *E. coli* have been identified in animals or foods (Johnson et al. 1996). Approximately 60 of these serotypes have been associated with disease in humans. Common non-O157:H7 serotypes causing disease in the United States include O111, O26, O121, and O103. Dr. David Acheson used an enzyme immunoassay to detect Stxs in ground beef samples purchased in cities throughout the United States and found that 15.6% of the samples contained Stx-producing *E. coli* (Acheson 2000). Interestingly, the serotypes detected in this screening were O133:Hu, O22NM, O82:H8, O8:H9, and O13:Hu, indicating that by limiting the screening of ground beef to the O157:H7 serotype, one may miss a significant fraction of contaminated and potentially harmful beef. Given the large number of *E. coli* serotypes that have been shown capable of producing Shiga toxins, screening cattle for specific O:H serotypes may be too labor intensive and cost restrictive. The use of sensitive assays for the detection of Stx1 and Stx2 may prove to be a more effective screening approach for cattle or beef products.

Outbreaks of bloody diarrhea caused by contaminated apple juice, apple cider, lettuce, alfalfa sprouts, and well water have been reported. Manure used as a fertilizer or leaking into irrigation water may be an important source of EHEC contamination of fruits and vegetables. *E. coli* O157:H7 has been shown to survive up to 21 months in nonaerated ovine manure (Kudva et al. 1998), but survival rates in bovine manure appear to be on the order of 40–50 days (Wang et al. 1996, Kudva et al. 1998). The potential of EHEC to readily contaminate vegetables was shown in a study in which bovine feces containing low numbers (1–10 cfu/g) of *E. coli* O157:H7 was applied to lettuce leafs. Even with storage at refrigerator temperatures, the bacteria were cultivated from the lettuce for 15 days (Beuchat 1999). The prolonged survival rates of EHEC in manure also indicate that in the absence of good farm management practices, fecal matter is certainly sufficient for the infection of multiple animals within a herd. Finally, it has been shown that insects may serve as vectors for the transmission of EHEC to plants (Janisiewicz et al. 1999). *E. coli* O157:H7 has been shown to survive for up to 6 days when introduced into bruised apple tissue (Dingman 2000).

Conclusions

The outbreak of bloody diarrhea in Sakai City, Japan, clearly indicates the potential of EHEC to cause massive and disruptive outbreaks of debilitating disease. The available epidemiologic data suggest that these food- and waterborne organisms continue to pose a significant health risk. Much has been learned about how EHEC cause disease. The key pathogenic mechanisms described to date are the ability of the organisms to adhere to the intestinal epithelium and cause A/E lesions and their ability to produce Stxs. Following adherence, EHEC proteins are secreted into the host cell membrane and cytosol, resulting in a complex array of changes in the target cell; these include major cytoskeletal rearrangements and the activation or disruption of host cell intracellular signaling cascades. An improved understanding of the cellular biology of bacteria–host cell interactions may lead to effective strategies to disrupt secretion of bacterial virulence factors and to counteract deleterious alterations in the host cell caused by EHEC. A number of putative adherence factors for EHEC have been characterized, and it is becoming clear that different EHEC serotypes may possess different adherence factors. Thus, a major focus of future experiments will be the characterization of the complete EHEC adherence armamentarium. The recent genomic sequencing of an *E. coli* O157:H7 strain revealed that EHEC have acquired additional genetic material lacking in non-pathogenic *E. coli* strains. It is likely that many of the *E. coli* O157:H7 unique genes encode factors important in pathogenesis, and these genes will serve as future targets for characterization. The Stxs are well characterized in terms of their genes and structure and are thought to target vascular endothelial cells for destruction. Although the toxin A subunit is essential for mediating protein synthesis inhibition, there is evidence that purified toxin B subunits induce apoptotic cell death in some cell types, suggesting that cross-linking membrane Gb_3 may activate programmed cell death in some cases. It has also become clear that different cell types may respond to the toxins in a manner that does not necessarily lead to cell death. For example, intestinal epithelial cells may actively transport the toxins across the gut epithelial barrier, neutrophils may bind and disseminate the toxins in the blood, and monocytes/macrophages may respond to toxins by producing inflammatory mediators. Initial attempts to protect patients from developing disease by oral administration of toxin receptor analogues did not prove to be effective. The development of injectable toxin receptor analogues containing multiple toxin binding sites holds promise as an effective interventional approach. Future development of antitoxin antibodies and toxin subunit vaccines also holds great promise. Advancements in our understanding of the host response to Stxs will be necessary to develop optimally effective therapeutic approaches to prevent and treat diseases caused by EHEC. Finally, recent in-plant screening studies suggest that the carriage rate of EHEC in cattle presented for slaughter may greatly exceed earlier estimates. These findings highlight the need to develop means to prevent EHEC carriage in herds and to detect EHEC in cattle before processing.

References

Acheson, D. W. K. 2000. How does *Escherichia coli* O157:H7 testing in meat compare with what we see clinically? *Journal of Food Protection* **63**:819–821.

Armstrong, G. D., Fodor, E. and Vanmaele, R. 1991. Investigation of Shiga-like toxin binding to chemically synthesized oligosaccharide sequences. *Journal of Infectious Diseases* **164**:1160–1167.

Armstrong, G. D., McLaine, P. N. and Rowe, P. C. 1998. Clinical trials of Synsorb-Pk in preventing hemolytic-uremic syndrome, pp. 374–384. *In Escherichia coli* O157:H7 and Other Shiga Toxin-Producing *E. coli* Strains, J. B. Kaper, A. D. O'Brien (Eds.); Washington, D.C.: American Society for Microbiology.

Banatvala, N., Griffin, P. M., Greene, K. D., Barrett, T. J., Bibb, W. F., Green, J. H., Wells, J. G. and the Hemolytic Uremic Syndrome Study Collaborators. 2001. The United States national prospective hemolytic uremic syndrome study: microbiologic, serologic, clinical, and epidemiologic findings. *Journal of Infectious Diseases* **183**:1063–1070.

Bäumler, A. J., Tsolis, R. M. and Heffron, F. 1996. The *lpf* fimbrial operon mediates adhesion of *Salmonella typhimurium* to murine Peyer's patches. *Proceedings of the National Academy of Sciences (USA)* **93**:279–283.

Bell, B. P., Goldoft, M., Griffin, P. M., Davis, M. A., Gordon, D. C., Tarr, P. I., Bartleson, C. A., Lewis, J. H., Barrett, T. J., Wells, J. G., Baron, R. and Kobayashi, J. 1994. A multistate outbreak of *Escherichia coli* O157:H7–associated bloody diarrhea and hemolytic uremic syndrome from hamburgers. *Journal of the American Medical Association* **272**:1349–1353.

Benjamin, M. M. and Datta, A. R. 1995. Acid tolerance of enterohemorrhagic *Escherichia coli*. *Applied and Environmental Microbiology* **61**:1669–1672.

Beuchat, L. R. 1999. Survival of enterohemorrhagic *Escherichia coli* O157:H7 in bovine feces applied to lettuce and the effectiveness of chlorinated water as a disinfectant. *Journal of Food Protection* **62**:845–849.

Brunder, W., Khan, A. S., Hacker, J. and Karch, H. 2001. Novel type of fimbriae encoded by the large plasmid of sorbitol-fermenting enterohemorrhagic *Escherichia coli* O157:H⁻. *Infection and Immunity* **69**:4447–4457.

Burland, V., Shao, Y., Perna, N. T., Plunkett, G., Sofia, H. J. and Blattner, F. R. 1998. The complete DNA sequence and analysis of the large virulence plasmid of *Escherichia coli* O157:H7. *Nucleic Acids Research* **26**:4196–4204.

Dingman, D. W. 2000. Growth of *Escherichia coli* O157:H7 in bruised apple (*Malus domestica*) tissue as influenced by cultivar, date of harvest, and source. *Applied and Environmental Microbiology* **66**:1077–1083.

Donnenberg, M. S., Tzipori, S., McKee, M. L., O'Brien, A. D., Alroy, J. and Kaper, J. B. 1993. The role of the *eae* gene of enterohemorrhagic *Escherichia coli* in intimate attachment *in vitro* and in a porcine model. *Journal of Clinical Investigation* **92**:1418–1424.

Dytoc, M. T., Ismaili, A., Philpott, D. J., Soni, R., Brunton, J. L. and Sherman, P. M. 1994. Distinct binding properties of *eae*-negative verocytotoxin-producing *Escherichia coli* of serotype O113:H21. *Infection and Immunity* **62**:3494–3505.

Elder, R. O., Keen, J. E., Siragusa, G. R., Barkocy-Gallagher G. A., Koohmaraie, M. and Laegreid, W. W. 2000. Correlation of enterohemorrhagic *Escherichia coli* O157 prevalence in feces, hides, and carcasses of beef cattle during processing. *Proceedings of the National Academy of Sciences (USA)* **97**:2999–3003.

Elliott, S. J., Yu, J. and Kaper, J. B. 1999. The cloned locus of enterocyte effacement from enterohemorrhagic *Escherichia coli* O157:H7 is unable to confer the attaching effacing phenotype upon *E. coli* K-12. *Infection and Immunity* **67**:4260–4263.

Endo, Y., Tsurugi, K., Yutsudo, T., Takeda, Y., Ogasawara, T. and Igarashi, K. 1987. Site of action of a Vero toxin (VT2) from *Escherichia coli* O157:H7 and of Shiga toxin on eukaryotic ribosomes. RNA *N*-glycosidase activity of the toxins. *European Journal of Biochemistry* **171**:45–50.

Frankel, G., Phillips, A. D., Rosenshine, I., Dougan, G., Kaper, J. B. and Knutton, S. 1998. Enteropathogenic and enterohemorrhagic *Escherichia coli*: more subversive elements. *Molecular Microbiology* **30**:911–921.

Fraser, M. E., Chernaia, M. M., Kozlov, Y. V. and James, M. N. G. 1994. Crystal structure of the holotoxin from *Shigella dysenteriae* at 2.5Å resolution. *Nature Structural Biology* **1**:59–64.

Goosney, D. L., DeVinney, R. and Finlay, B. B. 2001. Recruitment of cytoskeletal and signaling proteins to enteropathogenic and enterohemorrhagic *Escherichia coli* pedestals. *Infection and Immunity* **69**:3315–3322.

Griffin, P. M. 1998. Epidemiology of Shiga toxin-producing *Escherichia coli* infections in humans in the United States, pp. 15–22. *In Escherichia coli* O157:H7 and Other Shiga Toxin-Producing *E. coli* Strains, J. B. Kaper, A. D. O'Brien, (Eds.); Washington, D.C.: American Society for Microbiology.

Hurley, B. P., Jacewicz, M., Thorpe, C. M., Lincicome, L. L., King, A. J., Keusch, G. T. and Acheson, D. W. K. 1999. Shiga toxins 1 and 2 translocate differently across polarized intestinal epithelial cells. *Infection and Immunity* **67**:6670–6677.

Izumiya, H., Terajima, J., Wada, A., Inagaki, Y., Itoh, K. I., Tamura, K. and Watanabe, H. 1997. Molecular typing of enterohemorrhagic *Escherichia coli* O157:H7 isolates in Japan using pulsed-field gel electrophoresis. *Journal of Clinical Microbiology* **35**:1675–1680.

Jackson, M. P., Neill, R. J., O'Brien, A. D., Holmes, R. K. and Newland, J. W. 1987. Nucleotide sequence analysis and comparison of the structural genes for Shiga-like toxin I and Shiga-like toxin II encoded by bacteriophages from *Escherichia coli* 933. *FEMS Microbiology Letters* **44**:109–114.

Janisiewicz, W. J., Conway, W. S., Brown, M. W., Sapers, G. M., Fratamico, P. and Buchanan, R. L. 1999. Fate of *Escherichia coli* O157:H7 on fresh cut apple tissue and its potential for transmission by fruit flies. *Applied and Environmental Microbiology* **65**:1–5.

Jarvis, K. G. and Kaper, J. B. 1996. Secretion of extracellular proteins by enterohemorrhagic *Escherichia coli* via a putative type III secretion system. *Infection and Immunity* **64**:4826–4829.

Johnson, R. P., Clarke, R. C., Wilson, J. B., Read, S. C., Rahn, K., Renwick, S. A., Sandhu, K. A., Alves, D., Karmali, M. A., Lior, H., McEwen, S. A., Spika, J. S. and Gyles, C. A. 1996. Growing concerns and recent outbreaks involving non-O157:H7 serotypes of verotoxigenic *Escherichia coli*. *Journal of Food Protection* **59**:1112–1122.

Kitov, P. I., Sadowska, J. M., Mulvey, G., Armstrong, G. D., Ling, H., Pannu, N. S., Read, R. J. and Bundle, D. R. 2000. Shiga-like toxins are neutralized by tailored multivalent carbohydrate ligands. *Nature* **403**:669–672.

Knutton, S., Baldwin, T., Williams, P. H. and McNeish, A. S. 1989. Actin accumulation at sites of bacterial adhesion to tissue culture cells: basis for a new diagnostic test for enteropathogenic and enterohemorrhagic *Escherichia coli*. *Infection and Immunity* **57**:1290–1298.

Kresse, A. U., Guzman, C. A. and Ebel, F. 2001. Modulation of host cell signalling by enteropathogenic and Shiga toxin-producing *Escherichia coli*. *International Journal of Medical Microbiology* **291**:277–285.

Kudva, I. T., Blanch, K. and Hovde, C. J. 1998. Analysis of *Escherichia coli* O157:H7 survival in ovine or bovine manure and manure slurry. *Applied and Environmental Microbiology* **64**:3166–3174.

Levine, M. M. 1987. *Escherichia coli* that cause diarrhea: enterotoxigenic, enteropathogenic, enteroinvasive, enterohemorrhagic and enteroadherent. *Journal of Infectious Diseases* **155**:377–389.

Ling, H., Boodhoo, A., Hazes, B., Cummings, M. D., Armstrong, G. D., Brunton, J. L. and Read, R. J. 1998. Structure of the Shiga-like toxin I B-pentamer complexed with an analogue of its receptor Gb$_3$. *Biochemistry* **37**:1777–1788.

Lingwood, C. A. 1996. Role of verotoxin receptors in pathogenesis. *Trends in Microbiology* **4**:147–153.

Makino, K., Ishii, K., Yasunaga, T., Hattori, M., Yokoyama, K., Yutsudo, C. H., Kubota, Y., Yamaichi, Y., Iida, T., Yamamoto, K., Honda, T., Han, C. G., Ohtsubo, E., Kasamutsu, M., Hayashi, T., Kuhara, S. and Shinagawa, H. 1998. Complete nucleotide sequences of 93-kb and 3.3-kb plasmids of an enterohemorrhagic *Escherichia coli* O157:H7 derived from the Sakai outbreak. *DNA Research* **28**:1–9.

Mangeney, M., Lingwood, C. A., Taga, S., Caillou, B., Tursz, T. and Wiels, J. 1993. Apoptosis induced in Burkitt's lymphoma cells via Gb3/CD77, a glycolipid antigen. *Cancer Research* **53**:5314–5319.

Marcato, P., Mulvey, G., Read, R. J., van der Helm, K., Nation, P. N. and Armstrong, G. D. 2001. Immunoprophylactic potential of cloned Shiga toxin 2 B subunit. *Journal of Infectious Diseases* **183**:435–443.

McDaniel, T. K., Jarvis, K. G., Donnenberg, M. S. and Kaper, J. B. 1995. A genetic locus of enterocyte effacement conserved among diverse enterobacterial pathogens. *Proceedings of the National Academy of Sciences (USA)* **92**:1664–1668.

Mead, P. S., Slutsker, L., Dietz, V., McCaig, L. F., Bresee, J. S., Shapiro, C., Griffin, P. M. and Tauxe, R. V. 1999. Food-related illness and death in the United States. *Emerging Infectious Diseases* **5**:607–625.

Melton-Celsa, A. R. and O'Brien, A. D. 1998. Structure, biology and relative toxicity of Shiga toxin family members for cells and animals, pp. 121–128. *In Escherichia coli* O157:H7 and Other Shiga Toxin-Producing *E. coli* Strains, J. B. Kaper, A. D. O'Brien, (Eds.); Washington, D.C.: American Society for Microbiology.

Merritt, E. A. and Hol, W. G. J. 1995. AB$_5$ toxins. *Current Opinion in Structural Biology* **5**:165–171.

Michino, H., Araki, K., Minami, S., Takaya, S., Sakai, N., Miyazaki, M., Ono, A. and Yanagawa, H. 1999. Massive outbreak of *Escherichia coli* O157:H7 infection in school children in Sakai City, Japan, associated with consumption of white radish sprouts. *American Journal of Epidemiology* **150**:787–796.

Mukherjee, J., Chios, K., Fishwild, D., Hudson, D., O'Donnell, S., Rich, S. M., Donohue-Rolfe, A. and Tzipori, S. 2002. Human Stx2-specific monoclonal antibodies prevent systemic complications of *Escherichia coli* O157:H7 infection. *Infection and Immunity* **70**:612–619.

Nicholls, L., Grant, T. H. and Robins-Browne, R. M. 2000. Identification of a novel genetic locus that is required for *in vitro* adhesion of a clinical isolate of enterohemorrhagic *Escherichia coli* to epithelial cells. *Molecular Microbiology* **35**:275–288.

Nishikawa, K., Matsuoka, K., Kita, E., Okabe, N., Mizuguchi, M., Hino, K., Miyazawa, S., Yamasaki, C., Aoki, J., Takashima, S., Yamakawa, Y., Nishjima, M., Terunuma, D., Kuzuhara, H. and Natori, Y. 2002. A therapeutic agent with oriented carbohydrates for treatment of infections by Shiga toxin-producing *Escherichia coli* O157:H7. *Proceedings of the National Academy of Sciences (USA)* **99**:7669–7674.

Nyholm, P. G., Magnusson, G., Zheng, Z., Norel, R., Binnington-Boyd, B. and Lingwood, C. A. 1996. Two distinct binding sites for globotriaosyl ceramide on verotoxins: identification by molecular modelling and confirmation using deoxy analogues and a new glycolipid receptor for all verotoxins. *Chemistry and Biology* **3**:263–275.

Paton, A. W., Ratcliff, R. M., Doyle, R. M., Seymour-Murray, J., Davos, D., Lanser, J. A. and Paton, J. C. 1996. Molecular microbiological investigation of an outbreak of hemolytic-uremic syndrome caused by dry fermented sausage contaminated with Shiga-like toxin-producing *Escherichia coli*. *Journal of Clinical Microbiology* **34**:1622–1627.

Perna, N. T., Plunkett III, G., Burland, V., Mau, B., Glasner, J. D., Rose, D. J., Mayhew, G. F., Evans, P. S., Gregor, J., Kirkpatrick, H. A., Posfai, G., Hackett, J., Klink, S., Boutin, A., Shao, Y., Miller, S., Grotbeck, E. J., Davis, N. W., Lim, A., Dimalanta, E. T., Potamousis, K. D., Apodaca, J., Anantharaman, T. S., Lin, J., Yen, G., Schwartz, D. C., Welch, R. A. and Blattner, F. R. 2001. Genome sequence of enterohemorrhagic *Escherichia coli* O157:H7. *Nature* **409**:529–533.

Peterson, W. L., Mackowiak, P. A., Barnett, C. C., Marling-Cason, M. and Haley, M. L. 1989. The human gastric bactericidal barrier: mechanisms of action, relative antibacterial activity, and dietary influences. *Journal of Infectious Diseases* **159**:979–983.

Phillips, A. D., Navabpour, S., Hicks, S., Dougan, G., Wallis, T. and Frankel, G. 2000. Enterohaemorrhagic *Escherichia coli* O157:H7 target Peyer's patches in humans and cause attaching/effacing lesions in both human and bovine intestine. *Gut* **47**:377–381.

Ramegowda, B. and Tesh, V. L. 1996. Differentiation-associated toxin receptor modulation, cytokine production and sensitivity to Shiga-like toxins in human monocytes and monocytic cell lines. *Infection and Immunity* **64**:1173–1180.

Sandvig, K. and van Deurs, B. 1996. Endocytosis, intracellular transport, and cytotoxic action of Shiga toxin and ricin. *Physiology Review* **76**:949–966.

Saxena, S. K., O'Brien, A. D. and Ackerman, E. J. 1989. Shiga toxin, Shiga-like toxin II variant, and ricin are all single-site RNA *N*-glycosidases of 28S RNA when microinjected into *Xenopus* oocytes. *Journal of Biological Chemistry* **264**:596–601.

Schmidt, H. 2001. Shiga toxin-converting bacteriophages. *Research in Microbiology* **152**:687–695.

Selander, R. K., Caugant, D. A. and Whittam, T. S. 1987. Genetic structure and variation in natural populations of *Escherichia coli*, pp. 1625–1648. *In Escherichia coli* and *Salmonella typhimurium*: Cellular and Molecular Biology, Vol. 2, F. C. Neidhardt (Ed.); Washington, D.C.: American Society for Microbiology.

Siegler, R. L. 1994. Spectrum of extrarenal involvement in postdiarrheal hemolytic uremic syndrome. *Journal of Pediatrics* **125**:511–581.

Siegler, R. L. 1995. The hemolytic uremic syndrome. *Pediatric Clinics of North America* **42**:1505–1529.

Sixma, T. K., Stein, P. E., Hol, W. G. J. and Read, R. L. 1993. Comparison of the B-pentamers of heat-labile enterotoxin and verotoxin-1: two structures with remarkable similarity and dissimilarity. *Biochemistry* **32**:191–198.

Slutsker, L., Ries, A. A., Greene, K. D., Wells, J. G., Hutwagner, L., Griffin, P. M. and the *E. coli* O157:H7 Study Group. 1997. *Escherichia coli* O157:H7 diarrhea in the United States: clinical and epidemiological features. *Annals of Internal Medicine* **126**:505–513.

Stein, P. E., Boodhoo, A., Tyrrell, G. J., Brunton, J. L. and Read, R. J. 1992. Crystal structure of the cell-binding B oligomer of verotoxin-1 from *E. coli*. *Nature* **355**:748–750.

Stevens, M. P., van Diemen, P. M., Frankel, G., Phillips, A. D. and Wallis, T. S. 2002. Efa1 influences colonization of the bovine intestine by Shiga toxin–producing *Escherichia coli* serotypes O5 and O111. *Infection and Immunity* **70**:5158–5166.

Takatorige, T., Ida, O., Ikeda, K., Kimoto, K., Tatara, K., Kitani, T. and Okajima, S. (Eds.). 1997. Report on the Outbreak of *E. coli* O157:H7 Infection in Sakai City; Sakai City, Japan: Sakai City Medical Association.

Tarr, P. I., Bilge, S. S., Vary Jr., J. C., Jelacic, S., Habeeb, R. L., Ward, T. R., Baylor, M. R. and Besser, T. E. 2000. Iha: a novel *Escherichia coli* O157:H7 adherence-conferring molecule encoded on a recently acquired chromosomal island of conserved structure. *Infection and Immunity* **68**:1400–1407.

Tatsuno, I., Horie, M., Abe, H., Miki, T., Makino, K., Shinagawa, H., Taguchi, H., Kamiya, S., Hayashi, T. and Sasakawa, C. 2001. *toxB* gene on pO157 of enterohemorrhagic *Escherichia coli* O157:H7 is required for full epithelial cell adherence phenotype. *Infection and Immunity* **69**:6660–6669.

te Loo, D. M. W. M., Monnens, L. A. H., van der Velden, T. J. A. M., Vermeer, M. A., Preyers, F., Demacker, P. N. M., van den Heuvel, L. P. W. J. and van Hinsbergh, V. W. M. 2000. Binding and transfer of verocytotoxin by polymorphonuclear leukocytes in hemolyic uremic syndrome. *Blood* **95**:3396–3402.

Torres, A. G., Giron, J. A., Perna, N. T., Burland, V., Blattner, F. R., Avelino-Flores, F. and Kaper, J. B. 2002. Identification and characterization of *lpfABCC'DE*, a fimbrial operon of enterohemorrhagic *Escherichia coli* O157:H7. *Infection and Immunity* **70**:5416–5427.

van Setten, P. A., Monnens, L. A. H., Verstraten, R. G. G., van den Heuvel, L. P. W. J. and van Hinsbergh, V. W. M. 1996. Effects of verocytotoxin-1 on non-adherent human monocytes: binding characteristics, protein synthesis and induction of cytokine release. *Blood* **88**:174–183.

Wang, G., Zhao, T. and Doyle, M. P. 1996. Fate of enterohemorrhagic *Escherichia coli* O157:H7 in bovine feces. *Applied and Environmental Microbiology* **62**:2567–2570.

Wieler, L. H., McDaniel, T. K., Whittam, T. S. and Kaper, J. B. 1997. Insertion site of the locus of enterocyte effacement in enteropathogenic and enterohemorrhagic *Escherichia coli* differs in relation to the clonal phylogeny of the strains. *FEMS Microbiology Letters* **156**:49–53.

4 Bacterial Hazards in Fresh and Fresh-Cut Produce: Sources and Control

Alejandro Castillo and
M. Ofelia Rodríguez-García

Introduction

Several outbreaks of foodborne disease linked to the consumption of fresh fruits and vegetables make clear that these commodities constitute a serious threat to public health. Unique factors complicate the ability to assure safe fruits and vegetables: first, they are usually consumed without any previous antimicrobial treatment; second, they are highly perishable foods, usually providing no leftover food to test when an outbreak occurs; third, their consumption has increased in recent years (Gillman et al. 1995, Brownson et al. 1996); fourth, this group of foods includes a large variety of products with different physical structures, which makes it difficult to develop standard control measures applicable to all types of fruits and vegetables; and fifth, a large proportion of the produce consumed in the United States is grown and packed in countries in which there is not yet a fully developed on-farm food safety program. Some of these and other technical factors are also a burden in investigating produce-related outbreaks (Tauxe et al. 1997). Pathogen control during growing, packing, and shipping of fresh fruits and vegetables, as well as during processing and handling of fresh-cut produce, should have a positive effect on public health in the United States. Research on produce-related outbreaks has suggested that the contamination occurs at the growing or packing step. However, most of the time this contamination is superficial, and it is at the cutting step where the contamination is introduced into the flesh, increasing the probability that the pathogen will reach the consumer. This has been the case in different outbreaks in which contaminated fruits and vegetables were linked to outbreaks in the United States and Canada (Centers for Disease Control 1991, Ackers et al. 1998, Hedberg et al. 1999, Mohle-Boetani et al. 1999, California Department of Health Services 2000, 2001, Campbell et al. 2001, Dentinger et al. 2001, Naimi et al. 2003).

Pathogen control includes the application of measures for preventing, eliminating, or reducing pathogenic organisms from foods. However, different from processed foods, where a terminal treatment is applied to the food product, antimicrobial treatments applied to raw or fresh foods are not normally capable of totally eliminating microbial pathogens. This also makes the application of the Hazard Analysis and Critical Control Point System (HACCP) difficult to be successful in a production system in which significant pathogens are identified but no effective control measure is yet available. Product disinfection, if effective, can be used as a critical control point in a HACCP plan. However more research is needed to develop effective treatments.

In this chapter, information is presented about control measures applied at different steps from the field through fresh-cutting that may help reduce pathogens in produce. Research experiences from the authors are also presented. The objective of this chapter is to discuss the current status of research on pathogen control in fresh and fresh-cut fruits and vegetables as well as future trends in research in this area.

Effect of Produce Contamination on Public Health

Fruits and vegetables are among the five types of food most frequently linked to outbreaks of foodborne illness in the United States (Olsen et al. 2000). These outbreaks are associated with both imported and domestic produce (Tauxe et al. 1997). From the information presented in different reviews (Beuchat 1996, Nguyen-the and Carlin 2000), it is evident that as much as 80% of produce-related outbreaks have been caused by *Salmonella, Cyclospora,* pathogenic *Escherichia coli,* calicivirus, and *Shigella,* and that these pathogens have been associated with a variety of products.

Distribution of Pathogens in Produce

Fresh and fresh-cut produce have been reported to harbor different foodborne pathogens. The list of pathogens isolated from produce includes bacteria such as *Salmonella, Aeromonas* spp., *Campylobacter,* enterovirulent *E. coli, Shigella, Staphylococcus aureus, Listeria monocytogenes,* and *Yersinia enterocolitica* (Callister and Agger 1987, D'Aoust 2000, Nguyen-the and Carlin 2000) and protozoa such as *Cryptosporidium* or *Giardia* spp. (Monge and Chinchilla 1996, Nguyen-the and Carlin 2000). However, not all these pathogens have been linked consistently with outbreaks associated with fresh and fresh-cut produce.

The Food and Drug Administration (FDA) reported the incidence of *Salmonella* and *Shigella* to be 3.5% and 1.0%, respectively, for imported produce and 0.6% and 0.5%, respectively, for domestic produce (FDA/Center for Food Safety and Applied Nutrition [CFSAN], 2001). A comprehensive review revealed that the prevalence of pathogens in produce varies between 0% and 50% (FDA/CFSAN 2001). Different surveys for *L. mono-cytogenes* have shown incidences between 4.7% and 50% (Heisick et al. 1989, Prazak et al. 2002, Thunberg et al. 2002), whereas other pathogens, such as *Campylobacter jejuni, Bacillus cereus,* or *E. coli* O157:H7, have been sporadically isolated from different products (Beuchat 1996).

Salmonella seems to be the pathogen most commonly associated with fresh and fresh-cut produce. It has been isolated from artichokes, bean sprouts, beet leaves, broccoli, cabbage, carrots, cauliflower, celery, cilantro, endive, green onion, lettuce, parsley, pepper, radish, salad greens, and spinach (Beuchat 1996, FDA/CFSAN 2001). In a collaborative study between Texas A&M and the University of Guadalajara, Mexico, preharvest and postharvest cantaloupe samples were collected from farms located in South Texas and in the State of Colima, Mexico. *Salmonella* was isolated from five (0.5%) of 950 samples in South Texas and from one (0.3%) of 300 samples in Colima State, Mexico (Mercado-Uribe 2002). In Mexico, we have isolated *Salmonella* in 8% of various vegetables sampled at public markets (Rodríguez-García and Fernandez-Escartín 1993), and from 1% of cabbage samples and 6% of radish samples collected at the field immediately after harvest (Sevilla-Zamudio 2002).

Sources of Bacterial Pathogens in Fruits and Vegetables

The presence of bacterial pathogens in produce may be the result of exposure to fecal sources, either directly or indirectly through contaminated water, utensils, or field workers (Beuchat 1996), or of other non-fecal sources such as the environment, infected wounds in field or packing workers, or improperly sanitized equipment and utensils (Nguyen-the and Carlin 1994, De Roever 1998, Escartín 2000).

Different studies indicate that intestinal pathogens may be the organisms of greatest concern in fruit and vegetables (Beuchat 1996, D'Aoust 2000). Produce are exposed to fecal pathogens through contaminated water used to irrigate field crops, usually by using a poor irrigation technique such as spraying, which deposits the contaminants onto the product. Other mechanisms for bacterial contamination include fertilization of soil with untreated sludge, cross-contamination of crops by infected wildlife, manual handling of fresh fruits by workers with poor personal hygiene, washing of product by dumping it in contaminated washbasins, or using ice made from nonpotable water for cooling the product (D'Aoust 2001).

Knowing the major sources of pathogens and the mechanisms by which these pathogens contaminate fruits and vegetables enables the development of measures to prevent this contamination.

Survival and Growth of Pathogenic Bacteria in Produce

Bacteria present in fruits and vegetables can survive for long periods of time, or even grow, as there are enough nutrients and water available to support growth (Schena et al. 1999, Guo et al. 2002). Some of these organisms are psychrotrophs and therefore can multiply at refrigeration temperatures (Hotchkiss and Banco 1992, Bennik et al. 1995). This is especially important because under current commercial systems, produce may be shipped from remote places, frequently from overseas. When the product arrives in the United States, it is normally stored at warehouses until distribution. This system depends on refrigeration to preserve the product's freshness. Customs inspection may add to the storage time. The relatively long time of storage for produce may promote growth of psychrotrophic pathogens such as *L. monocytogenes* or *Aeromonas* spp.

Some reports demonstrate that *Salmonella* can multiply on the surface of fruits such as tomatoes (Zhuang et al. 1995, Weissinger et al. 2000), cut watermelons (Escartín et al. 1989), cantaloupe, and honeydew melon (Golden et al. 1993, Del Rosario and Beuchat 1995). According to Wells and Butterfield (1997), produce that has been affected by soft rot is more conducive to growth of *Salmonella* than non-diseased produce. These authors also reported that retail produce with soft rot had a higher incidence of *Salmonella* than non-damaged produce. *Shigella* has shown capability of growing when inoculated on fresh-cut produce (Escartín et al. 1989, Wu et al. 2000). *L. monocytogenes* has been reported to grow well on refrigerated avocado pulp (Arvizu-Medrano et al. 2001, Iturriaga et al. 2002), and Castillo and Escartín (1994) reported the ability of *C. jejuni* to survive on fresh-cut watermelon and papaya.

In sprouts, for which the major food safety problem seems to be the contamination of the seeds, pathogens such as *Vibrio cholerae* O1, *Salmonella* Typhi, and *E. coli* O157:H7 grew rapidly during the process of germination and sprouting, reaching counts of approximately 6.0 log CFU/g after 24 hours of incubation of the seeds, whereas no growth was observed when inoculation occurred 24 hours after germination. Large numbers of natural microbiota present after 24 hours probably prevented pathogen growth (Castro-Rosas and Escartín 1999).

Under favorable conditions, pathogens can be internalized in the product. Internalized pathogens have water and nutrients more readily available. Internalized pathogens have been reported to survive or grow in products such as tomato or lettuce (Seo and Frank 1999, Guo et al. 2002).

Pathogen Control

The control of bacterial hazards in fruits and vegetables was not given the importance that it deserves until several outbreaks of foodborne illness showed that these commodities were a common vehicle for pathogens. In addition, concerns were raised by the fact that a great amount of produce was being imported from tropical countries, where hygiene conditions during growth, harvest, and packing might not be optimal. This triggered different government initiatives to improve the safety of domestic and imported produce as part of the presidential Food Safety Initiative (Clinton 1997).

Control in the Field

The guidelines contained in the document "Guide to Minimize Microbial Food Safety Hazards for Fresh Fruits and Vegetables" (FDA, U.S. Department of Agriculture, Centers for Disease Control 1998) contain recommendations to reduce the risk for microbial hazards to contaminated plant foods. These include the Good Agricultural Practices (GAP) that should be followed during production, harvest, washing, selection, packing, and shipping of fresh or fresh-cut fruits and vegetables to prevent such contamination. The concepts included in the GAP are science based and can be applied by fruit and vegetable producers in the United States as well as other countries. As a result of the application of the GAP, the incidence of foodborne illness associated with fresh and fresh-cut produce is expected to be reduced.

Science-based procedures for reducing microbial pathogens in produce are not yet fully configured, and more investigation is necessary to develop strategies leading to the control of pathogens in fruits and vegetables. In fact, promoting and supporting research is one of the areas of interest within the Food Safety Initiative. Areas that need to be addressed include the detection of sources of pathogens in produce and the development of effective intervention strategies for pathogen reduction. The guide mentioned above focuses on reduction of microbial hazards, risk reduction instead of risk elimination, provision of scientifically based principles for controlling microbiological hazards, and constant updating or modification of the guidelines as new scientific information is available on detecting and reducing the microbial hazard in foods.

Control at the Packing Plant

In comparison with contamination occurring in the field (preharvest), the mechanisms for bacterial contamination occurring at the packing plant (postharvest) present slight differences. Human contamination gains greater importance as a source of pathogens postharvest. Equipment may be poorly sanitized, if at all, and the type of pests that may be in contact with the plant environment may be different from field fauna. The packing plant has been identified as the place where the product is at the highest risk of contamination, mainly because of the failure to apply good manufacturing practices and the common practice of dumping the product in a wash tank, where the contamination, if present in a few product units, is spread over all the lot. In a sampling of cantaloupe farms in Mexico, *Salmonella* or *E. coli* could not be detected in melons before or immediately after being harvested; however, as much as 52% of the cantaloupes were contaminated with *E. coli* contamination after being washed. *E. coli* was also found on the hands of the packing personnel, but on none of the field workers tested (Ponce de León et al. 2000, Mercado-Uribe 2002). Because *E. coli* is one of the organisms most commonly used as an index of fecal contamination, these results suggest that fecal contamination seems to occur more frequently during postharvest operations than during preharvest operations.

Pathogen Interventions

Several methods have been proposed for decontaminating fresh and fresh-cut fruits and vegetables. Most of these treatments are designed to reduce bacterial pathogens on the surface of the food product and are applied either by dipping or spraying an antimicrobial compound. Beuchat (1998) described a list of treatments recommended for postharvest disinfection of horticultural products. Treatments include chlorine, bromide, iodine, trisodium phosphate, quaternary ammonium compounds, acids, hydrogen peroxide, ozone, and ionizing irradiation. To avoid duplicating information that has been previously published, this chapter will focus on comparing the advantages and disadvantages of chlorine (by far the most common sanitizer used for fruit and vegetable disinfection) to other methods studied recently.

Chlorine

Chlorine is widely used for surface disinfection of fresh fruits and vegetables (Beuchat and Ryu 1997), usually by using chlorine gas or either sodium or calcium hypochlorite. Chlorine is also used for sanitizing contact surfaces in the food industry. The antibacterial activity of chlorine depends on the amount of free or available hypochlorous acid (HOCl) that is formed through a series of reactions between chlorine and the water. HOCl dissociation depends on the pH and the temperature of the water. In addition, organic matter, light, and metals interfere with the stability of chlorine and HOCl and therefore with the efficacy of chlorine as a sanitizer. The concentrations commonly used by the industry for surface disinfection of fruits and vegetables range between 50 and 200 mg/L, with a contact time of 1–2 minutes (Beuchat 1998). Although the greatest solubility of chlorine is reached at temperatures close to 4°C, adjusting the temperature of the solution 10°C higher than the temperature of the food product may create a positive pressure differential in the product. This minimizes the diffusion of wash water into the product tissues, which otherwise could affect the quality and shelf-life of the product (Bartz and Showalter 1981, Zhuang et al. 1995).

Contact of chlorine with open wounds, crevices, or other tissues of the fruits and vegetables may result in a decrease of the effectiveness of this compound at reducing pathogens in produce by exposure to increased amounts of organic matter, which consumes chlorine by being oxidized (Seo and Frank 1999, Takeuchi and Frank 2000). If the product was contaminated before harvest, the waxy cuticle present in many fruits and vegetables may be covering the pathogens, thus protecting the microorganisms from contact with the chlorine solution as well as any other sanitizer that may be used.

Surfactants such as ethanol or certain detergents reduce the hydrophobicity of the surface of fruits and vegetables. This enhances the contact of the sanitizer with the product surface, thereby increasing bacterial reduction, but at the same time, the quality of the product is sometimes affected during this process (Adams et al. 1989, Zhang and Farber 1996).

Use of chlorine to wash broccoli resulted in a reduction of total aerobic counts and coliform bacteria of 1 log cycle (Albrecht et al. 1995). In contrast, Wei et al. (1995) did not find a great reduction of *Salmonella* serovar Montevideo inoculated on the surface of tomatoes after washing with 100 mg/L chlorine for 0.5, 1, and 2 minutes. In fact, these authors reported that *S.* Montevideo survived for 20 hours on the tomato skin. This organism was able to grow in wounds of the whole fruits and in tomato slices, but decreased on the intact surface. These results underscore the great sensitivity of chlorine to inactivation by organic matter, making the concentration of the free chlorine a key factor in adjusting a chlorine treatment for pathogen reduction in fruits and vegetables.

Chlorine in the form of chlorine dioxide (ClO_2) has also been used to disinfect fruits and vegetables. Mason and Hicks (1985) reported that ClO_2 may be a good preservative for transport of packed fruits and vegetables. Reina et al. (1995) reported that total bacteria counts and *Enterobacteriaceae* were reduced in cucumbers by 2 and 6 logs, respectively, by washing the cucumbers with a solution containing 1.3 mg/L ClO_2. The authors did not observe any bacterial reduction after decreasing the ClO_2 concentration to 0.95 mg/L, whereas increasing the concentration to 2.8 or 5.1 mg/L was deleterious to product quality.

Organic Acids

Although widely used for decontamination of meat carcasses (Castillo et al. 2002), organic acids have not been commonly used for reducing bacterial pathogens on fruits and vegetables. The antimicrobial properties of organic acids are well documented (Baird-Parker 1980). The antimicrobial effect of organic acids increases with the temperature (Sorrells et al. 1989, Conner et al. 1990). Therefore, temperature adjustment is critical for achieving reduction of microorganisms with high acid tolerance, such as *E. coli* O157:H7.

Delaquis et al. (1999) tested the effectiveness of gaseous acetic acid for disinfecting bean seed inoculated with *Salmonella* Typhimurium, *E. coli* O157:H7 or *L. monocytogenes* (3.0–5.0 log CFU/g). After treating with 242 µL of acetic acid per liter of air for 12 hours at 45°C, *S.* Typhimurium or *E. coli* O157:H7 were not detected on the seed, and *L. monocytogenes* was recovered on two out of 10 samples. Fumigation with acetic acid was also lethal to natural microbiota on the bean seed, although no significant reduction in the sprouting rate or any physical damage was observed.

Richards et al. (1995) tested *p*-aminobenzoic acid (PABA) versus other organic acids against *L. monocytogenes*, *S.* Enteritidis, and *E. coli*. In their study, PABA showed greater antimicrobial activity than formic, propionic, citric, acetic, or lactic acids. PABA was also capable of reducing these pathogens at pH values higher than the values at which the other acids showed any bacterial reduction.

The use of organic acids creates concerns among packers about the potential presence of acid-adapted organisms. Ryu et al. (1999) compared different acids for their antimicrobial activity against acid-adapted and non-adapted *E. coli* O157:H7. They did observe that acid-adapted cells were able to survive and even grow in the presence of lactic (pH 3.9) or acetic (pH 4.2) acids. However, the acid resistance can be overcome by combining other factors such as the application of the acid at 55°C (Castillo et al. 2002).

Other Treatments

Several alternative treatments have been studied for reducing bacterial pathogens in fruits and vegetables. Lin et al. (2000a) studied the antimicrobial effect of allyl isothiocyanate (AITC) and methyl isothiocyanate against *S.* Montevideo, *E. coli* O157:H7, and *L. monocy-*

togenes in lettuce. Treatment of lettuce and tomatoes with AITC reduced the pathogens by 2–8 logs, depending on the concentration, time of contact, and product characteristics. These authors found that AITC was more effective against *S.* Montevideo and *E. coli* O157:H7 than against *L. monocytogenes*, whereas methyl isothiocyanate was more effective against *L. monocytogenes* than against the two Gram-negative organisms. Apparently, AITC antimicrobial activity is based on damage to the cell membrane, which is more extensive on Gram-negative compared with Gram-positive bacteria (Lin et al. 2000b). In another study, Ogawa et al. (2000) combined AITC (10–80 μg/mL) with hydrostatic pressure (200–250 Mpa) to enhance the reduction of *E. coli* O157 in Asazuke, a low-salt vegetable.

Hot-water washes also have been studied and recommended for reducing pathogens on fruits and vegetables. The key factor is to use a treatment that will increase the temperature at the product surface above 75°C. *E. coli* O157:H7 populations were reduced by 5 logs on apples by immersing in water at 80° or 95°C without significantly increasing the temperature in the interior of the fruit (Fleischman et al. 2001). Pao and Davis (1999) obtained a 5-log reduction in *E. coli* inoculated on the surface of oranges by immersing in water at 80°C for only 30 seconds. On the stem scar area, where microorganisms seem to be more difficult to reach, a 5-log reduction was achieved after immersion for 2 minutes in water at 80°C.

Hot-water treatments may be useful for reducing pathogens on products that have a stem scar or crevices where the pathogens may be confined and cannot be reached by a chemical sanitizer. Some superficial damage may occur on the product after hot-water treatment. Therefore, this treatment may be more appropriate for fruits or vegetables destined for further processing such as juice extraction or cutting, and not for cases where product quality is important, such as fresh or fresh-cut fruits and vegetables.

Other treatments may include electrolyzed acid water washes; immersion in ozonated water (Koseki et al. 2001); alkaline solutions such as sodium hydroxide, sodium bicarbonate, or trisodium phosphate (Zhuang and Beuchat 1996, Pao et al. 2000); different detergents and alcohols (Pao et al. 2000); ethanol (Martínez-Gonzalez 2003); pulsed electricity (Iu and Griffiths 2001); and even natural treatments such as garlic juice (Unal et al. 2001) or inoculation with lactic acid bacteria (Vescovo et al. 1996). All these treatments have been effective at reducing pathogens on the surface of produce by 1–5 logs. However, in-plant validations sometimes have indicated that surface treatments may not achieve bacterial reductions greater than 1 or 2 logs (Alvarado-Casillas et al. 2000). A combination of treatments may enhance the individual effect of different antimicrobials and should be further studied (Jordan et al. 1999).

Current Research on Pathogen Interventions

Reports on side-by-side comparison of different sanitizers are necessary to decide which sanitizers could be recommended to the produce industry for decontaminating fruits and vegetables at the packing level. Research conducted in our laboratory includes projects on the comparison of inexpensive treatments that would be attractive for packers. One of these projects included the comparison of water wash alone and combinations of water wash with waxing, disinfection with calcium hydroxide, hypochlorite, or lactic acid for reducing bacterial pathogens on cantaloupes (Alvarado-Casillas et al. 2000). These sanitizers were applied by immersion or spraying. In this study, we found a treatment consisting of spraying the product for 15 seconds with 2% L-lactic acid at 55°C to reduce populations of *Salmonella enterica* serovar Typhimurium and *Escherichia coli* O157:H7 by 3 and 2 logs, respectively. This treatment did not affect the sensorial characteristics or the shelf life of the product.

Spraying with, or submerging in, a solution of 200 ppm sodium hypochlorite resulted in a 1.9 log reduction of *S.* Typhimurium and only 0.2–0.5 log reduction for *E. coli* O157:H7. Bacterial reductions obtained by waxing or applying calcium hydroxide were no greater than those obtained by water wash alone. These comparisons are shown in Figure 4.1.

Martínez-Gonzales (2003) evaluated the effectiveness of a water wash and of water wash followed by hot water, 70% ethanol, 200 ppm hypochlorite, or 2% hot (55°C) *L*-lactic acid treatments to reduce *S.* Typhimurium, *E. coli* O157:H7 and *L. monocytogenes* inoculated on the surface of oranges. The water wash alone reduced these three pathogens by 0.9, 1.4, and 0.6 logs, respectively. When the water wash was followed by a sanitizing treatment, hot water and lactic acid showed the greatest reductions. However, the lactic acid treatment would need to be adjusted in a processing facility, as great variations were observed between lactic acid concentrations (2% vs. 4%), method of application (spray vs. immersion), and time of contact (1, 2, and 4 minutes by immersion and 15, 30, and 45 seconds by spraying).

From these two studies on comparison of methods it appears that *L*-lactic acid sprays may be a good alternative for fruit and vegetable disinfection. More work is needed to determine the type of products where this agent, as well as others, can be effectively applied.

Pathogen Internalization

Microorganisms can diffuse into the product through the water, especially during uncontrolled wash treatments, or through environmental contamination (FDA/CFSAN 1999). Physical damage and type of product can also promote bacterial internalization in the product. Microorganisms can become internalized by fruits and vegetables via different mechanisms. The physical structure of the product is an important factor. Bacteria can be localized in the stomata, the stem-scar, or the calyx areas of different fruits or vegetables (Annous et

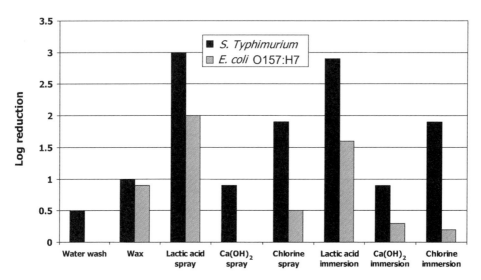

Figure 4.1 Reduction (log/cm²) by treatment of *S.* Typhimurium and *E. coli* O157:H7 inoculated onto the surface of fresh cantaloupes.

al. 2001, Guo et al. 2002, Solomon et al. 2002). Product injury due to insect wounds, hail damage, handling, etc., can promote bacterial diffusion (FDA/CFSAN 1999).

According to some studies, product washing before packing may contribute to the internal contamination of fruits and vegetables (Bartz and Showalter 1981, Janisiewicz et al. 1999, Merker et al. 1999). Washing with water at high pressure does not produce visible damage to the product; however, if the product has been bruised during harvesting, microorganisms can gain contact with the endocarp via different mechanisms. Product damage can provide a place for increased bacterial growth, which then can be released during washing. By this mechanism clean wash water can become contaminated, and then the microorganisms can spread to non-damaged product. If the contamination is limited to a small number of product units, dip washing can spread the contamination over the entire lot and even to later lots if the water is not changed between batches. Nonbruised product can also experience microbial internalization if the wash water is contaminated (Buchanan et al. 1999, FDA/CFSAN 1999, Merker et al. 1999).

The transmission of *E. coli* O157:H7 to apples by fruit flies was studied by Janisiewicz et al. (1999). The authors inoculated fruit flies with *E. coli* O157:H7 as well as with a surrogate *E. coli*–producing green fluorescent protein and freed them in the presence of wounded apples. All apples were contaminated with *E. coli* after 48 hours of storage. *E. coli* O157:H7 was able to grow by 3 logs inside the wounded apples in 48 hours of storage at 24°C. The non-inoculated fruit flies were allowed to come in contact with wounded, inoculated, and non-inoculated apples. *E. coli* was detected in the wounds of previously non-contaminated apples exposed to contaminated apples and non-inoculated fruit flies, indicating that *E. coli* could be transferred between apples by flies (Janisiewicz et al. 1999).

Current Research on Pathogen Internalization

If bacterial internalization of produce occurs, then sanitizing solutions can also penetrate the product, and depending on the sanitizing solutions' sensitivity to organic matter, the sanitizers may remain active inside products. This hypothesis was tested by Ibarra-Sánchez (2002), who studied the effect of product disinfection on populations of *S.* Typhimurium and *E. coli* O157:H7 internalized in tomatoes. The treatments compared were a water wash or water wash followed by lactic acid spray or a hypochlorite (200 ppm) spray. All treatments were applied at 5° and 55°C. Figure 4.2 shows the recovery of internalized pathogens from tomatoes. Water wash alone or chlorine spray did not significantly affect the internal bacterial populations of the tomatoes, whereas L-lactic acid spray reduced both pathogens to undetectable levels or, in one instance, to a level very close to the detection limit of the counting method. In this case it was hypothesized that, even though the hypochlorite solution probably penetrates the internal regions of the product, the organic matter inside the tomatoes might inactivate the free chlorine. In contrast, lactic acid is not as sensitive to inactivation by organic matter; thus, it remains active inside the product.

Pathogen Control after Minimal Processing

Minimal processing includes washing, trimming, shredding, peeling, cutting, packaging, and in general any treatment that does not modify the major structure of the food product. Fresh-cut produce is an example of a minimally processed food, and it is partially prepared so that no additional preparation is necessary for its use (Watada and Qi 1999). Because

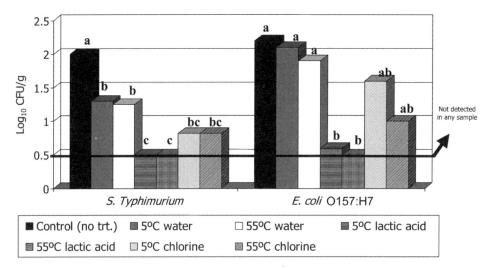

Figure 4.2 Detection of internalized pathogens in tomatoes after rinsing with different sanitizers. Bars with same letter are not significantly different ($P < 0.05$).

these products often involve tissue breakage resulting from cutting or peeling, fresh-cut products are highly perishable. Nutrients and water are readily available, and therefore these products need to be refrigerated. If present, pathogens can contaminate the product from the surface of the raw commodity, as has occurred in different outbreaks (Beuchat 1996). In some outbreaks, poor storage might have played a role in permitting pathogen multiplication.

Because fresh-cut produce is ready to use, the application of sanitizers is not always feasible, as the sanitizer might affect the color, odor, or taste of the product. Consumers usually rely on the level of food safety training among food handlers to follow good manufacturing practices during preparing the food product. Some sanitizers have been tested for effectiveness at reducing pathogens in fresh-cut produce. Zhang and Farber (1996) compared six different sanitizers for reducing *L. monocytogenes* on fresh-cut lettuce and cabbage. According to their results, chlorine compounds reduced the microorganism by 0.4–1.7 logs, depending on the food commodity as well as the concentration and temperature of the sanitizer applied. Lactic or acetic acid produced only a 0.2–0.5 log reduction at their highest concentration (1%) and longest time of contact (10 minutes), whereas trisodium phosphate had almost no effect on the *Listeria* populations, and Salmide (Bioxy Incorporated, Raleigh, NC), a chlorite-based halogenated disinfectant, led to maximum reductions of 0.6–1.8 logs.

Fresh-cut produce, especially greens that are not flavor sensitive, are usually treated with chlorine before cutting for safety and quality purposes; sometimes a final wash is applied after cutting to eliminate any chlorine residues. This final wash may change the flavor in the product, therefore affecting product quality (Guerzoni et al. 1996).

Other common practices with fresh produce involve adding acids that will enhance flavor. Some consumers add lime or lemon juice to the product before consumption. However, this must only be seen as a practice to enhance flavor, as research shows that lime juice did not completely eliminate the risk of infection through consumption of contaminated fruit (Escartín et al. 1989, Castillo and Escartín 1994).

Irradiation of Fresh-Cut Produce

There have been many articles published on irradiation of fruits and vegetables. However, most of these reports deal with disinfestation and quality issues, and few reports address food safety. Nevertheless, ionizing irradiation of fresh-cut fruits and vegetables seems to be a promising technology for microbial safety and quality (see Chapter 28). Irradiation seems to significantly reduce organisms that can grow on the product so that numbers do not increase over time, thereby enhancing product quality and perhaps increasing shelf life by reducing the amount of spoilage organisms (Nguyen-the and Carlin 1994). For example, Hagenmaier and Baker (1998) reported a reduction of 1.8 logs in total plate counts from shredded carrots stored for 2 days after irradiation with 0.5 kGy in comparison to non-irradiated carrots. This reduction remained stable even 9 days after irradiating. In another study, aerobic plate counts (APCs) from carrots irradiated with 1 kGy from a γ source were reduced by up to 3.5 logs after 1 day of storage; however, after 7 days the APCs were the same on both irradiated and non-irradiated carrots (Zagory 1999).

Current Research on Irradiation of Fresh-Cut Produce

The quality of different irradiated fresh-cut produce products after electron beam irradiation is under study in our laboratory to determine the effect on bacterial counts and sensory characteristics. In fresh-cut cantaloupes, electron beam irradiation at 0.7 kGy reduced the aerobic plate count by 2 logs in comparison with non-irradiated cantaloupe, and this difference remained after 9 days of storage at 4°C. After 13 days, the APCs were the same in irradiated and non-irradiated cantaloupe. When a chlorine wash of 200 mg/L (ppm) was applied to the cantaloupes before cutting, and the cantaloupes were then irradiated at 0.7 kGy, the APCs were reduced by 3.3 logs, and the counts remained significantly lower than those in the control during 15 days of storage. Increasing the dose to 1.4 kGy had a remarkable effect on bacterial counts, with reductions of 5–6 logs, which remained low even after 21 days of storage. Combining irradiation at 1.4 kGy with a pre-cut chlorine wash increased the log reduction by three more log cycles and did not significantly increase over 21 days of storage at 4°C. In our study, objective color and firmness values were not affected by any of the treatments applied, in contrast to the study referred by Zagory (1999). These results suggest that the microbiological and sensory quality of fresh-cut cantaloupe can be enhanced by low-dose electron beam irradiation and that applying effective sanitizing rinses such as a chlorine wash to whole melons may enhance the bactericidal effect of electron beam irradiation. These treatments may also reduce any bacterial pathogens present, and the safety of fresh-cut produce may be enhanced by low-dose irradiation.

Conclusions

Pathogen interventions have shown some reduction of bacterial pathogens on fresh and fresh-cut produce, some with less effect than others. In addition, the effectiveness of these interventions may be product dependant, as there are so many differences among commodities. More research is therefore needed to identify specific decontamination methods that are effective, as well as the type of products for which these interventions are applicable. After these treatments are designed, the implementation of HACCP plans may become a reality.

References

Ackers, M. L., Mahon, B. E., Leahy, E., Goode, B., Damrow, T., Hayes, P. S., Bibb, W. F., Rice, D. H., Barrett, T. J., Hutwagner, L., Griffin, P. M. and Slutsker, L. 1998. An outbreak of *Escherichia coli* O157:H7 infections associated with leaf lettuce consumption. *Journal of Infectious Diseases* **177**:1588–1593.

Adams, M. R., Hartley, A. D. and Cox, L. 1989. Factors affecting the efficiency of washing procedures used in the production of prepared salads. *Food Microbiology* **6**:69–77.

Albrecht, J., Jamamouz, H., Sumner, S. S. and Melch, V. 1995. Microbial evaluation of vegetable ingredients in salad bars. *Journal of Food Protection* **58**:683–685.

Alvarado-Casillas, S., Ibarra-Sánchez, L. S., Martínez-Ruiz, Y., Ponce de León-Torres, J. and Castillo, A. 2000. Evaluación de métodos de descontaminación de melón cantaloupe y chile morrón para exportación (Evaluation of methods for decontaminating cantaloupes and bell peppers). Proceedings of the 2nd International Food Safety Conference. November 10-11, Guadalajara, Mexico, p. 27.

Annous, B. A., Sapers, G. M., Mattrazzo, A. M. and Riordan, D. C. 2001. Efficacy of washing with a commercial flatbed brush washer, using conventional and experimental washing agents, in reducing populations of *Escherichia coli* on artificially inoculated apples. *Journal of Food Protection* **64**:159–163.

Arvizu-Medrano, S. M., Iturriaga, M. H. and Escartín, E. F. 2001. Indicator and pathogenic bacteria in guacamole and their behavior in avocado pulp. *Journal of Food Safety* **21**:233–244.

Baird-Parker, A. C. 1980. Organic acids, pp. 126–135. *In* Microbial Ecology of Foods, Vol. I. Factors affecting life and death of microorganisms. J. H. Silliker, R. P. Elliot, A. C. Baird-Parker, F. L. Bryan, J. H. B. Christian, D. S. Clark, J. C. Olson, T. A. Roberts (Eds.). International Commission on Microbiological Specifications for Foods, New York: Academic Press.

Bartz, J. A. and Showalter, R. K. 1981. Infiltration of tomatoes by aqueous bacterial suspensions. *Phytopathology* **71**:515–518.

Bennik, M. H. J., Smid, E. J., Rombouts, F. M. and Gorris, L. G. M. 1995. Growth of psychrotrophic foodborne pathogens in a solid surface model system under the influence of carbon dioxide and oxygen. *Food Microbiology* **12**:509–519.

Beuchat, L. R. 1998. Surface decontamination of fruits and vegetables eaten raw: a review. Food Safety Issues, World Health Organization. Geneva, WHO/FSF/FOS/98.2.

Beuchat, L. R. 1996. Pathogenic microorganisms associated with fresh produce. *Journal of Food Protection* **59**:204–216.

Beuchat, L. R. and Ryu, J. H. 1997. Produce handling and processing practices. *Emerging Infectious Diseases* **3**:459–465.

Brownson, R. C., Smith, C. A., Pratt, M., Mack, N. E., Jackson-Thompson, J., Dean, C. G., Dabney, S. and Wilkerson, J. C. 1996. Preventing cardiovascular disease through community-based risk reduction: the Bootheel Heart Health Project. *American Journal of Public Health* **86**:206–213.

Buchanan, R. L., Edelson, S. G., Miller, R. L. and Sapers, G. M. 1999. Contamination of intact apples after immersion in an aqueous environment containing *Escherichia coli* O157:H7. *Journal of Food Protection* **62**:444–450.

California Department of Health Services. 2000. State health director advises consumers to wash cantaloupe before eating, Number 29-00. California Department of Health Services, Office of Public Affairs Press Releases. Available at: http://www.applications.dhs.ca.gov/pressreleases/store/backup/29-00.html.

California Department of Health Services. 2001. State health director advises consumers to scrub cantaloupe before eating, Number 37-01. California Department of Health Services, Office of Public Affairs Press Releases. Available at: http://www.dhs.ca.gov/admin/ffdmb/templates/Misc/PressReleases/Store/PressReleases/37-01.html.

Callister, S. M. and Agger, W. A. 1987. Enumeration and characterization of *Aeromonas hydrophila* and *Aeromonas caviae* isolated from grocery store produce. *Applied and Environmental Microbioliology* **53**:249–253.

Campbell, J. V., Mohle-Boetani, J., Reporter, R., Abbott, S., Farrar, J., Brandl, M., Madrell, R. and Werner, S. B. 2001. An outbreak of Salmonella serotype Thompson associated with fresh cilantro. *Journal of Infectious Diseases* **183**:984–987.

Castillo, A. and Escartín, E. F. 1994. Survival of *Campylobacter jejuni* on sliced watermelon and papaya. *Journal of Food Protection* **57**:166–168.

Castillo, A., Hardin, M. D., Acuff, G. R. and Dickson, J. S. 2002. Reduction of microbial contamination on carcasses, pp. 351–381. *In* Control of Foodborne Microorganisms, V. K. Juneja, J. N. Sofos (Eds.). New York: Marcel Dekker.

Castro-Rosas, J. and Escartín, E. F. 1999. Incidence and germicide sensitivity of *Salmonella typhi* and *Vibrio cholerae* 01 in alfalfa sprouts. *Journal of Food Safety* **19**:137–146.

Centers for Disease Control. 1991. Epidemiologic notes and reports multistate outbreak of *Salmonella* Poona infections United States and Canada. *Morbidity and Mortality Weekly Report* **40**:549–552.

Clinton, W. J. 1997. Memorandum on the Food Safety Initiative. Administration of William J. Clinton, October 2, p. 1479. Available at: http://www.foodsafety.gov/~dms/fs-wh2.html.

Conner, D. E., Scott, V. N. and Bernard, D. T. 1990. Growth, inhibition, and survival of *Listeria monocytogenes* as affected by acidic conditions. *Journal of Food Protection* **53**:652–655.

D'Aoust, J. Y. 2000. *Salmonella*, pp. 1233–1299. *In* The Microbiological Safety and Quality of Food, Vol. II., B. M. Lund, A. C. Baird-Parker, G. W. Gould (Eds.). Gaithersburg, Md.: Aspen Publishers.

D'Aoust, J. Y. 2001. Salmonella, pp. 163–191. *In* Guide to Foodborne Pathogens, R. G. Labbe, S. García (Eds.). Hoboken, N.J.: Wiley Interscience.

Del Rosario, B. A. and Beuchat, L. R. 1995. Survival and growth of enterohemorrhagic *Escherichia coli* O157:H7 in cantaloupe and watermelon. *Journal of Food Protection* **58**:105–107.

De Roever, C. 1998. Microbiological safety evaluations and recommendations on fresh produce. *Food Control* **9**:321–347.

Delaquis, P. J., Sholberg, P. L. and Stanich, K. 1999. Desinfection of mung bean seed with gaseous acetic acid. *Journal of Food Protection* **62**:953–957.

Dentinger, C. M., Bower, W. A., Nainan, O. V., Cotter, S. M., Myers, G., Dubusky, L. M., Fowler, S., Salehi, E. D. P. and Bell, B. P. 2001. An outbreak of hepatitis A associated with green onions. *Journal of Infectious Diseases* **183**:1273–1276.

Escartín, E. F. 2000. Microbiología e Inocuidad de los Alimentos (Food microbiology and safety). Queretaro, Mexico: Universidad Autónoma de Querétaro Press.

Escartín, E. F., Castillo, A. and Lozano, J. S. 1989. Survival and growth of *Salmonella* and *Shigella* on sliced fresh fruit. *Journal of Food Protection* **52**:471–472.

Food and Drug Administration, Center for Food Safety and Applied Nutrition. 1999. Potential for infiltration, survival and growth of human pathogens within fruits and vegetables. Available at: http://www.cfsan.fda.gov/~comm/juicback.html.

Food and Drug Administration, Center for Food Safety and Applied Nutrition. 2001. Outbreaks associated with fresh produce: incidence, growth, and survival of pathogens in fresh and fresh-cut produce, Chapter IV. *In* Analysis and Evaluation of Preventive Control Measures for the Control and Reduction/Elimination of Microbial Hazards on Fresh and Fresh-Cut Produce. Available at: http://www.cfsan.fda.gov/~comm/ift3-4a.html.

Food and Drug Administration, U.S. Department of Agriculture, Centers for Disease Control. 1998. Guidance for industry: guide to minimize microbial food safety hazards for fresh fruits and vegetables. Available at: http://www.foodsafety.gov/~dms/prodguid.html.

Fleischman, G. J., Bator, C., Merker, R. and Keller, S. E. 2001. Hot water immersion to eliminate *Escherichia coli* O157:H7 on the surface of whole apples: thermal effects and efficacy. *Journal of Food Protection* **64**:451–455.

Gillman, M. W., Cupples, L. A., Gagnon, D., Posner, B. M., Ellison, R. C., Castelli, W. P. and Wolf, P. A. 1995. Protective effect of fruits and vegetables on development of stroke in men. *Journal of the American Medical Association* **273**:1113–1117.

Golden, D. A., Rhodehamel, E. J. and Kautter, D. A. 1993. Growth of *Salmonella* spp. in cantaloupe, watermelon, and honeydew melons. *Journal of Food Protection* **56**:194–196.

Guerzoni, M. E., Gianotti, A., Corbo, M. R. and Sinigaglia, M. 1996. Shelf-life modeling for fresh-cut vegetables. *Postharvest Biology and Technology* **9**:195–207.

Guo, X., Chen, J., Brackett, R. E. and Beuchat, L. R. 2002. Survival of *Salmonella* on tomatoes stored at high relative humidity, in soil, and on tomatoes in contact with soil. *Journal of Food Protection* **65**:274–279.

Hagenmaier, R. D. and Baker, R. A. 1998. Microbial population of shredded carrot in modified atmosphere packaging as related to irradiation treatment. *Journal of Food Science* **63**:162–164.

Hedberg, C. W., Angulo, F. J., White, K. E., Langkop, C. W., Schell, W. L., Stobierski, M. G., Schuchat, A., Besser, J. M., Dietrich, S., Helsel, L., Griffin, P. M., McFarland, J. W., Osterholm, M. T. and the investigation team. 1999. Outbreaks of salmonellosis associated with eating uncooked tomatoes: implications for public health. *Epidemiology and Infection* **122**:385–393.

Heisick, J. E., Wagner, D. E., Nierman, M. L. and Peeler, J. T. 1989. *Listeria* spp. found on fresh market produce. *Applied and Environmental Microbiology* **55**:1925–1927.

Hotchkiss, J. H. and Banco, M. J. 1992. Influence of new packaging technologies on the growth of microorganisms in produce. *Journal of Food Protection* **55**:815–820.

Ibarra-Sánchez, L. S. 2002. Infiltración microbiana en frutas sometidas a tratamientos de descontaminación (Microbial internalization in fruits subjected to decontamination treatments). Master's Thesis, Graduate Program in Biotechnological Processes, University of Guadalajara, Mexico.

Iturriaga, M. H., Arvizu-Medrano, S. M. and Escartín, E. F. 2002. Behavior of *Listeria monocytogenes* in avocado pulp and processed guacamole. *Journal of Food Protection* **65**:1745–1749.

Iu, J. and Griffiths, M. W. 2001. Reduction in levels of *Escherichia coli* O157:H7 in apple cider by pulsed electric fields. *Journal of Food Protection* **64**:964–969.

Janisiewicz, W. J., Conway, W. S., Brown, M. W., Sapers, G. M., Fratamico, P. and Buchanan, R. L. 1999. Fate of *Escherichia coli* O157:H7 on fresh-cut apple tissue and its potential for transmission by fruit flies. *Applied and Environmental Microbiology* **65**:1–5.

Jordan, S. L., Glover, J., Malcolm, L., Thomson-Carter, F. M., Booth, I. R. and Park, S. F. 1999. Augmentation of killing of *Escherichia coli* O157:H7 by combinations of lactate, ethanol, and low-pH conditions. *Applied and Environmental Microbiology* **65**:1308–1311.

Koseki, S., Yoshida, K., Isobe, S. and Itoh, K. 2001. Decontamination of lettuce using acidic electrolysed water. *Journal of Food Protection* **64**:652–658.

Lin, C. M., Kim, J., Du, W. and Wei, C. 2000a. Bactericidal activity of isothiocyanate against pathogens on fresh produce. *Journal of Food Protection* **63**:25–30.

Lin, C. M., Preston, J. F. and Wei, C. 2000b. Antibacterial mechanism of allyl isothiocyanate. *Journal of Food Protection* **63**:727–734.

Martínez-Gonzales, N. E. 2003. Riesgos a la salud por *Escherichia coli* O157:H7, *Salmonella* spp. y *Listeria monocytogenes* asociados a la preparacion y consumo de jugo de naranja fresco (Risk of disease caused by *Escherichia coli* O157:H7, *Salmonella* spp. and *Listeria monocytogenes* associated with preparation and consumption of fresh sqweezed orange juice). Ph.D. Thesis. Graduate program in Biological Sciences, Nacional Polytechnic Institute, Mexico City, Mexico.

Mason, J. Y. and Hicks, B. W. 1985. Dry compositions for the production of chlorine dioxide. U.S. Patent 86-08-V0014.

Mercado-Uribe, I. 2002. *Salmonella* contamination during production of domestic and imported cantaloupe. Master's Thesis, Texas A&M University, College Station, Tex.

Merker, R., Edelson-Mamel, S., Davis, V. and Beuchat, L. R. 1999. Preliminary experiments on the effect of temperature differences on dye uptake by oranges and grapefruit. U.S. Food and Drug Administration, Center for Food Safety and Applied Nutrition. Available at http://vm.cfsan.fda.gov/~comm/juicexp.html.

Mohle-Boetani, J. C., Reporter, R., Werner, B. S., Abbot, S., Farrar, J., Waterman, S. H. and Vugia, D. J. 1999. An outbreak of *Salmonella* serogroup Saphra due to cantaloupes from Mexico. *Journal of Infectious Diseases* **180**:1361–1364.

Monge, R. and Chinchilla, M. 1996. Presence of *Crisptosporidium* oocysts in fresh vegetables. *Journal of Food Protection* **59**:202–203.

Naimi, T. S., Wicklund, J. H., Olsen, S. J., Krause, G., Wells, J. G., Bartkus, J. M., Boxrud, D. J., Sullivan, M., Kassenborg, H., Besser, J. M., Mintz, E. D., Osterholm, M. T. and Herberg, C. W. 2003. Concurrent outbreaks of *Shigella sonnei* and enterotoxigenic *Escherichia coli* infections associated with parsley: implications for surveillance and control of foodborne illness. *Journal of Food Protection* **66**:535–541.

Nguyen-the, C. and Carlin, F. 1994. The microbiology of minimally processed fruits and vegetables. *Critical Reviews in Food Science and Nutrition* **34**:371–401.

Nguyen-the, C. and Carlin, F. 2000. Fresh and Processed Vegetables, pp. 620–984. *In* The Microbiological Safety and Quality of Food, Vol. I, B. M. Lund, A. C. Baird-Parker, G. W. Gould (Eds.). Gaithersburg, Md.: Aspen Publishers.

Ogawa, T., Nakatani, A., Matsukaki, H. and Isshiki. K. 2000. Combined effects of hydrostatic pressure, temperature, and the addition of allyl isothiocyanate on inactivation of *Escherichia coli*. *Journal of Food Protection* **63**:884–888.

Olsen, S. J., MacKinnon, L. C., Goulding, J. S., Bean, N. H. and Slutsker, L. 2000. Surveillance for foodborne-disease outbreaks—United States, 1993-1997, *In* CDC Surveillance Summaries, March 17, 2000. *Morbidity and Mortality Weekly Reports* **49(No. SS-1)**:1–64.

Pao, S. and Davis, C. L. 1999. Enhancing microbiological safety of fresh orange juice by fruit immersion in hot water and chemical sanitizers. *Journal of Food Protection* **62**:756–760.

Pao, S., Davis, C. L. and Kelsey, D. F. 2000. Efficacy of alkaline washing for the decontamination of orange fruit surfaces inoculated with *Escherichia coli*. *Journal of Food Protection* **63**:961–964.

Ponce de León, T. J., Martínez-Ruiz, Y., Ibarra-Sánchez, S., Alvarado-Casillas, S. and Castillo, y A. 2000. Estudio de la contaminación por *Salmonella* spp. y *Escherichia coli* en el melón cantaloupe de exportación (Study of *Sal-*

monella and *Escherichia coli* contamination in cantaloupe for export to the United States). Proceedings of the 2nd International Food Safety Conference. November 10-11, Guadalajara, Mexico.

Prazak, A. M., Murano, E. A., Mercado, I. and Acuff, G. R. 2002. Prevalence of *Listeria monocytogenes* during production and postharvest processing of cabbage. *Journal of Food Protection* **65**:1728–1734.

Reina, L. D., Fleming, H. P. and Humphries, E. G. 1995. Microbiological control of cucumber hydrocooling water with chlorine dioxide. *Journal of Food Protection* **58**:541–546.

Richards, R. M. E., Xing, D. K. L. and King, T. P. 1995. Activity of *p*-aminobenzoic acid compared with other organic acids against selected bacteria. *Journal of Applied Bacteriology* **78**:209–215.

Rodríguez-García, O. and Fernandez-Escartín, E. 1993. Bacterias indicadoras de contaminación fecal en verduras colectadas en mercados públicos (Fecal indicators in vegetables collected from public markets). Abstracts of the 24th Meeting of the Mexican Association for Microbiology. May 24-27, Guadalajara, Jalisco.

Ryu, J. H., Deng, Y. and Beuchat, L. R. 1999. Behavior of acid-adapted and unadapted *Escherichia coli* O157:H7 when exposed to reduced pH achieved with various organic acids. *Journal of Food Protection* **62**:451–455.

Schena, L., Ippolito, A., Zahavi, T., Cohen, L., Nigro, F. and Droby, S. 1999. Genetic diversity and biocontrol activity of *Aureobasidium pullulans* isolates against postharvest rots. *Postharvest Biology and Technology* **17**:189–199.

Seo, K. H. and Frank, J. F. 1999. Attachment of *Escherichia coli* O157:H7 to lettuce leaf surface and bacterial viability in response to chlorine treatment as demonstrated by using confocal scanning laser microscopy. *Journal of Food Protection* **62**:3–9.

Sevilla-Zamudio, S. V. 2002. Monitoreo de *Salmonella* y *E. coli* en hortalizas durante la cosecha en el campo de cultivo (Monitoring of *Salmonella* and *E. coli* during harvest at crop fields). Master's Thesis, Graduate Program in Food Sciences, University of Guadalajara, Mexico.

Solomon, E. B., Potenski, C. J. and Matthews, K. R. 2002. Effect of irrigation method on transmission to and persistence of *Escherichia coli* O157:H7 on lettuce. *Journal of Food Protection* **65**:673–676.

Sorrells, K. M., Enigl, D. C. and Hatfeld, J. H. 1989. Effect of pH, acidulant, time, and temperature on the growth and survival of *Listeria monocytogenes*. *Journal of Food Protection* **52**:571–573.

Takeuchi, K. and Frank, J. F. 2000. Penetration of *Escherichia coli* into lettuce tissues as affected by inoculum size and temperature and the effect of chlorine treatment on cell viability. *Journal of Food Protection* **63**:434–440.

Tauxe, R., Kruse, H., Hedberg, C., Potter, M., Madden, J. and Wachsmuth, K. 1997. Microbial hazards and emerging issues associated with produce: a preliminary report to the National Advisory Committee on Microbiologic Criteria for Foods. *Journal of Food Protection* **60**:1400–1408.

Thunberg, R. L., Tran, T. T., Bennett, R. W., Mattheus, R. N. and Belay, N. 2002. Microbial evaluation of selected fresh produce obtained at retail markets. *Journal of Food Protection* **65**:677–682.

Unal, R., Fleming, H. P., McFeeters, R. F., Thompson, R. L., Breidt, F. and Giesbrecht, F. G. 2001. Novel quantitative assays for estimating the antimicrobial activity of fresh garlic juice. *Journal of Food Protection* **64**:189–194.

Vescovo, M., Torriani, S., Orsi, C., Macchiarolo, F. and Scolari, G. 1996. Application of antimicrobial-producing lactic acid bacteria to control pathogens in ready-to-use vegetables. *Journal of Applied Bacteriology* **81**:113–119.

Watada, A. E. and Qi, L. 1999. Quality of fresh-cut produce. *Postharvest Biology and Technology* **15**:201–205.

Wei, C. I., Huang, T. S., Kim, J. M., Lin, W. F., Tamplin, M. L. and Bartz, J. A. 1995. Growth and survival of *Salmonella montevideo* on tomatoes and disinfection with chlorinated water. *Journal of Food Protection* **58**:829–836.

Weissinger, W. R., Chantaraparanont, W. and Beuchat, L. R. 2000. Survival and growth of *Salmonella baildon* in shredded lettuce and diced tomatoes, and effectiveness of chlorinated water as a sanitizer. *International Journal of Food Microbiology* **62**:123–131.

Wells, J. M. and Butterfield, J. E. 1997. *Salmonella* contamination associated with bacterial soft rot of fresh fruits and vegetables in the marketplace. *Plant Disease* **81**:867–872.

Wu, F. M., Doyle, M. P., Beuchat, L. R., Wells, J. G., Mintz, E. D. and Swaminathan, B. 2000. Fate of *Shigella sonnei* on parsley and methods of disinfection. *Journal of Food Protection* **63**:568–572.

Zagory, D. 1999. Effects of post-processing handling and packaging on microbial populations. *Postharvest Biology and Technology* **15**:313–321.

Zhang, S. and Farber, J. M. 1996. The effects of various desinfectants against *Listeria monocytogenes* on fresh-cut vegetables. *Food Microbiology* **13**:311–321.

Zhuang, R. Y. and Beuchat, L. R. 1996. Effectiveness of trisodium phosphate for killing *Salmonella montevideo* on tomatoes. *Letters in Applied Microbiology* **22**:97–100.

Zhuang, R.-Y., Beuchat, L. R. and Angulo, F. J. 1995. Fate of *Salmonella montevideo* on raw tomatoes as affected by temperature and treatment with chlorine. *Applied and Environmental Microbiology* **61**:2127–2131.

5 *Campylobacter* Species and Fresh Produce: Outbreaks, Incidence, and Biology

Robert E. Mandrell and Maria T. Brandl

Introduction

Many consumers in the United States and other developed countries have an increasing desire to eat fresh produce for the enhanced taste and healthful qualities it provides. Nevertheless, fresh produce and produce dishes have been implicated in 293 outbreaks and approximately 18,000 reported cases of illness in the United States since 1990 (Center for Science in the Public Interest 2002). Major recent outbreaks related to produce have been associated with *Salmonella enterica* or *Escherichia coli* O157:H7; sprouts, cantaloupe, lettuce, tomatoes, and cilantro have been implicated (Centers for Disease Control 2000).

Thermophilic *Campylobacter* species, *C. coli, C. lari,* and *C. jejuni,* cause more cases of human gastrointestinal illness in the United States, the United Kingdom, and other western countries than any other bacteria (World Health Organization 1998, Frost et al. 2002); over 90% of the cases are caused by *C. jejuni* (Friedman et al. 2000). However, in these and other countries that monitor for the etiology of diarrheal illnesses as part of public health surveillance systems, foodborne outbreaks caused by *Campylobacter* species have been reported infrequently compared to outbreaks caused by other enteric bacteria such as *S. enterica* or pathogenic *E. coli* (Centers for Disease Control 2000). The fact that most cases of *Campylobacter* infection appear to occur as sporadic illnesses emphasizes why point sources of contamination associated with sporadic illnesses usually are not pursued or identified (Friedman et al. 2000).

Campylobacter species are highly prevalent on both preharvest and postharvest chickens (Jacobs-Reitsma 2000, Stern et al. 2001). Poultry and poultry products are considered to be a major potential source for *Campylobacter* illness (Altekruse et al. 1999) and have been associated with *Campylobacter* outbreaks (Istre et al. 1984, Allerberger et al. 2003). *Campylobacter* species can also be prevalent in other animals such as cattle, swine, sheep (Stanley et al. 1998), horses, rabbits, domestic pets, rodents, and wild birds (Jacobs-Reitsma 2000).

Outbreaks of *Campylobacter* illness resulting from contaminated fruits, vegetables, or other produce-related products have been reported, but they occur infrequently compared with those associated with other enteric pathogens (Beuchat 1996, Jacobs-Reitsma 2000). In this review, we have summarized data from reported outbreaks and incidence studies related to *Campylobacter* species and produce. In addition, data from our laboratory on the fitness of *C. jejuni* in the plant environment are presented and discussed.

Mention of a trade name, proprietary product, or specific equipment does not constitute a guarantee or warranty by the U.S. Department of Agriculture and does not imply its approval to the exclusion of other products that may be suitable.

Outbreaks of *Campylobacter* Illness Associated with Food

Table 5.1 lists outbreaks of *Campylobacter* species-associated illness occurring in the United States between 1990 and 1999, selected outbreaks of *Campylobacter* illness related to water, and selected produce-related outbreaks of *Campylobacter* illness. The source of the outbreaks in some cases has been difficult to identify and, therefore, may not be the actual source for some outbreaks; nevertheless, these data provide a framework of information for discussion of comparative aspects related to *Campylobacter* species and produce.

The numbers of outbreaks and cases reported between 1990 and 1999 in the United States plus selected produce and waterborne outbreaks in other countries for dairy, produce, water, meat, poultry, and seafood as source categories were 16 outbreaks (totaling 286 cases), 18 outbreaks (totaling >961 cases), 12 outbreaks (totaling 5,161 cases), 10 outbreaks (totaling 111 cases), 8 outbreaks (totaling 69 cases), and 5 outbreaks (totaling 123 cases), respectively. Nine additional outbreaks causing at least 350 illnesses involving miscellaneous or mixed food sources were reported. A review of the epidemiology of outbreaks involving *Campylobacter* in water-related environments indicates that 12 major outbreaks have been reported (Table 5.1; Koenraad et al. 1997, Lee et al. 2002). These outbreaks reflect the ability of *Campylobacter* species to survive in aqueous environments, possibly even in a viable, but nonculturable, state that could be important in the cycling of *Campylobacter* species in the environment (Rollins and Colwell 1986). *Campylobacter* species survival in water could have implications for the potential contamination of crops with *Campylobacter* species by irrigation water.

Campylobacter species have been the most frequent bacterial cause of gastrointestinal illness in England and Wales since 1981 (Frost et al. 2002). However, they accounted for only 2.1% of the total number of outbreaks between 1995 and 1999 with an identified etiologic agent (50 of 2,374 total), a difference consistent with observations in the United States

Table 5.1   Selected outbreaks of *Campylobacter* and selected produce-related and waterborne outbreaks in the United States and other countries, 1990 to 2001.

Source	Number of outbreaks[a]	Total number of cases	Country
Seafood	5	123	United States
Poultry	8	69	United States
Miscellaneous, known	9	357	United States
Meat (beef, pork, other)	10	111	United States
Water[b]	12	5,161	Multiple
Produce[c]	18	961	Multiple
Dairy products	16	286	United States
Unknown[d]	43	646	United States

(a) Data were obtained from listings of U.S. outbreaks at the Centers for Disease Control (2000) and the Center for Science in the Public Interest (2002).

(b) Data summarized from Koenraad et al. 1997, Population and Public Health Branch 2000, and Lee et al. 2002; outbreaks occurred in the United States, Canada, New Zealand, Finland, and Norway.

(c) Data summarized from Centers for Disease Control Line Listings of outbreaks with a produce-related product as source (Centers for Disease Control 2000) and from Kirk et al. 1997, Wight et al. 1997, Frost et al. 2002, and Long et al. 2002; outbreaks occurred in the United States, Australia, United Kingdom, and France.

(d) Data summarized from Centers for Disease Control Line Listings of outbreaks with *Campylobacter* either identified or suspected as the causative agent (Centers for Disease Control 2000).

and other countries, indicating that *Campylobacter* species cause more cases of sporadic illness than do other food pathogens. In a recent survey in England and Wales between 1992 and 2000, five of 83 (6.0%) outbreaks of "infectious intestinal disease" associated with produce in England and Wales were caused by *Campylobacter* (Long et al. 2002). Most of the 83 outbreaks documented were associated with *Salmonella* (41%) or Norwalk-like virus (15.7%).

It is noteworthy that, with the exception of water-related outbreaks, produce is associated with more cases of *Campylobacter* illness (>961 cases) than any other food source and is second only to "unknown" sources in the total number of outbreaks in the United States (Table 5.1). Specific produce-related outbreaks caused by *Campylobacter* species are shown in Table 5.2. However, it is probable that produce was not the original source of *Campylobacter* contamination for most of these outbreaks; the food source listed for 12 of 18 outbreaks reported in Table 5.2 suggests that multiple ingredients may have been included in the food source. Although it cannot be determined from the outbreak data, a possible explanation for the relatively high number of outbreaks and cases associated with produce is that produce became cross-contaminated by another contaminated food during preparation (e.g., raw poultry or other meat; Acuff et al. 1986, De Boer and Hahné 1990). Such a scenario could result in sporadic illnesses caused by *Campylobacter*-contaminated produce without the food source being identified. This is in contrast to the identification of the food source and etiology occurring during an intensive traceback investigation during an outbreak. Therefore, an important area for consideration and study in determining the comparative risks associated with *Campylobacter* species is their overall fitness for survival and growth in preharvest environments and environmental conditions related to both pre- and postharvest produce (Mandrell et al. 2002).

For most produce-related outbreaks, few epidemiological details are available. Information is often difficult to obtain because of the perishable nature of the commodity, multiple points where contamination could occur (pre- and postharvest), multiple types of produce in servings or meals, and presence of other contaminated food during food preparation. However, even when the etiology of a produce outbreak is clear, the predominant hypothesis for *Campylobacter* contamination of produce is through cross-contamination by *Campylobacter* on poultry or meat (De Boer and Hahné 1990, Centers for Disease Control 1998) or from food-preparers. Consistent with this hypothesis, a study by Castillo and Escartín (1994) demonstrated that *C. jejuni* survived for 6 hours after inoculation onto sliced watermelon and papaya incubated at 25–29°C, indicating that *C. jejuni* can survive on these tissues for a period of time sufficient to pose a risk to the consumer if cross-contamination occurs. The potential for cross-contamination of raw produce by the food preparer is implied by a study reporting that *C. jejuni* on plates exposed to raw chicken products could be transferred to five of 54 (9%) samples of raw vegetables placed on the exposed plates (De Boer and Hahné 1990). *C. jejuni* was isolated also from 42 of 58 (72%) hands that had held raw chicken products (De Boer and Hahné 1990).

It should be emphasized that it is not possible to attribute any of the outbreaks noted in Table 5.2 to preharvest contamination of produce. The lack of any studies reporting the incidence of *Campylobacter* on field produce (i.e., preharvest) and the limited number of studies reporting the incidence of *Campylobacter* on postharvest produce obtained at markets (see next section) make it difficult to assess the potential risk of preharvest *Campylobacter* contamination of produce to public health. Multi-state outbreaks related to a produce product could indicate major contamination of crops in the field, during transport, or

Table 5.2 Produce-related outbreaks of *Campylobacter* in the United States and selected produce outbreaks in other countries, 1990 to 2001.

Produce source	Species	Number of cases	Country[a]	Date	References[b]
Melon or strawberries	*jejuni*	48	United States, Minn.	September 1993	CDC, CSPI
Lettuce	*jejuni*	5	United States, Wash.	June 1995	CDC, CSPI
Lettuce/lasagna	*jejuni*	14	United States, Okla.	August 1996	CDC, CSPI
Lettuce	*jejuni*	300	United States, Minn.	June 1998	CDC, CSPI
Sweet potatoes	ND[c]	17	United States, Conn.	August 1997	CDC, CSPI
Dish, fruit salad	*jejuni*	62	United States, Minn.	August 1994	CDC, CSPI
Dish, salad	*jejuni*	10	United States, Wash.	June 1994	CDC, CSPI
Dish, salad	*jejuni*	5	United States, Wash.	July 1994	CDC, CSPI
Dish, salad	*jejuni*	152	United States, Minn.	June 1998	CDC, CSPI
Dish, salad[d]	*jejuni*	70	United States, N.Y.	June 1996	CDC, CSPI
Cucumber	ND	78	Australia	October 1995	Kirk et al. 1997
Produce, salad/vegetables[e]/ meat fajitas	*jejuni*	8	England/ Wales	December 1993	Frost et al. 2002, Long et al. 2002[f]
Produce, lettuce/tomato	*jejuni*	16	England/ Wales	April 1996	Frost et al. 2002, Long et al. 2002[f]
Produce, salad vegetables[e]/ prawn/curred meat	*jejuni*	61	England/ Wales	April 1996	Frost et al. 2002, Long et al. 2002[f]
Produce, lettuce/mayonnaise/garlic	*jejuni*	30	England/ Wales	April 1996	Frost et al. 2002, Long et al. 2002[f]
Produce, lettuce	*jejuni*	18	England/ Wales	April 1996	Frost et al. 2002, Long et al. 2002[f]
Produce, orange juice/pasta salad	*jejuni*	30	England/ Wales	April 1996	Frost et al. 2002, Long et al. 2002[f]
Produce, Lettuce[g]	*jejuni*	37	France	November 1995	Wight et al. 1997

(a) For U.S. outbreaks, the state is noted also.
(b) CDC, see Centers for Disease Control 2000; CSPI, see Center for Science in the Public Interest 2002.
(c) ND, not determined.
(d) Suspected source.
(e) Salad composed of vegetables plus meat or seafood; the original contaminated food item was not determined.
(f) S. Meakins, 2003, personal communication. Outbreaks of Camplybacter illness in the United Kingdom associated with salad/vegetables/fruit.
(g) Both *Campylobacter* and *Escherichia coli* O111 were associated with this outbreak.

at a processing or distribution facility. Multi-state outbreaks of campylobacteriosis have been few or nonexistent regardless of the food source (Friedman et al. 2000).

Incidence of *Campylobacter* Species in Fresh Produce or Produce-Related Foods

Few studies have reported the incidence of *Campylobacter* species in fresh produce products. A summary of the results of 18 different studies, published between 1982 and 2003, of produce-related products for which isolation of *Campylobacter* was achieved or attempted is presented in Table 5.3. The cumulative data from these studies and the produce-associated outbreak source data (Table 5.1) are useful for assessing the potential risks associated with *Campylobacter* in produce relative to other foodborne pathogens.

Matsusaki and coworkers may have provided the first published report of the incidence of *Campylobacter* in produce products (Matsusaki et al. 1982). A total of 345 different foods obtained from markets, including 24 samples of vegetables, were tested for *C. coli*

Table 5.3   Studies of the incidence of *Campylobacter* in produce or produce-related food.

Source	Species[a]	Samples positive/total samples	References
Vegetables, retail, 24 types	Cc, Cj	0/NR[b]	Matsusaki et al. 1982
Mushrooms, retail	Cj	3/200	Doyle and Schoeni 1986
Vegetables, outdoor markets, and supermarkets	TC	9/1,564[c]	Park and Sanders 1992
Salad	TC	0/20	Karib and Seeger 1994
Preserved fruits, chutney, punch drinks	TC	0/NR	Adesiyun 1995
Vegetables, unprocessed	ND	0/65	Odumeru et al. 1997
Vegetables, freshcut	ND	0/296	Odumeru et al. 1997
Mixed salad vegetables, MAP[d]	ND	20/90	Phillips 1998
Salads	TC	0/12	Mosupye and von Holy 1999
Vegetables, RTE[e]	Cj	2/150[f]	Federighi et al. 1999
Cut tomato and onion	Cj	0/NR	Mosupye and von Holy 2000
Cereals, nuts, dried fruits, herbs, spices	ND	0/NR	Candlish et al. 2001
Vegetables, organic	ND	0/NR	McMahon and Wilson 2001
Vegetables	Cj	2/56[g]	Kumar et al. 2001
Vegetables, organic, multiple countries, RTE	ND	0/2,883	Sagoo et al. 2001
Vegetables, RTE	ND	0/NR	Moore et al. 2002
Fresh produce, retail	ND	0/127	Thunberg et al. 2002
Vegetables, RTE	ND	0/3,827	Sagoo et al. 2003

(a) Cc, *C. coli*; Cj, *C. jejuni*; TC, thermophilic *Campylobacters*; ND, *Campylobacter* species not determined.
(b) NR, not reported.
(c) Nine positives identified in samples of spinach, lettuce, radish, green onion, parsley, and potatoes.
(d) MAP, modified atmosphere packaged.
(e) RTE, ready-to-eat.
(f) Grated vegetables; carrots and coleslaw were positive.
(g) One sample each of spinach and fenugreek were positive.

and *C. jejuni*. *Campylobacter* strains were isolated from 14 of 34 chicken meat samples (41.2%) but not from any vegetable samples or other foods (e.g., beef, pork, frozen food, shellfish, and milk). The first published report of the isolation of *Campylobacter* from a "produce-related" product may be of the isolation of *C. jejuni* from fresh mushrooms (Doyle and Schoeni 1986). *C. jejuni* was isolated from three of 200 retail fresh mushroom samples (1.5%; Table 5.3). Even though mushrooms are not plants, but fungi, there are similarities between mushroom and vegetable production and processing practices.

Only five of the 18 incidence studies listed in Table 5.3 reported any *Campylobacter* species isolated from produce samples. In one of the most extensive and complete studies, Park and Sanders (1992) tested for the incidence of thermotolerant *Campylobacter* in 1,564 fresh produce samples obtained from farmers' outdoor markets (533 samples) and supermarkets (1,031 samples) in Ottawa, Canada. Of 10 different vegetables tested, *Campylobacter* species were isolated from two of 60, two of 67, two of 74, one of 40, one of 42, and one of 63 samples of spinach, lettuce, radish (with leaves), green onion, parsley, and potatoes, respectively (total incidence 1.7%). All of the positive samples originated from farmers' markets during the summer months. In contrast, summer and winter samples obtained from the supermarkets (1,031 samples) all were negative for *Campylobacter* species. In addition, all of the *Campylobacter* strains isolated from processed vegetable samples were obtained only after enrichment culture methods; *Campylobacter* were not isolated from any sample by direct plating. These results suggest that the culture methods are critical for accurate testing of the incidence of *Campylobacter* species in produce samples (discussed further later). Multiple species of *Campylobacter* were confirmed to be present in spinach (*C. jejuni, C. lari,* and *C. coli*), parsley (*C. jejuni* and *C. lari*), and radish samples. The presence of multiple species and strains of *Campylobacter* with different fitness levels in selective isolation media emphasizes the difficulty in obtaining accurate information for epidemiologic investigations of outbreaks. It also shows the need for developing methods for identifying more of the strains of the suspect species that may be present in a sample at a broad range of concentrations.

Campylobacter have also been isolated from modified atmosphere packaged (MAP) mixed salad vegetables (Phillips 1998). Samples purchased from two large food chain supermarkets in the United Kingdom were tested either on the "date of purchase" (<2 hours at 4°C) or were stored at 4°C and tested on the "use-by date." *Campylobacter* were isolated from 12 of 36 and 8 of 54 of the date-of-purchase and use-by-date samples, respectively. The difference in isolation rate from the two types of samples was significant, but the number of colony-forming units isolated from the use by date samples was not significantly different from the date of purchase samples. Nevertheless, the average rate of isolation of *Campylobacter* from MAP salad vegetables (22.2%) is much higher than that reported for any other study of produce or produce-related products. It is possible that the modified atmosphere used for packaging the salad vegetables (5%–8% CO_2:2%–5% O_2:87%–93% N_2) enhanced the survival of *Campylobacter* in these samples (Phillips 1998). The effect of MAP on the growth of different *Campylobacter* species is an important area for future studies.

In a study of 400 samples of ready-to-eat vegetable samples collected over a 2-year period in the Nantes region of France, two of 150 and zero of 250 grated ready-to-eat vegetables and salad samples, respectively, were positive for *C. jejuni* (Federighi et al. 1999). Differences were noted in the efficiency of isolation of *Campylobacter* on two different selection and three different plating media from all food samples tested, again indicating the importance of the media and culture methods for accurate estimates of *Campylobacter* incidence.

Finally, a recent study on the incidence of *Campylobacter* in produce reported that two of 56 vegetable samples collected in a local market in Northern India were positive: one of nine spinach and one of nine fenugreek samples (Table 5.3). *Campylobacter* strains were not isolated from cauliflower, cabbage, coriander, radish, or carrot samples in this study. The remainder of the incidence studies listed in Table 5.3 failed to isolate *Campylobacter* from a wide range of produce types and other plant-related foods.

It is worth noting that in the most recent *Campylobacter*-negative study, five strains of *S. enterica* were isolated from five of the 3,843 ready-to-eat salad vegetable samples tested. More important, however, was the fact that one of the *Salmonella* strains isolated from four-leaf bagged lettuce (S. Newport Phage Type 33) was the same type as that linked with a possible outbreak (19 cases) occurring in England and Wales at the same time (Sagoo et al. 2003). If this lettuce product was associated directly with the human cases, it could be the first documented report of a ready-to-eat produce-related product linked with an outbreak. The fact that *Campylobacter* species were not isolated from any of the ready-to-eat products, many of which were packaged presumably with modified atmospheres with CO_2 concentrations favorable for *Campylobacter* growth (5%–10%; see Chapter 6), emphasizes the need to determine whether there are differences in the methods used in studies of the same product types that yielded very different results (Phillips 1998).

The *Campylobacter* outbreak data related to produce suggests that between 1990 and 1999, produce was the second most common cause of human illness (18 outbreaks, approximately 800 illnesses; Table 5.1), even though incidence data demonstrate that *Campylobacter* is seldom isolated from produce (four of 17 studies; Table 5.3). The lack of appropriate culture methods for produce samples may explain some of the reported low-incidence data. The specific requirements of *Campylobacter* for growth and for isolation from food (e.g., nutrients, atmosphere, temperature, and osmolarity) indicate that current methods may need modification.

Comparison of Methods for Successful Isolation of *Campylobacter* Species from Produce

Fresh produce samples may require modified methods for isolation of *Campylobacter* species. Of the 18 studies reporting the incidences of *Campylobacter* species in produce (Table 5.3), only five studies reported a *Campylobacter*-positive (C[+]) isolation of any *Campylobacter* species from any sample. An examination of the major ingredients in the media for isolating *Campylobacter* in the five C[+] studies suggested that each of the five isolation protocols was different in the type of enrichment and plating media used. A few trends were apparent in comparing the C[+] and *Campylobacter*-negative (C[−]) methods. For example, the C[+] methods included an enrichment step (blood, four of five methods; antibiotics, four of five methods; and cycloheximide, five of five methods) and sodium pyruvate (four of five methods) in the plating medium; these ingredients are used commonly in other media described for isolation of *Campylobacter* (Corry et al. 1995). Two of the C[−] studies tested a wide variety of ready-to-eat organic and salad vegetables obtained from retail shops in the United Kingdom between May 1 and June 30 of 2000 (Sagoo et al. 2001) and 2001 (Sagoo et al. 2003), respectively. It is interesting that not a single positive sample was detected from the 6,710 total samples tested by the procedure used in these two studies. In addition, it is accepted that the commonly used methods for isolating *C. coli* and *C. jejuni* will prevent the isolation of many of the other *Campylobacter* species, some of which are emerging as human pathogens (Corry et al. 1995, Labarca et al. 2002). These results suggest the need to develop improved methods for isolating *Campylobacter* species from various food sources to obtain accurate assessments of the risks to public health after exposure to *Campylobacter* species.

A study of the effectiveness of a polymerase chain reaction method for detection of *C. jejuni* in foods involved testing both inoculated and noninoculated foods purchased from a retail market, including 13 different varieties of raw fruits and vegetables (Winters and Slavik 2000). This method failed to identify any *C. jejuni* in the noninoculated produce samples even after attempts to enrich for *C. jejuni* by incubation for 24 hours in Brain Heart Infusion medium overlaid on Brucella agar.

Campylobacter Species in the Plant Environment

The upper and lower surfaces of plants are considered hostile environments for microbes because of the repetitive and rapid fluctuations in physicochemical conditions (Handelsman and Stabb 1996, Hirano and Upper 2000). Ten different sugars and 25 amino acids, as well as organic acids, fatty acids, sterols, growth factors (vitamins), hormones, and nucleotides, have been detected in plant root exudates (Curl and Truelove 1986). Many of the nutrients present in root exudates have been detected also in leaf exudates (Lindow and Brandl 2003).

The ability of *Campylobacter* species to survive on and colonize plant tissue will depend on the physical and chemical conditions that exist in the plant environment. Listed in Table 5.4 are plant substances that are detected commonly on the surfaces of leaves and roots and that are also used by various strains of *C. coli* and *C. jejuni* on Biolog plates (Biolog Inc., Hayward, CA), either metabolically or for respiration. Alpha-hydroxybutyric acid, succinic acid, glutamic acid, *L*-asparagine, *L*-aspartic acid, glutamine, proline, and serine are potential nutrient sources for *Campylobacter* cells exposed to plants, because they are used by *C. coli* and *C. jejuni* (Table 5.4) and are spatially or temporally abundant on plants. Fumaric acid, which is present on the leaves and roots of many plant species (Morgan and Tukey 1964, Kraffczyk et al. 1984), can be used by some *Campylobacter* species as an electron acceptor for anaerobic growth (Holt et al. 1994).

Campylobacter cells are unable to ferment or oxidize sugars; rather, they obtain energy by metabolizing amino or organic acids (Elharrif and Megraud 1986). It was somewhat unexpected, therefore, that in the Biolog system most strains of *C. coli* and *C. jejuni* were respiring actively in *L*-arabinose and *L*-fucose, two sugars that are also present in plants (Kraffczyk et al. 1984, Zablackis et al. 1996; Table 5.4). These results are intriguing, but the nature of the responses of *C. coli* and *C. jejuni* species to these specific sugars remains unclear.

C. jejuni Survival and Growth on Plants

The ability of *C. jejuni* to colonize plant surfaces was tested in our laboratory. The upper parts of lettuce and spinach plants were inoculated by immersion in a suspension of *C. jejuni* (strain RM1221) cells that had been grown in Brucella broth, washed, and resuspended in 0.5× phosphate buffered saline (pH 7.0). The plants were then incubated at 28° or 33°C in a humid chamber that ensured the presence of a water film on parts of the leaf surface. *C. jeju-*

Table 5.4 Carbon substrates used by *Campylobacter coli* and *Campylobacter jejuni* strains on Biolog plates (Biolog Inc., Hayward, CA) and reported to be present on root or leaf surfaces.

	Activity[a]		Presence of substrate[b]	
Substrate	*C. jejuni*	*C. coli*	Root exudates	Leaf Rinses
L-arabinose	382 ± 33	191 ± 43	+	+
L-fucose	445 ± 41	789 ± 56	+[c]	+[c]
α-D-glucose	74 ± 22	5 ± 4	+	+
α-hydroxybutyric acid	871 ± 47	43 ± 9	+	NR[d]
Succinic acid	196 ± 20	68 ± 9	+	+
L-asparagine	300 ± 33	46 ± 13	+	+
L-aspartic acid	231 ± 27	47 ± 9	+	+
L-proline	68 ± 11	16 ± 5	+	+
L-serine	62 ± 21	3 ± 3	+	+
L-cysteine	564 ± 90[e]	1409[e]	+	−

(a) *C. coli* and *C. jejuni* were assayed for respiration induced by carbon substrates present in GN2 Biolog plates (Biolog Inc. Hayward, CA). The data are presented as the average ± standard error of the mean of the normalized optical density determined for all of the strains tested. Nine and 38 strains of *C. coli* and *C. jejuni*, respectively, were tested at least twice. Bacteria were grown on the medium recommended by the manufacturer and suspended in water, and the plates were incubated for 24 hours at 37°C.

(b) + or − indicates whether the compound was detected in soluble plant root exudates (Kraffczyk et al. 1984, Curl and Truelove 1986) or rinses of leaves (Morgan and Tukey 1964, Derridj 1996).

(c) Present in plant cell walls (Zablackis et al. 1996).

(d) NR, not reported to be present.

(e) Cysteine was tested separately with two strains of *C. jejuni* and one strain of *C. coli*.

ni populations on both lettuce and spinach leaves decreased over 100-fold at 28° and 33°C within the first 24 hours of incubation; populations on most lettuce and spinach leaves were below the detection threshold of 100 or 15 cells per leaf, respectively, 48 hours after inoculation (data not shown). In addition, enrichment in Preston broth (Uyttendaele and Debevere 1996) failed to recover *C. jejuni* from leaves, suggesting that *C. jejuni* cells were dead or nonculturable. *C. jejuni* survived for at least 6 days on the leaves of spinach plants incubated at 10°C under the same conditions as above (Figure 5.1). Thus, although *C. jejuni* was unable to grow on aerial plant surfaces, presumably because of the high levels of oxygen in that environment, it remained viable on leaves at cold temperatures considerably longer than at warm temperatures.

Campylobacter Species in Soil

Experiments were initiated in our laboratory to test the ability of *C. jejuni* to survive on roots and in soil. Using population studies, we determined that 6 days after inoculation, the population size of *C. jejuni* decreased 10^6-fold on spinach leaves but only 123-fold on spinach roots of plants incubated under similar conditions of high humidity and at 10°C; *C. jejuni* was culturable from the spinach rhizosphere for at least 28 days (data not shown). *C. jejuni* survived also at high levels and remained culturable for at least 23 days at 10°C on radish roots exposed to it in soil (Figure 5.1). Thus, the results of our radish and spinach experiments suggest that *C. jejuni* survives better in the root and soil environment than on

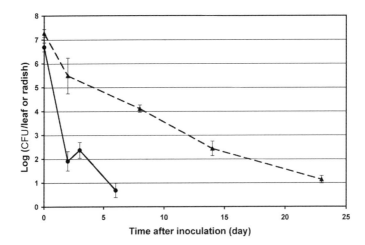

Figure 5.1　Survival of *C. jejuni* strain RM1221 on leaves of spinach plants (triangles) and on radish (circles). Spinach plants were inoculated by immersion of the upper plant part in a suspension of *C. jejuni* in 0.5× phosphte-buffered saline (5 mM sodium phosphate, 75 mM NaCl, pH 7.0). The radishes were grown in soil inoculated with a suspension of *C. jejuni* in 0.5× PBS. The spinach and radish plants were incubated at 10°C under humid conditions. Each data point represents the mean of the log value of the population size on six leaves, or six radishes; bars represent the standard error of the mean. The population size of *C. jejuni* was determined by plating onto selective agar.

leaf surfaces (Figure 5.1). To examine this further, we simulated the contamination of crops by water and soil splashing during rain and irrigation by inoculating the upper part of spinach plants with a suspension of *C. jejuni* mixed in organic soil. Four days after incubation of the plants under high humidity at 10°C, the population of *C. jejuni* was on average 1,000-fold higher on spinach leaves contaminated with both the pathogen and soil than on those with the pathogen alone (data not shown). It is noteworthy that the vegetables from which *Campylobacter* species have been isolated are either root crops or have a high probability of being contaminated with soil from the field (Table 5.3). Organic acids and chemicals in soil could be used by *Campylobacter* species as energy sources or electron acceptors (Elharrif and Megraud 1986). Humic acids, which are abundant in many soils, have been reported to increase the isolation of *Campylobacter* species when supplemented in growth agar, presumably because of their photo-protective effect (Weinrich et al. 1990). Conditions of low oxygen and high osmolality in wet and fertilized soil also could provide a suitable habitat for the enhanced survival of *C. jejuni*.

　　Although *C. jejuni* appeared to remain culturable at higher rates in the root environment than on leaf tissue, it is clear that, under the conditions of our studies, *C. jejuni* was unable to grow on plant surfaces at both cold temperatures and at temperatures that are conducive to its growth *in vitro*. Unlike *C. jejuni*, other enteric human pathogens such as *S. enterica*, *E. coli* O157:H7, and *Listeria monocytogenes* have been reported to survive well and to even colonize the surfaces of seeds, roots, and leaves of crop plants (Brandl and Mandrell 2002, Charkowski et al. 2002, Wachtel et al. 2002, Gorski et al. 2003). This weak fitness of *C. jejuni* on plants may explain the relatively rare campylobacteriosis outbreaks associated with produce, compared with the numerous and large epidemics of foodborne illness linked to the contamination of produce by *S. enterica* and pathogenic *E. coli*.

Conclusions and Perspectives

Raw vegetables, fruits, and mixed salads have become an increasingly important food source worldwide, and accordingly, outbreaks associated with consumption of contaminated produce continue to occur. However, outbreaks caused by *Campylobacter* species in produce are less frequent relative to those caused by other enteric pathogens, such as *Salmonella enterica*, even though *Campylobacter* species cause more total gastrointestinal illnesses in the United States and many European countries than any other bacteria. The fastidious growth characteristics and sensitivity of *Campylobacter* species in food and many food production environments would limit the concentration of *Campylobacter* cells encountered by humans consuming contaminated food. Low concentrations of cells in pre- or postharvest produce would result in limited isolation and, thus, low incidence of *Campylobacter* species in retail produce. *Campylobacter* association with sporadic rather than outbreak illness is consistent with a low incidence in food. Nevertheless, the occasional outbreaks and incidences of *Campylobacter* species associated with produce-related products, the prevalence of *Campylobacter* species in a variety of animals raised for food, and the unexplained high rates of sporadic human illness caused by *Campylobacter* all emphasize the need to assess carefully whether pre- and postharvest produce products are important sources of food leading to campylobacteriosis. The very high incidence of *Campylobacter* species in MAP salads (22%) noted in one study (Table 5.3) is an intriguing result and demonstrates the need for additional surveys of MAP products and studies of the survival and growth of *Campylobacter* species on produce under MAP conditions. It can be speculated, for example, that even though *Campylobacter* species may be less fit than other bacterial enteric pathogens on preharvest produce, certain processing and packaging conditions (e.g., 10–20°C, microaerophilic atmosphere, organic acids, and soil contaminants) could enhance the survival and perhaps growth of *Campylobacter* species, resulting in a limited number of produce items contaminated with a potentially infectious dose. Our preliminary studies indicate that *Campylobacter* species may find microsites with specific conditions that are conducive to its survival on plants (M. Brandl and R. Mandrell, unpublished observations). However, the limited number of sites available on plants for supporting survival of *Campylobacter* species and the lack of significant growth at these sites are consistent with limited infectious potential of, and outbreak illness caused by, *Campylobacter* species on produce.

To address the potential risks from *Campylobacter* contamination of produce products, including the increasingly popular ready-to-eat packaged salads, vegetables, and fruits, will require careful studies of the incidence and biology of *Campylobacter* related to fresh produce. Although *C. coli* and *C. jejuni* are the most common pathogens, other species of *Campylobacter* have been shown also to cause human disease (Klein et al. 1986, Goossens et al. 1990, Figura et al. 1993, Lastovica and Le Roux 2001). Many of these other species will not be identified in most clinical laboratories from diarrheal samples because of the highly selective media used for isolating thermophilic *Campylobacter*. The development of better culture; detection and identification methods for a variety of *Campylobacter* species, including nonthermophilic species; and careful analysis of the epidemiologic data are also needed. The results of preliminary studies in our laboratory (see Table 5.4 and Figure 5.1) suggest the types of nutrients and conditions that could be exploited to develop methods for better isolation and growth of *Campylobacter* species from produce environments. Recent knowledge gained from genomic and proteomic studies of *Campylobacter* species will provide information valuable for epidemiological and fundamental studies related to source tracking, metabolism, and stress responses of *Campylobacter* related to food. Recent data demonstrating that some enteric pathogens are relatively fit in the plant environment suggests that produce may be an important source of both outbreak and sporadic diarrheal illnesses. This new paradigm, in addition to an enhanced knowledge of the biology of *Campylobacter* species in non-animal environments, as well as improved detection methods, may lead to a reassessment of the importance of preharvest produce as a food source for campylobacteriosis.

Acknowledgments

We thank Aileen Haxo and Anna Bates for excellent technical assistance and Craig Parker for careful editing of the manuscript.

This work was supported by the U.S. Department of Agriculture, Agricultural Research Service CRIS projects 5325-42000-040 and 5325-42000-041, and supports a United States collaboration in the European Commission Fifth Framework Project QLK1-CT-2002-0220, "CAMPYCHECK."

References

Acuff, G. R., Vanderzant, C., Hanna, M. O., Ehlers, J. G. and Gardner, F. A. 1986. Effects of handling and preparation of turkey products on the survival of *Campylobacter jejuni*. *Journal of Food Protection* **49**:627–631.

Adesiyun, A. A. 1995. Bacteriologic quality of some Trinidadian ready-to-consume foods and drinks and possible health risks to consumers. *Journal of Food Protection* **58**:651–655.

Allerberger, F., Al-Jazrawi, N., Kreidl, P., Dierich, M. P., Feierl, G., Hein, I. and Wagner, M. 2003. Barbecued chicken causing a multi-state outbreak of *Campylobacter jejuni* enteritis. *Infection* **31**:19–23.

Altekruse, S. F., Stern, N. J., Fields, P. I. and Swerdlow, D. L. 1999. *Campylobacter jejuni*—an emerging foodborne pathogen. *Emerging Infectious Diseases* **5**:28–35.

Beuchat, L. R. 1996. Pathogenic microorganisms associated with fresh produce. *Journal of Food Protection* **59**:204–216.

Brandl, M. T. and Mandrell, R. E. 2002. Fitness of *Salmonella enterica* serovar Thompson in the cilantro phyllosphere. *Applied and Environmental Microbiology* **68**:3614–3621.

Candlish, A. A. G., Pearson, S. M., Aidoo, K. E., Smith, J. E., Kelly, B. and Irvine, H. 2001. A survey of ethnic foods for microbial quality and aflatoxin content. *Food Additives and Contaminants* **18**:129–136.

Castillo, A. and Escartín, E. F. 1994. Survival of *Campylobacter jejuni* on sliced watermelon and papaya. *Journal of Food Protection* **57**:166–168.

Centers for Disease Control. 1998. Outbreak of *Campylobacter enteritis* associated with cross-contamination of food—Oklahoma, 1996. *Morbidity and Mortality Weekly Report* **47**:129–131.

Centers for Disease Control. 2000. Centers for Disease Control and Prevention, U.S. Foodborne Disease Outbreak Line Listings, 1990-1999. Available at: http://www.cdc.gov/ncidod/dbmd/outbreak/us_outb.htm.

Center for Science in the Public Interest. 2002. Outbreak alert: closing the gaps in our federal food safety net. Available at: http://www.cspinet.org/.

Charkowski, A. O., Barak, J. D., Sarreal, C. Z. and Mandrell, R. E. 2002. Differences in growth of *Salmonella enterica* and *Escherichia coli* O157:H7 on alfalfa sprouts. *Applied Environmental Microbiology* **68**:3114–3120.

Corry, J. E., Post, D. E., Colin, P. and Laisney, M. J. 1995. Culture media for the isolation of campylobacters. *International Journal of Food Microbiology* **26**:43–76.

Curl, E. A. and Truelove, B. 1986. Root exudates, pp. 55–92. *In* The Rhizosphere, B. Yaron (Ed.); Berlin: Springer.

De Boer, E. and Hahné, M. 1990. Cross-contamination with *Campylobacter jejuni* and *Salmonella* spp. from raw chicken products during food preparation. *Journal of Food Protection* **53**:1067–1068.

Derridj, S. 1996. Nutrients on the leaf surface, pp. 25–42. *In* Aerial Plant Surface Microbiology, C. E. Morris, P. C. Nicot, C. Nguyen-The (Eds.); New York: Plenum Press.

Doyle, M. P. and Schoeni, J. L. 1986. Isolation of *Campylobacter jejuni* from retail mushrooms. *Applied and Environmental Microbiology* **51**:449–450.

Elharrif, Z. and Megraud, F. 1986. Characterization of thermophilic *Campylobacter*: I. Carbon substrate utilization tests. *Current Microbiology* **13**:117–122.

Federighi, M., Magras, C., Pilet, M. F., Woodward, D., Johnson, W., Jugiau, F. and Jouve, J. L. 1999. Incidence of thermotolerant Campylobacter in foods assessed by NF ISO 10272 standard: results of a two-year study. *Food Microbiology (London)* **16**:195–204.

Figura, N., Guglielmetti, P., Zanchi, A., Partini, N., Armellini, D., Bayeli, P. F., Bugnoli, M. and Verdiani, S. 1993. Two cases of *Campylobacter mucosalis* enteritis in children. *Journal of Clinical Microbiology* **31**:727–728.

Friedman, C. R., Neimann, J., Wegener, H. C. and Tauxe, R. V. 2000. Epidemiology of *Campylobacter jejuni* infections in the United States and other industrialized countries, pp. 121–138. *In* Campylobacter, I. Nachamkin, M. J. Blaser (Eds.); Washington, D.C.: American Society for Microbiology Press.

Frost, J. A., Gillespie, I. A. and O'Brien, S. J. 2002. Public health implications of campylobacter outbreaks in England and Wales, 1995–9: epidemiological and microbiological investigations. *Epidemiology and Infection* **128**:111–118.

Goossens, H., Vlaes, L., De Boeck, M., Pot, B., Kersters, K., Levy, J., De Mol, P., Butzler, J. P. and Vandamme, P. 1990. Is "*Campylobacter upsaliensis*" an unrecognised cause of human diarrhoea? *Lancet* **335**:584–586.

Gorski, L., Palumbo, J. D. and Mandrell, R. E. 2003. Attachment of *Listeria monocytogenes* to radish tissue is dependent upon temperature and flagellar motility. *Applied and Environmental Microbiology* **69**:258–266.

Handelsman, J. and Stabb, E. V. 1996. Biocontrol of soilborne plant pathogens. *The Plant Cell* **8**:1855–1869.

Hirano, S. S. and Upper, C. D. 2000. Bacteria in the leaf ecosystem with emphasis on *Pseudomonas syringae*: a pathogen, ice nucleus, and epiphyte. *Microbiology and Molecular Biology Reviews* **64**:624–653.

Holt, J. G., Krieg, N. R., Sneath, P. H. A., Staley, J. T. and Williams, S. T. (Eds). 1994. Group 2: aerobic/microaerophilic, motile, helical/vibroid Gram-negative bacteria, pp. 39–63. *In* Bergey's Manual of Determinative Bacteriology; Baltimore, Md.: Williams and Wilkins.

Istre, G. R., Blaser, M. J., Shillam, P. and Hopkins, R. S. 1984. *Campylobacter enteritis* associated with undercooked barbecued chicken. *American Journal of Public Health* **74**:1265–1267.

Jacobs-Reitsma, W. 2000. *Campylobacter* in the food supply, pp. 467–481. *In* Campylobacter, I. Nachamkin, M. J. Blaser (Eds.); Washington, D.C.: American Society for Microbiology Press.

Karib, H. and Seeger, H. 1994. Presence of *Yersinia* and *Campylobacter* spp. in foods. *Fleischwirtschaft* **74**:1104–1106.

Kirk, M., Waddell, R., Dalton, C., Creaser, A. and Rose, N. 1997. A prolonged outbreak of *Campylobacter* infection at a training facility. *Communicable Diseases Intelligence* **21**:57–61.

Klein, B. S., Vergeront, J. M., Blaser, M. J., Edmonds, P., Brenner, D. J., Janssen, D. and Davis, J. P. 1986. *Campylobacter* infection associated with raw milk. An outbreak of gastroenteritis due to *Campylobacter jejuni* and thermotolerant *Campylobacter fetus* subsp *fetus*. *Journal of the American Medical Association* **255**:361–364.

Koenraad, P. M. F. J., Rombouts, F. M. and Notermans, S. H. W. 1997. Epidemiological aspects of themophilic *Campylobacter* in water-related environments: a review. *Water and Environmental Research* **69**:52–63.

Kraffczyk, I., Trolldenier, G. and Beringer, H. 1984. Soluble root exudates of maize Zea-Mays influence of potassium supply and rhizosphere microorganisms. *Soil Biology and Biochemistry* **16**:315–322.

Kumar, A., Agarwal, R. K., Bhilegaonkar, K. N., Shome, B. R. and Bachhil, V. N. 2001. Occurrence of *Campylobacter jejuni* in vegetables. *International Journal of Food Microbiology* **67**:153–155.

Labarca, J. A., Sturgeon, J., Borenstein, L., Salem, N., Harvey, S. M., Lehnkering, E., Reporter, R. and Mascola, L. 2002. *Campylobacter upsaliensis*: another pathogen for consideration in the United States. *Clinical Infectious Diseases* **34**:E59–E60.

Lastovica, A. J. and Le Roux, E. 2001. Efficient isolation of *Campylobacter upsaliensis* from stools. *Journal of Clinical Microbiology* **39**:4222–4223.

Lee, S. H., Levy, D. A., Craun, G. F., Beach, M. J. and Calderon, R. L. 2002. Surveillance for waterborne-disease outbreaks—United States, 1999–2000. *Morbidity and Mortality Weekly Report* **51**:1–52.

Lindow, S. E. and Brandl, M. T. 2003. Microbiology of the phyllosphere. *Applied and Environmental Microbiology* **69**:1875–1883.

Long, S. M., Adak, G. K., O'Brien, S. J. and Gillespie, I. A. 2002. General outbreaks of infectious intestinal disease linked with salad vegetables and fruit, England and Wales, 1992–2000. *Communicable Disease and Public Health* **5**:101–105.

Mandrell, R., Barak, J., Brandl, M., Charkowski, A., Cooley, M., Gorski, L., Miller, W., Palumbo, J. and Wachtel, M. 2002. *Fitness of human enteric pathogens in models of pre-harvest and post-harvest produce contamination*, pp. A1-A14. 31st U.S. and Japan Natural Resources (UJNR) Protein Resources Panel Meeting, December 1-6, Monterey, CA.

Matsusaki, S., Katayama, A., Kawaguchi, N., Tanaka, K. and Goto, A. 1982. Incidence of *Campylobacter jejuni/coli* in foodstuffs. *Journal. Food Hygiene Society of Japan* **23**:434–437.

McMahon, M. A. and Wilson, I. G. 2001. The occurrence of enteric pathogens and *Aeromonas* species in organic vegetables. *International Journal of Food Microbiology* **70**:155–162.

Moore, J. E., Wilson, T. S., Wareing, D. R., Humphrey, T. J. and Murphy, P. G. 2002. Prevalence of thermophilic *Campylobacter* spp. in ready-to-eat foods and raw poultry in Northern Ireland. *Journal of Food Protection* **65**:1326–1328.

Morgan, J. V. and Tukey Jr., H. B. 1964. Characterization of leachate from plant foliage. *Plant Physiology* **39**:590–593.

Mosupye, F. M. and von Holy, A. 1999. Microbiological quality and safety of ready-to-eat street vended foods in Johannesburg, South Africa. *Journal of Food Protection* **62**:1278–1284.

Mosupye, F. M. and von Holy, A. 2000. Microbiological hazard identification and exposure assessment of street food vending in Johannesburg, South Africa. *International Journal of Food Microbiology* **61**:137–145.

Odumeru, J. A., Mitchell, S. J., Alves, D. M., Lynch, J. A., Yee, A. J., Wang, S. L., Styliadis, S. and Farber, J. M. 1997. Assessment of microbiological quality of ready-to-use vegetables for health-care food services. *Journal of Food Protection* **60**:954–960.

Park, C. E. and Sanders, G. W. 1992. Occurrence of thermotolerant campylobacters in fresh vegetables sold at farmers' outdoor markets and supermarkets. *Canadian Journal of Microbiology* **38**:313–316.

Phillips, C. A. 1998. The isolation of *Campylobacter* spp. from modified atmosphere packaged foods. *International Journal of Environmental Health Research* **8**:215–221.

Population and Public Health Branch. 2000. Waterborne outbreak of gastroenteritis associated with a contaminated municipal water supply, Walkerton, Ontario, May–June 2000. *Canada Communicable Disease Report* **26**:170–173.

Rollins, D. M. and Colwell, R. R. 1986. Viable but nonculturable stage of *Campylobacter jejuni* and its role in survival in the natural aquatic environment. *Applied and Environmental Microbiology* **52**:531–538.

Sagoo, S. K., Little, C. L. and Mitchell, R. T. 2001. The microbiological examination of ready-to-eat organic vegetables from retail establishments in the United Kingdom. *Letters in Applied Microbiology* **33**:434–439.

Sagoo, S. K., Little, C. L., Ward, L., Gillespie, I. A. and Mitchell, R. T. 2003. Microbiological study of ready-to-eat salad vegetables from retail establishments uncovers a national outbreak of Salmonellosis. *Journal of Food Protection* **66**:403–409.

Stanley, K. N., Wallace, J. S., Currie, J. E., Diggle, P. J. and Jones, K. 1998. Seasonal variation of thermophilic campylobacters in lambs at slaughter. *Journal of Applied Microbiology* **84**:1111–1116.

Stern, N. J., Fedorka-Cray, P., Bailey, J. S., Cox, N. A., Craven, S. E., Hiett, K. L., Musgrove, M. T., Ladely, S., Cosby, D. and Mead, G. C. 2001. Distribution of *Campylobacter* spp. in selected U.S. poultry production and processing operations. *Journal of Food Protection* **64**:1705–1710.

Thunberg, R. L., Tran, T. T., Bennett, R. W., Matthews, R. N. and Belay, N. 2002. Microbial evaluation of selected fresh produce obtained at retail markets. *Journal of Food Protection* **65**:677–682.

Uyttendaele, M. and Debevere, J. 1996. Evaluation of Preston medium for detection of *Campylobacter jejuni in vitro* and in artificially and naturally contaminated poultry products. *Food Microbiology (London)* **13**:115–122.

Wachtel, M. R., Whitehand, L. C. and Mandrell, R. E. 2002. Association of *Escherichia coli* O157:H7 with preharvest leaf lettuce upon exposure to contaminated irrigation water. *Journal of Food Protection* **65**:18–25.

Weinrich, V. K., Winkler, K. and Heberer, E. 1990. Studies concerning the use of selected humic acid products in media for isolation of thermophilic *Campylobacter* species. *Deutsche Tierarztliche Wochenschrift* **97**:511–515.

Wight, J. P., Rhodes, P., Chapman, P. A., Lee, S. M. and Finner, P. 1997. Outbreaks of food poisoning in adults due to *Escherichia coli* O111 and *Campylobacter* associated with coach trips to northern France. *Epidemiology and Infection* **119**:9–14.

Winters, D. K. and Slavik, M. F. 2000. Multiplex PCR detection of *Campylobacter jejuni* and *Arcobacter butzleri* in food products. *Molecular and Cellular Probes* **14**:95–99.

World Health Organization. 1998. Surveillance Programme for Control of Foodborne Infections and Intoxications in Europe, Seventh Report. Available at: http://www.bgvv.de/internet/7threport/7threp_ctryreps_fr.htm.

Zablackis, E., York, W. S., Pauly, M., Hantus, S., Reiter, W. D., Chapple, C. C. S., Albersheim, P. and Darvill, A. 1996. Substitution of L-fucose by L-galactose in cell walls of *Arabidopsis* mur1. *Science* **272**:1804–1808.

6 *Campylobacter* and Campylobacteriosis: What We Wish We Knew

Richard L. Ziprin

Introduction

Campylobacteria are medically important enteropathogens of man and are responsible for perhaps 2,000,000 incidents of enteritis per year within the United States (Smith 2002). These organisms are often present in the gastrointestinal tracts of swine, poultry, and beef. They are also recoverable from wild birds and field pests. From these sources, and sometimes from water, milk, fruit juice, or produce, they may find their way into the human gastrointestinal tract and cause a very unpleasant and often severe diarrhea. Sometimes infection with *Campylobacter* spp. can result in post-infection neurological damage, including either Guillain-Barré syndrome or Miller-Fisher syndrome (Smith 2002). Yet campylobacteria are relative "newcomers" to the world of microbiologists. They were probably initially isolated in the 1930s, and perhaps as early as the second decade of the twentieth century. However, less than three decades have elapsed since campylobacteria were fully associated with human illness and poultry were identified as a source of infection (Skirrow 1977, Skirrow and Butzler 2000). Poultry, of course, are not the only source of campylobacteria. Water, milk, fruit juice, and sometimes produce are all frequent sources. Poor hygiene or poor sanitation and improper food storage and handling are also contributors to the spread of infection.

Isolation and Cultivation

Campylobacters are perhaps the most fascinating microbes I have worked with throughout my long career. Before working with them I had erroneously considered campylobacteria to be obscure, even unimportant organisms. I had heard that campylobacteria were difficult to grow, and difficult to work with. I had the notion that very special complex methods were required for cultivation, but these preconceived notions were wrong.

In my laboratory we cultivate campylobacteria on "Campy-Cefex" agar plates (Stern et al. 1992). These plates are made of Brucella Agar (Becton Dickinson, Cockysville, MD). Lysed horse blood is added to the Brucella Agar, as are the antibiotics cefoperazine and cycloheximide. I often add other antibiotics, such as rifampicin, trimethoprim, and vancomycin, or omit cycloheximide. There is some art to deciding the appropriate mix of antimicrobials for this particular use, but that is unimportant. When Campy-Cefex plates have been inoculated with campylobacteria, the plates are then incubated in a gas mix consisting of 85% Nitrogen, 5% Oxygen, and 10% CO_2 (Stern et al. 1992). This gas is available

---

Mention of a trade name, proprietary product, or specific equipment does not constitute a guarantee or warranty by the U.S. Department of Agriculture and does not imply its approval to the exclusion of other products that may be suitable.

from commercial suppliers of industrial gases (Messer MG Industries, Houston, TX). I place inoculated plates in plastic storage bags, fill these with the gas mixture, and then incubate the plates at 42°C. There are other products for culturing *Campylobacter*. Some workers add tetrazolium salts to make the colonies more visible. Some use broth, such as Mueller Hinton Broth, with whole bovine blood added as a supplement. There are also enrichment broths that work reasonably well for isolating the organism from foods, but they are not as effective when isolating from heavily contaminated materials such as fecal or cecal material. Yet there is plenty of opportunity for clever improvements to *Campylobacter* isolation and cultivation methodology.

Identification

It is relatively easy to visually identify putative *Campylobacter* colonies on Campy-Cefex agar plates. Some caution is needed, as non-campylobacteria organisms do sometimes grow, notwithstanding the high-temperature and limited-oxygen growth conditions, but these organisms usually do not have the clear translucent pink appearance of typical *Campylobacter* colonies. Therefore, skill is required to visually identify typical *Campylobacter* colonies. It is a more difficult task to confirm that isolated colonies are *Campylobacter* and to determine which species has been isolated. In our laboratory we usually screen isolated colonies with a latex agglutination test that will identify an isolate as one of the following: *C. lari, C. coli,* or *C. jejuni.* The latex agglutination test does not respond to other *Campylobacter* species, nor does it help to differentiate among them. There is clearly room for improved secondary identification of putative *Campylobacter* isolates.

Biochemical tests can be used to determine the species isolated. Unfortunately, these rely on a test for hippurate hydrolysis, which is a notoriously inaccurate test. Commercial kits designed to identify *Campylobacter* are in my opinion equally unable to give fully reliable results. Thus, there is a need for improvements in the fundamental techniques of isolation, characterization, and identification of *Campylobacter,* especially from nonclinical yet heavily contaminated materials. Modern microbiology has some answers in the form of molecular techniques such as pulsed field gel electrophoresis and ribotyping. However, applicability of these tests are somewhat limited by cost, complexity, and the specialized equipment and knowledge needed. We certainly wish we knew of better ways to accomplish the basic tasks of *Campylobacter* microbiology.

The Organism

C. jejuni grows slowly, so at 18–24 hours after inoculation, it is often only possible to make out a slight reddish area on the portion of the plate that was most heavily seeded. By 36–48 hours of incubation, though sometimes earlier, very distinctive small translucent pink colonies appear. Once the microbiologist learns to recognize these colonies, particularly the translucence, he or she can easily select *Campylobacter* colonies from among other organisms that might also be growing on the plate. If the seeded plate is then regassed and placed back into the incubator and reexamined again at 3 or 4 days after incubation, the colonies will have lost their clear translucent appearance and will appear dark and opaque. Sometimes, distinct colonies are not seen. Instead, especially on freshly prepared plates, there are large areas of growth the size of a fingertip or thumbprint. The perimeter is not uniform or circular, but is somewhat irregular in appearance. The portion of the bacterial mass, between the perimeter and the center, appears to be a thin layer of growth that is sometimes difficult to see without holding the plate at an angle to the light. The center is a slightly thickened raised gray mass. The perimeter is slightly raised, with cells apparently piled on each other, in a manner reminiscent of the rings formed by *Bacillus circulans*, except that there are no concentric rings. Thus the organism can present itself in at least three different colony formats: distinct translucent colonies, dark or opaque colonies, and diffuse, spread-out colonies caused by motile swarming cells. The underlying bacterial physiology involved in these colony morphology variations is something worthy of basic research investigation.

The microbiologist will quickly want to make some microscopic observations in addition to noting the unusual growth characteristics and conditions that are typical of the campylobacteria. These organisms are Gram negative, but they do not counterstain well with safranin, as do most other organisms. Instead, it is best to counter stain with carbol-fuchsin. With light microscopy, the organism appears as a small curved rod. The observer can see "seagull" formations where two organisms lie side by side. However, the organism is actually spiral shaped. This spiral form is not the only morphological form that can be seen. In older cultures, the organisms appear coccoid, or round. These morphological changes are apparent by electron microscopy, and the coccoid form clearly occurs as the growth conditions worsen. Excellent photographs of these morphological variants have been published (Rollins and Colwell 1986, Lazaro et al. 1999). These variant colonial and cellular forms possibly play a role in transmission and spread through the environment. It is also possible that they are involved in bacterial survival within adverse environments. Much remains to be learned about them.

Survival and Spread in the Environment

Campylobacteria give the microbiologist a sense that they are temperamental and fragile, but in truth, they are clever survivors. Otherwise, they would not be so widely distributed as commensals in animals, nor would they be present in a wide variety of foods including meats, milk, dairy products, fresh fruits, and vegetables. This wide distribution suggests that contrary to the fragile nature of the organism in the laboratory, something quite different is happening in nature. We wish we understood more about how *Campylobacter* survive, thrive, and spread in the general environment. These issues are especially vexing as the organism seems so delicate in the laboratory, where desiccation or, conversely, suspension of the organism in water has rapid and profound effects on our ability to recover cells by conventional culture methods.

The Viable but Nonculturable State

Rollins and Colwell (1986) described a viable but nonculturable coccoid (VBNC) form of *C. jejuni* in natural aquatic environments. Coccoid cells form when campylobacteria are suspended in water. Although the cells are viable, they are nonculturable. The morphological change was thought to have been important for survival in aquatic environments. This idea may have been in part caused by the appearance of the coccoid cells, which are slightly reminiscent of bacterial spores. VBNC organisms are detected not by growth, but by their ability to reduce tetrazolium salts (Cappelier et al. 1997). The cells are viable, maintaining respiratory metabolism and DNA integrity (Lazaro et al. 1999), but they cannot be recovered when plated onto bacteriological media. The nature of the coccoid cells is unclear, as it now seems likely that the VBNC state occurs in the spiral form as well (Lazaro et al. 1999). Perhaps the coccoid cells are capable of elongation into a VBNC spiral form and reverting to a culturable form, but it is equally plausible that the VBNC coccoid forms are degenerative and not contributors to survival in the environment (Ziprin et al. 2003a). We wish we knew more about the purposes of coccoid cell formation. We need to determine whether the coccoid VBNC cells are indeed always degenerative or whether there are conditions under which they can revert to a spiral infective form. A method is needed to physically separate spiral forms from coccoid forms so that we can definitively determine the roles of each form in survival of campylobacteria within adverse environments. We should better understand the amazing bacterial physiology that enables the coccoid cells to remain viable for upward of 2 weeks while in nutrient deficient conditions, and also to remain viable in conditions that present the cells with a very high osmotic gradient; that is, suspended in reverse osmosis purified water.

The ubiquitous presence of campylobacteria among animals in nature indicates that the organism is in fact hardy, a good survivor, while paradoxically the organism appears quite delicate in the laboratory environment. We do not yet know whether either the morphological variations or the VBNC state are part of *Campylobacter*'s approach to enhancing their opportunity to survive in the environment and their opportunity to infect animals and man. Manuscripts by Chan et al. (2001), Federighi et al. (1998), Lazaro et al. (1999), Stern et al. (1994), and Tholozan et al. (1999) suggest that the VBNC state is part of the mechanism for survival in aquatic environments and low temperatures and, thus, may enhance the spread of campylobacteria. Other work (Medema et al. 1992, VAN DE Giessen et al. 1996, Hald et al. 2001, Ziprin et al. 2003a) demonstrates that VBNC cells may not recover from this state, are "degenerative," and are incapable of colonizing chickens. The biological purposes and nature of VBNC campylobacteria remain to be fully established.

How Do Campylobacter *Find Their Way into Chickens?*

Many cases of human campylobacteriosis arise from contamination of water, dairy products, juices, and sea foods with campylobacteria. However, live poultry and retail poultry products frequently are contaminated with campylobacteria, and contaminated retail poultry products are a major source of human infection (Jacobs-Reitsma 2000). It has been estimated that 75% of live broilers and 80% or more of retail poultry meat is contaminated (Jacobs-Reitsma 2000, Hiett et al. 2002). We do not yet fully understand how chickens become contaminated with campylobacteria.

Some authors have reported that commercial broiler flocks seem to become contaminated with *Campylobacter* after 2–3 weeks in grow-out facilities (Jacobs-Reitsma et al. 1995, Shane 2000). The reasons for the temporal lag between the time birds are placed into a rearing facility and the time *Campylobacter* may be isolated are unknown. Efforts to understand the apparent sudden appearance of campylobacteria within a broiler facility at approximately 2–3 weeks into the grow-out have proven futile. Some of the issues of source, time of colonization, and prevalence have been reviewed (Gregory et al. 1997). The roles of carry-over from previous flocks as new flocks are sequenced through broiler houses and the adequacy of typical decontamination procedures are debatable (Petersen and Wedderkopp 2001, Shreeve et al. 2002). Numerous other hypotheses have been proposed to explain the appearance of campylobacteria in production facilities. These include such obvious things as inadequate biosecurity, poor farm management practices, and *Campylobacter*-contaminated feed or water (Stern et al. 2002). Poor insect or rodent control, the droppings of wild birds (Stern et al. 1997, Craven et al. 2000), vertical transmission from the breeder birds (Petersen and On 2000, Cox et al. 2002a, Cox et al. 2002b), and contamination of chicks at the hatchery (Hiett et al. 2002) are potential sources of campylobacteria at the farm. However, none of these possibilities is fully capable of explaining the "mystery" appearance of campylobacteria within a grow-out facility at 2–3 weeks into the process. Although each of the foregoing possibilities provides a reasonable hypothetical identification of source, they do not adequately explain the timing. VBNC present in hatcheries, water, or newly hatched chicks might appear to be an attractive explanation, as it can be argued that it takes time for VBNC to recover and transform back into fully infective organisms. However, although this is an attractive theory, it is neither supported nor refuted by presently available information. Recent work has indicated that campylobacteria are able to survive within biofilms in water supply plumbing, which also has been proposed as a source of campylobacteria (Trachoo et al. 2002). There is a great deal more we need to learn about how *Campylobacter* species sustain themselves in nature and become part of the gut ecology of agricultural animals. A few citations of somewhat representative writings on this subject are Achen et al. 1998, Annan-Prah and Janc 1988, Berndtson et al. 1996, Evans 1992, Gibbens et al. 2001, Refregier-Petton et al. 2001, and van de Giessen et al. 1992.

Strain and Species Diversity

The ability of *C. jejuni* and *C. coli* strains to colonize poultry, other agricultural animals, and humans is variable. Strains isolated from chickens do not necessarily colonize other chickens (Glunder 1995, Stas et al. 1999), and *C. jejuni* strains have varying abilities to colonize chicks (Young et al. 1999, Ahmed et al. 2002). In addition to *Campylobacter* strain-determined "host preferences," each *Campylobacter* species shows some host specificity as well. Swine are the "preferred" host for *C. coli*, whereas only a small percentage of human campylobacteriosis incidents (in the United States) is caused by *C. coli*. Chickens in nature are also colonized by *C. coli*, and we have shown (Ziprin et al. 2002a) that swine isolates of *C. coli* have the ability to colonize chickens in the laboratory. Bovine isolates of *C. jejuni* will also colonize chicks in the laboratory (Ziprin et al. 2003b), but epidemiological evidence developed from antimicrobial resistance patterns shows that there are host-range preferences of bovine isolates for cattle (Aarestrup et al. 1997). Similarly, *C. lari* has a "preference" for nondomesticated birds, especially gulls. *C. lari* can also colonize chickens and then infect humans, especially immunocompromised individuals (Bezian et al. 1990, Chiu et al. 1995, Goudswaard et al. 1995, Krause et al. 2002, Werno et al. 2002). Some seafoods are a source of *C. lari* human infections (Endtz et al. 1997). Thus, campylobacteria move between animal species, from farm animals to humans, and probably from humans to agricultural animals. Yet unknown bacterial factors and host factors, and probably environmental factors as well, control the dispersion and distribution of the organism in nature. These factors are barely understood.

It is well known that there is a great deal of genetic diversity in *Campylobacter* isolates from poultry production facilities (Nielsen et al. 1997, Hanninen et al. 2000, Petersen and On 2000, Hanninen et al. 2001, Petersen et al. 2001a, Petersen et al. 2001b, Ahmed et al. 2002, Nylen et al. 2002). However, very little is known about the phenotypic properties and underlying bacterial genetics that enable the movement of campylobacteria between ecological niches in nature or among agricultural animals. Neither do we understand how campylobacteria are enabled to cause campylobacteriosis in humans, whereas the organism tends to be an inconsequential commensal in poultry, swine, and cattle.

Colonization of Chick Ceca

We do not yet understand, and I think we need to learn much more about, the bacterial factors that control the distribution of *Campylobacter* species and serotypes in nature, particularly with respect to colonizing ability. A recent review provides an excellent background on bacterial factors affecting intestinal colonization of poultry (Mead 2002). Very little is known about the genes and the gene products involved in colonization. Several studies have identified four genes (*ciaB*, *pldA*, *cadF*, and *dnaJ*) that possibly have some role in colonization. Strains containing mutations within these genes are rendered incapable of colonizing the chick cecum (Konkel et al. 1998, Ziprin et al. 1999, Ziprin et al. 2001, Ziprin et al. 2002b). However, although the laboratory evidence identifying the genes seems compelling, I still have some lingering doubts about the conclusions drawn from data reported in some of my publications, as follows. Each of four successive mutations examined interfered with cecal colonization. Mutant strains were chosen for colonization studies based on reasoning and on an understanding of what is likely involved in the physical act of maintaining campylobacteria within the gut environment (Konkel et al. 2000). It seems, however, extraordinarily good luck that each of the four mutant strains examined were poor colonizers. Therefore, I have begun to wonder whether something else besides the mutated gene is involved in the failure of these strains to colonize. Each strain carried a kanamycin resistance gene, and though there is no reason to suspect that this property could affect colonization, I still wonder whether introduction of the antimicrobial resistance gene is not somehow interfering with colonization. In addition to this concern, there are some other concerns I have that, though the organisms appear to be isogenic except for the mutated gene, there might be other unrecognized damage to the cells or genome that render them incapable of colonization. Others have also attempted to identify genes involved in colonization. Ahmed et al. (2002) used subtractive hybridization between a colonizing and a non-colonizing strain to identify 23 DNA inserts present in the colonizing strain that were absent from the non-colonizer. From the sequence data, the authors were able to determine either the gene or gene function for 17 of the sequences. It is, however, not clear how the putative genes actually act to promote colonizing ability. Flagellin genes (Nachamkin et al. 1993), superoxide dismutase genes (Purdy et al. 1999), and temperature-dependent genes (Bras et al. 1999), have also been indicated as playing roles in colonization. Given the power of modern molecular biology, there seems to be a lack of solid information on the molecular factors involved in colonization. This is somewhat understandable because the talent and laboratory facilities necessary for genetic manipulations are quite different from those needed for *in vivo* assessment of the effects of genetic manipulation on colonization. Laboratories with abilities to do live animal research need to better connect with microbiologists skilled in modern molecular genetics so that the two types of research can be joined to answer the core unknowns of *Campylobacter* colonization.

Just as we have too little knowledge of the bacterial factors needed for colonization, we also lack knowledge of host factors that enable colonization. We have some understanding of host barriers to colonization such as gastric acidity, but we do not have information about the role that things such as luminal secretions and undigested or unabsorbed nutrients might play. We also have little knowledge of the role native luminal microflora may play in both promoting and inhibiting colonization. There is a large body of literature on the general subject of competitive exclusion of *Campylobacter* (Achen et al. 1998, Hakkinen and Schneitz 1999, Chen and Stern 2001, Stern et al. 2001), but we do not know whether normal microflora contribute in any way to the creation of a permissive environment for colonization. Overall, our knowledge of the relevant gut microecology is relatively poor.

Campylobacteriosis, Disease in Humans

Human illness caused by *Campylobacter jejuni* and *C. coli* is usually an enteritis with acute diarrheal disease. Several related toxins seem to play a role. There is a cytolethal distending toxin (Hickey et al. 2000, Mooney et al. 2001, Hassane et al. 2003), and there is possibly an enterotoxin (Suzuki et al. 1994, Kanwar et al. 1995). There are also *C. jejuni* chromosomal sequences reported to hybridize with enterotoxin genes from *Vibrio cholerae* and *Escherichia coli* (Calva et al. 1989); however, such genes have not been found in *Campylobacter* laboratory strains after complete sequencing of the genome. The literature is inconclusive on this issue, and a recent review indicates a skeptical view as to the existence of *C. jejuni* enterotoxins (Pickett 2000).

The clinical aspects of campylobacteriosis in humans have been recently reviewed (Skirrow and Blaser 2000, Smith 2002). The disease tends to be self-resolving within 1–5 days, and treatment generally is usually limited to supportive therapy to maintain hydration. Deaths are rare, with an estimated death rate of 0.001%. Extraintestinal illness is rare and is most common among patients over 65 or under 4 years of age. Individuals in all age groups are susceptible to campylobacteriosis, but there may be an unusual bimodal age distribution in which peak infection incidence rates tend to cluster in children under 1 year of age and among older adolescents or young adults—people between 15 and 24 years old (Fenton et al. 2002). Not all epidemiological studies show this distribution clearly. The reasons for this age distribution are unclear. Certainly infants and toddlers are generally more susceptible to disease, so this aspect of the age distribution of campylobacteriosis in the population is not unexpected, nor is a higher incidence of extraintestinal infection and mortality among the aged. However, the excess prevalence among older teens and young adults is not understood. It has been suggested that the excess prevalence in young people might be a matter of lifestyle and poor hygiene, as young people have their initial experiences living away from home, or that institutional food hygiene issues on college campuses or nearby eateries might play a role. It is also possible that there is an underlying physiological phenomenon that is not understood.

We also do not know the answers to some fundamental questions about *Campylobacter* epidemiology. For example, why are most incidents of human campylobacteriosis in the United States caused by *C. jejuni* (95%), whereas a greater proportion of human infections in other parts of the world are caused by *C. coli*? The answer may lie in different distributions of *Campylobacter* species and variants within agricultural animals or in other subtle issues not yet discovered. I have often wondered whether perhaps the preponderant isolation of *C. jejuni* from human campylobacteriosis cases in the United States may be an artifact of clinical laboratory practice here in the United States.

Extraintestinal infections by *Campylobacter* and post-infection sequelae are rare, but are often severe. There are reports of maternal death, abortion and stillbirth, premature labor, diarrhea in newborn infants, and bacteremia in newborn infants. Appendicitis, colitis, toxic megacolon, intestinal hemorrhage, bacteremia, hepatitis, cholecystitis, pancreatitis, renal and urinary infection—including prostatitis, focal infections, myocarditis, and splenic rupture are among the occasional complications of campylobacteriosis. In addition, as with other bacterial infections, sequelae may include reactive arthritis, Reiter's Syndrome, and Guillain-Barré Syndrome (van Koningsveld et al. 2001).

Guillain-Barré Syndrome is an acute inflammatory demyelinating neuropathy that affects both motor and sensory fibers. There is a very similar disorder, Miller-Fisher Syndrome, in which the clinical presentation involves paralysis of eye muscles, loss of reflexes, and inability to coordinate voluntary muscle movement. The cell envelope of Gram-negative organisms contains lipopolysaccharides that contribute to their virulence (Fry et al. 2000). It is likely that *C. jejuni* lipopolysaccharides (and most Gram-negative cell envelope lipopolysaccharides) contain structures that mimic the structures present in neural tissue. Therefore, immunologic reaction to *C. jejuni* provokes an autoimmune-mediated attack on neural tissue (Jacobs et al. 1997, Ang et al. 2000a, Ang et al. 2000b, Wassenaar et al. 2000, Korzeniowska-Kowal et al. 2001, van Belkum et al. 2001). Unfortunately, there is no animal model in which to study the molecular and immunological foundation for *C. jejuni*–induced Guillain-Barré. An animal model of this disorder is desperately needed.

Guillain-Barré is certainly not the only demyelinating neuropathy, and as this chapter is intended to point toward subject matter that needs investigation, and to stimulate new lines of investigation, I will digress here briefly and talk about a seemingly unrelated viral disease, tropical spastic paraparesis. Then, I will return to campylobacteriosis and neuropathy and attempt to relate the two diseases to each other in a very speculative way.

Tropical spastic paraparesis is a myeloneuropathy of some individuals infected by a retrovirus, human T-cell leukemia (lymphoma) virus (HTLV) Type I, and possibly HTLV Type II. Molecular pathogenic mechanisms underlying the causation of tropical spastic paraparesis are probably from errant immune responses triggered by immunological responses to viral antigens and molecular mimicry (Plumelle 1999, Buczynski et al. 2001, Jacobson 2002, Levin et al. 2002a, Levin et al. 2002b, Osame 2002, Ribas and Melo 2002). Because the HTLV virus is transmitted through intravenous drug use, by sexual contact, and by blood transfusions, potential blood donors must be screened for antibodies to the virus. Unfortunately, the serological tests often yield indeterminate results. That is, donor sera react with some HTLV antigens in Western Blot Tests. Because the donated blood can not be used, there is a resultant diminishment of the potential blood donor pool (Ownby et al. 1997). The cause of these aberrant serological results is unknown. At one time there was a belief that influenza vaccination may have been a cause, but that theory probably is not valid (Simonsen et al. 1995). One study of 674,000 donors detected antibodies to HTLV-1 in 0.13% of the population (870 donors). Of these 870, 15 were true positives, 201 were true negatives, and 654 were indeterminate. A subsample of 234 seropositive individuals was studied further with polymerase chain reaction methods, and 206 of the samples that had been considered indeterminate by Western Blot immunoassays were negative by DNA methodology (Zaaijer et al. 1994). Thus, many individuals who are not infected have antibodies that react with HTLV antigens in Western Blot assays.

I was a blood donor during several years immediately before the time I initiated work with *Campylobacter*. However, approximately 6 months after starting work with campylobacteria I seroconverted to an "indeterminate" status and was appropriately asked to refrain from donating in the future. Polymerase chain reaction investigation failed to detect viral DNA, although additional studies clearly demonstrated that my serum contained antibodies that cross-react with tests specific for HTLV. As I have no relevant lifestyle issues, and in fact I am not infected, it occurred to me that there may have been a relationship between the onset of work with *Campylobacter* and seroconversion. I have been unable to find any citations that would support or even suggest that campylobacteriosis might result in such a seroconversion, and so far as I know I have never suffered a bout of campylobacteriosis. Nonetheless, I wonder whether molecular mimicry similar to that involved in reactive arthritis or Guillain Barré Syndrome might not also be responsible for the production of antibodies that recognize both neural tissue and HTLV. Recently, I stumbled on a report that simian immunodeficiency virus glycoprotein gp120 is an enterotoxin in the mouse ileal loop model (Swaggerty et al. 2000). This report has me wondering whether a *Campylobacter* enterotoxin (assuming one exists) might be a molecular mimic that could provoke antibody production to some retroviral proteins; hence, exposure to *Campylobacter* might account for the relatively large number of indeterminate Western Blot tests for retroviruses in blood donors. I realize that this is a highly speculative notion with no direct support. Still, I would like to see a study done to correlate incidents of indeterminate HTLV serological results with known incidents of campylobacteriosis.

Because *Campylobacter*-contaminated foods of animal origin very frequently cause human gastrointestinal illness, it would be desirable to have a means of vaccinating animals that would reduce carriage of the organism within their gastrointestinal tract. Similarly, it would be very advantageous to have a vaccine that could immunize humans against campylobacteriosis. Unfortunately, attempts at developing a *Campylobacter* vaccine have been unsuccessful. There have been a few good attempts, but little real success (Baqar et al. 1995a, Baqar et al. 1995b, Widders et al. 1996, Rice et al. 1997, Ziprin et al. 2002b). One report hints at success with a killed oral vaccine intended for eventual human use in the military (Scott 1997). There is also one report of successful vaccination of mice with a fusion protein of *Campylobacter* FlaA and the maltose binding protein of *E. coli*. An *E. coli* enterotoxin was used as an adjuvant (Lee et al. 1999). In any case, there are difficult practical and regulatory issues that surround any attempt at human use of a *Campylobacter* vaccine because of the potential for molecular mimicry and induction of autoimmunity, leading to neuropathology (Kopecko 1997, Pace et al. 1998).

Difficulties encountered in laboratory immunization experiments are somewhat confounding. Several studies of populations in underdeveloped countries have shown that maternal antibodies present in breast milk were somewhat protective. Also, studies conducted in geographic areas in which *Campylobacter* infections are endemic have shown that in these populations the levels of anti-*Campylobacter* antibodies naturally present in human sera increase with age; that is, with opportunity for exposure. The presence of antibodies to *Campylobacter* in older children correlates with a milder form of campylobacteriosis than is typically observed within the "developed" world (Oberhelman and Taylor 2000), and with asymptomatic carriage. Therefore, it is clear that some protective immunity can be developed through *in vivo* exposure to viable cells and through actual clinical infection. We need to find ways to translate that fact into a real-world efficacious vaccine that lacks any potential for inducing autoimmune disorders such as arthritis and demyelinating neuropathy.

Conclusion

Prevention of human campylobacteriosis through control of the natural distribution of campylobacteria within the farm and food production environment will require a much better scientific understanding of the organism than we presently have. There are many unanswered questions about the fundamental bacteriology of *Campylobacter* spp., about the strategies used by the organism to survive and spread within the natural environment, and about basic issues of host–pathogen interaction. There are many poorly understood aspects of campylobacteriosis in humans, especially with respect to sequelae and neuropathy, but also to the rare instances of frank extraintestinal systemic infection. Fortunately, although all of these issues are of very great scientific interest, there is one single overriding public health fact that needs mention and emphasis: Good food manufacturing practices, adequate plant hygiene, and sanitation practices, coupled with similarly good practices in the home kitchen (See Chapter 30), are the primary defenses available to food producers and consumers.

References

Aarestrup, F. M., Nielsen, E. M., Madsen, M. and Engberg, J. 1997. Antimicrobial susceptibility patterns of thermophilic *Campylobacter* spp. from humans, pigs, cattle, and broilers in Denmark. *Antimicrobial Agents and Chemotherapy* **41**:2244–2250.

Achen, M., Morishita, T. Y. and Ley, E. C. 1998. Shedding and colonization of *Campylobacter jejuni* in broilers from day-of-hatch to slaughter age. *Avian Diseases* **42**:732–737.

Ahmed, I. H., Manning, G., Wassenaar, T. M., Cawthraw, S. and Newell, D. G. 2002. Identification of genetic differences between two *Campylobacter jejuni* strains with different colonization potentials. *Microbiology* **148**:1203–1212.

Ang, C. W., Endtz, H. P., Jacobs, B. C., Laman, J. D., de Klerk, M. A., van der Meche, F. G. and van Doorn, P. A. 2000a. *Campylobacter jejuni* lipopolysaccharides from Guillain-Barré syndrome patients induce IgG anti-GM1 antibodies in rabbits. *Journal of Neuroimmunology* **104**:133–138.

Ang, C. W., van Doorn, P. A., Endtz, H. P., Merkies, I. S., Jacobs, B. C., de Klerk, M. A., van Koningsveld, R. and van der Meche, F. G. 2000b. A case of Guillain-Barré syndrome following a family outbreak of *Campylobacter jejuni* enteritis. *Journal of Neuroimmunology* **111**:229–233.

Annan-Prah, A. and Janc, M. 1988. The mode of spread of *Campylobacter jejuni/coli* to broiler flocks. *Journal of Veterinary Medicine* **35**:11–18.

Baqar, S., Applebee, L. A. and Bourgeois, A. L. 1995a. Immunogenicity and protective efficacy of a prototype *Campylobacter* killed whole-cell vaccine in mice. *Infection and Immunity* **63**:3731–3735.

Baqar, S., Bourgeois, A. L., Schultheiss, P. J., Walker, R. I., Rollins, D. M., Haberberger, R. L. and Pavlovskis, O. R. 1995b. Safety and immunogenicity of a prototype oral whole-cell killed *Campylobacter* vaccine administered with a mucosal adjuvant in non-human primates. *Vaccine* **13**:22–28.

Berndtson, E., Emanuelson, U., Engvall, A. and Dainelsson-Tham, M. L. 1996. A 1-year epidemiological study of campylobacters in 18 Swedish chicken farms. *Preventive Veterinary Medicine* **26**:167–185.

Bezian, M. C., Ribou, G., Barberis-Giletti, C. and Megraud, F. 1990. Isolation of a urease positive thermophilic variant of *Campylobacter lari* from a patient with urinary tract infection. *European Journal of Clinical Microbiology and Infectious Diseases* **9**:895–897.

Bras, A. M., Chatterjee, S., Wren, B. W., Newell, D. G. and Ketley, J. M. 1999. A novel *Campylobacter jejuni* two-component regulatory system important for temperature-dependent growth and colonization. *Journal of Bacteriology* **181**:3298–3302.

Buczynski, J., Yanagihara, R., Mora, C., Cartier, L., Verdugo, A., Araya, F., Castillo, L., Gibbs, C. J., Gajdusek, C. D., Rogers-Johnson, P. and Liberski, P. P. 2001. Tropical spastic paraparesis. *Folia Neuropathologica* **39**:265–269.

Calva, E., Torres, J., Vazquez, M., Angeles, V., de la Vega, H. and Ruiz-Palacios, G. M. 1989. *Campylobacter jejuni* chromosomal sequences that hybridize to *Vibrio cholerae* and *Escherichia coli* LT enterotoxin genes. *Gene* **75**:243–251.

Cappelier, J. M., Lazaro, B., Rossero, A., Fernandez-Astorga, A. and Federighi, M. 1997. Double staining (CTC-DAPI) for detection and enumeration of viable but non-culturable *Campylobacter jejuni* cells. *Veterinary Research* **28**:547–555.

Chan, K. F., Tran, H. L., Kanenaka, R. Y. and Kathariou, S. 2001. Survival of clinical and poultry-derived isolates of *Campylobacter jejuni* at a low temperature (4°C). *Applied and Environmental Microbiology* **67**:4186–4191.

Chen, H. C. and Stern, N. J. 2001. Competitive exclusion of heterologous *Campylobacter* spp. in chicks. *Applied Environmental Microbiology* **67**:848–851.

Chiu, C. H., Kuo, C. Y. and Ou, J. T. 1995. Chronic diarrhea and bacteremia caused by *Campylobacter lari* in a neonate. *Clinical Infectious Diseases* **21**:700–701.

Cox, N. A., Stern, N. J., Hiett, K. L. and Berrang, M. E. 2002a. Identification of a new source of *Campylobacter* contamination in poultry: transmission from breeder hens to broiler chickens. *Avian Diseases* **46**:535–541.

Cox, N. A., Stern, N. J., Wilson, J. L., Musgrove, M. T., Buhr, R. J. and Hiett, K. L. 2002b. Isolation of *Campylobacter* spp. from semen samples of commercial broiler breeder roosters. *Avian Diseases* **46**:717–720.

Craven, S. E., Stern, N. J., Line, E., Bailey, J. S., Cox, N. A. and Fedorka-Cray, P. 2000. Determination of the incidence of *Salmonella* spp., *Campylobacter jejuni*, and *Clostridium perfringens* in wild birds near broiler houses by sampling intestinal droppings. *Avian Diseases* **44**:715–720.

Endtz, H. P., Vliegenthart, J. S., Vandamme, P., Weverink, H. W., van den Braak, N. P., Verbrugh, H. A. and van Belkum, A. 1997. Genotypic diversity of *Campylobacter lari* isolated from mussels and oysters in The Netherlands. *International Journal of Food Microbiology* **34**:79–88.

Evans, S. J. 1992. Introduction and spread of thermophilic Campylobacters in broiler flocks. *Veterinary Record* **131**:574–576.

Federighi, M., Tholozan, J. L., Cappelier, J. M., Tissier, J. P. and Jouve, J. L. 1998. Evidence of non-coccoid viable but non-culturable *Campylobacter jejuni* in microcosm water by direct viable count, CTC-DAPI double staining, and scanning electron microscopy. *Food Microbiology* **15**:539–550.

Fenton, K., White, J., Gillespie, I. and Morgan, D. 2002. Quarterly communicable disease review: July to September 2001. *Journal of Public Health and Medicine* **24**:63–69.

Fry, B. N., Feng, S., Chen, Y. Y., Newell, D. G., Coloe, P. J. and Korolik, V. 2000. The galE gene of *Campylobacter jejuni* is involved in lipopolysaccharide synthesis and virulence. *Infection and Immunity* **68**:2594–2601.

Gibbens, J. C., Pascoe, S. J. S., Evans, S. J., Davies, R. H. and Sayers, A. R. 2001. A trial of biosecurity as a means to control *Campylobacter* infection of broiler chickens. *Preventive Veterinary Medicine* **48**:85–99.

Glunder, G. 1995. Infectivity of *Campylobacter jejuni* and *Campylobacter coli* in chickens. *Berler und Muenchener Tieraerztliche Wochenschrift* **108**:101–104.

Goudswaard, J., Sabbe, L. and te Winkel, W. 1995. Reactive arthritis as a complication of *Campylobacter lari* enteritis. *Journal of Infection* **31**:171.

Gregory, E., Barnhart, H., Dreesen, D. W., Stern, N. J. and Corn, J. L. 1997. Epidemiological study of *Campylobacter* spp. in broilers: source, time of colonization, and prevalence. *Avian Diseases* **41**:890–898.

Hakkinen, M. and Schneitz, C. 1999. Efficacy of a commercial competitive exclusion product against *Campylobacter jejuni*. *British Poultry Science* **40**:619–621.

Hald, B., Knudsen, K., Lind, P. and Madsen, M. 2001. Study of the infectivity of saline-stored *Campylobacter jejuni* for day-old chicks. *Applied and Environmental Microbiology* **67**:2388–2392.

Hanninen, M. L., Perko-Makela, P., Pitkala, A. and Rautelin, H. 2000. A three-year study of *Campylobacter jejuni* genotypes in humans with domestically acquired infections and in chicken samples from the Helsinki area. *Journal of Clinical Microbiology* **38**:1998–2000.

Hanninen, M. L., Perko-Makela, P., Rautelin, H., Duim, B. and Wagenaar, J. A. 2001. Genomic relatedness within five common Finnish *Campylobacter jejuni* pulsed-field gel electrophoresis genotypes studied by amplified fragment length polymorphism analysis, ribotyping, and serotyping. *Applied and Environmental Microbiology* **67**:1581–1586.

Hassane, D. C., Lee, R. B. and Pickett, C. L. 2003. *Campylobacter jejuni* cytolethal distending toxin promotes DNA repair responses in normal human cells. *Infection and Immunity* **71**:541–545.

Hickey, T. E., McVeigh, A. L., Scott, D. A., Michielutti, R. E., Bixby, A., Carroll, S. A., Bourgeois, A. L. and Guerry, P. 2000. *Campylobacter jejuni* cytolethal distending toxin mediates release of interleukin-8 from intestinal epithelial cells. *Infection and Immunity* **68**:6535–6541.

Hiett, K. L., Cox, N. A. and Stern, N. J. 2002. Direct polymerase chain reaction detection of *Campylobacter* spp. in poultry hatchery samples. *Avian Diseases* **46**:219–223.

Jacobs, B. C., Endtz, H. P., van der Meche, F. G., Hazenberg, M. P., de Klerk, M. A. and van Doorn, P. A. 1997. Humoral immune response against *Campylobacter jejuni* lipopolysaccharides in Guillain-Barré and Miller Fisher syndrome. *Journal of Neuroimmunology* **79**:62–68.

Jacobson, S. 2002. Immunopathogenesis of human T cell lymphotropic virus type I-associated neurologic disease. *Journal of Infectious Diseases* **186(Suppl. 2)**:S187–S192.

Jacobs-Reitsma, W. 2000. *Campylobacter* in the food supply, pp. 467–481. *In Campylobacter*, 2nd edn., I. Nachamkin, M. J. Blaser (Eds.); Washington, D.C.: American Society for Microbiology Press.

Jacobs-Reitsma, W. F., VAN DE Giessen, A. W., Bolder, N. M. and Mulder, R. W. A. 1995. Epidemiology of *Campylobacter* spp. at 2 Dutch broiler farms. *Epidemiology and Infection* **114**:413–421.

Kanwar, R. K., Ganguly, N. K., Kumar, L., Rakesh, J., Panigrahi, D. and Walia, B. N. 1995. Calcium and protein kinase C play an important role in *Campylobacter jejuni*–induced changes in Na$^+$ and Cl$^-$ transport in rat ileum in vitro. *Biochimica et Biophysica Acta* **1270**:179–192.

Konkel, M. E., Kim, B. J., Klena, J. D., Young, C. R. and Ziprin, R. 1998. Characterization of the thermal stress response of *Campylobacter jejuni*. *Infection and Immunity* **66**:3666–3672.

Konkel, M. E., Joens, L. A. and Mixter, P. F. 2000. Molecular characterization of *Campylobacter jejuni* virulence determinants, pp. 217–240. *In Campylobacteria*, 2nd edn., I. Nachamkin, M. J. Blaser (Eds.); Washington, D.C.: American Society for Microbiology Press.

Kopecko, D. J. 1997. Regulatory considerations for *Campylobacter* vaccine development. *Journal of Infectious Disease* **176(Suppl. 2)**:S189–S191.

Korzeniowska-Kowal, A., Witkowska, D. and Gamian, A. 2001. Molecular mimicry of bacterial polysaccharides and their role in etiology of infectious and autoimmune diseases. *Postepy Higieny i Medycyny Doswiadczalnej* **55**:211–232.

Krause, R., Ramschak-Schwarzer, S., Gorkiewicz, G., Schnedl, W. J., Feierl, G., Wenisch, C. and Reisinger, E. C. 2002. Recurrent septicemia due to *Campylobacter fetus* and *Campylobacter lari* in an immunocompetent patient. *Infection* **30**:171–174.

Lazaro, B., Carcamo, J., Audicana, A., Perales, I. and Fernandez-Astorga, A. 1999. Viability of DNA maintenance in nonculturable spiral *Campylobacter jejuni* cells after long-term exposure to low temperatures. *Applied and Environmental Microbiology* **65**:4677–4681.

Lee, L. H., Burg, E., Baqar, S., Bourgeois, A. L., Burr, D. H., Ewing, C. P., Trust, T. J. and Guerry, P. 1999. Evaluation of a truncated recombinant flagellin subunit vaccine against *Campylobacter jejuni*. *Infection and Immunity* **67**:5799–5805.

Levin, M. C., Lee, S. M., Kalume, F., Morcos, Y., Dohan Jr., F. C., Hasty, K. A., Callaway, J. C., Zunt, J., Desiderio, D. and Stuart, J. M. 2002a. Autoimmunity due to molecular mimicry as a cause of neurological disease. *Nature Medicine* **8**:509–513.

Levin, M. C., Lee, S. M., Morcos, Y., Brady, J. and Stuart, J. 2002b. Cross-reactivity between immunodominant human T lymphotropic virus type I tax and neurons: implications for molecular mimicry. *Journal of Infectious Diseases* **186**:1514–1517.

Mead, G. C. 2002. Factors affecting intestinal colonisation of poultry by campylobacter and role of microflora in control. *World Poultry Science Journal* **58**:169–178.

Medema, G. J., Schets, F. M., VAN DE Giessen, A. W. and Havelaar, A. H. 1992. Lack of colonization of 1 day old chicks by viable, non-culturable *Campylobacter jejuni*. *Journal of Applied Bacteriology* **72**:512–516.

Mooney, A., Clyne, M., Curran, T., Doherty, D., Kilmartin, B. and Bourke, B. 2001. *Campylobacter upsaliensis* exerts a cytolethal distending toxin effect on HeLa cells and T lymphocytes. *Microbiology* **147**:735–743.

Nachamkin, I., Yang, X. H. and Stern, N. J. 1993. Role of *Campylobacter jejuni* flagella as colonization factors for three-day-old chicks: analysis with flagellar mutants. *Applied and Environmental Microbiology* **59**:1269–1273.

Nielsen, E. M., Engberg, J. and Madsen, M. 1997. Distribution of serotypes of *Campylobacter jejuni* and *C. coli* from Danish patients, poultry, cattle and swine. *FEMS Immunology Medicine and Microbiology* **19**:47–56.

Nylen, G., Dunstan, F., Palmer, S. R., Andersson, Y., Bager, F., Cowden, J., Feierl, G., Galloway, Y., Kapperud, G., Megraud, F., Mølbak, K., Petersen, L. R. and Ruutu, P. 2002. The seasonal distribution of *Campylobacter* infection in nine European countries and New Zealand. *Epidemiology and Infection* **128**:383–390.

Oberhelman, R. A. and Taylor, D. N. 2000. *Campylobacter* infections in developing countries, pp. 139–153. *In Campylobacter*, 2nd edn., I. Nachamkin, M. J. Blaser (Eds.); Washington, D.C.: American Society for Microbiology Press.

Osame, M. 2002. Pathological mechanisms of human T-cell lymphotropic virus type I-associated myelopathy (HAM/TSP). *Journal of Neurovirology* **8**:359–364.

Ownby, H. E., Korelitz, J. J., Busch, M. P., Williams, A. E., Kleinman, S. H., Gilcher, R. O. and Nourjah, P. 1997. Loss of volunteer blood donors because of unconfirmed enzyme immunoassay screening results. Retrovirus Epidemiology Donor Study. *Transfusion* **37**:199–205.

Pace, J. L., Rossi, H. A., Esposito, V. M., Frey, S. M., Tucker, K. D. and Walker, R. I. 1998. Inactivated whole-cell bacterial vaccines: current status and novel strategies. *Vaccine* **16**:1563–1574.

Petersen, L. and On, S. L. 2000. Efficacy of flagellin gene typing for epidemiological studies of *Campylobacter jejuni* in poultry estimated by comparison with macrorestriction profiling. *Letters in Applied Microbiology* **31**:14–19.

Petersen, L., Nielsen, E. M., Engberg, J., On, S. L. and Dietz, H. H. 2001a. Comparison of genotypes and serotypes of *Campylobacter jejuni* isolated from Danish wild mammals and birds and from broiler flocks and humans. *Applied and Environmental Microbiology* **67**:3115–3121.

Petersen, L., Nielsen, E. M. and On, S. L. 2001b. Serotype and genotype diversity and hatchery transmission of *Campylobacter jejuni* in commercial poultry flocks. *Veterinary Microbiology* **82**:141–154.

Petersen, L. and Wedderkopp, A. 2001. Evidence that certain clones of *Campylobacter jejuni* persist during successive broiler flock rotations. *Applied and Environmental Microbiology* **67**:2739–2745.

Pickett, C. L. 2000. *Campylobacter* toxins and their role in pathogenesis, pp. 179–190. *In Campylobacter*, 2nd edn., I. Nachamkin, M. J. Blaser (Eds.); Washington, D.C.: American Society for Microbiology Press.

Plumelle, Y. 1999. HTLV-1-associated myelopathy/tropical spastic paraparesis (HAM/TSP) pathogenesis hypothesis. A shift of homologous peptides pairs, central nervous system (CNS)/HTLF-1, HTLV-1/thymus, thymus/CNS, in a thymus-like CNS environment, underlies the pathogenesis of HAM/TSP. *Medical Hypotheses* **52**:595–604.

Purdy, D., Cawthraw, S., Dickinson, J. H., Newell, D. G. and Park, S. F. 1999. Generation of a superoxide dismutase (SOD)-deficient mutant of *Campylobacter coli*: evidence for the significance of SOD in *Campylobacter* survival and colonization. *Applied and Environmental Microbiology* **65**:2540–2546.

Refregier-Petton, J., Rose, N., Denis, M. and Salvat, G. 2001. Risk factors for *Campylobacter* spp. contamination in French broiler-chicken flocks at the end of the rearing period. *Preventive Veterinary Medicine* **50**:89–100.

Ribas, J. G. and Melo, G. C. 2002. Human T-cell lymphotropic virus type 1(HTLV-1)-associated myelopathy. *Revista Sociedade Brasileira de Medicina Tropical* **35**:377–384.

Rice, B. E., Rollins, D. M., Mallinson, E. T., Carr, L. and Joseph, S. W. 1997. *Campylobacter jejuni* in broiler chickens: colonization and humoral immunity following oral vaccination and experimental infection. *Vaccine* **15**:1922–1932.

Rollins, D. M. and Colwell, R. R. 1986. Viable but nonculturable stage of *Campylobacter jejuni* and its role in survival in the natural aquatic environment. *Applied and Environmental Microbiology* **52**:531–538.

Scott, D. A. 1997. Vaccines against *Campylobacter jejuni*. *Journal of Infectious Diseases* **176(Suppl. 2)**:S183–S188.

Shane, S. M. 2000. *Campylobacter* infection of commercial poultry. *Revue Scentifique et Technique OIE (Office International des Epizooties)* **19**:376–395.

Shreeve, J. E., Toszeghy, M., Ridley, A. and Newell, D. G. 2002. The carry over of *Campylobacter* isolates between sequential poultry flocks. *Avian Diseases* **46**:378–385.

Simonsen, L., Buffington, J., Shapiro, C. N., Holman, R. C., Strine, T. W., Grossman, B. J., Williams, A. E. and Schonberger, L. B. 1995. Multiple false reactions in viral antibody screening assays after influenza vaccination. *American Journal of Epidemiology* **141**:1089–1096.

Skirrow, M. B. 1977. *Campylobacter* enteritis: a "new" disease. *British Medical Journal* **2**(6078):9–11.

Skirrow, M. B. and Blaser, M. J. 2000. Clinical aspects of *Campylobacter* infection, pp. 69–88. *In Campylobacter*, 2nd edn., I. Nachamkin, M. J. Blaser (Eds.); Washington, D.C.: American Society for Microbiology Press.

Skirrow, M. B. and Butzler, J. P. 2000. Foreword, xvii–xxiii. *In Campylobacter*, 2nd edn., I. Nachamkin, M. J. Blaser (Eds.); Washington, D.C.: American Society for Microbiology Press.

Smith, J. L. 2002. *Campylobacter jejuni* infection during pregnancy: long-term consequences of associated bacteremia, Guillain-Barré Syndrome, and reactive arthritis. *Journal of Food Protection* **65**:696–708.

Stas, T., Jordan, F. T. W. and Woldehiwet, Z. 1999. Experimental infection of chickens with *Campylobacter jejuni*: strains differ in their capacity to colonize the intestine. *Avian Pathology* **28**:61–64.

Stern, J. J., Wojton, B. and Kwiatek, K. 1992. A differential-selective medium and dry ice-generated atmosphere for recovery of *Campylobacter jejuni*. *Journal of Food Protection* **55**:514–517.

Stern, N. J., Jones, D. M., Wesley, I. and Rollins, D. M. 1994. Colonization of chicks by non-culturable *Campylobacter jejuni* spp. *Letters in Applied Microbiology* **18**:333–336.

Stern, N. J., Myszewski, M. A., Barnhart, H. M. and Dreesen, D. W. 1997. Flagellin A gene restriction fragment length polymorphism patterns of *Campylobacter* spp. isolates from broiler production sources. *Avian Diseases* **41**:899–905.

Stern, N. J., Cox, N. A., Bailey, J. S., Berrang, M. E. and Musgrove, M. T. 2001. Comparison of mucosal competitive exclusion and competitive exclusion treatment to reduce Salmonella and Campylobacter spp. colonization in broiler chickens. *Poultry Science* **80**:156–160.

Stern, N. J., Robach, M. C., Cox, N. A. and Musgrove, M. T. 2002. Effect of drinking water chlorination on *Campylobacter* spp. colonization of broilers. *Avian Diseases* **46**:401–404.

Suzuki, S., Kawaguchi, M., Mizuno, K., Takama, K. and Yuki, N. 1994. Immunological properties and ganglioside recognitions by *Campylobacter jejuni*–enterotoxin and cholera toxin. *FEMS Immunology and Medical Microbiology* **8**:207–211.

Swaggerty, C. L., Frolov, A. A., McArthur, M. J., Cox, V. W., Tong, S., Compans, R. W. and Ball, J. M. 2000. The envelope glycoprotein of simian immunodeficiency virus contains an enterotoxin domain. *Virology* **277**:250–261.

Tholozan, J. L., Cappelier, J. M., Tissier, J. P., Delattre, G. and Federighi, M. 1999. Physiological characterization of viable-but-nonculturable *Camppylobacter jejuni* cells. *Applied and Environmental Microbiology* **65**:1110–1116.

Trachoo, N., Frank, J. F. and Stern, N. J. 2002. Survival of *Campylobacter jejuni* in biofilms isolated from chicken houses. *Journal of Food Protection* **65**:1110–1116.

van Belkum, A., van den Braak, N., Godschalk, P., Ang, W., Jacobs, B., Gilbert, M., Wakarchuk, W., Verbrugh, H. and Endtz, H. 2001. A *Campylobacter jejuni* gene associated with immune-mediated neuropathy. *Nature Medicine* **7**:752–753.

VAN DE Giessen, A. W., Heuvelman, C. J., Abee, T. and Hazeleger, W. C. 1996. Experimental studies on the infectivity of non-culturable forms of *Campylobacter* spp. in chicks and mice. *Epidemiology and Infection* **117**:463–470.

VAN DE Giessen, A., Mazurier, S. I., Jacobsreitsma, W., Jansen, W., Berkers, P., Ritmeester, W. and Wernars, K. 1992. Study on the Epidemiology and Control of *Campylobacter Jejuni* in poultry broiler flocks. *Applied and Environmental Microbiology* **58**:1913–1917.

van Koningsveld, R., Rico, R., Gerstenbluth, I., Schmitz, P. I., Ang, C. W., Merkies, I. S., Jacobs, B. C., Halabi, Y., Endtz, H. P., van der Méche, F. G. and van Doorn, P. A. 2001. Gastroenteritis-associated Guillain-Barré syndrome on the Caribbean island Curaçao. *Neurology* **56**:1467–1472.

Wassenaar, T. M., Fry, B. N., Lastovica, A. J., Wagenaar, J. A., Coloe, P. J. and Duim, B. 2000. Genetic characterization of *Campylobacter jejuni* O:41 isolates in relation with Guillain-Barré syndrome. *Journal of Clinical Microbiology* **38**:874–876.

Werno, A. M., Klena, J. D., Shaw, G. M. and Murdoch, D. R. 2002. Fatal case of *Campylobacter lari* prosthetic joint infection and bacteremia in an immunocompetent patient. *Journal of Clinical Microbiology* **40**:1053–1055.

Widders, P. R., Perry, R., Muir, W. I., Husband, A. J. and Long, K. A. 1996. Immunisation of chickens to reduce intestinal colonisation with *Campylobacter jejuni*. *British Poultry Science* **37**:765–778.

Young, C. R., Ziprin, R. L., Hume, M. E. and Stanker, L. H. 1999. Dose response and organ invasion of day-of-hatch Leghorn chicks by different isolates of *Campylobacter jejuni*. *Avian Diseases* **43**:763–767.

Zaaijer, H. L., Cuypers, H. T., Dudok de Wit, C. and Lelie, P. N. 1994. Results of 1-year screening of donors in The Netherlands for human T-lymphotropic virus (HTLV) type I: significance of Western blot patterns for confirmation of HTLV infection. *Transfusion* **34**:877–880.

Ziprin, R. L., Young, C. R., Stanker, L. H., Hume, M. E. and Konkel, M. E. 1999. The absence of cecal colonization of chicks by a mutant of *Campylobacter jejuni* not expressing bacterial fibronectin-binding protein. *Avian Diseases* **43**:586–589.

Ziprin, R. L., Young, C. R., Byrd, J. A., Stanker, L. H., Hume, M. E., Gray, S. A., Kim, B. J. and Konkel, M. E. 2001. Role of *Campylobacter jejuni* potential virulence genes in cecal colonization. *Avian Diseases* **45**:549–557.

Ziprin, R. L., Hume, M. E., Young, C. R. and Harvey, R. B. 2002a. Cecal colonization of chicks by porcine strains of *Campylobacter coli*. *Avian Diseases* **46**:473–477.

Ziprin, R. L., Hume, M. E., Young, C. R. and Harvey, R. B. 2002b. Inoculation of chicks with viable non-colonizing strains of *Campylobacter jejuni*: evaluation of protection against a colonizing strain. *Current Microbiology* **44**:221–223.

Ziprin, R. L., Droleskey, R. E., Hume, M. E. and Harvey, R. B. 2003a. Viable nonculturable *Campylobacter jejuni* not colonizing the cecum of newly hatched leghorn chicks. *Avian Diseases* **47**:753–758.

Ziprin, R. L., Sheffield, C. L., Hume, M. E., Drinnon, D. L. J. and Harvey, R. B. 2003b. Cecal colonization of chicks by bovine-derived strains of *Campylobacter*. *Avian Diseases* **47**:1429–1433.

7 Global Analysis of the *Mycobacterium avium* subsp. *paratuberculosis* Genome and Model Systems Exploring Host–Agent Interactions

Thomas A. Ficht, L. Garry Adams,
Sangeeta Khare, Brian O'Shea, and
Allison C. Rice-Ficht

Introduction

The focus of this chapter will be on genetic and immunological methods to diagnose *Mycobacterium avium* subsp. *paratuberculosis* (MapTb) infections. Discussion will be limited to recent experimental results pertaining to the identification of infected animals, inconsistencies in organism identification, development of animal models of disease, and the genetic background of the host as a contributing factor to disease. We will also address growing concerns and continuing debate on MapTb as a source of foodborne illness in the human population, specifically Crohn's disease. For a comprehensive description of the organism and the disease it causes, the reader is referred to an earlier review (Harris and Barletta 2001).

Johne's is a chronic inflammatory bowel disease (enteritis) of ruminants caused by MapTb (Johne and Frothigham 1895). According to the U.S. Department of Agriculture's National Animal Health Monitoring System's (NAHMS) 1996 national dairy study, Johne's-positive herds experience an economic loss of almost $100–$200 per cow resulting from reduced milk production and increased cow-replacement costs. When averaged across all herds, Johne's disease costs the U.S. dairy industry $200–$250 million annually in reduced productivity (Ott et al. 1999).

Paratuberculosis in ruminants has a prolonged incubation period, and most animals remain subclinical. Infection is typically acquired via the fecal–oral route. Because of poor diagnosis, the progression from the subclinical to clinical state remains undefined, and subclinical animals spread the disease to other animals in the herd. Reports of survival of MapTb in pasteurized milk (Chiodini and Hermon-Taylor 1993, Sung and Collins 1998), isolation of the bacterium from wildlife, and the possible relationship to human disease (Crohn's) underscores the need for more accurate diagnostic testing (Kanazawa et al. 1999, Naser et al. 1999, Prantera and Scribano 1999, Suenaga et al. 1999, Collins et al. 2000).

The gold standard for diagnosis is detection of the organism, but the prolonged incubation period coupled with the extended *in vitro* growth rate complicates this form of diagnosis. The immune response is similar to other mycobacterial species in that protective immunity is characterized by a strong Th1 response. Animals with minimal disease have reduced antibody responses but elicit strong Th1-cell responses (Adams et al. 1996, Sweeney et al. 1998, Stabel 2000). Cell-mediated immunity is essential for protection and CD4+ Th1 cells are presumed to play a dominant role (Bassey and Collins 1997). These responses are measured using the IFN-γ enzyme immunoassay originally designed for diagnosis of bovine tuberculosis (Stabel 1996, McDonald et al. 1999). Unfortunately, being an indirect measure, this test has reduced specificity and may be confused by exposure to other pathogens including environmental mycobacteria (Gwozdz and Thompson 2002, Walravens et al. 2002). Animals with fulminating disease develop strong humoral and weak cellular responses (Bassey and Collins 1997). The progression of disease has been attributed to the shift from Th1-like to Th2-like response (Clarke and Little 1996, Navarro et al. 1998). A decrease in IFN-γ producing CD4+ cells has been reported (Bassey and Collins 1997, Zhao et al. 1997, Begara-McGorum et al. 1998) and may explain the limited *in vitro* evidence supporting antibacterial activity for IFN-γ. The host's defense mechanisms may be misdirected to favor pathogen survival. Antigen processing may also be suppressed, as indicated by the reduced presentation of exogenously added hemagglutin at a time when surface expressed costimulatory molecules are unaffected (Valentin-Weigand and Goethe 1999).

Similarity between Johne's disease in ruminants and Crohn's disease in humans has raised concerns regarding a causal relationship between the organism and human diseases. Although this contention awaits confirmation, it must be remembered that other explanations (autoimmunity) of Crohn's disease also await verification. Furthermore, description of the human disease is heterogeneous, so that correlation with and identification of the organism (itself a difficult task) is exceedingly complex. Of primary concern in this context are reports describing the resistance of MapTb to pasteurization as it is currently performed (Chiodini and Hermon-Taylor 1993, Sung and Collins 1998). With the elevated level of Johne's disease in dairy cattle, consumption of dairy products becomes a primary source of concern. There is a need to alter the pasteurization process as it is currently performed, but without a cost-effective approach there may be a lack of necessary support. Without this combined approach such decisions may be labeled alarmist in the absence of "proof," although proof against this contention might include the millions of people that have apparently consumed dairy products without effect. For this reason, the research approaches described aim to improve identification of infected animals as well as to characterize differences between isolates that may be related to the observed heterogeneity of the disease. Furthermore, the genetic background of the host or environmental cues may be important contributing factors to a disease that is most prevalent in well-developed nations.

Rationale for Current Research

Animal Health

The threat posed by MapTb to animal health originates with the prolonged incubation period associated with infection. Ruminants carry the organism for years without clinical signs, often remaining subclinical for years. During this time, the organism may be shed intermittently, resulting in intra-herd transmission, as well as vertical transmission to offspring via the fecal–oral route and transplacental infection of the fetus. The long-term effects include profound chronic granulomatous enteritis with extensive mucosal thickening, which is the underlying cause of associated chronic wasting (Clarke and Little 1996, Buergelt and Ginn 2000, Corpa et al. 2000). MapTb infection represents a problem to longer-lived animals such as dairy cattle, but it usually has a reduced effect in the beef cattle industry. The organism also infects other ruminant species including goats and bison. The disease in goats could represent a source of human infection worldwide, as such animals are often maintained as household pets in underdeveloped countries. In the United States, the disease has caused problems in conservation efforts to restore bison to pre–twentieth century levels.

The disease is primarily characterized by a chronic wasting that occurs gradually over several years. Disease progression is associated with a loss of T-cell mediated immune function, but there is little information available concerning intestinal immune activity. Experiments with severe combined immunodeficient mice have revealed increased infiltration of the lamina propria by mononuclear cells and significantly increased width of the villus (Mutwiri et al. 2001, Mutwiri et al. 2002). *In vitro* studies in Ussing chambers suggest significant abnormalities in intestinal transport functions (Mutwiri et al. 2001), all of which indicate that T-cell-independent mechanisms may be sufficient to produce clinical signs of disease.

Although immune protection is the ultimate goal consistent with prevention of disease, that possibility appears to be in the distant future at best. Hurdles include the lack of identified target immunogens, representing potential vaccine candidates, and the extended duration of the subclinical phase, which makes identification of strains with reduced virulence nearly impossible.

As a result, immediate research is required to address the development of diagnostic tests capable of identifying animals in the preclinical stages of infection, including tests to directly detect the presence of the organism or to indirectly identify animals infected with the organism based on changes in host gene expression. In the latter case, the use of biomarkers or biosignatures in detecting infection in agriculturally important animals may be impractical. However, because of the potential for human infection, characterization of such biomarkers appears to be readily justified.

Human Health

There is currently a lack of consensus among scientists concerning the potential link between Crohn's disease in humans and MapTb. Yet concern persists, primarily because of the unexplained increase in inflammatory bowel disease in developed countries and the identification of the organism as an inhabitant of the lumen of humans and other species (Beard et al. 2001a, Beard et al. 2001b). Recent isolations of the organism from individuals with Crohn's disease as well as from HIV-infected individuals (Mutwiri et al. 2001, Richter et al. 2002) emphasize the need for concern. Although experimental evidence also supports a link based on signs observed in MapTb-infected severe combined immunodeficient mice (Mutwiri et al. 2001), the organism is not consistently identified in human cases. Evidence indicating MapTb resistance to pasteurization as it is currently performed has also raised concerns regarding a potential cause of some forms of inflammatory bowel disease (i.e., Crohn's) in otherwise normal, healthy adults.

The link with human disease has and is still vigorously debated, and current consensus indicates that there is little proof of a causal role for the organism in human disease. This consensus is based on a lack of apparent adherence to Koch's postulates that includes the inability to demonstrate the presence of the organism in all cases of Crohn's disease, the difficulty of isolating the organism in a pure form, and the inability to reproduce the disease in healthy animals (Grimes 2003). However, this lack of adherence may be interpreted as a failure of modern medical research rather than proof that the organism is not associated with the disease. Until there is a paradigm to explain Crohn's disease that either excludes or includes MapTb conclusively, research must continue to explore any potential link.

Goals of Current Research

Improved Culture Techniques

Direct detection of the organism, as always, remains the gold standard for demonstrating infection status. Improvement to *in vitro* culture systems has been dramatic over the last few years, reducing culture times from months to weeks. The most promising approach employs the use of a modified substrate and indicator in conjunction with the Bactec system; however, even these enriched systems require a minimum of 2 weeks for sufficient growth (Eamens et al. 2000, Thornton et al. 2002). The practical use of such equipment in identifying infected animals is limited by apparent shedding of the organism only during the latter stages of infection (Pavlik et al. 2000).

Antibody-Based Detection Schemes

Enzyme-linked immunosorbent assay–based detection schemes have been evaluated in other laboratories, and although effective with heavily infected or shedding animals, these systems exhibit greatly reduced efficacy with preclinical or subclinical ruminant animals.

Lymphocyte blastogenesis in response to MapTb antigen has also been examined and has provided improved sensitivity over other testing methods (Williams et al. 1985). This of course gave rise to IFN-γ testing, which is a measure of antigen-specific T-cell activation (Stabel 1996, Gwozdz et al. 2000, Stabel and Whitlock 2001, Gwozdz and Thompson 2002). However, these methods are hampered by cross reactivity between antigens from closely related non-pathogenic mycobacterial organisms present in the environment. Increased specificity may be possible with the identification of specific antigens.

DNA-Based Detection Schemes

DNA-based detection schemes permit the detection of minimal numbers of organisms through the amplification of target sequences selected to enhance sensitivity and specificity. Current approaches exhibit increased sensitivity through enhanced organism recovery and elimination of impurities that may inhibit detection (Cavallini et al. 2000). Although all three features are vital to increased sensitivity, they are dependent on consistent shedding of the organism by the host. In the case of Johne's disease, it is not clear whether preclinical or even subclinical animals consistently shed the organism, and in fact, there is evidence to the contrary (Pavlik et al. 2000).

DNA amplification may be performed using cultured samples or directly; however, the latter typically requires removal of impurities. Polymerase chain reaction (PCR) amplification may be performed following immunomagnetic bead separation, bead beating, and DNA extraction. This technique offers the capacity to enrich the organism as well as remove impurities that interfere with amplification to consistently detect <10 MapTb organisms. An additional aspect of organism identification is the potential for variation among isolates. Recent evidence suggests that a great deal of genetic variation exists within the species classically identified based on mycobactin J growth dependence *in vitro* (Lambrecht and

Collins 1992, Aduriz et al. 1995). This variety could present difficulties in any detection scheme and must not be overlooked.

IS900 has been employed as the definitive diagnostic amplicon for identification of MapTb. Its presence in other mycobacterial species has complicated diagnosis, making it necessary to examine sequence variation through restriction fragment polymorphisms with the target amplicon (Englund et al. 2002). This has proven to be a valuable approach, but it does not lend itself to identification of the source of the organism or to additional epidemiological tracking. Additional loci have been identified that are unique to MapTb through genomic scale comparison with *M. avium* (Bannantine et al. 2002).

Population Structure

Variation among MapTb isolates has been described for surface antigens, virulence, and host range (Sugden et al. 1987, Roach et al. 1993). Sheep and bovine isolates exhibit variation that may reflect preferences for a particular host or differential gene expression in response to the host. The ability to identify markers for each of these variables is the first step in identifying the genes responsible for infection or persistence in a particular host. Recent genomic analysis has confirmed that as a group, MapTb can be distinguished from *M. avium* using several common genetic markers or globally based genomic comparisons (Bannantine et al. 2003, Motiwala et al. 2003). However, these approaches did not explore in detail differences between MapTb isolates that could reflect important genomic variability.

Our laboratories have explored the use of amplified fragment length polymorphisms (AFLPs) to characterize the genetic diversity among MapTb isolates. Although it is also possible to identify single nucleotide polymorphisms using this technique, AFLPs may be used to analyze the entire genome at various levels of selectivity. Differences in sequence appear as polymorphic bands during electrophoresis. This approach has been used to identify a number of genetic differences between MapTb and *M. avium* corresponding to deleted sequences that may represent significant biological differences. In addition, this approach has provided the added benefit of distinguishing between isolates of MapTb that could prove useful in tracking the origin of infection.

Identification of Isolates

Isolates were obtained from our collaborators at the following facilities: Dr. Melissa Libal, Texas Veterinary Medical Diagnostic Laboratory; Dr. Gilberto Chavez-Gris, Department of Pathobiology, Universidad Nacional Autónoma de México, University of California, Davis, and the University of Wisconsin Diagnostic Laboratory. DNA was extracted by bead beating as described elsewhere (Khare et al. 2003).

AFLP-based fingerprinting requires only small amounts of DNA (500 ng or less), and depth of analysis can be varied depending on primer design. AFLP technology is based on digestion of DNA with two different restriction enzymes (i.e., EcoRI and MseI) followed by ligation of different adapters to each type of restriction site (Vos et al. 1995). Subsequent amplification of DNA fragments with primers complementary to the two different adapters results in preferential amplification of DNAs containing both adapters because of suppression PCR (Siebert et al. 1995). Following this pre-amplification, a selective PCR is performed in which different numbers of DNAs can be amplified using primers complementary to the adapters but containing additional selective bases ($n \geq 1$). For example, amplification of DNAs from complex plant genomes with primers containing three selective bases results in the amplification of approximately 50 different DNAs that are <1 kbp different in size. This is an ideal number of DNA fragments for analysis on gels and provides a unique fingerprint of the source DNA. The number of selective bases used can be adjusted depending on genome complexity, allowing fingerprints of bacteria, fungi, viruses, insects, animals, or plants to be collected using the same technology. Often, with bacteria, the pre-amplification step provides an ideal 30–50 bands because of the small genome size, and a second more definitive amplification step is not required. Examples of the success of this approach in bacterial epidemiology include *Vibrio cholera*, *Salmonella enterica* subsp. *enterica*, and *Campylobacter* spp. (de Boer et al. 2000, Jiang et al. 2000a, Jiang et al. 2000b, Lindstedt et al. 2000). The use of fluorescent labeled primers permits examination of multiple fragments simultaneously (Lindstedt et al. 2000). In some cases a single, 30-cycle PCR amplification may produce the ideal banding pattern; in other cases selective PCR is necessary (de Boer et al. 2000).

Texas A&M University researchers have automated most steps in the AFLP-fingerprinting procedure to allow rapid and reproducible collection of DNA patterns using double LI-COR DNA sequencers (Lincoln, NE). Data from the DNA sequencers are directly downloaded into the software program, Bionumerics (BioSystemactia, Devon), for analysis of images. This system together with fingerprint analysis software from the Sanger Centre allows rapid collection, archiving, and analysis of DNA fingerprints from any organism. The data may be stored for comparison with other isolates.

Initial experiments established the efficacy of AFLP-based fingerprinting for distinguishing MapTb from *M. avium* and for distinguishing pathotypes and biotypes among MapTb isolates. Fingerprints were determined using up to 200 different primer combinations in an initial scan for DNA polymorphisms. Because each primer combination amplifies approximately 50 DNA bands, this analysis had the potential to examine approximately 10,000 loci, and the data for one such analysis are collected on four gels in 4 hours. Each DNA band scored represents a search for polymorphisms in 10–16-bp (two restriction sites plus selective bases), plus insertions and deletions in each amplified DNA (averaging 250 bp). Once polymorphisms were identified with specific primer sets, subsequent analyses were performed with these informative primer combinations.

Data have been collected on local isolates from the veterinary diagnostic laboratory, and useful primer sets have been established for diagnostics and epidemiology of MapTb outbreaks. This approach permits tracking of disease transmission in and between herds and possibly to human populations. Reference databases will be collected containing gene patterns from a variety of isolates collected from the field in a number of geographic locations and will allow hypothesis-driven analysis of the data and graphical display of results.

Comparison of M. avium *and MapTb*

Variation among MapTb isolates has been described for surface antigens, virulence, and host range (Sugden et al. 1987, Roach et al. 1993). Sheep and bovine isolates exhibit variation in their lipoarabinomannan that may reflect preferences for a particular host or changes in response to a host. The ability to identify markers for each of these variables is the first step in identifying the genes responsible for infection or persistence in a particular host. We propose the use of AFLP and PFGE (pulsed field gel electrophoresis) to characterize the genetic diversity among MapTb isolates.

Our labs have had extensive experience in the molecular fingerprinting of microorganisms and have employed PFGE, restriction fragment length polymorphism, PCR, and immunomagnetic bead separation techniques in our studies of *M. bovis*, the causative agent of bovine tuberculosis (Perumaalla et al. 1996, Sreevatsan et al. 1996, Perumaalla et al. 1999). In the experiments described below we implemented the use of a relatively new DNA fingerprinting method, AFLP, that provides greater genetic resolution and is less sensitive to DNA quality than PFGE (Zhao et al. 2000). In collaboration with members of the Johne's working group at TAMU and the assistance of the diagnostic bacteriology lab at the Texas Veterinary Medical Diagnostic Lab and their collaborators at the University of Wisconsin at Madison and the University of California at Davis, we have access to isolates that represent a broad geographic distribution. This analysis has the potential to identify useful epidemiological markers.

Twenty-one MapTb clinical isolates were compared with *M. avium* isolates obtained from the American Type Culture Collection (Manassas, VA) and from a human infection. A total of six PstI- and 16 MseI-specific primers were employed for 2,300 AFLP reactions. Although most of the PCR products are shared by these closely related organisms, several bands were produced in most reactions that were unique to either MapTb or *M. avium*. Because the aim of this work was the development of diagnostics capable of positively identifying MapTb-infected animals, the bands specific for MapTb were excised and their DNA sequence was determined (O'Shea et al. 2003). Diagnostic primers were designed for PCR amplification of four independent regions based on the interior sequence of what was ultimately identified as deleted segments. These primers were used with 20 MapTb field strain isolates positive for IS900 and 25 isolates obtained from NAHMS. The 20 MapTb field isolates were relatively homogeneous in their banding patterns. Three of the four amplicons applied were positive for all Wisconsin-derived field isolates. A fourth amplicon divided the samples into two groups displaying four negative amplifications within the 20 samples. A very different result was observed with the 25 isolates obtained from the USDA/NAHMS. The four amplicons delineated a number of genotypes within the National Veterinary Services Laboratory test set. For example, one primer set produced an amplification product from 15 of 17 positive samples, whereas another produced a product from only 11 of 17 samples. These results suggest that there are differences among the isolates that are not identified using classical biotyping methods and that may be used to epidemiologically track outbreaks.

Application of this approach to field isolates received at the Texas Veterinary Medical Diagnostic Lab has already provided interesting results. Two isolates made from one bull suffering from Johne's and exhibiting a strong positive serum enzyme-linked immunosorbent assay for MparaTb exposure were identified by NVSL as fast-growing *M. avium* subsp. *avium*. The isolates were distinguished by their mycobactin J growth requirement. However, both organisms tested positive for the presence of IS900 including restriction digestion. Using the four loci described above and in detail elsewhere (O'Shea et al. 2003), only the mycobactin J–independent isolate had all four loci consistent with identification of MparaTb. The mycobactin J–dependent isolate had only one of the four loci. One interpretation of these results is that *M. avium* complex organisms represent a continuum of genotypes exhibiting a spectrum of pathogenesis. Complete genomic characterization will be necessary to reveal genes required for pathogenesis in specific hosts.

Host Gene Expression in Response to MapTb Infection

Ligated Ileal Loops

The ligated ileal loop assay permits the study of early events in the pathogenesis of disease (Frost et al. 1997). For this non-survival ligated ileal loop procedure, 4–6-week-old male Holstein/Friesian calves are anesthetized and maintained analgesic for the course of the 12-hour experiment. The abdominal wall is opened and the distal ileum exteriorized. Each calf has up to 24 8-cm ileal and eight colonic loops, half of which are injected directly into the lumen with media (negative time matched controls) MapTb, MapTb mutants, or *M. avium* (10^9 CFU per ligated Peyer's patch ileal loop). At 0.5, 1, 4, 8, and 12 hours post-inoculation, four 3-mm skin biopsy punch disks from each loop of Peyer's patch, colonic mucosa, and local drainage mesenteric lymph nodes are collected directly into cold Tri-reagent (Molecular Research Center, Inc., Cincinnati, OH) for mRNA extractions and subsequent microarray analysis, using commercially available bovine chips. Enteropathogenesis is assessed by measuring fluid accumulation and leukocyte infiltration compared with control loops inoculated with sterile growth medium. Tissue samples from each loop are also examined ultrastructurally and histopathologically for the location of organisms and development of lesions.

Presumably because of the slow progression of Johne's disease, we have not observed gross changes over the course of the experiment; however, we have detected the MapTb organisms within macrophages and mononuclear infiltrates into the lamina propria at 1 hour post-infection. We also expect to detect changes in host gene expression, particularly cytokines and chemokines, consistent with invasion. Thus, virulent MapTbs are expected to stimulate a different set of genes than *M. avium*. Because *M. avium* is not known to cause any disease in ruminants, one may expect that the organism is either cleared from this host or that it is engulfed and destroyed by the ruminant immune system. The more critical step lies with the interpretation of the differences between cytokine and chemokine expression stimulated by MapTb and *M. avium* and their potential role in clearance of MapTb. Probably the most interesting potential outcome would be the identification of host genes activated in response to MapTb infection, which could serve diagnostically as cytokine and chemokine expression for infection. Identification of such genes may provide additional insight into Johne's pathogenesis and natural disease resistance.

Initial experiments have focused on detection of differential gene expression in infected (MapTb vs. *M. avium*) versus control (uninfected) cells. Controls for this analysis include primers for amplification of cDNA of glyceraldehyde-3-phosphate dehydrogenase (GAPDH), a gene expressed at constant levels, which is included in each PCR reaction as an internal control for variation in template. Early experiments indicate that MapTb or *M. avium* infection causes differential induction of cytokine expression in calves. We will correlate the ability to elicit cytokine production with their ability to invade intestinal tissue and cause tissue injury and diarrhea.

Host gene expression is evaluated in ileal Peyer's patches and in the mesenteric lymph nodes. We have had extensive experience with these host pathogen systems, mRNA extractions, and differential-display reverse transcription PCR analysis (Gutierrez-Pabello et al. 2002). In addition, the Peyer's patches, colonic loops, and lymph nodes have been cultured quantitatively and examined histopathologically for lesions and compared with samples from uninfected tissues.

Reverse transcription PCR was performed on temporal samples from the Peyer's patch mucosa after intraluminal challenge. The ligated ileal loop assay measures intestinal fluid accumulation and transmigration of leukocytes as well as the histomorphogenesis of the lesions. Early response to MapTb has not been documented, but it is presumed to be essential to the development of disease. Chemokines involved in the recruitment and activation of leukocytes at the site of infection have been examined. During this 12-hour assay, there was no apparent change in fluid accumulation, in contrast to observations with *Salmonella*, and the level of bacteria inoculated into the loops did not change appreciably during this time. Conserved cysteines separated by some other amino acid cytokines are involved in chemoattraction and activation of neutrophils (family of growth-related oncogenes, granulocyte chemotactic proteins), and CC (a chemokine subgroup in which cysteine residues are contiguous) cytokines are involved in the chemoattraction of monocytes, eosinophils, lymphocytes, and basophils (family of monocytes and macrophages chemotactic proteins).

Real-time PCR was used to determine the cytokine expression level. Primers were designed using Primer Express software (Applied Biosystems, Foster City, Calif.). Real-time PCR was run on SDS5700, and the results were expressed as the fold increase in gene expression in infected tissue over non-infected tissue. All results were normalized to the GAPDH target mRNA. There was a 14-fold stimulation of growth-related oncogene (GRO)-α transcription by 4 hours post-inoculation and a 12-fold increase in GRO-γ transcription by 8 hours post-inoculation. Increases in transcription of GCP1 (fourfold) and GCP2 (sixfold) were also observed. Increases in MCP1 (sixfold) and MCP2 (11-fold) transcription reached an optimum at 4 hours post-infection. These were accompanied by a 14-fold increase in IL1β (14-fold) transcription as reported elsewhere (Lee et al. 2001).

A variety of signals may initiate the intestinal response to MapTb infection, including the transient expression of chemokine genes. Expression of GRO may be under the control of interleukin 1 or tumor necrosis factor α. Our results suggest involvement of interleukin 1 on GRO expression. These results support the contention that differential activation of granulocyte and monocyte chemotactic cytokines may play a major role in the pathogenesis of MapTb infection. It will be interesting to compare these results with that obtained using the closely related organism *M. avium*.

Host Genetic Factors

Evaluation of differences between cattle is a potential pitfall of the ileal loop experiments that may complicate interpretation of the host response. We have established techniques to assay for differential gene expression from resistant and susceptible cattle to identify new candidate genes; to quantify kinetic mRNA expression in Peyer's patches stimulated by infection; and to identify cognate families of activating nuclear transcription factors. Clearly there are major genes in the host that control critical points in this complex network of host antimicrobial defense. We propose to define the genetic mechanisms active in host–agent interaction of MapTb-infected bovine ileal Peyer's patches and macrophage. Initial trials will employ the methods described in the previous section (i.e., real-time PCR to characterize expression from target genes). However, long-term experimental goals necessitate the use of more global analysis such as that provided by microarray technology to better analyze the host pathogen interaction.

All cattle employed will be genotyped for resistance versus susceptibility (Nramp1) to intracellular pathogens (Feng et al. 1996). The outcome of this research may also be used to advise producers of the importance of selective breeding and the need for early diagnosis before the appearance of disease. These experiments also include an examination of mutant MapTb constructed via transposon mutagenesis to identify individual gene products essential for disease progression.

Conclusions

Paratuberculosis in ruminants has a prolonged incubation period, and most animals remain subclinical for prolonged periods. Infection is typically acquired via the fecal–oral route. Because of poor-quality diagnostic tests, the rate of progression from the subclinical to clinical state remains unknown, and subclinical animals spread the disease to other animals within the herd. Perhaps more important, concern has arisen regarding the possible survival of MapTb in pasteurized milk and a potential link between MapTb and Crohn's disease or ulcerative colitis. Improved pasteurization and detection systems are under investigation, and this technology will rely on the development of diagnostic reagents that are specific for MapTb. However, knowledge of the genomic structure of this group of organisms is incomplete, and evidence suggests that *M. avium* and MapTB may represent only two forms from a diverse continuum of *M. avium* complex isolates.

Research regarding MapTb infection and disease has also been hampered by a lack of suitable animal models. Recently developed animal models now offer the opportunity to reproduce the symptoms of disease and to examine specific interactions between the host and agent with the ultimate goal of defining interactions that represent potential targets for improved diagnosis and therapeutic treatment (Beard et al. 2001b, Lee et al. 2001, Mutwiri et al. 2001, Mutwiri et al. 2002). Toward this goal we have described a model system offering one way to explore host–agent interactions *in situ* and to establish the biological signature (bio-signature) of MapTb infection.

References

Adams, J. L., Collins, M. T. and Czuprynski, C. J. 1996. Polymerase chain reaction analysis of TNF-alpha and IL-6 mRNA levels in whole blood from cattle naturally or experimentally infected with Mycobacterium paratuberculosis. *Canadian Journal of Veterinary Research* **60**:257–262.

Aduriz, J. J., Juste, R. A. and Cortabarria, N. 1995. Lack of mycobactin dependence of mycobacteria isolated on Middlebrook 7H11 from clinical cases of ovine paratuberculosis. *Veterinary Microbiology* **45**:211–217.

Bannantine, J. P., Baechler, E., Zhang, Q., Li, L. and Kapur, V. 2002. Genome scale comparison of *Mycobacterium avium* subsp. *paratuberculosis* with *Mycobacterium avium* subsp. *avium* reveals potential diagnostic sequences. *Journal of Clinical Microbiology* **40**:1303–1310.

Bannantine, J. P., Zhang, Q., Li, L. L. and Kapur, V. 2003. Genomic homogeneity between *Mycobacterium avium* subsp. a*vium* and *Mycobacterium avium* subsp. *paratuberculosis* belies their divergent growth rates. *BioMed Central Microbiology* **3**:10–17.

Bassey, E. O. and Collins, M. T. 1997. Study of T-lymphocyte subsets of healthy and *Mycobacterium avium* subsp. *paratuberculosis*-infected cattle. *Infection and Immunity* **65**:4869–4872.

Beard, P. M., Daniels, M. J., Henderson, D., Pirie, A., Rudge, K., Buxton, D., Rhind, S., Greig, A., Hutchings, M. R., McKendrick, I., Stevenson, K. and Sharp, J. M. 2001a. *Paratuberculosis* infection of nonruminant wildlife in Scotland. *Journal of Clinical Microbiology* **39**:1517–1521.

Beard, P. M., Stevenson, K., Pirie, A., Rudge, K., Buxton, D., Rhind, S. M., Sinclair, M. C., Wildblood, L. A., Jones, D. G. and Sharp, J. M. 2001b. Experimental *paratuberculosis* in calves following inoculation with a rabbit isolate of *Mycobacterium avium* subsp. *paratuberculosis*. *Journal of Clinical Microbiology* **39**:3080–3084.

Begara-McGorum, I., Wildblood, L. A., Clarke, C. J., Connor, K. M., Stevenson, K., McInnes, C. J., Sharp, J. M. and Jones, D. G. 1998. Early immunopathological events in experimental ovine *paratuberculosis*. *Veterinary Immunology and Immunopathology* **63**:265–287.

Buergelt, C. D. and Ginn, P. E. 2000. The histopathologic diagnosis of subclinical Johne's disease in North American bison (Bison bison). *Veterinary Microbiology* **77**:325–331.

Cavallini, A., Notarnicola, M., Berloco, P., Lippolis, A. and De Leo, A. 2000. Use of macroporous polypropylene filter to allow identification of bacteria by PCR in human fecal samples. *Journal of Microbiological Methods* **39**:265–270.

Chiodini, R. J. and Hermon-Taylor, J. 1993. The thermal resistance of *Mycobacterium paratuberculosis* in raw milk under conditions simulating pasteurization. *Journal of Veterinary Diagnostic Investigation* **5**:629–631.

Clarke, C. J. and Little, D. 1996. The pathology of ovine *paratuberculosis*: gross and histological changes in the intestine and other tissues. *Journal of Comparative Pathology* **114**:419–437.

Collins, M. T., Lisby, G., Moser, C., Chicks, D., Christensen, S., Reichelderfer, M., Hoiby, N., Harms, B. A., Thomsen, O. O., Skibsted, U. and Binder, V. 2000. Results of multiple diagnostic tests for *Mycobacterium avium* subsp. *paratuberculosis* in patients with inflammatory bowel disease and in controls. *Journal of Clinical Microbiology* **38**:4373–4381.

Corpa, J. M., Garrido, J., Garcia Marin, J. F. and Perez, V. 2000. Classification of lesions observed in natural cases of *paratuberculosis* in goats. *Journal of Comparative Pathology* **122**:255–265.

de Boer, P., Duim, B., Rigter, A., van Der Plas, J., Jacobs-Reitsma, W. F. and Wagenaar, J. A. 2000. Computer-assisted analysis and epidemiological value of genotyping methods for *Campylobacter jejuni* and *Campylobacter coli*. *Journal of Clinical Microbiology* **38**:1940–1946.

Eamens, G. J., Whittington, R. J., Marsh, I. B., Turner, M. J., Saunders, V., Kemsley, P. D. and Rayward, D. 2000. Comparative sensitivity of various faecal culture methods and ELISA in dairy cattle herds with endemic Johne's disease. *Veterinary Microbiology* **77**:357–367.

Englund, S., Bolske, G. and Johansson, K. E. 2002. An IS900-like sequence found in a Mycobacterium sp. other than *Mycobacterium avium* subsp. *paratuberculosis*. *FEMS Microbiology Letters* **209**:267–271.

Feng, J., Li, Y., Hashad, M., Schurr, E., Gros, P., Adams, L. G. and Templeton, J. W. 1996. Bovine natural resistance associated macrophage protein 1 (Nramp1) gene. *Genome Research* **6**:956–964.

Frost, A. J., Bland, A. P. and Wallis, T. S. 1997. The early dynamic response of the calf ileal epithelium to *Salmonella typhimurium*. *Veterinary Pathology* **34**:369–386.

Grimes, D. S. 2003. *Mycobacterium avium* subspecies *paratuberculosis* as a cause of Crohn's disease. *Gut* **52**:155.

Gutierrez-Pabello, J. A., McMurray, D. N. and Adams, L. G. 2002. Upregulation of thymosin beta-10 by *Mycobacterium bovis* infection of bovine macrophages is associated with apoptosis. *Infection and Immunity* **70**:2121–2127.

Gwozdz, J. M. and Thompson, K. G. 2002. Antigen-induced production of interferon-gamma in samples of peripheral lymph nodes from sheep experimentally inoculated with *Mycobacterium avium* subsp. *paratuberculosis*. *Veterinary Microbiology* **84**:243–252.

Gwozdz, J. M., Thompson, K. G., Murray, A., Reichel, M. P., Manktelow, B. W. and West, D. M. 2000. Comparison of three serological tests and an interferon-gamma assay for the diagnosis of *paratuberculosis* in experimentally infected sheep. *Austrian Veterinary Journal* **78**:779–783.

Harris, N. B. and Barletta, R. G. 2001. *Mycobacterium avium* subsp. *paratuberculosis* in Veterinary Medicine. *Clinical Microbiological Reviews* **14**:489–512.

Jiang, S. C., Louis, V., Choopun, N., Sharma, A., Huq, A. and Colwell, R. R. 2000a. Genetic diversity of *Vibrio cholerae* in Chesapeake Bay determined by amplified fragment length polymorphism fingerprinting. *Applied Environmental Microbiology* **66**:140–147.

Jiang, S. C., Matte, M., Matte, G., Huq, A. and Colwell, R. R. 2000b. Genetic diversity of clinical and environmental isolates of *Vibrio cholerae* determined by amplified fragment length polymorphism fingerprinting. *Applied Environmental Microbiology* **66**:148–153.

Johne, H. A. and Frothigham, L. 1895. Ein eigenthumlicher Fall von Tuberkulose beim Rind. *Deutsche Zeitschrift fuer Thiermedizin und vergleichende Pathologie* **21**:438–454.

Kanazawa, K., Haga, Y., Funakoshi, O., Nakajima, H., Munakata, A. and Yoshida, Y. 1999. Absence of *Mycobacterium paratuberculosis* DNA in intestinal tissues from Crohn's disease by nested polymerase chain reaction. *Journal of Gastroenterology* **34**:200–206.

Khare, S., Ficht, T. A., Santos, R. L., Romano, J., Rice-Ficht, A. C., Zhang, S., Grant, I. R., Libal, M., Hunter, D. and Adams, L. G. 2003. Rapid and sensitive detection of *Mycobacterium avium* subsp. *paratuberculosis* in bovine milk and feces samples by an immunomagnetic bead-conventional and real-time PCR. *Journal of Clinical Microbiology* (in press).

Lambrecht, R. S. and Collins, M. T. 1992. *Mycobacterium paratuberculosis*. Factors that influence mycobactin dependence. *Diagnostic Microbiology and Infectious Disease*. **15**:239–246.

Lee, H., Stabel, J. R. and Kehrli Jr., M. E. 2001. Cytokine gene expression in ileal tissues of cattle infected with *Mycobacterium paratuberculosis*. *Veterinary Immunology and Immunopathology* **82**:73–85.

Lindstedt, B. A., Heir, E., Vardund, T. and Kapperud, G. 2000. Fluorescent amplified-fragment length polymorphism genotyping of *Salmonella enterica* subsp. *enterica* serovars and comparison with pulsed-field gel electrophoresis typing. *Journal of Clinical Microbiology* **38**:1623–1627.

McDonald, W. L., Ridge, S. E., Hope, A. F. and Condron, R. J. 1999. Evaluation of diagnostic tests for Johne's disease in young cattle. *Australian Veterinary Journal* **77**:113–119.

Motiwala, A. S., Strother, M., Amonsin, A., Byrum, B., Naser, S. A., Stabel, J. R., Shulaw, W. P., Bannantine, J. P., Kapur, V. and Sreevatsan, S. 2003. Molecular epidemiology of *Mycobacterium avium* subsp. *paratuberculosis*: evidence for limited strain diversity, strain sharing, and identification of unique targets for diagnosis. *Journal of Clinical Microbiology* **41**:2015–2026.

Mutwiri, G. K., Kosecka, U., Benjamin, M., Rosendal, S., Perdue, M. and Butler, D. G. 2001. *Mycobacterium avium* subspecies *paratuberculosis* triggers intestinal pathophysiologic changes in beige/SCID mice. *Comparative Medicine* **51**:538–544.

Mutwiri, G. K., Rosendal, S., Kosecka, U., Yager, J. A., Perdue, M., Snider, D. and Butler, D. G. 2002. Adoptive transfer of BALb/c mouse splenocytes reduces lesion severity and induces intestinal pathophysiologic changes in the *Mycobacterium avium* Subspecies *paratuberculosis* beige/scid mouse model. *Comparative Medicine* **52**:332–341.

Naser, S., Shafran, I. and El-Zaatari, F. 1999. *Mycobacterium avium* subsp. *paratuberculosis* in Crohn's disease is serologically positive [letter]. *Clinical Diagnostic Laboratory Immunology* **6**:282.

Navarro, J. A., Ramis, G., Seva, J., Pallares, F. J. and Sanchez, J. 1998. Changes in lymphocyte subsets in the intestine and mesenteric lymph nodes in caprine *paratuberculosis*. *Journal of Comparative Pathology* **118**:109–121.

O'Shea, B., Khare, S., Bliss, K., Klein, P., Ficht, T. A., Adams, L. G. and Rice-Ficht, A. C. 2003. Genotyping *Mycobacterium avium* subsp. *paratuberculosis* using amplified fragment length polymorphism in the development of diagnostic reagents. *Journal of Clinical Microbiology* (Submitted).

Ott, S. L., Wells, S. J. and Wagner, B. A. 1999. Herd-level economic losses associated with Johne's disease on US dairy operations. *Preventive Veterinary Medicine* **40**:179-192.

Pavlik, I., Matlova, L., Bartl, J., Svastova, P., Dvorska, L. and Whitlock, R. 2000. Parallel faecal and organ *Mycobacterium avium* subsp. *paratuberculosis* culture of different productivity types of cattle. *Veterinary Microbiology* **77**:309–324.

Perumaalla, V. S., Adams, L. G., Payeur, J., Baca, D. and Ficht, T. A. 1999. Molecular fingerprinting confirms extensive cow-to-cow intra-herd transmission of a single *Mycobacterium bovis* strain. *Veterinary Microbiology* **70**:269–276.

Perumaalla, V. S., Adams, L. G., Payeur, J. B., Jarnagin, J. L., Baca, D. R., Suarez Guemes, F. and Ficht, T. A. 1996. Molecular epidemiology of *Mycobacterium bovis* in Texas and Mexico. *Journal of Clinical Microbiology* **34**:2066–2071.

Prantera, C. and Scribano, M. L. 1999. Crohn's disease: the case for bacteria. *Italian Journal of Gastroenterology & Hepatology* **31**:244–246.

Richter, E., Wessling, J., Lugering, N., Domschke, W. and Rusch-Gerdes, S. 2002. *Mycobacterium avium* subsp. *paratuberculosis* infection in a patient with HIV, Germany. *Emerging Infectious Diseases* **8**:729–731.

Roach, T. I., Barton, C. H., Chatterjee, D. and Blackwell, J. M. 1993. Macrophage activation: lipoarabinomannan from avirulent and virulent strains of *Mycobacterium tuberculosis* differentially induces the early genes c-fos, KC, JE, and tumor necrosis factor-alpha. *Journal of Immunology* **150**:1886–1896.

Siebert, P. D., Chenchik, A., Kellogg, D. E., Lukyanov, K. A. and Lukyanov, S. A. 1995. An improved PCR method for walking in uncloned genomic DNA. *Nucleic Acids Research* **23**:1087–1088.

Sreevatsan, S., Escalante, P., Pan, X., Gillies 2nd, D. A., Siddiqui, S., Khalaf, C. N., Kreiswirth, B. N., Bifani, P., Adams, L. G., Ficht, T., Perumaalla, V. S., Cave, M. D., van Embden, J. D. and Musser, J. M. 1996. Identification of a polymorphic nucleotide in oxyR specific for *Mycobacterium bovis*. *Journal of Clinical Microbiology* **34**:2007–2010.

Stabel, J. R. 1996. Production of gamma-interferon by peripheral blood mononuclear cells: an important diagnostic tool for detection of subclinical *paratuberculosis*. *Journal of Veterinary Diagnostic Investigation* **8**:345–350.

Stabel, J. R. 2000. Cytokine secretion by peripheral blood mononuclear cells from cows infected with *Mycobacterium paratuberculosis*. *American Journal of Veterinary Research* **61**:754–760.

Stabel, J. R. and Whitlock, R. H. 2001. An evaluation of a modified interferon-gamma assay for the detection of *paratuberculosis* in dairy herds. *Veterinary Immunology and Immunopathology* **79**:69–81.

Suenaga, K., Yokoyama, Y., Nishimori, I., Sano, S., Morita, M., Okazaki, K. and Onishi, S. 1999. Serum antibodies to *Mycobacterium paratuberculosis* in patients with Crohn's disease. *Digestive Disease Science* **44**:1202–1207.

Sugden, E. A., Samagh, B. S., Bundle, D. R. and Duncan, J. R. 1987. Lipoarabinomannan and lipid-free arabino-mannan antigens of *Mycobacterium paratuberculosis*. *Infection and Immunity* **55**:762–770.

Sung, N. and Collins, M. T. 1998. Thermal tolerance of *Mycobacterium paratuberculosis*. *Applied and Environmental Microbiology* **64**:999–1005.

Sweeney, R. W., Jones, D. E., Habecker, P. and Scott, P. 1998. Interferon-gamma and interleukin 4 gene expression in cows infected with *Mycobacterium paratuberculosis*. *American Journal of Veterinary Research* **59**:842–847.

Thornton, C. G., MacLellan, K. M., Stabel, J. R., Carothers, C., Whitlock, R. H. and Passen, S. 2002. Application of the C(18)-carboxypropylbetaine specimen processing method to recovery of *Mycobacterium avium* subsp. *paratuberculosis* from ruminant tissue specimens. *Journal of Clinical Microbiology* **40**:1783–1790.

Valentin-Weigand, P. and Goethe, R. 1999. Pathogenesis of *Mycobacterium avium* subspecies *paratuberculosis* infections in ruminants: still more questions than answers. *Microbes and Infection* **1**:1121–1127.

Vos, P., Hogers, R., Bleeker, M., Reijans, M., van de Lee, T., Hornes, M., Frijters, A., Pot, J., Peleman, J., Kuiper, M. and Zambeau, M. 1995. AFLP: a new technique for DNA fingerprinting. *Nucleic Acids Research* **23**:4407–4414.

Walravens, K., Marche, S., Rosseels, V., Wellemans, V., Boelaert, F., Huygen, K. and Godfroid, J. 2002. IFN-gamma diagnostic tests in the context of bovine mycobacterial infections in Belgium. *Veterinary Immunology and Immunopathology* **87**:401–406.

Williams, E. S., DeMartini, J. C. and Snyder, S. P. 1985. Lymphocyte blastogenesis, complement fixation, and fecal culture as diagnostic tests for *paratuberculosis* in North American wild ruminants and domestic sheep. *American Journal of Veterinary Research* **46**:2317–2321.

Zhao, B., Collins, M. T. and Czuprynski, C. J. 1997. Effects of gamma interferon and nitric oxide on the interaction of *Mycobacterium avium* subsp. *paratuberculosis* with bovine monocytes. *Infection and Immunity* **65**:1761–1766.

Zhao, S., Mitchell, S. E., Meng, J., Kresovich, S., Doyle, M. P., Dean, R. E., Casa, A. M. and Weller, J. W. 2000. Genomic typing of *Escherichia coli* O157:H7 by semi-automated fluorescent AFLP analysis. *Microbes and Infection* **2**:107–113.

8 Viruses in Food

Sagar M. Goyal

Introduction

Human enteric viruses are shed in the feces of infected individuals in large numbers and are transmitted mainly by the fecal–oral route or by person-to-person contact. Fecal contamination of food or water may also result in outbreaks of viral diseases. Essentially any virus that is excreted in the feces has the potential to cause foodborne infections. The viruses that can be transmitted via food include rotavirus, enteric adenoviruses, noroviruses (human caliciviruses known previously as Norwalk and Norwalk-like viruses), hepatitis virus type A (HAV), hepatitis virus type E (HEV), astrovirus, and small round structured viruses (SRSVs; Bosch et al. 2001). As opposed to bacteria, viruses do not multiply in foods but are capable of surviving for long periods of time in food and on food contact surfaces, which may serve as vehicles of virus transmission to consumers.

Although a number of foodborne viral disease outbreaks are reported each year, the problem is much bigger because all foodborne outbreaks are not reported. For example, sporadic cases of foodborne illness are sometimes treated as common stomach ailments and go unreported for want of specific diagnostic tests. Often it is difficult to prove that a particular outbreak was caused by viruses because of the paucity of virus detection methods and because of the inadequacy of epidemiological procedures. If the cause of a particular foodborne outbreak is not determined, the outbreak is labeled as "acute gastrointestinal illness" (AGI) of unknown etiology. Of the 2,423 reported foodborne disease outbreaks in the United States during 1988–1992, the causative agent was not determined in 1,422 (59%) outbreaks. It is widely believed that a majority of the AGI outbreaks are actually caused by viruses (Hedberg et al. 1994, Bean et al. 1996, Glass et al. 2000, Parashar and Monroe 2001).

Mead et al. (1999) compiled and analyzed data from multiple surveillance systems and concluded that AGI of unknown etiology accounted for more than 81% of illnesses and hospitalizations and 64% of deaths. Of the foodborne outbreaks for which an agent was known, noroviruses accounted for more than 66% of cases, 33% of hospitalizations, and 7% of deaths. The results of this study demonstrated that foodborne diseases cause more illnesses but fewer deaths than previously reported.

Although proper handling and hygienic preparation of foods has resulted in the elimination of some diseases, new causes of foodborne illnesses have also been identified. During the last three decades, more than 20 new bacterial, parasitic, and viral agents have been recognized as etiological agents of gastroenteritis (Glass et al. 2000). Epidemiologists, clinical investigators, and public health officials were called on to classify and characterize new patterns of disease outbreaks and to develop rapid and sensitive methods for the detection of these new agents (Glass et al. 2000).

The purpose of this chapter is not to comprehensively review all the information available on the occurrence of viruses in food, but to point out gaps in our knowledge and to make some recommendations for new research that can be helpful in prevention and control of foodborne disease outbreaks.

Pathogens

As mentioned above, many different viruses including HAV, HEV, noroviruses (NVs), rotaviruses, adenoviruses, parvoviruses and astroviruses may be transmitted via foods (Appleton 2000, Bosch et al. 2001). Of these, NVs are estimated to be the most common causes of foodborne diseases in the United States, accounting for two-thirds of all food-related illnesses (Bresee et al. 2002). In fact, NVs have been cited as the most frequent cause of viral gastroenteritis in humans throughout the world (Monroe et al. 2000). Certain animal enteric viruses may also cause human infections. Such viruses are known as zoonotic virus-es and may include rotavirus, NV, and HEV (Enriquez et al. 2001). The most common set-tings for NV-related outbreaks include nursing homes, hospitals, restaurants, summer camps, conferences, cruise ships, and day care centers (Frankhauser et al. 1998, Frankhauser et al. 2002). Noroviruses can cause repeated infections because they are anti-genically diverse and provide short-lived immunity to infection (Glass et al. 2001). Infec-tions with NV are reported throughout the year, but most outbreaks are seen in winter months (Frankhauser et al. 1998, Mounts et al. 2000).

Foods Involved

A variety of food items and modes of transmission have been implicated in foodborne out-breaks. Outbreaks have commonly been associated with foods that are served raw or only lightly cooked, such as molluscan shellfish, fresh produce (fruits, vegetables, and salads), and ice, or products contaminated after cooking, such as frosted bakery products (Frankhauser et al. 1998, Richards 2001).

Produce

Fresh produce has been implicated in several foodborne viral disease outbreaks. The source of contamination for fresh produce may either be sewage-contaminated irrigation water or an infected food handler. The potential for water to serve as the vehicle of viral contamina-tion of fresh produce is an emerging concern for the following reasons: first, produce is grown and harvested in many areas of the world that lack adequate waste disposal and water treatment facilities; second, untreated surface water is often used to irrigate produce; third, current water treatment may be incapable of removing viruses; fourth, enteric viruses can survive on produce for long periods of time; and fifth, no virological criteria exist for water used in produce production (Ooi et al. 1997, Parashar et al. 1998, Deneen et al. 2000, Glass et al. 2001, Grabow et al. 2001, Croci et al. 2002, Solomon et al. 2002).

Contaminated food-contact surfaces may also be an important source of viruses for food. To reduce the risks of cardiac diseases and cancers, the public has increased their consump-tion of fresh fruits and produce, which has resulted in increased international trade in these commodities (Hedberg 2000). Most of the fresh fruits and vegetables are ready-to-eat, field-grown commodities, with an inherent risk of contamination by irrigation water. As a consequence, the incidence of outbreaks of foodborne diseases associated with fresh fruits and produce has increased in the United States. The mean number of produce-associated outbreaks tripled from 4.0 per year in 1973–1982 to 11.8 per year in 1993–1997. From 1973 to 1997, 164 foodborne outbreaks caused by fresh produce were reported to Centers for Disease Control and Prevention (Tauxe et al. 1997). In addition to these 164 outbreaks, consumption of salads resulted in an additional 202 outbreaks, two-thirds of which were caused by unknown agents, most probably by NVs (Hedberg 2000).

In a gastroenteritis outbreak aboard a cruise ship, consumption of fresh-cut fruit was found to be significantly associated with illness, and a significant dose-dependent relationship was evident between illness and number of various fresh fruit items eaten (Herwaldt et al. 1994). Celery, a component of chicken salad, was responsible for a large outbreak of NV affecting 48% of 3,000 cadets at the U.S. Air Force Academy. It was discovered that the celery had been exposed to nonpotable water and had acted as a vehicle for transmission (Warner et al. 1991). Imported, frozen raspberries were responsible for an outbreak of gastroenteritis in Helsinki. In a retrospective cohort study, it was found that employees of the company who had eaten raspberry dressing had an attack rate of 65% compared with 15% for those who had not. Calicivirus was detected in stool samples of four affected employees by reverse-transcriptase polymerase chain reaction (RT-PCR; Ponka et al. 1999). A common trend that can be followed from these episodes is that the implicated foods are cold, have come in contact with water, or are handled by sick food handlers.

Seafood

Seafood is an important source for human nutrition. However, several human pathogens can be transmitted by seafood, especially shellfish. Estuarine and costal marine waters are subject to human fecal contamination because of sewage discharges, and bivalve mollusks, such as clams, cockles, mussels, and oysters, may bioaccumulate viruses from surrounding polluted waters during the course of their filter-feeding activity and subsequently transmit these to humans (Gerba and Goyal 1978). Although human viruses do not multiply in shellfish, they are harbored for days or weeks in the digestive tract of shellfish and are apparently more difficult to remove than bacteria during depuration (Grohmann et al. 1981), a process by which shellfish cleanse themselves of microbial contamination when placed in clean waters. Unlike other types of seafood, shellfish are usually eaten raw or lightly cooked, along with their digestive tracts, thus increasing the risk of viral transmission (Sobsey et al. 1978). In rare instances, even cooking may not be able to prevent gastroenteritis outbreaks. For example, in one particular outbreak, the attack rate among persons who had eaten steamed oysters increased with the number of oysters eaten (Kirkland et al. 1996).

Several shellfish-associated outbreaks of viral diseases have been documented, and the presence of human enteric viruses has been demonstrated in shellfish harvested from both clean and polluted waters (Goyal et al. 1979). Recently, Formiga-Cruz et al. (2002) detected human adenovirus, Norwalk-like virus, and enteroviruses in shellfish from Greece, Spain, Sweden, and the United Kingdom. Similarly, Simmons et al. (2001) detected NV in two batches of commercially farmed Pacific oysters from different growing areas that were implicated in four different outbreaks.

From 1991 to 1998, more than 2,100 shellfish-related illnesses have occurred in the United States, of which 1,266 were attributed to enteric viruses, predominantly NVs. Almost 78% of the illnesses caused by NVs were associated with oysters harvested during the months of November to January from the Gulf Coast in the 1990s (Burkhardt and Calci 2000). Oyster-related outbreaks can be widespread, may affect a large number of people over many states, and may pose a great challenge for trace-back investigators (Berg et al. 2000). In one such outbreak, 60 clusters (comprising of 493 persons from Alabama, Florida, Georgia, Louisiana, and Mississippi) of gastroenteritis caused by SRSVs were reported following the consumption of raw oysters. In all clusters, ill persons had eaten oysters harvested from Louisiana waterways. On inspection of eight oyster-harvesting boats, it was found that seven boats had inadequate sewage-collection facilities and indulged in the practice of throwing garbage overboard, which may have contributed to contamination of the oyster harvesting areas (Berg et al. 2000). A similar outbreak occurred in southeast Queensland and northern New South Wales over a 4-week period (Stafford et al. 1997). Ninety-two of the 97 cases were confirmed as having consumed raw oysters within 3 days before developing the illness. No other food items or beverages were significantly associated with the illness.

During 1982, outbreaks of gastroenteritis associated with eating raw shellfish reached epidemic proportions in New York State. From May to December, 103 well-documented outbreaks occurred in which 1,017 persons became ill. Of these, 813 cases were related to eating clams and 204 to eating oysters. Norwalk virus was implicated as the predominant etiologic agent on the basis of clinical features, seroconversion, and presence of anti-NV IgM antibody in the sera of affected persons in some of the outbreaks. In addition, NV virus was identified by radioimmunoassay in clam and oyster specimens from two of the outbreaks. The magnitude, persistence, and widespread nature of these outbreaks raised questions about the safety of consuming raw shellfish (Morse et al. 1986).

Other Foods

Foods other than shellfish and fresh produce have also been implicated in foodborne viral disease outbreaks. In one outbreak, 35 members of a wedding party developed gastroenteritis following consumption of pasta and spring rolls. Descriptive epidemiology, based on interviews of affected guests, suggested that the outbreak was caused by SRSVs (Hicks et al. 1996), which were indeed identified subsequently by RT-PCR. Cream cakes were the source of infection in an outbreak of acute gastroenteritis in a small Norwegian community in which 250 persons were affected (Andersen et al. 1996). Earlier in 1984, two outbreaks of gastrointestinal illness occurred in two parties attending banquets on consecutive evenings at a large hotel. SRSVs resembling NV were demonstrated by electron microscopy (EM) in the stools of affected persons. Food-history analysis showed that the illness was significantly associated with eating cooked ham (Riordan et al. 1984).

Sources

Foods may be contaminated preharvest if they come into contact with sewage-polluted irrigation water. They may also be contaminated postharvest when handled by infected food handlers or by contact with contaminated surfaces. Produce may be contaminated by improper irrigation or fertilization practices, by the hands of infected pickers or processors, or as the result of adulteration during any stage of handling (Richards 2001).

Infected Food Handlers

Ill or asymptomatic food handlers can very easily cause contamination of food and water. A high proportion of gastroenteritis outbreaks has been associated with the consumption of food items handled by ill food workers. Of the 120 confirmed foodborne outbreaks of viral gastroenteritis, 53 were associated with consumption of cold food items handled by ill food workers. In an additional 21 (18%) outbreaks, food handlers denied being ill but confirmed illness among their family members, suggesting person-to-person transmission (Deneen et al. 2000).

In a restaurant-related outbreak of NV in Alaska, 191 of 343 persons attending a luncheon became sick. Eating potato salad was associated strongly with illness. The potato salad was prepared 2 days before the luncheon by two food handlers, one of whom was ill. The ill food handler used bare hands to mix the ingredients in a 12-gallon plastic container. When tested by RT-PCR, the virus from fecal samples of one luncheon attendee, one restaurant patron, and the implicated food handler had identical nucleotide sequences (Monroe et al. 2000).

The food handler does not necessarily have to be sick to transmit viral infections. Even in the presymptomatic phase of infection, the potential exists for contamination. In one outbreak, the implicated food handler prepared salad and appetizers 30 minutes before showing symptoms of disease (Gaulin et al. 1999). Other preinfectious, asymptomatic episodes of NV gastroenteritis have been documented (Hedberg and Osterholm 1993). Persons recovered from a disease episode can also be responsible for food contamination (Patterson et al. 1993, Parashar et al. 1998). During an outbreak of gastroenteritis among employees of a manufacturing company, an association was found between disease and eating sandwiches prepared by food handlers, one of whom had recovered from gastroenteritis days earlier. Norwalk-like viruses with a similar genetic sequence were detected in stool specimens of several employees and the sick food handler 10 days after the resolution of the illness. This indicates that virus can be shed for up to 10 days after illness while the infected person is free from symptoms.

In an outbreak affecting students who had eaten at a cafeteria, NV RNA was detected in a deli sandwich. A food handler who prepared the sandwiches for lunch reported that her infant had been sick with diarrhea just before the outbreak. A stool sample from the infant was positive for NV by RT-PCR, and the sequence of the amplified product was identical to that amplified from deli ham and from the affected student's stool (Daniels et al. 2000).

Person-to-person transmission of NV in an unusual manner was recently documented at a football game. Members of a team who had eaten box lunches developed vomiting and diarrhea, and on the next day some members of the opposing team developed similar symptoms without sharing any food or drinks. Norwalk-like virus of genogroup G I was detected in stool specimens from both teams by RT-PCR (Becker et al. 2000), strongly suggesting transmission through personal contact.

Water and Ice

Waterborne disease outbreaks are often reported separately from foodborne disease outbreaks, but almost all food items come in contact with water at some point. If the water is contaminated, it may contaminate foods that come in contact with it. In August 1994, 30 of 135 bakery plant employees and over 100 people from South Wales and Bristol in the United Kingdom were affected by an outbreak of gastroenteritis caused by SRSVs. Epidemiological studies of employees and three community clusters that used a dried custard mix may have inadvertently reconstituted the mix with contaminated water (Brugha et al. 1999). Ice has also been shown to act as a vehicle for NVs. In 1987, NV outbreaks occurred among persons who had attended a museum fundraiser in Wilmington, Delaware, and an intercollegiate football game in Philadelphia. Persons who consumed ice were 12 times more likely to experience vomiting or diarrhea than those who did not. Ice consumed at these events was traced to a manufacturer in southeastern Pennsylvania whose wells became contaminated when flooded by a nearby creek after a torrential rainfall (Cannon et al. 1991).

Surfaces

Environmental surfaces and other fomites may also act as vehicles of infection. In 1994, an outbreak of SRSV occurred in a 28-bed long-stay ward for the mentally infirm. The outbreak lasted for 17 days. SRSV was detected by RT-PCR in feces, throat swabs, and vomitus of patients. Viral RNA was also detected from 11 of 36 environmental swabs from lockers, curtains, and commodes located in the immediate environment of symptomatic patients (Green et al. 1998).

Methods

Although methods to detect viruses in clinical specimens have improved significantly over the last 10 years, their application to detect viruses in food and environmental samples has progressed more slowly (Sair et al. 2002). Based on methods for the concentration and detection of enteric viruses from large volumes of water and shellfish (Gerba et al. 1978, Sobsey et al. 1978, Gerba and Goyal 1982, Goyal and Gerba 1983), a few studies have been undertaken to develop simple methods for virological examination of foods other than shellfish (Leggitt and Jaykus 2000). Because of low-level contamination of food with viruses, it is important to have methods available that can detect small numbers of viruses in large quantities of food. In most methods, viruses are first eluted from food in 5–7 volumes of a suitable buffer at an alkaline pH. The volume of the eluate is then reduced by organic flocculation or polyethylene glycol (PEG) precipitation. The next step consists of virus detection in the processed concentrate by virus isolation or by molecular methods such as PCR or RT-PCR.

Dix and Jaykus (1998) modified an adsorption-elution-precipitation method to extract and concentrate human enteric viruses from hard-shell clams. Fifty-gram samples of clams were seeded with poliovirus type 1 (PV1) or HAV. Seeded viruses were purified by fluorocarbon (Freon; DuPont, Wilmington, DE) extraction and concentrated by PEG precipitation. Virus recovery ranged from 45% to 95% and was found to be dependent on PEG concentration and volume of the elution buffer. To further concentrate viruses, remove inhibitors, and reduce sample volumes, the protein-precipitating agent Pro-Cipitate (CPG, Lincoln Park, N.J.) was used in an adsorption-elution-precipitation scheme. RT-PCR and subsequent oligoprobe hybridization detected as low as 450 RT-PCR amplifiable units of Norwalk virus in 50-g samples of clams (Dix and Jaykus 1998).

An elution-concentration approach was used to extract and detect human enteric viruses from lettuce and hamburger (Leggitt and Jaykus 2000). Samples of lettuce or hamburger were artificially contaminated with PV1, HAV, or the NV virus and processed by sequential steps of homogenization, filtration, Freon extraction, and PEG precipitation. To reduce sample volume and remove RT-PCR inhibitors, a secondary PEG precipitation step was included. Using this method, 50-g samples of lettuce and hamburger were concentrated to final volumes of 3–5 mL, with virus recoveries ranging from 10% to 70% for PV1 and from 2% to 4% for HAV. Total RNA from PEG concentrates was extracted to a small volume (30–40 μL) and subjected to RT-PCR amplification. In studies with NV, 1.5×10^3 PCR-amplifiable units per 50 g of sample were detectable and PV1 and HAV were consistently detected at initial inoculum levels of $\geq 10^2$ plaque-forming units (PFU) and $\geq 10^3$ PFU per 50 g sample of food, respectively (Leggitt and Jaykus 2000).

Because of the importance of NVs in food, recent studies have focused on the development of methods for the detection of caliciviruses in food. Attempts to develop methods using NV have not been successful because human caliciviruses cannot be propagated *in vitro*. Hence, feline calicivirus (FCV), which grows well in cell culture, producing distinct cytopathic effects, has been used as a surrogate of human caliciviruses in methods development and survival studies (Slomka and Appleton 1998, Doultree et al. 1999, Gulati et al. 2001, Taku et al. 2002, Bidawid et al. 2003). We have recently developed methods for the concentration and detection of FCV from stainless steel bench tops, cutting boards, and sinks. Recovery of FCV from artificially contaminated surfaces ranged from 60% to 100% (Gulati et al. 2001, Taku et al. 2002).

Dubois et al. (2002) modified an elution-concentration method to detect PV1, HAV, and NV from fresh and frozen berries and from fresh vegetables. The method consisted of virus elution from fruit or vegetable surfaces by rinsing them with a buffer containing 100 mM Tris-HCl, 50 mM glycine, 50 mM $MgCl_2$, and 3% beef extract (pH 9.5) followed by volume reduction by PEG precipitation. Viruses from 100-g portions of produce were recovered in a final volume of 3–5 mL. The percentage of viral RNA recovery was estimated by RT-PCR to be 13%, 17%, and 45%–100% for NV, HAV, and PV1, respectively.

An alternate method using zirconium hydroxide was described by D'Souza and Jaykus (2002) to concentrate human enteric viruses from food. The recovery of PV1 ranged from 16% to 59%, with minimal loss to the supernatant (1%–5%). For both HAV and NV, RT-PCR amplicons of appropriate sizes were detected and confirmed in the pellet fraction with no visible amplicons from the supernatant. This rapid and inexpensive method was believed to hold promise as an alternative means to concentrate enteric viruses.

Schwab et al. (2000) developed a method to detect NV and HAV in food samples. The method involved washing of food samples with a guanidinium-phenol-based reagent, extraction with chloroform, and precipitation with isopropanol. Recovered viral RNA was amplified by RT-PCR, using a viral RNA internal standard control to identify potential sample inhibition. By this method, 10–100 PCR units (estimated to be equivalent to from 10^2 to 10^3 viral genome copies) of HAV and NV seeded onto ham, turkey, and roast beef were detected. When this method was used in an actual NV-associated outbreak at a university cafeteria, genogroup II NV was detected in sliced deli ham. Sequence analysis of a PCR-amplified genome indicated that the sequence was identical to that detected in the stools of ill students, thus proving the efficacy of the procedure under field conditions.

Virus Detection in Sample Concentrates

The classical method for the detection of viruses in a sample is by isolating the suspect virus in cell cultures or embryonating chicken eggs. However, most of the foodborne viruses (e.g., HAV, HEV, NV, and SRSV) cannot be propagated *in vitro* (Richards 1999). The presence of such viruses in foods and environmental samples, therefore, is detected by molecular biological tools (e.g., PCR and RT-PCR). Although molecular techniques have been very useful in clinical and environmental virology, their limitations should be recognized. These limitations may include the failure of molecular techniques to discriminate between viable and inactivated viruses, lack of test sensitivity and specificity, high assay costs, potential presence of RT-PCR inhibitors in sample concentrates, and the need for a sophisticated laboratory (Richards 1999).

In spite of these limitations, the use of molecular methods has permitted outbreak strains to be traced back to their common source and has led to the identification of viruses in implicated vehicles (e.g., water, shellfish, and foods) contaminated both at their source and by food handlers (De Leon et al. 1992, Ando et al. 1995, Lees et al. 1995, Le Guyader et al. 1996, Glass et al. 2000, Le Guyader et al. 2000). Using nested PCR and DNA hybridization, Schvoerer et al. (2000) studied environmental samples from bathing areas and sewage treatment plants in southwestern France and found five of 26 bathing water samples to be positive for viruses although no bacterial contamination was detected in these samples.

A multiplex RT-PCR method was developed for the simultaneous detection of the human enteric viruses. Three different sets of primers were used to produce three size-specific amplicons of 435, 270, and 192 bp for PV1, NV, and HAV, respectively. When tested on mixed virus suspensions, the multiplex method achieved detection of one infectious unit (PV1 and HAV) or RT-PCR-amplifiable unit (NV) for all viruses (Rosenfield and Jaykus 1999). With the use of single-tube amplification and liquid hybridization, multiplex PCR has potential advantages over monoplex PCR in terms of rapidity and cost effectiveness.

In an effort to improve the sensitivity and speed of virus detection from non-shellfish food commodities by RT-PCR, Sair et al. (2002) compared multiple RNA extraction methods and NV primer sets. Hamburger and lettuce samples were artificially contaminated with HAV or NV, followed by their concentration using a filtration-extraction-precipitation procedure. The use of TRIzol (Gibco-BRL, Gaithersburg, Md.) with the QIAshredder (Qiagen, Valencia, Calif.) Homogenizer (TRIzol/Shred) yielded the best RT-PCR detection limits (<1 RT-PCR amplifiable unit per reaction for NV), and the NVp110/NVp36 primer set was the most efficient for detecting NV from seeded food samples. Subsequently, a one-step RT-PCR protocol using the TRIzol/Shred extraction method and the NVp110/NVp36 or HAV3/HAV5 primer sets demonstrated improved sensitivity (>10-fold) over the routinely used two-step method. The ability to detect viral RNA in food concentrates without prior dilution was taken as evidence that RT-PCR inhibitors were effectively removed by the use of the above method.

Many authors have addressed the problem of co-concentration of RT-PCR inhibitors with sample concentrates. To remove RT-PCR inhibitors from oyster extracts, Burkhardt et al. (2002) compared a compartmentalized tube-within-a-tube device for RT-PCR-nested PCR with conventional RT-PCR-nested PCR. In the presence of 100 mg of shellfish tissue extract, the tube-within-a-tube device was able to increase the sensitivity of calicivirus detection 10-fold over that of conventional RT-PCR-nested PCR. Shieh et al. (1999) demonstrated that changes in elution conditions (glycine buffer at pH 7.5 rather than pH 9.5) and RNA extraction (using a silica gel membrane rather than single-step RNA precipitation) may aid in the removal of RT-PCR inhibitors from oyster extracts.

Often a single method is not sufficient to detect the presence of viruses. For example, the role of NV in three successive gastroenteritis outbreaks in a Mediterranean-style restaurant was evaluated by EM and RT-PCR. NV was detected in each of the three outbreaks. However, RT-PCR failed to detect NV in a specimen that was found positive by EM.

Efforts have been made to develop enzyme immunoassays for the detection of NV. Vipond et al. (2000) compared the performance of EM, RT-PCR, and an enzyme-linked immunosorbent assay (ELISA) using recombinant capsid protein from an SRSV. Of the 213 SRSV positive outbreaks, 71% were identified by enzyme immunoassay, 63% with EM, and 84% by RT-PCR. Thus, SRSV ELISA may provide a simple cost-effective assay for the detection of currently circulating SRSV strains and may allow rapid identification of emerging SRSV strains (Vipond et al. 2000). Using recombinant capsid protein of NV and its purified antisera in an ELISA test, Yoda et al. (2000) could detect both antibodies and the virus from outbreak cases, suggesting that the immunologic detection of NV antigens is possible.

Baculovirus-expressed capsid protein immunoassays have also been described to detect both antigens and antibodies (Jiang et al. 1992). Whether such immunological tests will detect all genetically diverse strains of human caliciviruses is not known, however.

Indicators

Routine virological monitoring of food and water would help improve the safety of food (Barrett et al. 2001, Grabow et al. 2001, Beuret et al. 2002, Campos et al. 2002). However, methods for routine detection of all viruses in all types of food are currently not available, and even if they were available, they may be too expensive and time consuming to be used on a regular basis. At present, the presence or absence of indicator bacteria (such as fecal coliforms and *Escherichia coli*) in food and water is taken as evidence of the presence or absence of fecal (and hence pathogen) contamination. Shellfish-growing waters are generally monitored for fecal contamination by testing for the fecal coliform group of bacteria. Although these bacteria are good indicators for the presence of bacterial pathogens, they do not adequately reflect the presence of human enteric viruses (Goyal et al. 1979, Dore et al. 2000, Kator and Rhodes 2001, Wait and Sobsey 2001, Duran et al. 2002). Hence, the use of alternate indicators to determine the virological quality of food and water has been suggested. Among others, these alternate indicators include adenoviruses (Pina et al. 1998), enteric viruses such as poliovirus, and bacteriophages (Zanetti et al. 2003).

Although many authors have advocated the use of bacteriophage as indicators of human enteric viruses, Leclerc et al. (2000) pointed out that bacteriophage data collected by various laboratories are not directly comparable because of lack of uniformity in bacteriophage detection methods. Differences exist in bacterial hosts used, method of phage concentration, determination of endpoints, volume of sample used (varying from 1 to 400 L), type of filter used for viral adsorption, types of beef extract used for elution, conditions of centrifugation, and the delivery of the concentrate to the host cells. In addition, factors affecting the ability of the host to continue to be receptive to the bacteriophage after continued subculture may also influence test results.

Hsu et al. (2002) compared three different eluents for recovering F+ RNA coliphage, somatic coliphage, and Salmonella phage from ground beef and chicken breast meat. After elution with glycine, threonine, or 3% beef extract, the eluates were concentrated by PEG precipitation and assayed on appropriate bacterial hosts (*E. coli* Famp, *E. coli* C, and *Salmonella* Typhimurium for F+ RNA coliphage, somatic coliphage, and Salmonella phage, respectively).

The highest recoveries of the three phage groups were obtained with 0.5 M threonine (pH 9.0), and the overall detection sensitivity was 3 PFU per 100 g of meat. The authors suggested that F+ RNA coliphage, and perhaps other enteric bacteriophage, may be effective candidates indicators for monitoring the microbiological quality of meat, poultry, and perhaps other foods during processing.

Dore et al. (2000) monitored the levels of F+ RNA coliphage and *E. coli* over a 2-year period in oysters (*Crassostrea gigas*) harvested from three polluted and one non-polluted site. All of the shellfish tested were found to be in compliance with the mandatory European Community *E. coli* standard of <230 organisms per 100 g of shellfish flesh. However, as many as 1,000 PFU of F+ RNA coliphage were frequently detected in 100-g samples of shellfish, and their presence was found to be strongly associated with harvest area fecal pollution and with shellfish-associated disease outbreaks. On the basis of these results, it was suggested that F+ RNA coliphage should be used as viral indicators for market-ready oysters (Dore et al. 2000).

To determine whether F+ RNA coliphage can be used as indicators of fecal contamination for fresh carrots, Endley et al. (2003) tested 25 samples each from a farm, truck, and processing shed. The number of samples positive for *E. coli*, Salmonella, and F+ RNA coliphage was 6, 3, and 19, respectively. It was suggested that F+ RNA coliphage could be used as a conservative indicator for feces-associated viruses on carrots.

One of the criteria for an ideal indicator is that it should survive longer than the pathogen itself. The decay rates of F+ RNA phage, enteric viruses, and coliform bacteria are known to vary with salinity, pH, temperature, turbidity, and dissolved oxygen concentration (Stenström and Carlandar, 2001, Wait and Sobsey 2001, Sinton et al. 2002). Despite this variability, levels of F+ RNA coliphage in seafood harvested from sewage-contaminated waters have consistently been higher than those of fecal-indicator bacteria (Dore et al. 2000, Hernroth et al. 2002). In addition, F+ RNA coliphage have been reported to survive for up to 30 days in groundwater (Yates et al. 1985).

In a recent study, we compared the survival of F+ RNA coliphage, FCV, and *E. coli* in water at different temperatures (Allwood et al. 2003). The results of this study suggested that both *E. coli* and F+ RNA phage would survive for as long as, or longer, than NV in clean water free of disinfectants, and that the latter could be used as a conservative indicator of NV in clean water. Because bacterial indicators are generally less resistant to disinfection than enteric viruses, it is possible that in the presence of disinfectants such as chlorine, *E. coli* survival would likely differ from the two viral organisms (Dore et al. 2000, Sinton et al. 2002). It is conceivable, therefore, that F+ RNA phage may even be a better indicator for monitoring food, water, and surfaces that have been sanitized/disinfected.

Although many authors have suggested the use of F+ RNA coliphage as an alternate indicator of enteric viruses (Havelaar et al. 1993, Hsu et al. 1995, Woody and Cliver 1995, Dore et al. 2000), F+ RNA coliphage proposed use as indicators is not without controversy. For example, although a strong correlation has been observed between the seasonal accumulation of F+ RNA coliphage in oysters and the incidence of NV diseases associated with oysters (Burkhardt and Calci 2000), some investigators have reported that F+ RNA coliphage are rarely detected in human feces (F+ RNA coliphage are present in the feces of approximately 3% of human population), suggesting that the presence of these coliphage in water does not necessarily indicate human fecal pollution (Havelaar et al. 1990, Schaper et al. 2002a, Schaper et al. 2002b) and that *Bacteroides fragilis* HSP40 phage be considered as an alternate indicator. However, Bacteroides is not an easy host for use in routine laboratory analysis and, hence, has not gained much popularity.

Somatic phage, those that do not require the F+ for attachment, have also been considered as indicators of viruses because the methods for the enumeration of somatic phage are simpler and results can be available in 4–6 hours. Clearly, more data are needed on the ecology and lifespan of phage in the environment to ascertain their usefulness as potential viral indicators in food and water.

Virus Survival, Disinfection, and Sanitization

In general, enteric viruses are much more resistant to environmental conditions than are bacteria, including fecal-indicator bacteria. Kurdziel et al. (2001) conducted a series of studies to ascertain the potential for survival of PV1 on various foodstuffs. The studies were mostly performed using fresh produce stored at refrigeration temperature for 2 weeks or so, which was considered to represent the maximum time elapsing between purchase and consumption. Each food sample was inoculated with PV1, and samples were analyzed immediately and at intervals throughout the experiment. The decimal reduction times or number of days after which the initial virus numbers had declined by 90% were as follows: lettuce, 11.6 days; green onion, no decline; white cabbage, 14.2 days; fresh raspberries, no decline; and frozen strawberries, 8.4 days. These results demonstrated that enteric viruses may persist on fresh fruit and vegetables for several days under conditions commonly used for storage in households and may pose a risk of infection to consumers.

Petterson and Ashbolt (2001) constructed a model for virus decay on lettuce and carrot following irrigation with secondarily treated sewage. The presence of a very persistent subpopulation of viruses was observed, as evidenced by an initial rapid phase of decay followed by a very slow phase. Because virus counts fitted a negative binomial distribution rather than Poisson distribution, it was deduced that viruses were not uniformly distributed over the surfaces of either lettuce or carrot. The presence of over-dispersion and a persistent subpopulation of viruses contributed to a significant increase in the heterogeneity of the risk estimates. In addition, predicted infection rates were significantly underestimated if the presence of a persistent subpopulation of viruses was not considered in the decay kinetics of the risk model. It was suggested that both viral clumping and persistence should be accounted for in future risk assessments of enteric viruses associated with wastewater reuse (Petterson and Ashbolt 2001).

To determine the mechanism of inactivation of enteric viruses in the environment, Nuanualsuwan and Cliver (2003) studied the capsid functions of PV1, HAV, and FCV after exposing them to ultraviolet light, hypochlorite, high temperature (72°C), and physiological temperature (37°C). The viral RNA of 37°C-inactivated viruses was protected by their capsids, but not that of the viruses inactivated at 72°C, suggesting that viral capsids play a role in virus survival. Virus inactivation almost always resulted in the loss of virus attachment further indicating that the primary target of ultraviolet light, hypochlorite, and 72°C inactivation is the viral capsid.

Food such as salads, cold food items, and frosted confectionery items may either be contaminated by ill or asymptomatic infected food handlers (Blacklow et al. 1987, Ponka et al. 1999, Schwab et al. 2000) or be contaminated by coming in contact with contaminated surfaces during harvest, processing, or handling in restaurants. Chemical disinfection of food contact surfaces and rinsing food items with sanitizers is generally relied on to prevent and control foodborne outbreaks, but the efficacy of disinfectants for the inactivation of viruses is generally not known.

Using FCV as a surrogate of NV, we conducted a study to determine the anti-FCV efficacy of commonly used disinfectants and food sanitizers on artificially contaminated stainless steel surfaces and fresh produce (strawberry and lettuce). Discs of stainless steel, strawberry, and lettuce were contaminated with known amounts of FCV. The disinfectants were applied at one, two, and four times the manufacturer's recommended concentrations for contact times of 1 and 10 minutes. The action of the disinfectants was stopped by dilution, and the number of surviving FCV was determined by titration in cell cultures. An agent was considered effective if it reduced the virus titer by at least 3 logs from an initial level of 10^7 TCID$_{50}$. Only one of seven disinfectants was found to be effective when used at the manufacturer's recommended concentration for 10 minutes, indicating that most of the sanitizing and disinfecting agents used in restaurants are not capable of removing calicivirus contamination from work surfaces or from produce at recommended use levels. Thus, preventing outbreaks of foodborne calicivirus depends on maintaining hand-washing practices and excluding ill food workers from the establishment.

Conclusions

This brief review has attempted to highlight the importance of human enteric viruses in different foods. Foodborne viral infections are prevalent worldwide, and steps are needed to contain or reduce their prevalence. Human enteric viruses can persist for a long time in the environment and can contaminate growing shellfish or irrigation waters. The latter may result in contamination of various foods such as fresh fruits and vegetables. Restaurants, nursing homes, hospitals, cruise ships, and conferences are the likely starting points for some of the foodborne disease outbreaks through meals prepared or handled by ill food workers or through foods coming in contact with contaminated surfaces or water.

Control measures should focus on ensuring that food does not come in contact with contaminated water, food handlers, or surfaces. Contamination via food handlers can be minimized by exclusion of ill food handlers and by the maintenance of strict personal hygiene. In outbreak situations, not only should the source of contamination be removed, but attempts should also be made to prevent further spread of infection by interrupting person-to-person transmission. Viral disease outbreaks associated with shellfish can be controlled by preventing contamination of shellfish growing waters.

The need for proper education of the food industry in this regard cannot be overemphasized. Food handlers and consumers should be made aware of the risks associated with contaminated food products and of the steps needed to minimize such contamination. Food handlers should be not only educated but re-educated on a regular basis about the importance of practicing good personal and environmental hygiene, including correct food-handling practices. Workers in countries from which food is imported should also be included in such training. Random monitoring of food and water, especially of ready-to-eat items, for the presence of viruses and their indicators should be mandated.

It is clear from this review that further research in food virology is badly needed. This need has become more acute in light of the potential threats of agro-terrorism, in which food commodities may be a target of miscreants. There is an urgent need for the development of simple, sensitive, and rapid methods for the elution and concentration of small amounts of different viruses in large amounts of food. Attempts should be made to develop easy-to-use diagnostic methods for virus detection following elution-concentration from foods. Simple methods to remove inhibitors of PCR and RT-PCR from food concentrates are also needed. Because epidemiologic investigation and detection of viruses in food and water may be cumbersome and expensive, research on finding a suitable viral indicator should continue. Research should also focus on developing and testing effective sanitizers/disinfectants for food, food contact surfaces, and human hands. In addition, the role of worker immunization on interruption of foodborne infections should also be studied.

References

Allwood, P. B., Malik, Y. S., Hedberg, C. W. and Goyal, S. M. 2003. Comparative survival of F-specific RNA coliphage, feline calicivirus, and *Escherichia coli* in water. *Applied and Environmental Microbiology* **69**:5707–5710.

Andersen, F. R., Birkeland, F. G., Bo, G., Eidsto, A. and Bruu, A. L. 1996. Outbreak of food-borne gastroenteritis caused by a Norwalk-like virus. Evaluation of methods for confirmation of the etiology in suspected viral gastroenteritis. *Tidsskr Nor Laegeforen* **20**:3325–3338.

Ando, T., Jin, Q., Gentsch, J. R., Monroe, S. S., Noel, J. S., Dowell, S. F., Cicirello, H. G., Kohn, M. A. and Glass, R. I. 1995. Epidemiologic applications of novel molecular methods to detect and differentiate small round structured viruses (Norwalk-like viruses). *Journal of Medical Virology* **47**:145–152.

Appleton, H. 2000. Control of food-borne viruses. *British Medical Bulletin.* **56**:172–183.

Barrett, E. C., Sobsey, M. D., House, C. H. and White, K. D. 2001. Microbial indicator removal in onsite constructed wetlands for wastewater treatment in the southeastern U.S. *Water Science and Technology* **44**:177–182.

Bean, N. H., Goulding, J. S., Lao, C. and Angulo, F. J. 1996. Surveillance of foodborne disease outbreaks—United States, 1988–1992. *Morbidity Mortality Weekly Reports* **45**:1–55.

Becker, K. M., Moe, C. L., Southwick, K .L. and MacCormack, J. N. 2000. Transmission of Norwalk virus during football game. *New England Journal of Medicine* **26**:1223–1227.

Berg, D. E., Kohn, M. A., Farley, T. A. and McFarland, L. M. 2000. Multi-state outbreaks of acute gastroenteritis traced to fecal-contaminated oysters harvested in Louisiana. *Journal of Infectious Diseases* **181**:S381–S386.

Beuret, C., Kohler, D., Baumgartner, A. and Luthi, T. M. 2002. Norwalk-like virus sequences in mineral waters: one-year monitoring of three brands. *Applied and Environmental Microbiology* **68**:1925–1931.

Bidawid, S., Malik, N., Adgburin, O., Sattar, S. A. and Farber, J. M. 2003. A feline kidney cell line based plaque assay for FCV, a surrogate for Norwalk virus. *Journal of Virological Methods* **107**:163–167.

Blacklow, N. R., Herrmann, J. E. and Cubitt, W. D. 1987. Immunobiology of Norwalk virus. *Ciba Foundation Symposium* **128**:144–161.

Bosch, A., Sanchez, G., Le Guyader, F., Vanaclocha, H., Haugarreau, L. and Pinto, R. M. 2001. Human enteric viruses in Coquina clams associated with a large hepatitis A outbreak. *Water Science and Technology* **43**:61–65.

Bresee, J. S., Widdowson, M. A., Monroe. S. S. and Glass, R. I. 2002. Foodborne viral gastroenteritis: challenges and opportunities. *Clinical Infectious Disease* **35**:748–753.

Brugha, R., Vipond, I. B., Evans, M. R., Sandifer, Q. D., Roberts, R. J., Salmon, R. L., Caul, E. O. and Mukerjee, A. K. 1999. A community outbreak of food-borne small round-structured virus gastroenteritis caused by a contaminated water supply. *Epidemiology and Infection* **122**:145–154.

Burkhardt, W., Blackstone, G. M., Skilling, D. and Smith, A. W. 2002. Applied technique for increasing calicivirus detection in shellfish extracts. *Journal of Applied Microbiology* **93**:235–240.

Burkhardt, W. and Calci, K. R. 2000. Selective accumulation may account for shellfish-associated viral illness. *Applied and Environmental Microbiology* **66**:1375–1378.

Campos, C., Guerrero, A. and Cardenas, M. 2002. Removal of bacterial and viral faecal indicator organisms in a waste stabilization pond system in Choconta, Cundinamarca (Colombia). *Water Science and Technology* **45**:61–66.

Cannon, R. O., Poliner, J. R., Hirschhorn, R. B., Rodeheaver, D. C., Silverman, P. R., Brown, E. A., Talbot, G. H., Stine, S. E., Monroe, S. S. and Dennis, D. T. 1991. A multistate outbreak of Norwalk virus gastroenteritis associated with consumption of commercial ice. *Journal of Infectious Diseases* **164**:860–863.

Croci, L., De, M. D., Scalfaro, C., Fiore, A. and Toti, L. 2002. The survival of hepatitis A virus in fresh produce. *International Journal of Food Microbiology* **73**:29–34.

Daniels, N. A., Bergmire-Sweat, D. A., Schwab, K. J., Hendricks, K. A., Reddy, S., Rowe, S. M., Fankhauser, R. L., Monroe, S. S., Atmar, R. L., Glass, R. I. and Mead, P. 2000. A food borne outbreak of gastroenteritis associated with Norwalk-like viruses: first molecular traceback to deli sandwiches contaminated during preparation. *Journal of Infectious Diseases* **181**:1467–1470.

De Leon, R., Matsui, S. M., Baric, R. S., Herrmann, J. E., Blacklow, N. R., Greenberg, H. B. and Sobsey, M. D. 1992. Detection of Norwalk virus in stool specimens by reverse transcriptase-polymerase chain reaction and nonradioactive oligoprobes. *Journal of Clinical Microbiology* **30**:3151–3157.

Deneen, V. C., Hunt, J. M., Paule, C. R., James, R. I., Johnson, R. G., Raymond, M. J. and Hedberg, C. W. 2000. The impact of food borne calicivirus disease: the Minnesota experience. *Journal of Infectious Diseases* **181**:S281–S283.

Dix, A. B. and Jaykus, L. A. 1998. Virion concentration method for the detection of human enteric viruses in extracts of hard-shelled clams. *Journal of Food Protection* **61**:458–465.

Dore, W. J., Henshilwood, K. and Lees, D. N. 2000. Evaluation of F-specific RNA bacteriophage as a candidate human enteric virus indicator for bivalve molluscan shellfish. *Applied and Environmental Microbiology* **66**:1280–1285.

Doultree, J. C., Druce, J. D., Birch, C. J., Bowden, D. S. and Marshall, J. A. 1999. Inactivation of feline calicivirus, a Norwalk virus surrogate. *Journal of Hospital Infection* **41**:51–57.

D'Souza, D. H. and Jaykus, L. A. 2002. Zirconium hydroxide effectively immobilizes and concentrates human enteric viruses. *Letters Applied Microbiology* **35**:414–418.

Dubois, E., Agier, C., Traore, O., Hennechart, C., Merle, G., Cruciere, C. and Laveran, H. 2002. Modified concentration method for the detection of enteric viruses on fruits and vegetables by reverse transcriptase-polymerase chain reaction or cell culture. *Journal of Food Protection* **65**:1962–1969.

Duran, A. E., Muniesa, M., Mendez, X., Valero, F., Lucena, F. and Jofre, J. 2002. Removal and inactivation of indicator bacteriophages in fresh waters. *Journal of Applied Microbiology* **92**:338–347.

Endley, S., Lu, L., Vega, E., Hume, M. E. and Pillai, S. D. 2003. Male-specific coliphages as an additional fecal contamination indicator for screening fresh carrots. *Journal of Food Protection* **66**:88–93.

Enriquez, C., Nwachuku, N. and Gerba, C. P. 2001. Direct exposure to animal enteric pathogens. *Reviews of Environmental Health* **16**:117–131.

Frankhauser, R. L., Monroe, S. S., Noel, J. S., Humphrey, C. D., Bresee, J. S., Parashar, U. D., Ando, T. and Glass, R. I. 2002. Epidemiologic and molecular trends of "Norwalk-like viruses" associated with outbreaks of gastroenteritis in the United States. *Journal of Infectious Diseases* **186**:1–7.

Frankhauser, R. L., Noel, J. S., Monroe, S. S., Ando, T. and Glass, R. I. 1998. Molecular epidemiology of "Norwalk-like viruses" in outbreaks of gastroenteritis in the United States. *Journal of Infectious Diseases* **178**:1571–1578.

Formiga-Cruz, M., Tofino-Quesada, G., Bofill-Mas, S., Lees, D. N., Henshilwood, K., Allard, A. K., Conden-Hansson, A. C., Hernroth, B. E., Vantarakis, A., Tsibouxi, A., Papapetropoulou, M., Furones, M. D. and Girones, R. 2002. Distribution of human virus contamination in shellfish from different growing areas in Greece, Spain, Sweden, and the United Kingdom. *Applied and Environmental Microbiology* **68**:5990–5998.

Gaulin, C., Frigon, M., Poirier, D. and Fournier, C. 1999. Transmission of calicivirus by a foodhandler in the presymptomatic phase of illness. *Epidemiology and Infection* **123**:475–478.

Gerba, C. P., Farrah, S. R., Goyal, S. M., Wallis, C. and Melnick, J. L. 1978. Concentration of enteroviruses from large volumes of tap water, treated sewage, and seawater. *Applied and Environmental Microbiology* **35**:540–548.

Gerba, C. P. and Goyal, S. M. 1978. Detection and occurrence of enteric viruses in shellfish: a review. *Journal of Food Protection* **41**:743–754.

Gerba, C. P. and Goyal, S. M. (Eds.). 1982. Methods in Environmental Virology. New York: Marcel Dekker.

Glass, R. I., Bresee, J., Jiang, B., Gentsch, J., Ando, T., Fankhauser, R., Noel, J., Parashar, U., Rosen, B. and Monroe, S. S. 2001. Gastroenteritis viruses: an overview. *Novartis Foundation Symposium* **238**:5–19.

Glass, R. I., Noel, J., Ando, T., Fankhauser, R., Belliot, G., Mounts, A., Parashar, U. D., Bresee, J. S. and Monroe, S. S. 2000. The epidemiology of enteric caliciviruses from humans: a reassessment using new diagnostics. *Journal of Infectious Diseases* **181**:S254–S261.

Goyal, S. M. and Gerba, C. P. 1983. Viradel method for detection of rotavirus from seawater. *Journal of Virological Methods* **7**:279–285.

Goyal, S. M., Gerba, C. P. and Melnick, J. L. 1979. Human enteroviruses in oysters and their overlying waters. *Applied and Environmental Microbiology* **37**:572–581.

Grabow, W. O., Taylor, M. B. and de Villiers, J. C. 2001. New methods for the detection of viruses: call for review of drinking water quality guidelines. *Water Science and Technology* **43**:1–8.

Green, J. K., Wright, P. A. and Gallimore, C. I. 1998. The role of environmental contamination with small round structured viruses in a hospital outbreak investigated by reverse-transcriptase polymerase chain reaction assay. *Journal of Hospital Infection* **39**:39–45.

Grohmann, G. S., Murphy, A. M., Christopher, P. J., Auty, E. and Greenberg, H. B. 1981. Norwalk virus gastroenteritis in volunteers consuming depurated oysters. *Australian Journal of Experimental Biology and Medical Sciences* **59**:219–228.

Gulati, B. R., Allwood, P. B., Hedberg, C. W. and Goyal, S. M. 2001. Efficacy of commonly used disinfectants for the inactivation of calicivirus on strawberry, lettuce, and a food-contact surface. *Journal of Food Protection* **64**:1430–1434.

Havelaar, A. H., Pot-Hogeboom, W. M., Furuse, K., Pot, R. and Hormann, M. P. 1990. F-specific RNA bacteriophages and sensitive host strains in faeces and wastewater of human and animal origin. *Journal Applied Bacteriology* **69**:30–37.

Havelaar, A., van Olphen, M. and Drost, Y. 1993. F-specific RNA bacteriophages are adequate model organisms for enteric viruses in fresh water. *Applied and Environmental Microbiology* **59**:2956–2962.

Hedberg, C. W. 2000. Global surveillance needed to prevent food borne diseases. *California Agriculture* **54**:54–61.

Hedberg, C. W., MacDonald, K. L. and Osterholm, M. T. 1994. Changing epidemiology of foodborne disease: a Minnesota perspective. *Clinical Infectious Disease* **18**:671–682.

Hedberg, C. W. and Osterholm, M. T. 1993. Outbreaks of food-borne and waterborne viral gastroenteritis. A Minnesota perspective. *Clinical Microbiology Reviews* **6**:199–210.

Hernroth, B. E., Conden-Hansson, A. C., Rehnstam-Holm, A. S., Girones, R. and Allard, A. K. 2002. Environmental factors influencing human viral pathogens and their potential indicator organisms in the blue mussel, *Mytilus edulis*: the first Scandinavian report. *Applied and Environmental Microbiology* **68**:4523–4533.

Herwaldt, B. L., Lew, J. F., Moe, C. L., Lewis, D. C., Humphrey, C. D., Monroe, S. S., Pon, E. W. and Glass, R. I. 1994. Characterization of a variant strain of Norwalk virus from a food-borne outbreak of gastroenteritis on a cruise ship in Hawaii. *Journal of Clinical Microbiology* **32**:861–866.

Hicks, N. J., Beynon, J. H., Bingham, P., Soltanpoor, N. and Green, J. 1996. An outbreak of viral gastroenteritis following a wedding reception. *Communicable Disease Report. CDR Review* **6**:R136–R139.

Hsu, F. C., Shieh, Y. S. and Sobsey, M. D. 2002. Enteric bacteriophages as potential fecal indicators in ground beef and poultry meat. *Journal of Food Protection* **65**:93–99.

Hsu, F. C., Shieh, Y. S., van Duin, J., Beekwilder, M. J. and Sobsey, M. D. 1995. Genotyping male-specific RNA coliphages by hybridization with oligonucleotide probes. *Applied and Environmental Microbiology* **61**:3960–3966.

Jiang, X., Wang, J., Graham, D. Y. and Estes, M. K. 1992. Detection of Norwalk virus in stool by polymerase chain reaction. *Journal of Clinical Microbiology* **30**:2529–2534.

Kator, H. and Rhodes, M. 2001. Elimination of fecal coliforms and F-specific RNA coliphage from oysters (*Crassostrea virginica*) relaid in floating containers. *Journal of Food Protection* **64**:796–801.

Kirkland, K. B., Meriwether, R. A., Leiss, J. K. and MacKenzie, W. R. 1996. Steaming oysters does not prevent Norwalk-like gastroenteritis. *Public Health Reports* **111**:527–530.

Kurdziel, A. S., Wilkinson, N., Langton, S. and Cook, N. 2001. Survival of poliovirus on soft fruit and salad vegetables. *Journal of Food Protection* **64**:706–709.

Leclerc, H., Edberg, S., Pierzo, V. and Delattre, J. M. 2000. Bacteriophages as indicators of enteric viruses and public health risk in groundwaters. *Journal of Applied Microbiology* **88**:5–21.

Le Guyader, F., Estes, M. K., Hardy, M. E., Neill, F. H., Green, J., Brown, D. W. and Atmar, R. L. 1996. Evaluation of a degenerate primer for the PCR detection of human caliciviruses. *Archives of Virology* **141**:2225–2235.

Le Guyader, F., Haugarreau, L., Miossec, L., Dubois, E. and Pommepuy, M. 2000. Three-year study to assess human enteric viruses in shellfish. *Applied and Environmental Microbiology* **66**:3241–3248.

Lees, D. N., Henshilwood, K., Green, J., Gallimore, C. I. and Brown, D. W. 1995. Detection of small round structured viruses in shellfish by reverse transcription-PCR. *Applied and Environmental Microbiology* **61**:4418–4424.

Leggitt, P. R. and Jaykus, L. A. 2000. Detection methods for human enteric viruses in representative foods. *Journal of Food Protection* **63**:1738–1744.

Mead, P. S., Slutsker, L., Dietz, V., McCaig, L. F., Bresee, J. S., Shapiro, C., Griffin, P. M. and Tauxe, R. V. 1999. Food-related illness and death in the United States. *Emerging Infectious Disease* **5**:607–625.

Monroe, S. S., Ando, T. and Glass, R. I. 2000. Introduction: human enteric caliciviruses—an emerging pathogen whose time has come. *Journal of Infectious Diseases* **181**:S249–S251.

Morse, D. L., Guzewich, J. J., Hanrahan, J. P., Stricof, R., Shayegani, M., Deibel, R., Grabau, J. C., Nowak, N. A., Herrmann, J. E. and Cukor, G. 1986. Widespread outbreaks of clam- and oyster-associated gastroenteritis. Role of Norwalk virus. *New England Journal of Medicine* **314**:678–681.

Mounts, A. W., Ando, T., Koopmans, M., Bresee, J. S., Noel, J. and Glass, R. I. 2000. Cold weather seasonality of gastroenteritis associated with Norwalk-like viruses. *Journal of Infectious Diseases* **181**:S284–S287.

Nuanualsuwan, S. and Cliver, D. O. 2003. Capsid functions of inactivated human picornaviruses and feline calicivirus. *Applied and Environmental Microbiology* **69**:350–357.

Ooi, P. L., Goh, K. T., Neo, K. S. and Ngan, C. C. 1997. A shipyard outbreak of salmonellosis traced to contaminated fruits and vegetables. *Annals of the Academy of Medicine (Singapore)* **26**:539–543.

Parashar, U. D., Dow, L., Fankhauser, R. L., Humphrey, C. D., Miller, J., Ando, T., Williams, K. S., Eddy, C. R., Noel, J. S., Ingram, T., Bresee, J. S., Monroe, S. S. and Glass, R. I. 1998. An outbreak of viral gastroenteritis associated with consumption of sandwiches: implications for the control of transmission by food handlers. *Epidemiology and Infection* **121**:615–621.

Parashar, U. D. and Monroe, S. S. 2001. "Norwalk-like viruses" as a cause of foodborne disease outbreaks. *Reviews in Medical Virology* **11**:243–252.

Patterson, T., Hutchings, P. and Palmer, S. 1993. Outbreak of SRSV gastroenteritis at an international conference traced to food handled by a post symptomatic caterer. *Epidemiology and Infection* **111**:157–162.

Petterson, S. R. and Ashbolt, N. J. 2001. Viral risks associated with wastewater reuse: modeling virus persistence on wastewater irrigated salad crops. *Water Science and Technology* **43**:23–26.

Pina, S., Puig, M., Lucena, F., Jofre, J. and Girones, R. 1998. Viral pollution in the environment and in shellfish: human adenovirus detection by PCR as an index of human viruses. *Applied and Environmental Microbiology* **64**:3376–3382.

Ponka, A., Maunula, L., von Bonsdorff, C. H. and Lyytikainen, O. 1999. An outbreak of calicivirus associated with consumption of frozen raspberries. *Epidemiology and Infection* **123**:469–474.

Richards, G. P. 1999. Limitations of molecular biological techniques for assessing the virological safety of foods. *Journal of Food Protection* **62**:691–697.

Richards, G. P. 2001. Enteric virus contamination of foods through industrial practices: a primer on intervention strategies. *Journal of Industrial Microbiology and Biotechnology* **27**:117–125.

Riordan, T., Craske, J., Roberts, J. L. and Curry, A. 1984. Food borne infection by a Norwalk like virus (small round structured virus). *Journal of Clinical Pathology* **37**:817–820.

Rosenfield, S. I. and Jaykus, L. A. 1999. A multiplex reverse transcription polymerase chain reaction method for the detection of foodborne viruses. *Journal of Food Protection* **62**:1210–1214.

Sair, A. I., D'Souza, D. H., Moe, C. L. and Jaykus, L. A. 2002. Improved detection of human enteric viruses in foods by RT-PCR. *Journal of Virological Methods* **100**:57–69.

Schaper, M., Duran, A. E. and Jofre, J. 2002a. Comparative resistance of phage isolates of four genotypes of F-specific RNA bacteriophages to various inactivation processes. *Applied and Environmental Microbiology* **68**:3702–3707.

Schaper, M., Jofre, J., Uys, M. and Grabow, W. O. 2002b. Distribution of genotypes of F-specific RNA bacteriophages in human and non-human sources of faecal pollution in South Africa and Spain. *Journal of Applied Microbiology* **92**:657–667.

Schvoerer, E., Bonnet, F., Dubois, V., Cazaux, G., Serceau, R., Fleury, H. J. and Lafon, M. E. 2000. PCR detection of human enteric viruses in bathing areas, waste waters and human stools in Southwestern France. *Research in Microbiology* **151**:693–701.

Schwab, K. J., Neill, F. H., Fankhauser, R. L., Daniels, N. A., Monroe, S. S., Bergmire-Sweat, D. A., Estes, M. K. and Atmar, R. L. 2000. Development of methods to detect "Norwalk-like" viruses (NLVs) and hepatitis A virus in delicatessen foods: application to a food-borne NLV outbreak. *Applied and Environmental Microbiology* **66**:213–218.

Shieh, Y. C., Calci, K. R. and Baric, R. S. 1999. A method to detect low levels of enteric viruses in contaminated oysters. *Applied and Environmental Microbiology* **65**:4709–4714.

Simmons, G., Greening, G., Gao, W. and Campbell, D. 2001. Raw oyster consumption and outbreaks of viral gastroenteritis in New Zealand: evidence for risk to the public's health. *Australian and New Zealand Journal of Public Health* **25**:234–240.

Sinton, L. W., Hall, C. H., Lynch, P. A. and Davies-Colley, R. J. 2002. Sunlight inactivation of fecal indicator bacteria and bacteriophages from waste stabilization pond effluent in fresh and saline waters. *Applied and Environmental Microbiology* **68**:1122–1131.

Slomka, M. J. and Appleton, H. 1998. Feline calicivirus as a model system for heat inactivation studies of small round structured viruses in shellfish. *Epidemiology and Infection* **121**:401–407.

Sobsey, M. D., Carrick, R. J. and Jensen, H. R. 1978. Improved methods for detecting enteric viruses in oysters. *Applied and Environmental Microbiology* **36**:121–128.

Solomon, E. B., Potenski, C. J. and Matthews, K. R. 2002. Effect of irrigation method on transmission to and persistence of *Escherichia coli* O157:H7 on lettuce. *Journal of Food Protection* **65**:673–676.

Stafford, R., Strain, D., Heymer, M., Smith, C., Trent, M. and Beard, J. 1997. An outbreak of Norwalk virus gastroenteritis following consumption of oysters. *Communicable Diseases Intelligence* **21**:317–320.

Stenström, T. A. and Carlander, A. 2001. Occurrence and die-off of indicator organisms in the sediment in two constructed wetlands. *Water Science and Technology* **44**:223–230.

Taku, A., Gulati, B. R., Allwood, P. B., Palazzi, K., Hedberg, C. W. and Goyal, S. M. 2002. Concentration and detection of caliciviruses from food-contact surfaces. *Journal of Food Protection* **65**:999–1004.

Tauxe, R., Kruse, H., Hedberg, C. W., Potter, M., Madden, J. and Wachsmuth, K. 1997. Microbial hazards and emerging issues associated with produce. A preliminary report to the National Advisory Committee on Microbiologic Criteria for Foods. *Journal of Food Protection* **60**:1400–1408.

Vipond, I. B., Pelosi, E., Williams, J., Ashley, C. R., Lambden, P. R., Clarke, I. N. and Caul, E. O. 2000. A diagnostic EIA for detection of the prevalent SRSV strain in United Kingdom outbreaks of gastroenteritis. *Journal of Medical Virology* **61**:132–137.

Wait, D. A. and Sobsey, M. D. 2001. Comparative survival of enteric viruses and bacteria in Atlantic Ocean seawater. *Water Science and Technology* **43**:139–142.

Warner, R. D., Carr, R. W., McCleskey, F. K., Johnson, P. C., Elmer, L. M. and Davison, V. E. 1991. A large nontypical outbreak of Norwalk virus. Gastroenteritis associated with exposing celery to nonpotable water and with *Citrobacter freundii*. *Archives of Internal Medicine* **151**:2419–2424.

Woody, M. A. and Cliver, D. O. 1995. Effects of temperature and host cell growth phase on replication of F-specific RNA coliphage Q beta. *Applied and Environmental Microbiology* **61**:1520–1526.

Yates, M. V., Gerba, C. P. and Kelley, L. M. 1985. Virus persistence in groundwater. *Applied and Environmental Microbiology* **49**:778–781.

Yoda, T., Terano, Y., Shimada, A., Suzuki, Y., Yamazaki, K., Sakon, N., Oishi, I., Utagawa, E. T., Okuno, Y. and Shibata, T. 2000. Expression of recombinant Norwalk-like virus capsid proteins using a bacterial system and the development of its immunologic detection. *Journal of Medical Virology* **60**:475–481.

Zanetti, S., Deriu, A., Manzara, S., Cattani, P., Mura, A., Molicotti, P., Fadda, G. and Sechi, L. A. 2003. A molecular method for the recovery and identification of enteric virus in shellfish. *New Microbiology* **26**:157–162.

Part II

Ecology, Distribution, and Spread of Foodborne Hazards

9 Microbial Ecology: Poultry Foodborne Pathogen Distribution

J. Allen Byrd II

Introduction

In 1994, a report entitled "Foodborne Pathogens: Risk and Consequences" estimated that as many as 9,000 deaths and 6.5–33 million cases of illness each year are caused by ingestion of contaminated food products (Council for Agricultural Science and Technology 1994). In 2002, the Foodborne Diseases Active Surveillance Network (FoodNet) collected data within the United States and found that *Campylobacter* (5,006 cases) and *Salmonella* (6,028) accounted for over 66% of the confirmed foodborne-related diseases (Centers for Disease Control 2003). New requirements from the U.S. federal government mandate stringent regulations for the reduction of foodborne pathogens in commercially processed poultry in the United States. Reasons for this include increasing pressure from public interest groups, economic costs for the U.S. population, and restrictions for poultry export to some countries. Control of foodborne pathogens continues to receive media attention and continues to influence research preformed by governmental agencies (U.S. Department of Agriculture-Food Safety Inspection Service 1997). The initiation of the U.S. Department of Agriculture-Food Safety Inspection Service hazard analysis and critical control point program (HACCP) has renewed industry interest for the control of foodborne pathogens. The HACCP concept has brought about intense record keeping and sampling programs and has influenced the development of innovative reduction strategies across the nation (see Chapter 20). Although implementation of new intervention strategies to reduce foodborne pathogens is the essential goal, large numbers of variables and sources of foodborne pathogens may limit the ability to produce pathogen-free broiler chickens. Therefore, to understand where these critical control points can be implemented, scientists and industry personnel must review the sites of potential foodborne pathogen contamination within the poultry continuum.

The normal gastrointestinal microflora for an animal contains over 2,000 species and over 10^{12} organisms per gram of digesta. These bacteria must adapt to survive in the harsh environment of the gastrointestinal tract. Many pathogens can invade different animal species and can be spread in a cyclic manner. Others may not be shed into the environment by the host for days or weeks. Therefore, the best detection methods may not always demonstrate whether an animal or sample is pathogen positive during a single sampling period. These are only a few of the things that must be considered in evaluating a pathogen detection program.

In the gastrointestinal tract, both beneficial and pathogenic bacteria compete for nutrients, space, and intestinal attachment sites. If a bacterium does not compete well for its

---

position in the ecosystem, then it will be eliminated from the gastrointestinal tract. Bacteria undergo a constant battle for survival in the intestinal environment. Each organism must maintain a constant internal homeostasis while being exposed to multiple environmental factors such as pH, salt concentrations, or temperature changes (Montville and Matthews 2001). These external factors stimulate genetic coding mechanisms for up-regulation of protective virulence factors that are influenced by gene expression (Montville and Matthews 2001). Thus, environmental conditions play critical roles in organism survivability in nature. Many organisms have specific mechanisms that help them to survive for many years in adverse conditions. Such mechanisms include the formation of resistant spores by bacteria, molds, fungi, and protozoa. Organisms without such protective mechanisms are therefore frequently disadvantaged. There are several abiotic factors that may affect these organisms, including relative humidity, pH, temperature, oxygen toxicity, ultraviolet radiation, chemical pollutants, and many other environmental factors.

Relative humidity has different effects on different bacteria depending on their Gram-staining characteristics. Gram-negative aerosol bacteria tend to survive longer at lower relative humidity, whereas Gram-positive bacteria tend to remain viable in higher relative humidities (Theunissen et al. 1993). One possible reason for this difference is that as water leaves the bacteria, the lipid bilayer undergoes structural changes to a gel phase, which ultimately leads to inactivation of the cell (Maier et al. 2000). Another closely linked factor is environmental temperature. High temperatures usually lead to cell inactivation and protein denaturation, whereas lower temperatures increase the survival times of organisms. Many organisms lose their viability when frozen because of intracellular ice crystal formation. Ultraviolet radiation and ionizing radiation such as X-rays have detrimental effects on bacteria. These physiological disruptions usually cause inhibition or alteration of biological activities such as transcription and translation.

Changes in pH can benefit particular bacteria, or they may be detrimental in certain circumstances. For example, a 1-pH unit change can cause death, cessation of growth, or a reduced rate of bacterial growth, which may be advantageous to other bacteria with different pH tolerance in terms of competition for a particular environment (Hill et al. 1995). For any environmental extreme, there are groups of bacteria that have protective measures to ensure their survival under specific harsh conditions. Acid-tolerant bacteria can adapt to manage low-pH conditions quite effectively. There are a variety of responses possibly elicited by *Salmonella* when faced with nutrient depravation. Many of these responses cannot take place in a single generation, and to be phenotypically expressed, they may take several generations for complete adaptation to occur. Bacteria placed in adverse conditions show very significant die-offs until a few of the bacteria adapt to the environment (Morita 1997). Reduction in cell size is the first noticeable sign of starvation, and it has been shown to occur within the first 2 days of food deprivation (Novitsky and Morita 1976). The cell size depreciates in the magnitude of 10- to 100-fold reductions. However the intrinsic characteristics of the cells are maintained, as can be seen following the addition of nutrients to the environment. In this event, cells will regain there normal morphological and biological activity (Morita 1997). All of these functions take place to ensure the survival of the species. If bacteria were unable to perform some of these metabolic activities, then they would have been eliminated from the environment long ago.

Campylobacter is an example of a foodborne pathogen that adapts very well to changing environments. *Campylobacter* is a bacterium that is poorly understood. In several investigations, poultry have been fed with *Campylobacter* with little or no recoverability (Shreeve et al. 2002). Still other investigators have recovered *Campylobacter* from experimental fed chicks (Ziprin et al. 2002). An explanation for this observation was provided by Talibart and co-workers (2000), who found that two-thirds of *Campylobacter* spp. were nonculturable after 14–21 days in the laboratory. *Campylobacter* changes the type of protein synthesis profile during starvation, which enhances heat resistance in starved cells and allows the cells to survive under these conditions (Cappelier et al. 2000). These are important factors that must be considered when reviewing the microbial ecology of the gastrointestinal tract (see Chapter 6 and Chapter 17).

A good starting point in understanding the microbial ecology of the gastrointestinal tract or external environment is through the use of competitive exclusion (CE) cultures. The concept of CE was first demonstrated by provision of microflora from healthy adult chickens to neonatal chickens for the prevention of intestinal colonization by *Salmonella* (Nurmi and Rantala 1973). In the commercial poultry industry, beneficial bacteria (intestinal microflora) may be unavailable for young chicks to ingest from adult chickens because broiler production is based on an all-in, all-out system. This process prevents or delays the transfer of microorganisms from one batch of birds to the next (Nurmi et al. 1996).

The precise mechanisms used by CE cultures to exclude *Salmonella* are not fully understood. It has been postulated that CE cultures provide protection against pathogens in several ways. One mechanism involves competition for intestinal attachment sites on the mucosa of the intestine (Nurmi et al. 1992). If the bacteria of the CE culture can fill intestinal attachment sites, it is believed that the pathogen will not have a site to bind and, thus, will pass through the animal. Another method involves the competition for nutrients. Pathogens that do not have the appropriate nutrients available for growth will not be established in the host (Nurmi et al. 1992). Another method postulated to prevent pathogen establishment is through the production of compounds that may be toxic to invading pathogens, such as volatile fatty acids or bacteriocins. Regardless of mechanisms involved, some researchers believe that CE cultures function as a bacteriostatic intervention strategy. Thus, pathogens can increase in the intestines if the beneficial microflora are disrupted because of disease, changes in nutrition, weather changes, and so forth (Lafont et al. 1983). This is the same scenario that every bacterium must endure to survive in the animal's gastrointestinal tract or in the external environment.

Breeder Poultry

In an ideal situation, pathogen-free breeder birds would be placed into a pathogen-free environment to produce pathogen-free eggs for commercial broiler production. In the United States, poultry producers attempt to provide adequate biosecurity to maintain healthy parent stocks. In 1996, Davies and Wray investigated *Salmonella* in three broiler breeder flocks and found that no matter how much the animal caretakers cleaned and disinfected the houses, no farm could totally eliminate this pathogen (Davies and Wray 1996). In another investigation, in which 111 breeder flocks were surveyed, the study showed that a footbath could reduce the incidence of *Salmonella* 46.1-fold (Henken et al. 1992). It has also been shown that the feed mill is another possible source of contamination. A Dutch laboratory found that feed ingredients were capable of contaminating a breeder farm (Veldman et al. 1995).

Normally, *Campylobacter* prevalence is lower in the spring of the year, when it is associated with breeder flocks (Berndtson et al. 1996). These researchers also found that the

Campylobacter infection rate increases as the size of the flock increases. Their studies showed that in 77 out of 287 flocks evaluated, it was possible to have a farm test negative for *Campylobacter* throughout the entire 1-year study (Berndtson et al. 1996). Most flocks start to become positive for *Campylobacter* after 2–3 weeks of age, and by 7 weeks of age 90% of the flocks are positive (Evans and Sayers 2000).

With high biosecurity associated with breeder flocks, how does *Salmonella* or *Campylobacter* enter these flocks? It is possible that chicks are contaminated at the hatchery or through a contaminated environment. Jones and co-workers (1991b) found that 3.9% of the environmental samples were contaminated with *Salmonella*, whereas feed samples were only sometimes contaminated with *Salmonella* (Veldman et al. 1995, Jones et al. 1991b). *Salmonella* has been found in the breeder fluff, eggshell, and feces of breeders, suggesting that *Salmonella* can be transported to the hatchery via the eggs (Bailey et al. 2002). *Campylobacter* has been difficult to detect in the environment. This is most likely because of desiccation of the bacteria or possibly because of bacterial assumption of a viable but nonculturable state. The most recent information suggests that *Campylobacter* may gain entrance into a new breeder flock through the chicks. Buhr and co-workers (2002) found that essentially all sections of the oviduct and the cloaca of hens could be contaminated with *Campylobacter*. Similar studies have been performed in roosters, suggesting that *Campylobacter* has the potential of entering the flocks at day of hatch but not being detected until the fowl are 2–4 weeks of age (Cox et al. 2002).

Hatchery

Hatcheries are the first site at which most intervention strategies are implemented. Although hatcheries are commonly fumigated with formaldehyde, studies have been performed with ozone as an alternative fumigant (Whistler and Sheldon 1989). Ozone was effective in reducing bacteria 4 to 7 logs, and greater than 4 logs for fungi. Although ozone was effective, it still did not reduce bacterial counts to the same extent as formaldehyde.

Chicks undergo a series of stressors during the first few days of life. First, chicks must hatch out of the "egg-friendly" environment into the new world, a world in which there is a dramatic change in temperature and exposure to pathogens even before the chick has exited the egg. Bailey and co-workers (1992) demonstrated that the presence of a single *Salmonella*-contaminated egg inside a hatching cabinet can lead to a significant spread of the pathogenic organism within the hatching cabinet. Further research has indicated that within the hatchery, eggshells are the best indicators for *Salmonella* contamination (Bailey et al. 1994). A review of the literature found that 5%–12% of paper pads were contaminated with *Salmonella* (Bailey et al. 2001). On further examination it was found that 42% of the individual flocks were positive for *Salmonella*. This finding is significant, because as little as 5% of the birds leaving a hatchery may be exposed to 100 colony-forming units of *Salmonella* and may subsequently infect over half of the uninfected birds leaving the hatchery. This leads to grower-house contamination (Byrd et al. 1998). In this scenario, the entire grow-out house would be contaminated with *Salmonella*, providing a potential reservoir for future flocks. Little has been published on *Campylobacter* regarding the hatchery because of the low level of detection. Broiler hatcheries surveyed from 1984 to 1989 had *Campylobacter*-positive samples ranging from 17.6% to 42.9% (Pearson et al. 1996). The authors of this study reported that the most common mode of *Campylobacter jejuni* contamination occurs by vertical transmission rather than hatchery contamination; this finding agrees with the work by Stern and co-workers (2001).

Grow-Out Houses

In terms of time in production, the grow-out period is the longest time of potential pathogen contact that directly affects the consumer; broilers are raised in this environment for 6–7 weeks. The environment is considered to be the primary source of contamination for many different bacteria, including *Salmonella* (Blankenship et al. 1993). There are conflicting reports regarding hatcheries versus environmental sources of *Salmonella*, some of which have already been discussed. To further add to the debate, in 1988, an 8-million-bird survey reported that *Salmonella* serotypes on processed carcasses were more closely associated with the hatchery-related serotypes than with the feed-related serotypes (Goren et al. 1988). A survey over the period from 1994 to 1996 reported that the *Salmonella* serotypes found before and at the end of grow-out were sporadic and could not be directly attributed to either the environment or the hatchery (Byrd et al. 1999). Workers surveying turkey brooder houses found that box-liners, litter, and drinkers were all negative for *Salmonella* before placement of a flock, with 53.8% of the poult's ceca positive for *Salmonella* on entry to the house (Hoover et al. 1997). The following flock had 39.1% of the feed pans positive for *Salmonella* before poult placement. Similar to this study, a recent large-scale survey found that *Salmonella*-positive broilers within a grow-out house were followed by a second *Salmonella*-positive flock in 26% of the flocks evaluated (Bailey et al. 2001). Moreover, houseflies trapped on fly strips were the best indicator of *Salmonella* status of the broiler flock.

There are seasonal trends regarding the detection of both *Campylobacter* and *Salmonella* (Skov et al. 1999, Stern et al. 2001). For example, *Campylobacter* is normally detected more frequently in the spring and summer, with slight increases during the fourth week of grow-out (Stern et al. 2001). Flocks are consistently high in *Campylobacter* positives (approximately 76%) at the end of the grow-out period (Humphrey et al. 1993). These numbers do not correlate well with environmental samples taken from broiler houses, in which 5.8% of the total samples were positive for *Campylobacter* (Jones et al. 1991a). In grow-out houses poultry tend to be *Campylobacter* positive before the appearance of positive drag swabs, mouse carcass rinses, insects, wild bird feces, or domestic animal feces, indicating that birds are positive as they enter farms or shortly after placement (Stern et al. 2001). However, after each house is cleaned and disinfected, *Campylobacter* could not be recovered (Shreeve et al. 2002). A possible explanation could be desiccation, as *Campylobacter* enters the viable but nonculturable state to enhance survivability within a biofilm in water lines (Cox et al. 2001, Tranchoo et al. 2002).

Both beneficial bacteria and pathogens must face a changing environment. Antimicrobials and diet continually shift the microflora within the gastrointestinal tract and the external environment (Cox et al. 2003). The pH of the external environment, such as litter, can be manipulated to a basic pH (12) or an acidic pH (3–4), which reduces both *Salmonella* and *Campylobacter* (Line 2002, Bennett et al. 2003). There needs to be further investigation to fully understand the benefits of litter treatment.

Rose and co-workers (1999) reviewed the literature regarding some known risk factors. These risk factors were associated with *Salmonella* but may also apply to *Campylobacter*. Some studies have shown that feed trucks that park near the entrances to feed mills or grow-out facilities were at increased risk of contamination when the feed was unpelleted, as compared with pelleted feed. Biosecurity should always be enforced with regard to grow-out houses, with proper clean-up between flocks to reduce the spread of pathogens to new flocks (Henken et al. 1992, Fris and van den Bos 1995). Contaminated day-old chicks and feed could affect the pathogen status of the entire farm (Vaughn et al. 1974, Oystein et al. 1996, Christensen et al. 1997, Davies et al. 1997). The number of houses on the farm (>3) increased the potential of positive *Salmonella* flocks (Oystein et al. 1996).

Conclusions

Reviewing the research and applying this information to an individual unique system is the first step to controlling foodborne pathogens. The use of intervention strategies at specific locations and time points may help to reduce foodborne pathogenic bacteria in the live animal. However, the use of these strategies to eliminate pathogens from live animals should be part of a complete management program. Additional research is needed to understand how strategies that alter the intestinal microbial ecosystem affect the health and productivity of the host animal. Further research is obviously needed to determine which intervention strategies in which combinations may provide synergistic effects in the reduction of foodborne pathogens. Further areas for future study by microbial ecologists also include understanding the succession of pathogenic and non-pathogenic bacterial species in breeder animals and farm wastes.

References

Bailey, J. S., Carson, J. A. and Cox, N. A. 1992. Ecology and implications of *Salmonella*-contaminated hatching eggs. *In* Proceedings of the XIX World's Poultry Congress, September 20–24, Vol. 3, Amsterdam, The Netherlands, p. 72–75.

Bailey, J. S., Cox, N. A. and Berrang, M. E. 1994. Hatchery-acquired *Salmonella* in broiler chicks. *Poultry Science* **73:**1153.

Bailey, J. S., Cox, N. A., Craven, S. E. and Cosby, D. E. 2002. Serotype tracking of *Salmonella* through integrated broiler chicken operations. *Journal of Food Protection* **65**:742–745.

Bailey, J. S., Stern, N. J., Fedorka-Cray, P., Craven, S. E., Cox, N. A., Cosby, D. E., Ladely, S. and Musgrove, M. T. 2001. Sources and movement of *Salmonella* through integrated poultry operations: a multistate epidemiological investigation. *Journal of Food Protection* **64**:1690–1697.

Bennett, D. D., Higgins, S. E., Moore, R. W., Beltran, R., Caldwell, D. J., Byrd, J. A. and Hargis, B. M. 2003. Effect of lime on *Salmonella enteritidis* survival in vitro. *Journal of Applied Poultry Research* **12**:65–68.

Berndtson, E., Emanuelson, U., Engvall, A. and Danielsson-Tham, M.-L. 1996. A 1-year epidemiological study of campylobacters in 18 Swedish chicken farms. *Preventive Veterinary Medicine* **26**:167–185.

Blankenship, L. C., Bailey, J. S., Cox, N. A., Stern, N. J., Brewer, R. and Williams, O. 1993. Two-step mucosal competitive flora treatment to diminish salmonellae in commercial broiler chickens. *Poultry Science* **72**:1667–1672.

Buhr, R. J., Cox, N. A., Stern, N. J., Musgrove, M. T., Wilson, J. L. and Hiett, K. L. 2002. Recovery of *Campylobacter* from segments of the reproductive tract of broiler breeder hens. *Avian Diseases* **46**:919–924.

Byrd, J. A, Corrier, D. E., DeLoach, J. R., Nisbet, D. J. and Stanker, L. H. 1998. Horizontal transmission of *Salmonella* Typhimurium in broiler chicks. *Journal of Applied Poultry Research* **7**:75–80.

Byrd, J. A., DeLoach, J. R., Corrier, D. E., Nisbet, D. J. and Stanker, L. H. 1999. Evaluation of *Salmonella* serotype distribution from commercial broiler hatcheries and growout houses. *Avian Diseases* **43**:39–47.

Cappelier, J. M., Rossero, A. and Federighi, M. 2000. Demonstration of a protein synthesis in starved *Campylobacter jejuni* cells. *International Journal of Food Microbiology* **55**:63–67.

Centers for Disease Control. 2003. Preliminary FoodNet data on the incidence of foodborne illness—selected sites United States 2002. *Morbidity and Mortality Weekly Report* **52**:340–343.

Christensen, J. P., Brown, D. J., Madsen, M., Olsen, J. E. and Bisgaard, M. 1997. Hatchery-borne *Salmonella enterica* serovar Tennessee infections in broilers. *Avian Pathology* **26**:155–158.

Council for Agricultural Science and Technology. 1994. *Foodborne Pathogens: Risks and Consequences*. Council for Agricultural Science and Technology, Ames, Iowa. Available at: http://www.cast-science.org/cast-science.lh/path_is.htm. Accessed: June 12, 2003.

Cox, N. A., Berrang, M. E., Stern, N. J. and Musgrove, M. T. 2001. Difficulty in recovering inoculated *Campylobacter jejuni* from dry poultry-associated samples. *Journal of Food Protection* **64**:252–254.

Cox, N. A., Craven, S. E., Musgrove, M. T., Berrang, M. E. and Stern, N. J. 2003. Effect of sub-therapeutic levels of antimicrobials in feed on the intestinal carriage of *Campylobacter* and *Salmonella* in turkeys. *Journal of Applied Poultry Research* **12**:32–36.

Cox, N. A., Stern, N. J., Wilson, J. L., Musgrove, M. T., Buhr, R. J. and Hiett, K. L. 2002. Isolation of *Campylobacter* spp. from semen samples of commercial broiler breeder roosters. *Avian Diseases* **46**:717–720.

Davies, R. H., Nicholas, R. A. J., McLaren, I. M., Corkish, J. D., Lanning, D. G. and Wray, C. 1997. Bacteriological and serological investigation of a persistent *Salmonella enteritidis* infection in an integrated poultry organization. *Veterinary Microbiology* **58**:277–293.

Davies, R. H. and Wray, C. 1996. Studies of contamination of three broiler breeder houses with *Salmonella enteritidis* before and after cleansing and disinfection. *Avian Diseases* **40**:626–633.

Evans, S. J. and Sayers, A. R. 2000. A longitudinal study of *Campylobacter* infection of broiler flocks in Great Britain. *Preventive Veterinary Medicine* **46**:209–223.

Fris, C. and van den Bos, J. 1995. A retrospective case-control study of risk factors associated with *Salmonella enterica* subsp. enterica serovar Enteritidis infections on Dutch broiler breeder farms. *Avian Pathology* **24**:255–272.

Goren, E., DeJong, W. A., Doornebal, P., Bolder, N. W., Mulder, R. W. A. W. and Jansen, A. 1988. Reduction of salmonellae infection of broilers by spray application of intestinal microflora: a longitudinal study. *Veterinary Quarterly* **10**:249–255.

Henken, A. M., Frankena, K., Goelema, J. O., Graat, E. A. M. and Noordhuizen, J. P. T. M. 1992. Multivariate epidemiological approach to salmonellosis in broiler breeder flocks. *Poultry Science* **71**:838–843.

Hill, C., O'Drisoll, B. and Booth, I. 1995. Acid adaptation and food poisoning microorganisms. *International Journal of Food Microbiology* **28**:245–254.

Hoover, N. J., Kenny, P. B., Amick, J. D. and Hypes, W. A. 1997. Preharvest sources of *Salmonella* colonization in turkey production. *Poultry Science* **76**:1232–1238.

Humphrey, T. J., Henley, A. and Lanning, D. G. 1993. The colonization of broiler chickens with *Campylobacter jejuni*: some epidemiological investigations. *Epidemiology Infection* **110**:601–609.

Jones, F., Axtell, R. C., Rives, D. V., Scheideler, S. E., Tarver, F. R., Walker, R. L. and Wineland, M. J. 1991a. A survey of *Campylobacter jejuni* contamination in modern broiler production and processing systems. *Journal of Food Protection* **54**:259–262.

Jones, F., Axtell, R. C., Rives, D. V., Scheideler, S. E., Tarver, F. R., Walker, R. L. and Wineland, M. J. 1991b. A survey of *Salmonella* contamination in modern broiler production. *Journal of Food Protection* **54**:502–507.

Lafont, J. P., Brée, A., Naciri, M., Yvoré, P., Guillot, J. F. and Chaslus-Dancla, E. 1983. Experimental study of some factors limiting "competitive exclusion" of *Salmonella* in chickens. *Research in Veterinary Science* **34**:16–20.

Line, J. E. 2002. *Campylobacter* and *Salmonella* populations associated with chickens raised on acidified litter. *Poultry Science* **81**:1473–1477.

Maier, R. M., Pepper, I. L. and Gerba, C. P. 2000. Environmental Microbiology. San Diego, Calif.: Academic Press.

Montville, T. J. and Matthews, K. R. 2001. Principles which influence microbial growth, survival, and death in foods, p. 13–32. *In* Food Microbiology: Fundamentals and Frontiers, 2nd edn., M. P. Doyle, L. R. Beuchat, T. J. Montville (Eds.). Washington, D.C.: American Society for Microbiology Press.

Morita, R. Y. 1997. Bacteria in Oigotrophic Environments: Starvation-Survival Lifestyle. New York: Chapman and Hall.

Novitsky, J. A. and Morita, R. Y. 1976. Morphological characterization of small cells resulting from nutrient starvation in a psychrophilic marine *Vibrio*. *Applied and Environmental Microbiology* **32**:617–622.

Nurmi, E. and Rantala, M. 1973. New aspects of *Salmonella* infection in broiler production. *Nature* **241**:210–211.

Nurmi, E., Nuotio, L. and Schneitz, C. 1992. The competitive exclusion concept development and future. *Journal of Food Microbiology* **15**:237–240.

Nurmi, E., Hakkinen, M. and Nuotio, L. 1996. Competitive exclusion. Paper read at the 31st National Meeting on Poultry Health and Processing, October 16–18, 1996, Ocean City, Ocean City, Md., pp. 72–77.

Oystein, A., Skov, M. N., Chriel, M., Affer, J. F. and Bisgaard, M. 1996. A retrospective study on *Salmonella* infection in Danish broiler flocks. *Preventative Veterinary Medicine* **26**:223–237.

Pearson, A. D., Greenwood, M. H., Feltman, R. K. A., Healing, T. D., Donaldson, J., Jones, D. M. and Colwell, R. R. 1996. Microbial ecology of *Campylobacter jejuni* in a United Kingdom chicken supply chain: intermittent common source, vertical transmission, and amplification by flock propagation. *Applied Environmental Microbiology* **62**:4614–4620.

Rose, N., Beaudeau, F., Drouin, P., Toux, J. Y., Rose, V. and Colin, P. 1999. Risk factors for *Salmonella enterica* subsp. enterica contamination in French broiler-chicken flocks at the end of the rearing period. *Preventive Veterinary Medicine* **39**:265–277.

Shreeve, J. E., Toszeghy, M., Ridley, A. and Newell, D. G. 2002. The carry-over of *Campylobacter* isolates between sequential poultry flocks. *Avian Diseases* **46**:378–385.

Skov, M. N., Angen, Ø., Chriél, M., Olsen, J. E. and Bisgaard, M. 1999. Risk factors associated with *Salmonella enterica* serovar *typhimurium* infected in Danish flocks. *Poultry Science* **78**:848–854.

Stern, N. J., Fedorka-Cray, P., Bailey, J. S., Cox, N. A., Craven, S. E., Hiett, K. L., Musgrove, M. T., Ladely, S., Cosby, D. E. and Mead, G. C. 2001. Distribution of Campylobacter spp. in selected U.S. poultry protection and processing operations. *Journal of Food Protection* **64**:1705–1710.

Talibart, R., Denis, M., Castillo, A., Cappelier, J. M. and Ermel, G. 2000. Survival and recovery of viable but non-cultivable forms of *Campylobacter* in aqueous microcosm. *International Journal of Food Microbiology* **55**:263–267.

Theunissen, H. J., Lemmens-den Toom, N. A., Burggraf, A., Stolz, E. and Michel, M. F. 1993. Influence of temperature and relative humidity on the survival of *Chlamydia pneumoniae* in aerosols. *Applied and Environmental Microbiology* **59**:2589–2593.

Tranchoo, N., Frank, J. F. and Stern, N. J. 2002. Survival of *Campylobacter jejuni* in biofilms isolated from chicken houses. *Journal of Food Protection* **65**:1110–1116.

U.S. Department of Agriculture-Food Safety Inspection Service. 1997. FSIS/CDC/FDA Sentinel Site Study: the establishment and implementation of an active surveillance for bacterial foodborne disease in the United States. Report to Congress. Available at: http://www.fsis.usda.gov/OPHS/fsisrep1.htm. Accessed: June 12, 2003

Vaughn, J. B., Williams, L. P., LeBlanc, D. R., Helsdon, H. L. and Taylor, C. 1974. *Salmonella* in a modern broiler operation: a longitudinal study. *American Journal of Veterinary Research* **35**:737–741.

Veldman, A., Vahl, H. A., Borggreve, G. J. and Fuller, D. C. 1995. A survey of the incidence of *Salmonella* species and Enterobacteriaceae in poultry feeds and feed components. *Veterinary Record* **136**:169–172.

Whistler, P. E. and Sheldon, B. W. 1989. Biocidal activity of ozone versus formaldehyde against poultry pathogens inoculated in a prototype setter. *Poultry Science* **68**:1068–1073.

Ziprin, R. L., Harvey, R. B., Hume, M. E. and Kubena, L. F. 2002. *Campylobacter* colonization of the crops of newly hatched Leghorn chicks. *Avian Diseases* **46**:985–988.

10 Microbial Ecological Principles Underlying Preharvest Intervention Strategies

Todd R. Callaway, Robin C. Anderson, Thomas S. Edrington, Kenneth M. Bischoff, Kenneth J. Genovese, Toni L. Poole, and David J. Nisbet

Introduction

The gastrointestinal tract of food animals is populated by over 2,000 known species of bacteria and can contain total bacterial populations $>10^{12}$ cells per gram of digesta. Because of the highly competitive nature of the intestinal microbial environment, a constant, intense competitive pressure always affects members of this ecosystem. Changes in the environmental conditions of the intestinal tract can profoundly alter the microbial community and even result in the elimination of specific members of the consortium.

In spite of the resources devoted to ensuring the safety of the food supply, more than 76 million U.S. citizens are made ill each year by consuming foods contaminated with pathogenic bacteria (Mead et al. 1999). Many of these food-associated illnesses have been linked to consumption of meat products or to direct contact with animals, their feces, or fecal runoff. Consumer confidence in a safe and wholesome food supply has been further damaged following repeated large-scale recalls of meat products in recent years (Sanchez et al. 2002). Losses in productivity and health costs attributed to the five most economically important foodborne pathogens (i.e., *Campylobacter*, *Salmonella*, *Escherichia coli* O157:H7, non-O157:H7 Enterohemorrhagic *E. coli*, and *Listeria*) cost the U.S. economy approximately $7 billion each year (Economic Research Service/United States Department of Agriculture [ERS/USDA] 2001). All of these pathogenic bacterial species can be found, possibly only as transient members, in the gastrointestinal microbial population of food animals.

Although postharvest antimicrobial treatments do reduce carcass contamination levels (Elder et al. 2000), some animals still carry potential foodborne pathogenic bacteria into the slaughterhouse. These pathogens are frequently not detected on the farm because they often do not affect animal health or production and are often shed sporadically or at low concentrations. Because fecal shedding levels are correlated with the incidence of carcass contamination (Elder et al. 2000), it appears that the status of live animals is crucial to our safe and wholesome food supply. Therefore, strategies that decrease foodborne pathogenic bacterial populations in the live animal on-farm, or just before they enter the abattoir, could produce "the most significant reduction in human exposures to organisms, and therefore reduction

---

Mention of a trade name, proprietary product, or specific equipment does not constitute a guarantee or warranty by the U.S. Department of Agriculture and does not imply its approval to the exclusion of other products that may be suitable.

in related illnesses and deaths" (Hynes and Wachsmuth 2000). An additional benefit to strategies that reduce pathogenic bacterial carriage in live animals could decrease the incidence of outbreaks associated with exposure to pathogens via petting zoos and open farm visits (Sanchez et al. 2002).

Pathogen Distribution

Pathogenic bacteria are fairly widespread in food animals, but the incidence of colonization can vary dramatically. For example, the incidence of *E. coli* O157:H7 in cattle varies seasonally, but it can be found in up to 80% of animals during the summer months (Elder et al. 2000). In addition, animals may shed pathogens in their feces on a given day, but not again for days or even weeks. This sporadic shedding poses a severe limitation on detection methodologies. *Salmonella* can be commonly found in poultry, cattle, swine, sheep, and goats, and *Campylobacter* can be found in nearly all food animal species as well. The ubiquitous nature of these pathogenic bacteria and the variability of shedding make detection of "carrier" animals and their subsequent elimination from the food chain a non-viable technique. Therefore, methods to eliminate pathogens from all animals before they enter the food chain might hold the most promise for food safety improvement.

Competitive Exclusion

One strategy for reducing foodborne pathogenic bacteria from the food chain directly uses non-pathogenic enteric bacteria to prevent or reduce colonization by pathogenic species. Competitive exclusion (CE) is the exclusion of enteric pathogens from the gastrointestinal tract through preferential colonization with commensal or beneficial bacteria indigenous to a particular animal species. The aim of this technology is to facilitate in the young animal the natural succession of a healthy and protective gut microflora, which otherwise may take up to a week (or longer) to become established (Schaedler 1973, Muralidhara et al. 1977). The use and understanding of CE exemplifies the need for further understanding of the microbial ecology of the gastrointestinal tract.

Early research reported that inoculating young chicks with gut bacteria obtained from healthy *Salmonella*-free adult chickens resulted in an increased resistance of the young chicks to subsequent colonization by *Salmonella* (Nurmi and Rantala 1973). Other research has shown that similar treatments with defined or undefined mixtures of healthy gut bacteria provide a measure of protection to newly hatched avian species (Lloyd et al. 1977, Barnes et al. 1979, Weinack et al. 1982, Nisbet et al. 1994). The extensive use of CE outside the United States to enhance colonization resistance of avian species to *Salmonella* has been well documented (Stavric and Kornegay 1995). The effectiveness of CE in protecting swine against *Salmonella* colonization has also been reported in recently developed CE mixtures (Fedorka-Cray et al. 1999, Nisbet et al. 1999, Genovese et al. 2000).

Although weaning of suckling pigs has been proposed as a critical control point for *Salmonella* control (Fedorka-Cray et al. 1997), evidence suggests that other critical control points may also exist during the grower and finisher phase of production (Davies et al. 1998). Because CE, at least in its present state, is less effective when applied to mature animals or to those harboring preexisting *Salmonella* infections (Corrier et al. 1998), strategies that prevent enteric infections occurring in older animals are still needed. For example, development of a lactose-adapted CE culture for young poultry has shown great promise in controlling *Salmonella* infections (Corrier et al. 1993, Corrier et al. 1997). Dietary lactose enhanced resistance to *Salmonella enteritidis* colonization by stimulating growth of beneficial anaerobes at the expense of the non-lactose-fermenting *Salmonella* in molting hens and market-age broilers. Because *Salmonella* infections in U.S. poultry operations are asymptomatic, the extra cost of providing dietary lactose made the complimentary use of lactose-adapted CE cultures too costly for use in the poultry industry. However, *Salmonella* infections in swine often result in increased morbidity and mortality (leading to decreased profitability), particularly those infections caused by the host-adapted serotype Choleraesuis (Schwartz 1991). Young swine diets often contain lactose; therefore, the use of lactose-adapted CE cultures combined with effective quantities of small intestinal bypass-lactose may likely be cost effective in controlling *Salmonella* infections of swine.

Several mechanisms have been proposed to describe how the indigenous flora excludes pathogenic bacteria from the gut: competition for nutrients or attachment sites within the gut, production of antibacterial substances or conditions (e.g., volatile fatty acids, bacteriocins, or anaerobiosis), and immunostimulation (Stavric and Kornegay 1995). The ability of enteric pathogens to persist within the competitive gut environment undoubtedly depends on their ability to secure available nutrients against a highly competitive microbial population and to survive inhospitable conditions (Falk et al. 1998). Although considerable information is known regarding the nutritional physiology of *E. coli* and *Salmonella* in pure culture, little is known regarding their ability to compete for specific nutrients within mixed microbial ecosystems, and this area of research has only recently begun to be investigated (Duncan et al. 1999).

Bacteriophage, Natural Predators of the Intestinal Microbial Ecosystem

Other natural members of the microbial ecosystem can also be used to reduce foodborne pathogenic bacteria. Bacteriophage are viruses that specifically affect bacteria and that are common members of the intestinal microbial flora of food animals. Phage have been repeatedly isolated from the bovine and ovine rumen and intestinal contents, and they vary with dietary changes (Klieve and Bauchop 1988, Klieve and Swain 1993). Phage play an important role in ruminal nutrient cycling and appear to allow the intestinal microflora to quickly adapt to dietary changes. In recent results, 46% of the sheep were found to be naturally infected with *E. coli* O157:H7–specific bacteriophage (T. Callaway, unpublished data).

Bacteriophage are highly specific and can be active against a single strain of a certain species (Barrow and Soothill 1997). Because phage have such a high degree of "target" specificity, it has been suggested that selected bacteriophage could be introduced to the intestinal microbial population to eliminate specific microorganisms from a mixed microbial population (Merril et al. 1996, Summers 2001).

Experimentally induced diarrhea in calves, caused by enteropathogenic *E. coli* (EPEC), was inhibited by treatment with a combination of EPEC-specific phage (Smith and Huggins 1983, Smith and Huggins 1987). Calves challenged with EPEC died, but phage-treated calves remained healthy when challenged and contained no EPEC in their spleens compared with high-EPEC concentrations in control calf spleens (Smith and Huggins 1987). Phage were also successfully used recently to reduce respiratory infections (*E. coli*) in broiler chickens (Huff et al. 2002).

Some researchers have examined the use of phage to reduce *E. coli* O157:H7 in cattle (Kudva et al. 1999). Several O157-specific phage were isolated but were only active under highly aerated conditions. These highly aerated conditions could be easily achieved during the treatment of foods (e.g., sprouts), but not within the gastrointestinal tract of food animals. Thus, it appears that a better understanding of the effects of phage in more gastrointestinal environmentally relevant conditions is crucial. With this consideration, the role of phage in the intestinal tract of animals and the potential for using phage to remove specific foodborne pathogens from animals before slaughter may be better understood (Lederberg 1996).

Dietary Effects on Pathogen Shedding in Food Animals

The microbial ecology of the ruminant gastrointestinal tract is well characterized; however, in spite of this large base of knowledge, the effect of diet on intestinal microbial populations is still not fully understood. In the last half-century, ruminants have been fed increasing amounts of grain (an energy-dense feedstuff) to increase growth efficiency, instead of the forage-based diet that they evolved to consume (Huntington 1997, Russell and Rychlik 2001). Feeding cattle high-grain diets results in more efficient growth but can cause serious changes in the ruminal and intestinal microbial ecosystem and can lead to severe intestinal dysfunction that impacts overall animal health and performance (Russell and Rychlik 2001). It has been demonstrated that abruptly switching cattle from a predominantly grain diet to a hay diet for 5 days prior to slaughter reduces fecal *E. coli* populations (Diez-Gonzalez et al. 1998), but this remains a highly contentious suggestion because of conflicting experimental data (Hancock et al. 2000, Jarvis and Russell 2001).

Studies have demonstrated that altering either the forage-to-grain ratio or the type or quality of forage used in cattle diets could alter *E. coli* populations. Early studies demonstrated that reducing the amount of forage in a ration, inducing starvation, or switching from high-quality forage to poorer-quality forage caused an increase in generic *E. coli* or O157:H7 populations (Brownlie and Grau 1967, Allison et al. 1975, Kudva et al. 1996). Experimentally inoculated calves that were fed a high-grain ration consistently shed more *E. coli* O157:H7 in their feces (Tkalcic et al. 2000).

Cattle fed a high-grain (feedlot-type) ration contained generic *E. coli* populations that were 3 logs higher than cattle fed only hay (Diez-Gonzalez et al. 1998). When cattle were switched abruptly from a high-grain diet to a ration composed of 100% hay, fecal *E. coli* populations declined 1,000-fold, and the population of *E. coli* resistant to an acid shock similar to that of the human stomach declined more than 100,000-fold within a 5-day period (Diez-Gonzalez et al. 1998). On the basis of these results it was suggested that feedlot cattle might be switched from high-grain diets to hay for the 5 days immediately before slaughter to help reduce *E. coli* populations before cattle enter the food chain (Diez-Gonzalez et al. 1998). Unfortunately, Diez-Gonzalez and coworkers only measured *E. coli*, and thus only generalized inferences could be made about the effect of the diet on *E. coli* O157:H7 or other EHEC. In a similar follow-up study, naturally colonized cattle were fed a high-grain ration and were maintained on the high-grain diet or were abruptly switched to a hay diet; 52% of the cattle that continued to be fed grain were *E. coli* O157:H7–positive, compared with only 18% of the hay-fed cattle (Keen et al. 1999). In other related studies, feeding cattle conserved forages (e.g., hay) was hypothesized to potentially improve food safety by reducing the risk of carcass contamination with intestinal/fecal bacteria (Jacobson et al.

2002). However, other research groups have produced contradictory results that suggest forage feeding had no effect or even increased *E. coli* O157:H7 shedding (Hovde et al. 1999, Buchko et al. 2000a, Buchko et al. 2000b). Therefore, although it appears that dietary composition and changes in diet affect *E. coli* populations (Callaway et al. 2003), the debate over this controversial concept is not settled.

Use of Bacterial Physiological Systems to Reduce Foodborne Pathogenic Bacteria

Although pathogenic bacteria can be common members of the microbial ecosystem, many of these species share physiological traits that the rest of the microbial population does not. Certain bacteria (e.g., *E. coli* and *Salmonella*) can respire under anaerobic conditions by converting nitrate to nitrite via a dissimilatory nitrate reductase (Stewart 1988). Interestingly, the intracellular bacterial enzyme nitrate reductase does not differentiate between nitrate and its analog, chlorate. In these bacteria, chlorate is reduced to chlorite in the cytoplasm, which accumulates and kills the bacteria (Figure 10.1; Stewart 1988).

The addition of chlorate to the diet reduced the intestinal and fecal populations of *Salmonella* and *E. coli* O157:H7 in experimentally challenged pigs (Anderson et al. 2001a, Anderson et al. 2001b). Further studies demonstrated that chlorate administered in the drinking water reduced ruminal, intestinal, and fecal populations of *E. coli* O157:H7 in both cattle and sheep (Callaway et al. 2002). Initial work with chlorate in broilers and turkeys has yielded promising results (J. Byrd, personal communication). Chlorate does not appear to have any effect on the ruminal or cecal/colonic fermentation in ruminants or monogastrics (Callaway et al. 2002). In addition, selection of chlorate-resistant bacteria is unlikely because these bacteria are unable to compete effectively with the normal intestinal microbial flora (Callaway et al. 2001). The ability of chlorate to quickly reduce pathogenic bacterial populations in food-producing animals has led to the idea that chlorate should be given to animals as a supplement immediately before shipment to the slaughterhouse (Anderson et al. 2000). The use of chlorate in food animals is currently under review by the U.S. Food and Drug Administration (FDA).

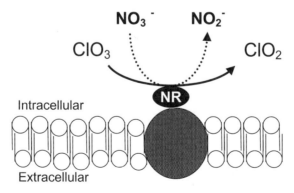

Figure 10.1   The mechanism of action of chlorate in bacteria equipped with nitrate reductase (NR).

Use and Effects of Antimicrobials on the Gastrointestinal Microecology

For over 50 years, antimicrobials have been used in agriculture to manage clinically important diseases in food animals. When individual animals display clinical symptoms, often the entire herd is treated prophylactically to prevent spread of the disease and to treat asymptomatic animals (Gustafson and Bowen 1997). This practice is essential to maintain herd health and to increase animal productivity, but producers have increasingly relied on antibiotics to treat disease rather than to emphasize good animal husbandry and hygiene practices that prevent disease. As a consequence, we are now seeing an increase in the incidence of drug-resistant bacteria in animals (Coates and Hoopes 1980, Wondwossen et al. 2000).

Another controversial practice is the use of antimicrobials at sub-therapeutic levels for growth promotion (Gustafson and Bowen 1997). The growth-promoting effects of antibiotics were first noted when streptomycin added to the diet of chickens improved their growth (Moore et al. 1946). Shortly afterward, residual antibiotic in the spent by-products of chlortetracycline fermentations was shown to improve weight gain and reduce the amount of feed needed to bring chickens and other livestock to market weight (Stokstad et al. 1949, Stokstad and Jukes 1950). Later, other antibiotics were shown to have similar effects, and at present, there are at least nine antimicrobial classes that are approved for use in feed as growth promoters. The mechanism of action behind growth promotion is not clearly understood; however, antimicrobials may modulate the metabolic activity of certain intestinal microorganisms or shift the balance of the microbial ecosystem in a manner that promotes more efficient digestion (Chopra and Roberts 2001). There is also evidence that animals fed antibiotics have increased serum levels of growth factors such as insulin-like growth factor 1 (Hathaway et al. 1996, Hathaway et al. 1999). Both broad-spectrum and narrow-spectrum antibiotics (see Chapter 17) have been shown to enhance growth performance. For example, ionophores (monensin), ionophore-like compounds (flavomycin), and macrolides (tylosin) generally target Gram-positive bacteria, but it is not clear that the antibacterial action of antibiotics is solely responsible for the growth-promoting effect.

The public health community is deeply concerned that enteric bacteria resistant to the growth-promoting antibiotics, particularly the foodborne pathogens *Salmonella*, *Campylobacter*, and *E. coli* O157:H7, may enter the food chain via contaminated meat products and produce. Although most cases of human illness are self-limiting, serious human infections caused by these pathogens may require treatment with antibiotics. In addition, there are concerns about the selection of antibiotic-resistant bacteria that have the ability to readily exchange genetic elements with many diverse microbial species. When food animals are harvested and processed, commensal bacteria that carry antibiotic resistance genes may contaminate meat products. Once in the human gastrointestinal tract, genetic transfer or "deck shuffling" can occur. There is significant evidence that *Bacteroides* spp. are involved in the transfer of tetracycline and erythromycin resistance genes among *Bacteroides* and other genera in the human colon (Shoemaker et al. 2001).

The potential for antibiotic-resistant bacteria entering the food chain through contaminated meat has put pressure on the FDA to propose tighter controls on antibiotic use in food animal production. Several political action groups are lobbying for the FDA to ban the sub-therapeutic use of antimicrobials for growth promotion—actions that will affect current production practices and that may have significant implications for animal health. Finally, as many European nations are banning the use of antimicrobials in animal feed, the use of such drugs in food animals may pose potential trade barriers.

In spite of these potential drawbacks to antibiotic treatment, recent research has found that some antibiotics do have the potential to improve food safety. Neomycin sulfate is approved for use in cattle and has a 24-hour withdrawal period. Cattle fed neomycin sulfate for 48 hours followed by a 24-hour withdrawal period shed significantly lower generic *E. coli* and *E. coli* O157:H7 populations in their feces, and this difference continued 5 days after neomycin withdrawal (Elder et al. 2002). In other research, tilmicosin increased shedding, whereas ceftiofur decreased the shedding of *E. coli* O157:H7 in calves (S. Price, personal communication).

Ionophores are growth-promoting antimicrobials unrelated to antibiotics used in human medicine that are therefore unlikely to lead to an increase in antibiotic resistance. Monensin is the most widely used ionophore in the livestock industry; it is used primarily to enhance performance of growing ruminants and also as a coccidiostat in poultry (Russell and Strobel 1989). Because the physiology of some common pathogenic bacteria renders themselves insensitive to the action of ionophores (the Gram-negative membrane structure prevents ionophores from reaching the cell membrane), ionophores do not appear to have any direct effect on gut populations of Gram-negative bacteria (Busz et al. 2002, Edrington et al. 2003a). However, because ionophores alter ruminal and intestinal microbial populations, it was hypothesized that they could provide a competitive advantage to pathogenic bacteria. *In vitro* and *in vivo* studies have demonstrated that the use of ionophores does not, in fact, provide such a competitive advantage to *E. coli* O157:H7 or *Salmonella* species in ruminants (Edrington et al. 2003a, Edrington et al. 2003b).

Water Systems as a Transmission Route?

In addition to foodstuffs, other recent human outbreaks of gastroenteritis caused by *E. coli* O157:H7 have been linked to cattle manure or the runoff from dairies infiltrating municipal water supplies (e.g., in Walkerton, ON; Population and Public Health Branch 2000). Cattle, as humans, can be infected by pathogens via water (Jackson et al. 1998, Shere et al. 2002); thus, water troughs can be reservoirs of pathogenic bacteria and are thought to be responsible for some of the dissemination of pathogenic bacteria (e.g., EHEC) within a herd (LeJeune et al. 2001). Although the actual significance of the water troughs in horizontal or vertical transmission has not been definitively established, intervention strategies focused on water troughs (e.g., ozonation and chlorination) offer potential to reduce the number of animals colonized by recycling pathogenic bacteria between pen- or herd-mates.

In addition, water runoff from farms contaminated with pathogenic bacteria from feces may be used to irrigate feed crops, where it can be consumed by animals or people (Maule 2000, Sanchez et al. 2002). Further research into reducing pathogen survival and reproduction in the water supply and runoff can potentially increase food safety by reducing the risk of foodborne pathogen horizontal transmission via drinking water (see Chapter 13).

Conclusions

The use of several of the individual strategies described above to reduce foodborne pathogenic bacteria in the live animal shows great promise. Furthermore, the use of these strategies to eliminate pathogens from live animals could significantly reduce illnesses caused by direct animal contact. However, additional research is needed to understand how these strategies that alter the intestinal microbial ecosystem affect the health and productivity of the host animal. The implementation of several of these intervention strategies in tandem (e.g., vaccination coupled with CE in the early phases of production, feeding a "pathogen-reducing diet" during the grow-out phase, addition of chlorate or antibiotics targeted against specific pathogens immediately before slaughter) could introduce "multiple hurdles" to the entry of foodborne pathogenic bacteria to the human food chain. Further research is obviously needed to determine the complementarity of these strategies to enhance synergistic effects. Further areas for future study by microbial ecologists also include understanding the succession of pathogenic and non-pathogenic bacterial species in water systems, farm wastes (manure), and water runoff from farms.

References

Allison, M. J., Robinson, I. M., Dougherty, R. W. and Bucklin, J. A. 1975. Grain overload in cattle and sheep: changes in microbial populations in the cecum and rumen. *American Journal of Veterinary Research* **36**:181–185.

Anderson, R. C., Buckley, S. A., Callaway, T. R., Genovese, K. J., Kubena, L. F., Harvey, R. B. and Nisbet, D. J. 2001a. Effect of sodium chlorate on *Salmonella typhimurium* concentrations in the pig gut. *Journal of Food Protection* **64**:255–259.

Anderson, R. C., Buckley, S. A., Kubena, L. F., Stanker, L. H., Harvey, R. B. and Nisbet, D. J. 2000. Bactericidal effect of sodium chlorate on *Escherichia coli* O157:H7 and *Salmonella typhimurium* DT104 in rumen contents *in vitro*. *Journal of Food Protection* **63**:1038–1042.

Anderson, R. C., Callaway, T. R., Buckley, S. A., Anderson, T. J., Genovese, K. J., Sheffield, C. L. and Nisbet, D. J. 2001b. Effect of oral sodium chlorate administration on *Escherichia coli* O157:H7 in the gut of experimentally infected pigs. *International Journal of Food Microbiology* **71**:125–130.

Barnes, E. M., Impey, C. S. and Stevens, B. J. H. 1979. Factors affecting the incidence and anti-*Salmonella* activity of the anerobic cecal flora of the chick. *Journal of Hygiene* **82**:263–283.

Barrow, P. A. and Soothill, J. S. 1997. Bacteriophage therapy and prophylaxis: rediscovery and renewed assessment of potential. *Trends in Microbiology* **5**:268–271.

Brownlie, L. E. and Grau, F. H. 1967. Effect of food intake on growth and survival of Salmonellas and *Escherichia coli* in the bovine rumen. *Journal of General Microbiology* **46**:125–134.

Buchko, S. J., Holley, R. A., Olson, W. O., Gannon, V. P. J. and Veira, D. M. 2000a. The effect of different grain diets on fecal shedding of *Escherichia coli* O157:H7 by steers. *Journal of Food Protection* **63**:1467–1474.

Buchko, S. J., Holley, R. A., Olson, W. O., Gannon, V. P. J. and Veira, D. M. 2000b. The effect of fasting and diet on fecal shedding of *Escherichia coli* O157:H7 by cattle. *Canadian Journal of Animal Science* **80**:741–744.

Busz, H. W., McAllister, T. A., Yanke, L. J., Olson, M. E., Morck, D. W. and Read, R. R. 2002. Development of antibiotic resistance among *Escherichia coli* in feedlot cattle. *Journal of Animal Science* **80(Suppl. 1)**:102.

Callaway, T. R., Anderson, R. C., Anderson, T. J., Poole, T. L., Kubena, L. F. and Nisbet, D. J. 2001. *Escherichia coli* O157:H7 becomes resistant to sodium chlorate addition in pure culture but not in mixed culture or in vivo. *Journal of Applied Microbiology* **91**:427–434.

Callaway, T. R., Anderson, R. C., Genovese, K. J., Poole, T. L., Anderson, T. J., Byrd, J. A., Kubena, L. F. and Nisbet, D. J. 2002. Sodium chlorate supplementation reduces *E. coli* O157:H7 populations in cattle. *Journal of Animal Science* **80**:1683–1689.

Callaway, T. R., Elder, R. O., Keen, J. E., Anderson, R. C. and Nisbet, D. J. 2003. Forage feeding to reduce pre-harvest *E. coli* populations in cattle, a review. *Journal of Dairy Science* **86**:852–860.

Chopra, I. and Roberts, M. 2001. Tetracycline antibiotics: mode of action, applications, molecular biology, and epidemiology of bacterial resistance. *Microbiology and Molecular Biology Reviews* **65**:232–260.

Coates, S. R. and Hoopes, K. H. 1980. Sensitivities of *Escherichia coli* isolated from bovine and porcine enteric infections to antimicrobial antibiotics. *American Journal of Veterinary Research* **41**:1882–1883.

Corrier, D. E., Byrd, J. A., Hume, M. E., Nisbet, D. J. and Stanker, L. H. 1998. Effect of simultaneous or delayed competitive exclusion treatment on the spread of *Salmonella* in chicks. *Journal of Applied Poultry Research* **7**:132–137.

Corrier, D. E., Hargis, B. M., Hinton, A. and DeLoach, J. R. 1993. Protective effect of used poultry litter and lactose in the feed ration on *Salmonella enteritidis* colonization of leghorn chicks and hens. *Avian Diseases* **37**:47–52.

Corrier, D. E., Nisbet, D. J., Hargis, B. M., Holt, P. S. and DeLoach, J. R. 1997. Provision of lactose to molting hens enhances resistance to *Salmonella enteriditis* colonization. *Journal of Food Protection* **60**:10–15.

Davies, P. R., Bovee, F. G. E. M., Funk, J. A., Morrow, W. E. M., Jones, F. T. and Deen, J. 1998. Isolation of *Salmonella* serotypes from feces of pigs raised in a multiple-site production system. *Journal of the American Veterinary Medical Association* **212**:1925–1929.

Diez-Gonzalez, F., Callaway, T. R., Kizoulis, M. G. and Russell, J. B. 1998. Grain feeding and the dissemination of acid-resistant *Escherichia coli* from cattle. *Science* **281**:1666–1668.

Duncan, S. H., Scott, K. P., Flint, H. J. and Stewart, C. S. 1999. Commensal-pathogen interactions involving *Escherichia coli* O157:H7 and the prospects for control, pp. 71–89. *In Escherichia Coli* O157:H7 in Farm Animals, C. S. Stewart, H. J. Flint (Eds.). New York: CABI Publishing.

Economic Research Service/United States Department of Agriculture. 2001. ERS estimates foodborne disease costs at $6.9 billion per year. Economic Research Service/United States Department of Agriculture. Available at: http://www.ers.usda.gov/Emphases/SafeFood/features.htm.

Edrington, T. S., Callaway, T. R., Bischoff, K. M., Genovese, K. J., Elder, R. O., Anderson, R. C. and Nisbet, D. J. 2003a. Effect of feeding the ionophores monensin and laidlomycin propionate and the antimicrobial bambermycin to sheep experimentally infected with *E. coli* O157:H7 and *Salmonella* Typhimurium. *Journal of Animal Science* **81**:553–560.

Edrington, T. S., Callaway, T. R., Varey, P. D., Jung, Y. S., Bischoff, K. M., Elder, R. O., Anderson, R. C., Kutter, E., Brabban, A. D. and Nisbet, D. J. 2003b. Effects of the antibiotic ionophores monensin, lasalocid, laidlomycin propionate and bambermycin on *Salmonella* and *E. coli* O157:H7 in vitro. *Journal of Applied Microbiology* **94**:207–213.

Elder, R. O., Keen, J. E., Siragusa, G. R., Barkocy-Gallagher, G. A., Koohmaraie, M. and Lagreid, W. W. 2000. Correlation of enterohemorrhagic *Escherichia coli* O157 prevalence in feces, hides, and carcasses of beef cattle during processing. *Proceedings of the National Academy of Sciences (USA)* **97**:2999–3003.

Elder, R. O., Keen, J. E., Wittum, T. E., Callaway, T. R., Edrington, T. S., Anderson, R. C. and Nisbet, D. J. 2002. Intervention to reduce fecal shedding of enterohemorrhagic *Escherichia coli* O157:H7 in naturally infected cattle using neomycin sulfate. *Journal of Animal Science* **80(Suppl. 1)**:151.

Falk, P. G., Hooper, L. V., Midtvedt, T. and Gordon, J. I. 1998. Creating and maintaining the gastrointestinal ecosystem: what we know and need to know from gnotobiology. *Microbiology and Molecular Biology Reviews* **62**:1157–1170.

Fedorka-Cray, P. J., Bailey, J. S., Stern, N. J., Cox, N. A., Ladely, S. R. and Musgrove, M. 1999. Mucosal competitive exclusion to reduce *Salmonella* in swine. *Journal of Food Protection* **62**:1376–1380.

Fedorka-Cray, P. J., Harris, D. L., and Whipp, S. C. 1997. Using isolated weaning to raise *Salmonella*-free swine. *Veterinary Medicine* **April**:375–382.

Genovese, K. J., Anderson, R. C., Harvey, R. B. and Nisbet, D. J. 2000. Competitive exclusion treatment reduces the mortality and fecal shedding associated with enterotoxigenic *Escherichia coli* infection in nursery-raised pigs. *Canadian Journal of Veterinary Research* **64**:204–207.

Gustafson, R. H. and Bowen, R. E. 1997. Antibiotic use in animal agriculture. *Journal of Applied Microbiology* **83**:531–541.

Hancock, D. D., Besser, T. E., Gill, C. and Bohach, C. H. 2000. Cattle, hay, and *E. coli. Science* **284**:49–50.

Hathaway, M. R., Dayton, W. R., White, M. E., Henderson, T. L. and Hennington, T. B. 1996. Serum insulin-like growth factor I (IGF-I) concentrations are increased in pigs fed antimicrobials. *Journal of Animal Science* **74**:1541–1547.

Hathaway, M. R., Dayton, W. R., White, M. E., Henderson, T. L., Young, T. A. and Doan, T. A. 1999. Effect of feed intake on antimicrobially induced increases in porcine serum insulin-like growth factor. *Journal of Animal Science* **77**:3208–3214.

Hovde, C. J., Austin, P. R., Cloud, K. A., Williams, C. J. and Hunt, C. W. 1999. Effect of cattle diet on *Escherichia coli* O157:H7 acid resistance. *Applied and Environmental Microbiology* **65**:3233–3235.

Huff, W. E., Huff, G. R., Rath, N. C., Balog, J. M., Xie, H., Moore, P. A. and Donoghue, A. M. 2002. Prevention of *Escherichia coli* respiratory infection in broiler chickens with bacteriophage (SPR02). *Journal of Poultry Science* **81**:437–441.

Huntington, G. B. 1997. Starch utilization by ruminants: from basics to the bunk. *Journal of Animal Science* **75**:852–867.

Hynes, N. A. and Wachsmuth, I. K. 2000. *Escherichia coli* O157:H7 risk assessment in ground beef: a public health tool, p. 46. Proceedings of the 4th International Symposium on Shiga Toxin-Producing *Escherichia coli* Infections, October 29–November 2, Kyoto, Japan.

Jackson, S. G., Goodbrand, R. B., Johnson, R. P., Odorico, V. G., Alves, D., Rahn, K., Wilson, J. B., Welch, M. K. and Khakhria, R. 1998. *Escherichia coli* O157:H7 diarrhoea associated with well water and infected cattle on an Ontario farm. *Epidemiology and Infection* **120**:17–20.

Jacobson, L. H., Nagle, T. A., Gregory, N. G., Bell, R. G., Le Roux, G. and Haines, J. M. 2002. Effect of feeding pasture-finished cattle different conserved forages on *Escherichia coli* in the rumen and faeces. *Meat Science* **62**:93–106.

Jarvis, G. N. and Russell, J. B. 2001. Differences in *Escherichia coli* culture conditions can have a large impact on the induction of extreme acid resistance. *Current Microbiology* **43**:215–219.

Keen, J. E., Uhlich, G. A. and Elder, R. O. 1999. Effects of hay- and grain-based diets on fecal shedding in naturally-acquired enterohemorrhagic *E. coli* (EHEC) O157 in beef feedlot cattle, p. 71. 80th Conference of Research Workers in Animal Diseases, November 9–11, Chicago, IL.

Klieve, A. V. and Bauchop, T. 1988. Morphological diversity of ruminal bacteriophages from sheep and cattle. *Applied and Environmental Microbiology* **54**:1637–1641.

Klieve, A. V. and Swain, R. A. 1993. Estimation of ruminal bacteriophage numbers by pulsed-field electrophoresis and laser densitometry. *Applied and Environmental Microbiology* **59**:2299–2303.

Kudva, I. T., Hatfield, P. G. and Hovde, C. J. 1996. *Escherichia coli* O157:H7 in microbial flora of sheep. *Journal of Clinical Microbiology* **34**:431–433.

Kudva, I. T., Jelacic, S., Tarr, P. I., Youderian, P. and Hovde, C. J. 1999. Biocontrol of *Escherichia coli* O157 with O157-specific bacteriophages. *Applied and Environmental Microbiology* **65**:3767–3773.

Lederberg, J. 1996. Smaller Fleas . . . ad infinitum: therapeutic bacteriophage redux. *Proceedings of the National Academy of Sciences (USA)* **93**:3167–3168.

LeJeune, J. T., Besser, T. E. and Hancock, D. D. 2001. Cattle water troughs as reservoirs of *Escherichia coli* O157. *Applied and Environmental Microbiology* **67**:3053–3057.

Lloyd, A. B., Cumming, R. B. and Kent, R. D. 1977. Prevention of *Salmonella typhimurium* infection in poultry by pre-treatment of chickens and poults with intestinal extracts. *Australian Veterinary Journal* **53**:82–87.

Maule, A. 2000. Survival of vero cytotoxigenic *Escherichia coli* O157:H7 in soil, water and on surfaces. *Journal of Applied Microbiology* **88**:71S–78S.

Mead, P. S., Slutsker, L., Dietz, V., McCraig, L. F., Bresee, J. S., Shapiro, C., Griffin, P. M. and Tauxe, R. V. 1999. Food-related illness and death in the United States. *Emerging Infectious Diseases* **5**:607–625.

Merril, C. R., Biswas, B., Carlton, R., Jensen, N. C., Creed, G. J., Zullo, S. and Adhya, S. 1996. Long-circulating bacteriophage as antibacterial agents. *Proceedings of the National Academy of Sciences (USA)* **93**:3188–3192.

Moore, P. R., Evanson, A., Luckey, T. D., McCoy, E., Elvehjen, C. A. and Hart, E. B. 1946. Use of sulfasuxidine, stretothricin, and streptomycin in nutritional studies with the chick. *Journal of Biological Chemistry* **165**:437–441.

Muralidhara, K. S., Sheggeby, G. G., Elliker, P. R., England, D. C. and Sandine, W. E. 1977. Effect of feeding lactobacilli on the coliform and *Lactobacillus* flora of intestinal tissue and feces from piglets. *Journal of Food Protection* **40**:288–295.

Nisbet, D. J., Anderson, R. C., Harvey, R. B., Genovese, K. J., DeLoach, J. R. and Stanker, L. H. 1999. Competitive exclusion of *Salmonella* serovar *Typhimurium* from the gut of early weaned pigs, pp. 80–82. 3rd International Symposium on the Epidemiology and Control of *Salmonella* in Pork, October 8–14, Washington, D.C.

Nisbet, D. J., Ricke, S. C., Scanlan, C. M., Corrier, D. E., Hollister, A. G. and DeLoach, J. R. 1994. Inoculation of broiler chicks with a continuous-flow derived bacterial culture facilitates early cecal bacterial colonization and increases resistance to *Salmonella typhimurium*. *Journal of Food Protection* **57**:12–15.

Nurmi, E. and Rantala, M. 1973. New aspects of *Salmonella* infection in broiler production. *Nature* **24**:210–211.

Population and Public Health Branch. 2000. Waterborne outbreak of gastroenteritis associated with a contaminated municipal water supply, Walkerton, Ontario, May–June 2000. *Canadian Communicable Disease Reports* **26**:170–173.

Russell, J. B. and Rychlik, J. L. 2001. Factors that alter rumen microbial ecology. *Science* **292**:1119–1122.

Russell, J. B. and Strobel, H. J. 1989. Effect of ionophores on ruminal fermentation. *Applied and Environmental Microbiology* **55**:1–6.

Sanchez, S., Lee, M. D., Harmon, B. G., Maurer, J. J. and Doyle, M. P. 2002. Animal issues associated with *Escherichia coli* O157:H7. *Journal of the American Veterinary Medical Association* **221**:1122–1126.

Schaedler, R. W. 1973. The relationship between the host and its intestinal microflora. *Proceedings of the Nutrition Society* **32**:41–47.

Schwartz, K. 1991. Salmonellosis in swine. *Compendium of Continuing Education for the Practicing Veterinarian* **13**:139–147.

Shere, J. A., Kaspar, C. W., Bartlett, K. J., Linden, S. E., Norrell, B., Francey, S. and Schaefer, D. M. 2002. Shedding of *Escherichia coli* O157:H7 in dairy cattle housed in a confined environment following waterborne inoculation. *Applied and Environmental Microbiology* **68**:1947–1954.

Shoemaker, N. B., Vlamakis, H., Hayes, K. and Salyers, A. A. 2001. Evidence for extensive resistance gene transfer among *Bacteroides* spp. and among *Bacteroides* and other genera in the human colon. *Applied and Environmental Microbiology* **67**:561–568.

Smith, H. W. and Huggins, M. B. 1983. Effectiveness of phages in treating experimental *Escherichia coli* diarrhoea in calves, piglets and lambs. *Journal of General Microbiology* **129**:2659–2675.

Smith, H. W. and Huggins, R. B. 1987. The control of experimental *E. coli* diarrhea in calves by means of bacteriophage. *Journal of General Microbiology* **133**:1111–1126.

Stavric, S. and Kornegay, E. T. 1995. Microbial probiotics for pigs and poultry, pp. 205–230. *In* Biotechnology in Animal Feeds and Animal Feeding, R. J. Wallace, A. Chesson (Eds.); New York: VCH.

Stewart, V. J. 1988. Nitrate respiration in relation to facultative metabolism in enterobacteria. *Microbiology Reviews* **52**:190–232.

Stokstad, E. L. R. and Jukes, T. H. 1950. Further observations on the animal protein factor. *Proceedings of the Society of Experimental Biology and Medicine* **73**:523–528.

Stokstad, E. L. R., Jukes, T. H., Pierce, J., Page, A. C. and Franklin, A. L. 1949. The multiple nature of the animal protein factor. *Journal of Biological Chemistry* **180**:647–654.

Summers, W. C. 2001. Bacteriophage therapy. *Annual Review of Microbiology* **55**:437–451.

Tkalcic, S., Brown, C. A., Harmon, B. G., Jain, A. V., Mueler, E. P. O., Parks, A., Jacobsen, K. L., Martin, S. A., Zhao, T. and Doyle, M. P. 2000. Effects of diet on rumen proliferation and fecal shedding of *Escherichia coli* O157:H7 in calves. *Journal of Food Protection* **63**:1630–1636.

Weinack, O. M., Snoeyenbos, G. H., Smyser, C. F. and Soerjadi, A. S. 1982. Reciprocal competitive exclusion of *Salmonella* and *Escherichia coli* by native intestinal microflora of the chicken and turkey. *Avian Diseases* **26**:585–595.

Wondwossen, A. G., Davies, P. R., Morgan-Morrow, W. E., Funk, J. A. and Altier, C. 2000. Antimicrobial resistance of *Salmonella* isolates from swine. *Journal of Clinical Microbiology* **38**:4633–4636.

11   Limiting Avian Gastrointestinal Tract *Salmonella* Colonization by Cecal Anaerobic Bacteria, and a Potential Role for Methanogens

Steven C. Ricke, Casendra L. Woodward, Young Min Kwon, Leon F. Kubena, and David J. Nisbet

Introduction

In the poultry chick, native adult microflora becomes established in the small intestine within the first 2 weeks (Barnes et al. 1980), but it takes over 4 weeks to develop in the ceca (Mead and Adams 1975). Natural resistance increases with age as the adult microflora become more established. As a consequence, adult birds possess a stable gut microbial ecosystem that is relatively resistant to infection by salmonellae. There are two circumstances in which the presence of an adult cecal microflora has been identified as a critical factor required to limit colonization and spread of *Salmonella* spp. in poultry. The first circumstance involves newly hatched chicks and the addition of competitive exclusion cultures consisting of adult avian cecal microorganisms capable of preventing *Salmonella* spp. colonization. A second possible role for the cecal microflora derives from the fact that hens undergoing feed deprived induced molt are more susceptible to intestinal infection by *Salmonella enteritidis* and shed more organisms (Holt and Porter 1992, Holt et al. 1994, Holt 1995). However, because the understanding of cecal microbial ecology is incomplete, there is still considerable debate as to what mechanisms are involved.

Mechanistically, how avian gastrointestinal microbial populations serve as a barrier to transient pathogen establishment will not be resolved until a more complete picture of microbial ecology in the adult chicken gut is established. Although the microbial populations in the avian gastrointestinal tract have been periodically studied for over a half century and an extensive number of organisms have been identified, little has been done to assign ecological roles for groups of identified organisms. As a consequence, only minimal understanding is possible regarding metabolic and functional niches that various indigenous microorganisms may occupy in the different gastrointestinal regions of both young chicks and adult birds. It has become apparent that the less stable, but more transient, facultative microbial population in the young chick does shift to a more stable strict anaerobic fermentative population in certain sections of the gastrointestinal tract as the bird matures. In particular, the formation of methane may be a key microbial fermentation process that has been observed in adult birds and that therefore could be considered an indicator of the stabilized mature cecal population in chickens. However, because of the difficulties associated with their isolation, methanogens have been overlooked when the cecal microflora have been characterized. This review is focused on the importance of developing a better general understanding of the avian gastrointestinal microbial population and why some subpopulations of organisms such as the methanogens may have an important role to play in stabilizing that population.

Biological Additives and the Importance of Gut Ecology

The practice of supplementing prebiotic or probiotic biological additives to domestic animals during growth is becoming widespread in food animal production. Prebiotics involve some sort of selective additive that is not usable by the host animal but that can be selectively used, and therefore is stimulatory to a portion of the gastrointestinal population that is then presumed beneficial to the host (Orban et al. 1997, Berg 1998, Patterson and Burkholder 2003). A probiotic involves the supplementation of actual viable microorganisms that potentially can establish in the gastrointestinal tract and elicit benefits to the host (Patterson and Burkholder 2003). Reported beneficial effects, particularly in cattle, pigs and poultry, include improved general health, more efficient food use, faster growth rate and increased milk and egg production (Fuller 1992, Fuller 1997, Berg 1998). Obviously, these benefits have fueled tremendous commercial growth in marketing such products, but considerable controversy exists as to how direct the influence of addition of biological additives is in exercising these benefits. Most mechanisms suggested to explain apparent success in animals are too simplistic ecologically when the complexities of the gastrointestinal tract are considered. Thus, the problem remains that even when such biological additives are consistently effective, the mechanisms remain elusive.

Probiotics have made their largest commercial inroads in poultry production. This is because probiotic addition has been shown to be consistently successful in minimizing *Salmonella* colonization of the ceca in baby chicks (Stavric 1987, Stavric 1992). Competitive exclusion by the normal bacterial flora is considered to be the main mode of prevention of various enteropathogen colonizations in the intestinal tracts of man and animals (Freter et al. 1983a, Freter et al. 1983b, Ushijima and Seto 1991). Natural infection of poultry by *Salmonella* occurs via the oral route, and salmonellae colonize the intestinal tract, with the ceca being a primary site of colonization (Brownwell et al. 1970, Soerjadi et al. 1981, Stavric 1987, Impey and Mead 1989). The newly hatched chick has a nearly sterile cecum that lacks competitive microflora and that therefore is quite susceptible to colonization by bacteria such as foodborne *Salmonella* spp. (Stavric 1987). Modern hatchery practices that remove the egg from the mature hen, plus the use of disinfection regimes in grower houses, ensure further delay in the development of a protective cecal microflora and extend the window of susceptibility to colonization by *Salmonella* spp. and other foodborne pathogens (Bailey 1987, Stavric 1987). This has been demonstrated in experimental housing, where chicks introduced to environments without previous exposure to chickens exhibited a delay in the appearance of intestinal lactobacilli until 3 weeks after feeding (Bare and Wiseman 1964). However, several research groups demonstrated, by providing newly hatched chicks with cecal microflora from *Salmonella*-free adult chickens, that the incidence of salmonellae cecal colonization could be decreased (Nurmi and Rantala 1973, Schleifer 1985, Stavric 1987, Stavric 1992). Such results would suggest that probiotics are a means to accelerate the establishment of a maximally protective microflora that approximates the pathogen colonization resistance observed in adult chickens.

The more effective probiotic cultures usually consist of mixtures of facultative and anaerobic bacteria that require administration as intact metabolic consortia to be efficacious (Nisbet et al. 1996a, Nisbet et al. 1996b). It also appears that the more successful probiotic cultures tend to closely mimic the complex anaerobic microflora found in adult birds. As a consequence, undefined cultures that have been taken directly from cecal or fecal extracts are usually more effective at limiting colonization of pathogens than are probiotic cultures consisting of combinations of defined microorganisms (Stavric 1992). This would suggest that there are missing organisms in the defined cultures that may have not been isolated. Simply put, relative success of various probiotic cultures in poultry may partially be the consequence of their respective effectiveness in speeding up the establishment of the dominant anaerobic microflora characteristic of the adult bird gastrointestinal tract. Thus, the primary goal for assessing probiotic capabilities and *in vivo* potential for success may require identifying the characteristics that signify complete establishment of a fully developed gastrointestinal microflora in the mature chicken.

Influence of Feed Removal on Cecal Microbial Ecology

Feed withdrawal is an additional circumstance in which the role of the fully developed cecal microflora in the mature chicken is believed to be important in minimizing colonization and spread of *Salmonella* spp. in poultry. Feed removal is commonly practiced in the poultry industry in broiler production to minimize fecal contamination on carcasses during processing. Removal of feed is also the primary method used in the layer industry to induce molting and stimulate multiple egg-laying cycles in hens (Brake 1993, Holt 1995). Invariably, increased *Salmonella* contamination occurs during feed withdrawal in both laying hens and broilers (Humphrey et al. 1993, Ramirez et al. 1997, Corrier et al. 1997, Corrier et al. 1999, Durant et al. 1999). Emptying of the gastrointestinal tract by feed removal appears to be a consistent factor favoring increased levels of *Salmonella* in a variety of animal species including fasted chickens (Moran and Bilgili 1990, Humphrey et al. 1993, Corrier et al. 1999), mice (Miller and Bohnhoff 1962) and ruminants (Brownlie and Grau 1967, Grau et al. 1968). However in fasted birds, based on the cecal measurements taken and microbial populations examined, the changes in the cecal microflora are not obvious (Holt et al. 1994, Corrier et al. 1997). Consequently, what is really needed is an understanding of the key processes during establishment of anaerobes in the avian gastrointestinal system that leads to the subsequent fermentation characteristics and ecological balance exhibited by the highly protective microflora (Ricke and Pillai 1999). This requires a much more complete picture of cecal microbial ecology as well as application of molecular techniques to identify microbial species that are difficult to cultivate (Ricke and Pillai 1999, Zhu et al. 2002).

Development and Characteristics of the Avian Cecal Microflora

All components of the avian digestive tract (crop, proventriculus, gizzard, small and large intestine, and ceca) can serve as potential sites for microbial colonization, but in commercially raised poultry, only the ceca can be considered a stable enough environment to serve as the primary site for establishment of a complex microbial ecosystem consisting predominantly of obligate anaerobes (Mead 1997). In domestic poultry, paired ceca are blind pouches 15–18 cm in length, consisting of a narrow, constricted open end connected to the colon and a dilated, thinner-walled blind component (McNab 1973). Fermentation is the primary microbial activity, with CH_4 and CO_2 being the principal gases produced and acetate the primary volatile fatty acid, along with some propionate and butyrate (Annison et al. 1968, Beattie and Shrimpton 1958, Shrimpton 1963, McNab 1973), and with some isobutyric, valeric, and isovaleric acids (Hinton et al. 1990, Kubena et al. 2001). The ceca are the primary site for microbial fermentation, with a lumen population estimated to be at from 10^{10} to 10^{11} organisms per gram wet weight, thus representing the region of the gastrointestinal tract with the greatest concentration of intestinal microorganisms in mature chickens (Johansson et al. 1948, Barnes and Shrimpton 1957, McNab 1973, Mead 1989, Vispo and Karasov 1997).

A total count of organisms that approximates that found in mature birds can be reached within 2 days after hatching and can remain similar thereafter (Lev and Briggs 1956, McNab 1973). However, the complexity of the cecal microflora composition dramatically changes as the bird ages. The majority of the organisms present during the initial few days after hatch are usually facultatively anaerobic bacteria (Shapiro and Sarles 1949, Huhtanen and Pensack 1965, Smith 1965, Mead and Adams 1975). At 24 hours posthatch, lactobacilli begin to appear, and by the third day they are the predominant part of the cecal microflora (Barnes et al. 1980, Mead and Adams 1975, Schleifer 1985). After this, many of the organisms can only be cultured under anaerobic incubation laboratory conditions, and the overall microbial diversity continues to increase for the next 5–6 weeks (Barnes et al. 1972, Fuller and Brooker 1974). In the chick, gut microflora characteristic of adult birds become established in the small intestine within the first 2 weeks after the chick has hatched out (Barnes et al. 1980). However, in the ceca it takes over 4 weeks after the chick has hatched out for the microbial population characteristic of adult birds to develop, and for establishment of the predominant cecal microorganisms, it may require up to 6 weeks (Barnes et al. 1972, Mead and Adams 1975).

Determining Cecal Microbial Composition

The microbial composition of the cecal microflora has been difficult to estimate not only because of the changes that occur as the bird ages but because the isolation methods used in some studies do not necessarily support the growth of the more oxygen-sensitive anaerobes (Mead 1997). Anaerobic isolation methods and media developed for enumeration of the rumen anaerobic microflora (Hungate 1950) have been shown to be the best for studying predominant anaerobic chicken cecal microflora, and more than 200 strains of bacteria have been isolated from the ceca (Barnes et al. 1972, Barnes et al. 1979). Several groups of anaerobic cocci, streptococci, and Gram-negative and Gram-positive nonsporeforming anaerobes, including species of *Peptostreptococcus*, *Clostridium*, *Bacteroides*, *Fusobacterium*, *Eubacterium*, *Propionibacterium*, and *Bifidobacterium*, have been isolated from chicken and turkey ceca (Barnes and Impey 1970, Barnes and Impey 1972, Bedbury and Duke 1983, Salanitro et al. 1974a, Salanitro et al. 1974b).

Unfortunately, knowledge of the cecal ecosystem is incomplete because even when strict anaerobic methodology is practiced, only a portion of the total viable counts can be recovered (Mead 1997). Supplementation with rumen fluid supports optimal growth for some cecal isolates (Salanitro et al. 1974a, Salanitro et al. 1974b, Salanitro et al. 1978). This has been used as an indication of chicken cecal bacteria possessing nutritional requirements similar to rumen organisms (Salanitro et al. 1974a, Salanitro et al. 1974b, Salanitro et al. 1978, Mead 1997). Based on approaches developed for rumen bacteria (Leedle and Hespell 1980), rumen fluid-based and non-rumen fluid-based carbohydrate differential selective media for inoculation of plate media in an anaerobic glove box have been examined (Fan et al. 1995, Nisbet et al. 1994, Ricke et al. 1993, Shermer et al. 1998). Use of carbohydrate differential media has allowed the enumeration of total anaerobes on plates and given us the ability to distinguish specific bacterial populations based on energy and nutrient requirements that respond to specific dietary changes such as lactose.

Evidence for Methanogens in the Chicken Cecum

Most conventional anaerobic methods only account for the fermentative saccharolytic microbial population; important components of the cecal microbial anaerobic consortium may not be isolated and, therefore, may be overlooked (Barrow 1992, Mead 1989, Mead 1997). A primary example of this is methane gas, which was detected in chicken cecal contents over 40 years ago, yet methanogens were not isolated until some 20 years later, when the methanogenic bacterium (genus *Methanogenium*) was isolated from chicken and turkey feces and identified as the primary type associated with chickens and turkeys (Beattie and Shrimpton 1958, Shrimpton 1966, Miller and Wolin 1986, Miller et al. 1986, Miller 1991, Jensen 1996, Mead 1997, Lin and Miller 1998). Only recently has a methanogen also been tentatively isolated directly from chicken ceca (C. Woodward and S. Ricke, unpublished data; L. Daniels, personal communication).

Methane formation in the avian cecum by indigenous methanogens is not surprising because methanogens commonly inhabit the gastrointestinal tract of most food animals including ruminants and a variety of nonruminant animals (Hungate 1966, Bryant 1979, Lin and Miller 1998), and the ceca in hens should be sufficiently anaerobic (Corrier et al. 1997) to sustain methanogens as well. Even so, because retention time in the intestine is relatively short, methane generated in the intestine is the product of a phylogenetically limited range of H_2-CO_2-using methanogens (Bryant 1979, Miller and Wolin 1986). However, use of H_2 by methanogens is of central importance in the gastrointestinal ecosystem because generation of methane maintains a very low partial pressure of H_2 that thermodynamically favors production of specific fermentation products (acetate, CO_2, and H_2), rather than more reduced products (propionate, lactate, or ethanol), by the carbohydrate-using anaerobes (Hungate 1966, Thauer et al. 1977, Wolin and Miller 1982, Bryant 1979).

Ecology of Methanogens in the Chicken Cecum

The ecological importance of methanogens in chicken ceca is difficult to ascertain and may be based on several factors including presence of other potentially competitive hydrogen users, diet, and age of bird. The presence of H_2- or CO_2-using acetogenic bacteria (reductive acetogenesis) has been demonstrated in other gastrointestinal ecosystems including in the rumens of cattle, in the large intestines of pigs and humans, and in termites (Bernalier et al. 1993, Breznak and Switzer 1986, De Graeve et al. 1994, Genthner et al. 1981, Greening and Leedle 1989, Le Van et al. 1998). In batch cultures of cecal microorganisms from laying hens, Marounek et al. (1996) observed a shift from acetate and methane to propionate production when salinomycin was added. However, when the specific methanogen inhibitor 2-bromoethanesulphonic acid was added to these cecal cultures, hydrogen recovery was not increased, suggesting possible H_2 consumption by H_2-dependent acetogenesis (Marounek et al. 1996).

Dietary changes may also influence methane production in chickens. When dietary fiber was increased in the diet in the form of wheat, barley, or oats, methane was increased nearly

10-fold compared with corn-fed hens (Shrimpton 1966). However, when cecal contents of 7-week-old chickens were incubated as batch cultures in the presence of different carbohydrates including pectin, xylan, and cellulose, methane production was not significantly changed, and an average of 11 moles of volatile fatty acid were produced per mole of methane (Marounek et al. 1999). Such differences between *in vitro* and *in vivo* studies indicate that some adaptation to diet by the cecal microflora is required and that age of the bird may be a factor. Most studies have only involved detection of methane from hens (Miller and Wolin 1986, Miller et al. 1986, Miller 1991, Shrimpton 1963, Shrimpton 1966), but in preliminary experiments, Marounek et al. (1996) were unable to detect production of methane in young chickens, although observing significant methanogenesis in laying hens. However, when ceca from chicks were examined during their initial 4 weeks after hatch, 11% were positive for methane by week 2 (27 chicks total) and 100% (29 chicks) were methane positive by week 3 (C. Woodward and S. Ricke, unpublished data).

The age at which methane becomes detectable in the cecum may be dependent on the time course for establishment and stabilization of the adult anaerobic microflora. The composition and complexity of the cecal microflora is in a dynamic state for the first 5–6 weeks of life (Barnes et al. 1972, Fuller and Brooker 1974, Fuller and Turvey 1971, Stavric 1987), until the definitive microbial community is stabilized. Barnes et al. (1972) observed that the types of predominant organisms continued to change up to 6 weeks after hatching. This slow development of adult microflora has been attributed to the sanitary rearing conditions of the hatcheries, which lead to the delay of normal gut flora establishment (Bailey 1987, Stavric 1987). Chicks hatched by sitting hens would be considered much more likely to be rapidly inoculated by contact with the adult hen, whereas no such contact is possible in the relatively sterile hatchery (Bailey 1987, Stavric 1987). This isolation of hatched chicks from potential contact with adult microflora is continued in the grower houses because the houses are usually sanitized and fresh litter is frequently added to the floors before new groups of chickens are brought in (Bailey 1987).

An understanding of the key processes during establishment of anaerobes in the avian gastrointestinal system that lead to the subsequent fermentation characteristics and ecological balances exhibited by the highly protective microflora is necessary. This requires a much more complete picture of cecal microbial ecology. Only then can some of the variability of experimental results for probiotic mechanisms and susceptibility during feed deprivation be resolved. Although formation of methane may be a key microbial process that seems to be present in adult birds, methanogens are generally overlooked when extensive studies on the characterization of the chicken cecal microflora have been conducted. However, before this can be proven, extensive in-depth studies are required to quantitatively detect fermentation patterns and concomitant microbial profiles that are most likely to support methanogen colonization. This includes experimental approaches that link specific microbial populations with ecological characteristics that are thermodynamically favorable for methane formation.

Conclusions

Avian gastrointestinal microorganisms have received periodic interest over the past 40–50 years, with numerous isolations of individual organisms usually accompanied by identification and characterization studies. Relating these findings back to the ecology of the avian gastrointestinal tract and the potential effect on the nutrition and health of the bird has been the focus of most of these studies. However, in commercial poultry production the successful development and application of effective competitive exclusion microbial cultures that prevent colonization in the avian gastrointestinal tract by foodborne pathogens has resulted in a renewed interest in understanding the key factors that influence the establishment of indigenous microbial populations in chickens. The recent successful applications of these competitive exclusion or probiotic cultures ensure continued commercialization and widespread use of such cultures (Ricke and Pillai 1999).

It has become evident that the more effective probiotic cultures usually consist of mixtures of facultative and anaerobic bacteria that require administration as intact metabolic consortia to be efficacious (Nisbet et al. 1996a, Nisbet et al. 1996b). It also appears that the more successful probiotic cultures tend to closely mimic the complex anaerobic microflora found in adult birds. Given that the fermentation response and ecological balance of the probiotic consortium appears to be essential for the effectiveness of such cultures, an understanding of the microbial ecosystem in the cecum of the adult chicken is required. This should provide an understanding of the key processes during establishment of anaerobes in the avian gastrointestinal system that leads to the subsequent fermentation characteristics and ecological balance exhibited by the highly protective microflora. However, this clearly requires a much more complete picture of cecal microbial ecology than exists at present, which may be accomplished using a variety of cultural and molecular approaches.

Any characterization of the cecal microbial profile must include methanogens. Specifically, there is a need to determine when methanogens become a part of the cecal microflora, identify what characteristics in the cecal microenvironment are associated with establishment of methanogens, and quantify the methanogen populations. In addition, the question remains as to how stable the cecal methanogen populations are under these conditions. Answering this question will require examining whether feed deprivation alters methanogen activity while birds are undergoing fasting and whether key characteristics in the chicken cecal microenvironment can be linked accordingly with increases and decreases in methanogen populations.

Acknowledgements

This chapter was supported by Hatch grant H8311, administered by the Texas Agriculture Experiment Station, and U.S. Department of Agriculture National Research Initiative grants 2001-35201-09946 and 2002-35201-12585.

References

Annison, E. F., Hill, K. J. and Kenworthy, R. 1968. Volatile fatty acids in the digestive tract of the fowl. *British Journal of Nutrition* **22**:207–216.

Bailey, J. S. 1987. Factors affecting microbial competitive exclusion in poultry. *Food Technolology* **41**:88–92.

Bare, L. N. and Wiseman, R. F. 1964. Delayed appearance of lactobacilli in the intestines of chicks reared in a "new" environment. *Applied Microbiology* **12**:457–459.

Barnes, E. M. and Impey, C. S. 1972. Some properties of the nonsporing anaerobes from poultry ceca. *Journal of Applied Bacteriology* **35**:241–251.

Barnes, E. M. and Impey, C. S. 1970. The isolation and properties of the predominant anaerobic bacteria in the ceca of chickens and turkeys. *British Poultry Science* **11**:467–481.

Barnes, E. M., Impey, C. S. and Cooper, D. M. 1980. Manipulation of the crop and intestinal flora of the newly hatched chick. *The American Journal Clinical Nutrition* **33**:2426–2433.

Barnes, E. M., Impey, C. S. and Stevens, B. J. H. 1979. Factors affecting the incidence and anti-salmonella activity of the anaerobic caecal flora of the young chick. *Journal of Hygiene (Cambridge)* **82**:263–283.

Barnes, E. M., Mead, G. C., Barnum, D. A. and Harry, E. G. 1972. The intestinal flora of the chicken in the period 2 to 6 weeks of age, with particular reference to the anaerobic bacteria. *British Poultry Science* **13**:311–326.

Barnes, E. M. and Shrimpton, D. H. 1957. Causes of greening of unvisterated poultry carcasses during storage. *Journal of Applied Bacteriology* **20**:273–285.

Barrow, P. A. 1992. Probiotics for chickens, pp. 225–257. *In* Probiotics—The Scientific Basis, R. Fuller (Ed.). London: Chapman and Hall.

Beattie, J. and Shrimpton, D. H. 1958. Surgical and chemical techniques for *in vivo* studies of the metabolism of the intestinal microflora of domestic fowls. *Quarterly Journal of Experimental Physiology* **43**:399–407.

Bedbury, H. P. and Duke, G. E. 1983. Cecal microflora of turkeys fed low or high fiber diets: enumeration, identification, and determination of cellulolytic activity. *Poultry Science* **62**:765–682.

Berg, R. D. 1998. Probiotics, prebiotics or "conbiotics"? *Trends in Microbiology* **6**:89–92.

Bernalier, A., Doisneau, E., Cordelet, C., Beaumatin, P., Durand, M. and Grivet, J. P. 1993. Competition for hydrogen between methanogenesis and hydrogenotrophic acetogenesis in human colonic flora studied by ^{13}C NMR. *The Proceedings of the Nutrition Society* **52**:118A.

Brake, J. 1993. Recent advances in induced molting. *Poultry Science* **72**:929–931.

Breznak, J. A. and Switzer, J. M. 1986. Acetate synthesis from H_2 plus CO_2 by termite gut microbes. *Applied and Environmental Microbiology* **52**:623–630.

Brownlie, L. E. and Grau, F. H. 1967. Effect of food intake on growth and survival of salmonellas and *Escherichia coli* in the bovine rumen. *Journal of General Microbiology* **46**:125–134.

Brownwell, J. R., Sadler, W. W. and Fanelli, M. J. 1970. Role of the ceca in intestinal infections of chickens with *Salmonella typhimurium*. *Avian Diseases* **14**:106–116.

Bryant, M. P. 1979. Microbial methane production—theoretical aspects. *Journal of Animal Science* **48**:193–201.

Corrier, D. E., Byrd, J. A., Hargis, B. M., Hume, M. E., Bailey, R. H. and Stanker, L. H. 1999. Presence of Salmonella in the crop and ceca of broiler chickens before and after preslaughter feed withdrawal. *Poultry Science* **78**:45–49.

Corrier, D. E., Nisbet, D. J., Hargis, B. M., Holt, P. S. and DeLoach, J. R. 1997. Provision of lactose to molting hens enhances resistance to *Salmonella enteritidis* colonization. *Journal of Food Protection* **60**:10–15.

De Graeve, K. G., Grivet, J. P., Durand, M., Beaumatin, P., Cordelet, C., Hannequart, G. and Demeyer, D. 1994. Competition between reductive acetogenesis and methanogenesis in the pig large-intestinal flora. *Journal of Applied Bacteriology* **76**:55–61.

Durant, J. A., Corrier, D. E., Byrd, J. A., Stanker, L. H. and Ricke S. C. 1999. Feed deprivation affects crop environment and modulates *Salmonella enteritidis* colonization and invasion of Leghorn hens. *Applied and Environmental Microbiology* **65**:1919–1923.

Fan, Y.-Y., Ricke, S. C., Scanlan, C. M., Nisbet, D. J., Vargas-Moskola, A. A., Corrier, D. E. and DeLoach, J. R. 1995. Use of a differential rumen fluid-based carbohydrate agar media for culturing lactose selected cecal bacteria from chickens. *Journal of Food Protection* **58**:361–367.

Freter, R., Stauffer, E., Cleven, D., Holdeman, L. V. and Moore, W. E. C. 1983a. Continuous-flow cultures as *in vitro* models of the ecology of large intestinal flora. *Infection and Immunity* **39**:666–675.

Freter, R., Brickner, H., Botney, M., Cleven, D. and Aranki, A. 1983b. Mechanisms that control bacterial populations in continuous-flow culture models of mouse large intestinal flora. *Infection and Immunity* **39**:676–685.

Fuller, R. 1992. History and development of probiotics, pp. 1–8. *In* Probiotics—the Scientific Basis, R. Fuller (Ed.). London: Chapman and Hall.

Fuller, R. 1997. Introduction, pp. 1–9. *In* Probiotics 2: Applications and Practical Aspects, R. Fuller (Ed.). New York: Chapman and Hall.

Fuller, R. and Brooker, B. E. 1974. Lactobacilli which attach to the crop epithelium of the fowl. *The American Journal of Clinical Nutrition* **27**:1305–1312.

Fuller, R. and Turvey, A. 1971. Bacteria associated with the intestinal wall of the fowl (*Gallus domesticus*). *Journal of Applied Bacteriology* **34**:617–622.

Genthner, B. R. S., Davis, C. L. and Bryant, M. P. 1981. Features of rumen and sewage sludge strains of *Eubacterium limosum*, a methanol- and H_2-CO_2-utilizing species. *Applied and Environmental Microbiology* **42**:12–19.

Grau, F. H., Brownlie, L. E. and Roberts, E. A. 1968. Effect of some preslaughter treatments on the *Salmonella* population in the bovine rumen and faeces. *Journal of Applied Bacteriology* **31**:157–163.

Greening, R. C. and Leedle, J. A. Z. 1989. Enrichment and isolation of *Acetitomaculum ruminis*, gen. Nov., sp. nov: acetogenic bacteria from the bovine rumen. *Archives of Microbiology* **151**:399–406.

Hinton Jr., A., Corrier, D. E., Spates, G. E., Norman, J. O., Ziprin, R. L., Beier, R. C. and DeLoach, J. R. 1990. Biological control of *Salmonella typhimurium* in young chickens. *Avian Diseases* **34**:626–633.

Holt, P. S. 1995. Horizontal transmission of *Salmonella enteritidis* in molted and unmolted laying chickens. *Avian Diseases* **39**:239–249.

Holt, P. S., Buhr, R. J., Cunningham, D. L. and Porter Jr., R. E. 1994. Effects of two different molting procedures on a Salmonella enteritidis infection. *Poultry Science* **73**:1267–1275

Holt, P. S. and Porter Jr., R. E., 1992. Microbiological and histopathological effects of an induced-molt fasting procedure on a *Salmonella enteritidis* infection in chickens. *Avian Diseases* **36**:610–618.

Huhtanen, C. N. and Pensack, J. M. 1965. The development of the intestinal flora in the young chick. *Poultry Science* **44**:825–830.

Humphrey, T. J., Baskerville, A., Whitehead, A., Rowe, B. and Henly, A. 1993. Influence of feeding patterns on the artificial infection of laying hens with *Salmonella enteritidis* phage type 4. *The Veterinary Record* **132**:407–409.

Hungate, R. E. 1966. The Rumen and Its Microbes. New York: Academic Press.

Hungate, R. E. 1950. The anaerobic, mesophilic, cellulolytic bacteria. *Bacteriological Reviews* **14**:1–49.

Impey, C. S. and Mead, G. C. 1989. Fate of salmonellas in the alimentary tract of chicks pre-treated with a mature caecal flora to increase colonization resistance. *Journal of Applied Bacteriology* **66**:469–475.

Jensen, B. B. 1996. Methanogenesis in monogastric animals. *Environmental Monitoring and Assessment* **42**:99–112.

Johansson, K. R., Sarles, W. B. and Shapiro, S. K. 1948. The intestinal microflora of hens as influenced by various carbohydrates in a biotin deficient ration. *Journal of Bacteriology* **56**:619–634.

Kubena, L. F., Bailey, R. H., Byrd, J. A., Young, C. R., Corrier, D. E., Stanker, L. H. and Rottinghaus, G. E. 2001. Cecal volatile fatty acids and broiler chick susceptibility to Salmonella typhimurium colonization as affected by aflatoxins and T-2 toxin. *Poultry Science* **80**:411–417.

Leedle, J. A. Z. and Hespell, R. B. 1980. Differential carbohydrate media and anaerobic replica plating techniques in delineating carbohydrate-utilizing subgroups in rumen bacterial populations. *Applied and Environmental Microbiology* **39**:709–719.

Lev, M. and Briggs, C. A. E. 1956. The gut microflora of the chick. II. The establishment of the flora. *Journal of Applied Bacteriology* **19**:224–230.

Le Van, T. D., Robinson, J. A., Ralph, J., Greening, R. C., Smolenski, W. J., Leedle, J. A. Z. and Schaefer, D. M. 1998. Assessment of reductive acetogenesis with indigenous ruminal bacterium populations and *Acetitomaculum ruminis*. *Applied and Environmental Microbiology* **64**:3429–3436.

Lin, C. and Miller, T. L. 1998. Phylogenetic analysis of *Methanobrevibacter* isolated from feces of humans and other animals. *Archives of Microbiology* **169**:397–403.

Marounek, M., Rada, V. and Bendra, V. 1996. Effect of ionophores and 2-bromoethanesulphonic acid in hen caecal methanogenic cultures. *Journal of Animal and Feed Sciences* **5**:425–431.

Marounek, M., Suchorska, O. and Savka, O. 1999. Effect of substrate and feed antibiotics on *in vitro* production of volatile fatty acids and methane in caecal contents of chickens. *Animal and Feed Science Technology* **80**:223–230.

McNab, J. M. 1973. The avian caeca: a review. *World's Poultry Science Journal* **29**:251–263.

Mead, G. C. 1997. Bacteria in the gastrointestinal tract of birds, pp. 216–240. *In* Gastrointestinal Microbiology: Gastrointestinal Microbes and Host Interactions, Vol. 2, R. I. Mackie, B. A. White, R. E. Isaacson (Eds.). New York: Chapman and Hall.

Mead, G. C. 1989. Microbes of the avian cecum: types present and substrates utilized. *The Journal of Experimental Zoology* **3(Suppl.)**:48–54.

Mead, G. C. and Adams, B. W. 1975. Some observations on the caecal microflora of the chick during the first two weeks of life. *British Poultry Science* **16**:169–176.

Miller, C. P. and Bohnhoff, M. 1962. A study of experimental *Salmonella* infection in the mouse. *The Journal of Infectious Diseases* **111**:107–116.

Miller, T. L. 1991. Biogenic sources of methane, pp. 175–187. *In* Microbial Production and Consumption of Greenhouse Gases, Methane, Nitrogen Oxide and Halomethanes, J. E. Rogers and W. B. Whitman (Eds.). Washington, D.C.: American Society for Microbiology.

Miller, T. L. and Wolin, M. J. 1986. Methanogens in human and animal intestinal tracts. *Systematic and Applied Microbiology* **7**:223–229.

Miller, T. L., Wolin, M. J. and Kusel, E. A. 1986. Isolation and characterization of methanogens from animal feces. *Systematic and Applied Microbiology* **8**:234–238.

Moran Jr., E. T., and Bilgili, S. F. 1990. Influence of feeding and fasting broilers prior to marketing on cecal access of orally administered *Salmonella*. *Journal of Food Protection* **53**:205–207.

Nisbet, D. J., Corrier, D. E., Ricke, S. C., Hume, M. E., Byrd II, J. A. and DeLoach, J. R. 1996a. Maintenance of the biological efficacy in chicks of a cecal competitive-exclusion culture against *Salmonella* by continuous-flow fermentation. *Journal of Food Protection* **59**:1279–1283.

Nisbet, D. J., Corrier, D. E., Ricke, S. C., Hume, M. E., Byrd II, J. A. and DeLoach, J. R. 1996b. Cecal propionic acid as a biological indicator of the early establishment of a microbial ecosystem inhibitory to *Salmonella* in chicks. *Anaerobe* **2**:345–350.

Nisbet, D. J., Ricke, S. C., Scanlan, C. M., Corrier, D. E., Hollister, A. G. and DeLoach, J. R. 1994. Inoculation of broiler chicks with a continuous-flow derived bacterial culture facilitates early cecal bacterial colonization and increases resistance to *Salmonella typhimurium. Journal of Food Protection* **57**:12–15.

Nurmi, E. and Rantala, M. 1973. New aspects of *Salmonella* infection in broiler production. *Nature* **241**:210–211.

Orban, J. I., Patterson, J. A., Sutton, A. L. and Richards, G. N. 1997. Effect of sucrose thermal oligosaccharide caramel, dietary vitamin-mineral level, and brooding temperature on growth and intestinal bacterial populations of broiler chickens. *Poultry Science* **76**:482–490.

Patterson, J. A. and Burkholder, K. M. 2003. Application of prebiotics and probiotics in poultry production. *Poultry Science* **82**:627–631.

Ramirez, G. A., Sarlin, L. L., Caldwell, D. J, Yezak Jr., C. R., Hume, M. E., Corrier, D. E., DeLoach, J. R. and Hargis, B. M. 1997. Effect of feed withdrawal on the incidence of Salmonella in the crops and ceca of market age broiler chickens. *Poultry Science* **76**:654–656.

Ricke, S. C., Nisbet, D. J., Shermer, C. L., Corrier, D. E. and DeLoach, J. R. 1993. Utilization of a rumen fluid differential carbohydrate media for the selective enumeration of the cecal bacterial population of broiler chicks. *Poultry Science* **72(Suppl. 1)**:187.

Ricke, S. C. and Pillai, S. D. 1999. Conventional and molecular methods for understanding probiotic bacteria functionality in gastrointestinal tracts. *Critical Reviews in Microbiology* **25**:19–38.

Salanitro, J. P., Fairchilds, I. G. and Zgornicki, Y. D. 1974a. Isolation, culture characteristics, and identification of anaerobic bacteria from the chicken cecum. *Applied Microbiology* **27**:678–687.

Salanitro, J. P., Blake, I. G. and Muirhead, P. A. 1974b. Studies on the cecal microflora of commercial broiler chickens. *Applied Microbiology* **28**:439–447.

Salanitro, J. P., Blake, I. G., Muirhead, P. A., Maglio, M. and Goodman, J. R. 1978. Bacteria isolated from the duodenum, ileum, and cecum of young chicks. *Applied and Environmental Microbiology* **35**:782–790.

Schleifer, J. H. 1985. A review of the efficacy and mechanism of competitive exclusion for the control of *Salmonella* in poultry. *World's Poultry Science Journal* **41**:72–83.

Shapiro, S. K. and Sarles, W. B. 1949. Microorganisms in the intestinal tract of normal chickens. *Journal of Bacteriology* **58**:531–544.

Shermer, C. L., Maciorowski, K. G., Bailey, C. A., Byers, F. M. and Ricke, S. C. 1998. Caecal metabolites and microbial populations in chickens consuming diets containing a mined humate compound. *Journal of the Science of Food and Agriculture* **77**:479–486.

Shrimpton, D. H. 1963. Some volatile products of microbial metabolism in the caeca of fowl. *Journal of Applied Bacteriology* **26**:i–ii.

Shrimpton, D. H. 1966. Metabolism of the intestinal microflora in birds and its possible influence on the composition of flavour precursors in their muscles. *Journal of Applied Bacteriology* **29**:222–230.

Smith, H. W. 1965. The development of the flora in the alimentary tract in young animals. *The Journal of Pathology and Bacteriology* **90**:495–513.

Soerjadi, A. S., Stehman, S. M., Snoeyenbos, G. H., Weinack, O. M. and Smyser, C. F. 1981. Some measurements of protection against paratyphoid *Salmonella* and *Escherichia coli* by competitive exclusion in chickens. *Avian Diseases* **25**:706–712.

Stavric, S. 1987. Microbial colonization control of chicken intestine using defined cultures. *Food Technology* **41**:93–98.

Stavric, S. 1992. Defined cultures and prospects. *International Journal of Food Microbiology* **15**:245–263.

Thauer, R. K., Jungermann, K. and Decker, K. 1977. Energy conservation in chemotrophic anaerobic bacteria. *Bacteriological Reviews* **41**:100–180.

Ushijima, T. and Seto, A. 1991. Selected faecal bacteria and nutrients essential for antagonism of *Salmonella typhimurium* in anaerobic continuous flow cultures. *The Journal of Medical Microbiology* **35**:111–117.

Vispo, C. and Karasov, W. H. 1997. The interaction of avian gut microbes and their host: an elusive symbiosis, pp. 116–155. *In* Gastrointestinal Microbiology: Gastrointestinal Ecosystems and Fermentations, Vol. 1, R. I. Mackie, B. A. White (Eds.). New York: Chapman and Hall.

Wolin, M. J. and Miller, T. L. 1982. Interspecies hydrogen transfer: 15 years later. *American Society for Microbiology News* **48**:561–565.

Zhu, X. Y., Zhong, T., Pandya, Y. and Joerger, R. D. 2002. 16S rRNA-based analysis of microbiota from the cecum of broiler chickens. *Applied and Environmental Microbiology* **68**:124–137.

12 Distribution and Spread of Enteric Pathogens in Swine: Outlook for the Future

Roger B. Harvey and H. Morgan Scott

Introduction

Consumers in the United States have become increasingly concerned about the safety of their food supplies. They are bombarded daily by the media about new and dangerous diseases that are associated with the food chain. A few examples would include O157:H7 *Escherichia coli* in ground beef; *Listeria monocytogenes* in dairy products; *Salmonella* in beef, poultry, pork, fruits, and vegetables; *Campylobacter jejuni* in poultry; hepatitis A in raw oysters; and hormone, antibiotic, and pesticide residues in meat. Yet the U.S. food supply is safer than it has ever been. A few short years ago, most food was prepared in the home, and home cooks were aware of the necessity of proper food handling and storage. This knowledge was obtained through home economic classes in public schools in addition to mothers teaching their children that food safety was an important part of food preparation (see Chapter 30). However, during the 1960s and 1970s, major changes occurred in Americans' lifestyles. Young families became more mobile, and many were far removed from the previous generations. For the younger generation, this led to a loss in institutional memory for cooking and food handling. Changes in the workforce, lifestyles, and demographics; the advent of fast foods; and the increase in immunocompromised individuals have all compounded the issue (Collins 1997, Morris and Potter 1997).

At present, few households cook meals on a regular daily basis, and knowledge of home food safety is at low ebb. Because of the real and perceived danger of bacterial contamination in the home, consumers want assurances that their food is totally safe, even to the point of food sterilization. This has led to increased pressure on food processors to produce, package, and deliver products free of bacterial contamination. To help resolve some of these issues and maintain consumers' confidence, most of the food industry processors have adopted the Hazard Analysis Critical Control Point (HACCP) programs (see Chapter 20). However, because the food chain is a long one, food processors are now looking farther "upstream" and are demanding quality assurance or on-farm HACCP programs for food-producing animals (Tauxe 1997).

This chapter will discuss the past and present effects of selected enteropathogens on pork production and the swine industry. On the basis of our present knowledge, future trends and an outlook for the distribution, spread, and control of enteropathogens will be discussed.

Mention of a trade name, proprietary product, or specific equipment does not constitute a guarantee or warranty by the U.S. Department of Agriculture and does not imply its approval to the exclusion of other products that may be suitable.

Salmonella, Campylobacter, and E. coli in Swine

Historical Perspective and Present Control Efforts

Salmonellosis in swine (swine paratyphoid) caused by *Salmonella* serovar Choleraesuis at one time was considered the only disease associated with salmonellae; however, now it is known that *Salmonella* serovar Typhimurium may induce clinical disease (enterocolitis) in young pigs (Schwartz 1991). Swine can be colonized by many of the serovars of *Salmonella* but may remain as undiagnosed asymptomatic carriers of the organisms, thereby increasing the risk of food chain contamination. With the exception of serovars Choleraesuis and Typhimurium, the remaining serovars of *Salmonella* do not cause economic losses to the swine industry, and other than food safety concerns, there are few economic incentives for the industry to reduce carriage rates of the other *Salmonella* serovars in swine.

Most of the food safety research in the United States for *Salmonella* in swine has centered on identification of on-farm prevalence compared with slaughter prevalence, improved sanitation/disinfection procedures on farms and at processing plants, and the effects of feed withdrawal, transport, and lairage on carcass contamination. Commercial and autogenous bacterins have been used with limited success on farms, with recurring clinical disease induced by serovars of Choleraesuis or Typhimurium.

Denmark, however, has embarked on an ambitious plan for *Salmonella* reduction in swine. In that country, there is a national serological (enzyme-linked immunosorbent assay–based) program that monitors swine at slaughter. *Salmonella*-positive herds are identified, and farms are categorized with a numerical score as to the rate of infection. The farms with the highest prevalence of *Salmonella* are required to cooperate in a *Salmonella*-reduction plan. *Salmonella*-reduction procedures may include quarantine, farm depopulation, restricted access to slaughter facilities, increased on-farm testing, improved biosecurity methods, increased disinfection procedures, and price reduction for pork products from those farms. The use of this program has resulted in reduced numbers of *Salmonella*-positive pigs presented at slaughter (Christensen et al. 1999, Dahl 1999).

In recent years, *Campylobacter* has emerged as one of the most common causes of human enteric disease in developed countries. In the United States, *Campylobacter* is considered by many to be the primary cause of enteric disease (Tauxe 1992). The number of cases of campylobacteriosis in Denmark was doubled from 1992 to 1996 (Nielsen et al. 1997), and *Campylobacter* was reported as responsible for 12% of all cases of acute gastroenteritis in the Netherlands (Weijtens et al. 1993).

Swine at slaughter may have an intestinal prevalence of 70%–100% of *Campylobacter*, with the predominant species being *C. coli*. However, on some farms, swine can have a relatively high prevalence of *C. jejuni* (Harvey et al. 1999). *C. coli*, *C. jejuni*, or *C. lari* do not appear to cause clinical disease in swine but are of concern for foodborne disease risk. Human campylobacteriosis in the United States is primarily caused by infection with *C. jejuni*, whereas in Europe both *C. jejuni* and *C. coli* are routinely isolated from human cases. Present control procedures center on random bacteriological sampling of carcasses at slaughter, on carcass washes with acidic solutions, on steam pasteurization of carcass surfaces, and on bacteriological monitoring of retail meats.

Colibacillosis is the most frequently reported condition of suckling pigs, and *E. coli*-associated diarrhea is the number three reported disease of weaned pigs in the United States (National Animal Health Monitoring System Report 2000). Enterotoxigenic strains of *E. coli* such as F18 and K88 can cause gut edema and mortality in nursery-age pigs.

Although these strains are not pathogenic to humans, they annually cost the swine industry millions of dollars. Present control practices for *E. coli* disease include enhanced sanitation and disinfection practices, increased biosecurity measures, feed additives such as spray-dried plasma and zinc oxide, use of antibiotics to treat clinical disease, and vaccination with autogenous bacterins.

E. coli O157:H7 is a human pathogen that has been associated with foodborne disease linked to undercooked ground beef. Although not commonly isolated from pigs in the United States, *E. coli* O157:H7 has been isolated from pigs at slaughter in South America (Borie et al. 1997, Rios et al. 1999). The U.S. pork industry is understandably very concerned about the potential colonization of swine with O157:H7. As a measure of precaution in the United States, swine and pork products are randomly sampled to monitor for the presence of *E. coli* O157:H7.

Current Trends in the U.S. Swine Industry: Influence on Enteropathogens

What are some of the factors and production practices in the U.S. swine industry that could facilitate the increase of enteropathogens? First and foremost is the tendency of the industry to become more vertically integrated. This has caused the average swine farm to increase the total number of animals produced annually, yet farms have not increased dramatically in acreage. The only way to achieve these increased production capabilities is through intense confinement rearing. Although intense rearing can increase efficiency of facilities and labor, it can initiate its own set of problems. In intense rearing, there is an increased likelihood that bacterial populations, including enteropathogens, could become more prevalent. Once a disease or infection begins, it can spread at an alarming rate throughout a nursery or grower/finisher barn. Therefore, improved biosecurity and monitoring of herd health status are essential.

Traditional methods of bacterial disease control have relied on the use of antibiotics; however, enteropathogens are becoming resistant to many of our available antibiotics, and it does not appear that new antibiotics are on the horizon. In fact, some regulatory agencies have proposed that antibiotics (sub-therapeutic) be removed from animal production. In Sweden, antibiotics in animal feeds have been outlawed, and there has been a subsequent increase in bacterial disease conditions. Apramycin is a feed/water antibiotic that was approved in the United States for treatment of coliform infections in pigs. Recently, the manufacturer voluntarily removed it from the market because of regulatory issues and pressures. It is unknown whether this removal will become a trend for other antibiotics.

In today's atmosphere of environmental awareness, odor control and manure management have become key issues and problems for large confinement operations. Because of offensive odors, reduced air quality, and groundwater contamination associated with animal production, some states have enacted "no growth" restrictions for the swine industry. Along with those problems are the public concerns about animal well-being and animal welfare in confinement operations. To address these issues, some European countries have banned confinement rearing for livestock and poultry and have gone to "free-range" rearing. On the surface, this may appear to be an easy fix for welfare and well-being, yet these practices can create another set of problems. For example, parasite loads, hypothermia, and foot rot, although minimal in confinement rearing, can increase dramatically in animals raised outdoors on dirt. *Salmonella* and *E. coli* organisms can survive for long periods in mud and damp soil, so free-range animals could actually have an increase in their bacterial carriage rates compared with their confinement counterparts.

Future Directions and Potential Intervention Strategies to Reduce the Prevalence of Enteropathogens

What changes are likely to occur, and what could be some mitigation/remediation strategies to counteract antibiotic resistance of enteropathogens, carriage rates of enteropathogens in swine, problems associated with intense confinement rearing, and the potential loss of sub-therapeutic antibiotics? Let us address each of these issues individually.

Enteropathogen Carriage Rates

If enteropathogen carriage rates in swine could be reduced, overall health in live animals would improve, and reduced bacterial loads at slaughter could potentially reduce risk of carcass contamination. Carriage rates of enteropathogens are influenced by various production practices in growing, rearing, and pre-slaughter. Neonatal pigs acquire their normal gut microflora from the sow; however, they also can acquire enteropathogens from the sow. In addition, improperly sanitized and disinfected farrowing barn facilities can lead to a carry-over of pathogens from previous batches of pigs. To counteract these effects, high health status of the sow and strict cleaning and disinfection procedures are mandatory. Coupled with these procedures, an enforced biosecurity program can decrease disease exposure to growing pigs. In an effort to counteract the exposure of young pigs to enteropathogens, several innovative management techniques appear promising. It has been shown experimentally that if *Campylobacter*-positive piglets are removed from the sow within 24 hours of birth, they will revert to a *Campylobacter*-negative status within days (Harvey et al. 2000). Likewise, when piglets are weaned early (12–14 days of age) from their *Salmonella*-positive sows and reared on farms separate from their dams, *Salmonella* in growing pigs will be significantly reduced (Fedorka-Cray et al. 1997). It is possible that these findings will lead to practical and cost-effective management practices for pathogen control.

Pre-slaughter handling procedures of market-age pigs can also affect pathogen shedding. *Salmonella*, *Campylobacter*, and *E. coli* concentrations and shedding are reported to increase following feed withdrawal and transport (Williams and Newell 1970, Berends et al. 1996, Isaacson et al. 1999, Nattress and Murray 2000, Harvey et al. 2001, Hurd et al. 2002). Feed withdrawal is desirable to the processors because reduced intestinal fill decreases the chance of gut rupture during the evisceration process. To counteract this dilemma, it is possible that a high-calorie liquid diet could be fed up to the time for transport. This would theoretically keep the gut microflora intact but would reduce the mass of the intestinal tract presented at slaughter. Another procedure that anecdotally appears to have merit is the addition of organic acids to drinking water to lower intestinal pH before transport. Reduced intestinal pH should decrease intestinal pathogen loads. There are no apparent remedies to the transport problem except that reduced intestinal concentrations and shedding of enteropathogens are directly related to reduced transit times (i.e., distances). The obvious choice would be to select slaughter plants within a shorter transport distance. Transport and feed withdrawal can act additively for bacterial shedding, so that beneficially affecting one could positively affect the other.

Other pre-slaughter factors that can increase the intestinal load of enteropathogens at slaughter are contaminated transportation vehicles, time spent in holding pens at the slaughter plant, and level of pen contamination at slaughter plants. *Salmonella* prevalence was compared in swine from the same farms that were slaughtered either on-farm or at an abattoir. The abattoir group had a sevenfold-higher *S. enterica* isolation rate compared with the on-farm group. That study showed that rapid *Salmonella* infection occurred during transport and holding and implicated the holding pen as an important *S. enterica* control point for pork production (Hurd et al. 2002). The solution to these problems is to first determine the extent of the problem, then incorporate corrective measures. Obviously, increased sanitation and disinfection efforts should be employed on the vehicles and at the slaughter plant holding pens, and holding times in the pens should be reduced.

Antibiotic Resistance and Discontinuation of Sub-Therapeutic Antibiotics

It is generally perceived by the medical community that increased antibiotic resistance in human infections has been generated by continuous use of sub-therapeutic antibiotics as growth promoters in animal production. Although this might sound good in theory, no definitive studies have established this as fact. Nevertheless, the impetus is there to remove antibiotics from animal feeds. If this were to include therapeutic antibiotics, the swine industry could see an increase in clinical disease from enteropathogens. The question arises: What will happen if sub-therapeutic antibiotics are removed from all swine feeds? Will enteropathogens increase in prevalence? If there is an increase in prevalence, will that expansion lead to increased carcass contamination at slaughter? Answers to these questions are unknown; however, there is a possibility that carcass contamination will increase. Will removal of antibiotics in feeds change the resistance pattern of enteropathogens and promote more susceptible microorganisms? Maybe and maybe not. On one hand, the acquisition of resistance gives bacteria a competitive advantage for survival, and resistance patterns are hard to reverse. On the other hand, a study showed that when sub-therapeutic antibiotics were discontinued in swine and poultry feeds in Denmark, *Enterococcus faecium* isolated from those species became less resistant to antibiotics within 5 years of the ban (Aarestrup et al. 2001).

What are some alternatives to antibiotics for enteropathogen control? One way to improve resistance to disease organisms is to enhance the immune system and increase its response to pathogens. This can be done in a number of ways. For example, through improved genetics, pigs could be produced that are naturally resistant to pathogens. This could be done through traditional breeding programs, by selection for those traits, or it could be done through genetic manipulation such as cloning or transgenic animals. Indeed, in 2003 a major swine breeding firm (Pig Improvement Corp., Franklin, Ky.) is planning to introduce a pig line that is genetically more resistant to enterotoxigenic *E. coli*. Another method of increased immune response could be induced by the use of genetically modified vaccines or bacterins or by development of new classes of adjuvants that could be administered with vaccines or bacterins.

Several new developments hold great promise for alternatives to antibiotics in the control of enteropathogens of swine. One such alternative is the use of competitive exclusion (CE) cultures. The theory of CE is to colonize the neonatal gastrointestinal tract with beneficial/commensal bacteria that make up the normal flora of healthy adult animals (Nurmi and Rantala 1973, Lloyd et al. 1977). The use of CE establishes a healthy adult gut flora in neonates sooner than would occur naturally, thereby reducing the window of opportunity for pathogens to colonize the gastrointestinal tract (Smith and Crabb 1961, Schaedler 1973). An experimental, porcine-derived, continuous-flow CE culture has been developed that inhibited *Salmonella* Choleraesuis and *S.* Typhimurium and *E. coli* O157:H7 and *E. coli* strain F-18 colonization *in vitro* (Harvey et al. 2002a). On the basis of *in vitro* results, piglets were treated with the CE culture within 24 hours of birth, and 7 days later (when they were 8 days old) they were challenged with virulent *Salmonella* or *E. coli*. CE-treated pigs had decreased colonization, shedding, morbidity, and mortality associated with *Salmonella* and *E. coli* compared with untreated controls (Anderson et al. 1999, Genovese et al. 2000, Genovese et al. 2001). Field trial experiments have demonstrated that CE treatments can significantly reduce mortality and medication costs associated with *E. coli* disease in nursery-age pigs (Harvey et al. 2002b, Harvey et al. 2003).

A second experimental method of preharvest reduction of *Salmonella* and *E. coli* is the addition of sodium chlorate to drinking water of swine within 24 hours of slaughter. This concept is based on the fact that both *Salmonella* and *E. coli*, like most members of the family *Enterobacteriaceae*, possess respiratory nitrate reductase activity that also catalyses the intracellular reduction of chlorate, an analogue of nitrate, to cytotoxic chlorite (Anderson et al. 2000, Anderson et al. 2001). Because most gastrointestinal anaerobes lack respiratory nitrate reductase, sodium chlorate can be administered without affecting most of the beneficial microbiota of the gastrointestincal tract. *In vitro* studies showed that sodium chlorate (5 mM) cleared *E. coli* O157:H7 and *Salmonella* Typhimurium DT104 from rumen contents within 24 hours of treatment (Anderson et al. 2000). On the basis of these results, sodium chlorate (100 mM) was administered to *Salmonella*-infected pigs. Significant ($P < 0.05$) reductions of cecal concentrations of *Salmonella* were observed within 16 hours of treatment (Anderson et al. 2001). With the dosages used, there is a broad margin of safety between the therapeutic dose and the published toxic dose of sodium chlorate to the animal. Negligible intestinal absorption and residue accumulation of sodium chlorate occur; so with these factors in mind, this compound could potentially be a safe, practical, and cost-effective pre-slaughter treatment for reduction of *Salmonella* and *E. coli* in food-producing animals.

If CE cultures and sodium chlorate become commercially available, it is possible that they could become viable alternatives to antibiotics for preharvest reduction of *Salmonella* and *E. coli* in swine. However, these treatments would be administered to swine at different ages. For example, CE cultures would be administered to neonates, whereas sodium chlorate would be used in market-age swine within 24 hours of slaughter.

Miscellaneous Factors that Influence Swine Production Procedures

Some additional factors come into play that could affect current methods of swine production and enteropathogen control. Earlier in this chapter, it was mentioned that with HACCP programs, there are tendencies for food processors to look farther "upstream" for quality control. This could include a national animal identification program. In that way, if retailers wanted to sell meat that came from organically grown animals, from animals raised under strict animal-welfare conditions, or for a brand-name specialty product, they would have a verification process. In addition, if there were major recalls because of meat contamination problems, a trace-back process would be available for packers and processors. Unfortunately, the cost of such programs falls directly on the swine producer, yet little or no incentives such as premium sales price come into play. As can be seen in the scenarios above, product liability may be another downside for the individual farmer. How all of this would affect animal rearing procedures and enteropathogen prevalence is unknown at this time.

Although somewhat outside the objectives of this chapter, another major factor could come into play for future animal production practices: bioterrorism (see Afterword). The threat of a disrupted food supply in this country is unthinkable; however, if conditions were right, it could happen. This chapter has focused on preharvest food safety, and if a bioterrorist incident occurred, it probably would not depend on postharvest adulteration of meat. Instead, it would be disruption of animal agriculture as we know it; that is, the introduction of exotic swine diseases. In that scenario, there could be high mortality and morbidity in U.S. swine herds. The subsequent outcome would be reduced pork supplies, and the economic impact would be tremendous. If anything like this should occur, biosecurity in swine production would take on a whole new meaning. Current production practices of rearing, transporting, and marketing would be dramatically changed forever.

Finally, one new aspect of food safety for meat products needs to be discussed. Although it concerns postharvest rather than preharvest food safety, it could have major implications on how we view control of enteropathogens in the future. That aspect is food irradiation (see Chapter 28, Food Irradiation). If all of the packaging, regulatory, and organoleptic considerations can be overcome, it appears to be a procedure that could dramatically improve food safety. If retail meat products could be sterilized, then some of the preharvest control measures we have discussed may be of little consequence. Other than for animal health reasons, enteropathogen control in swine may be a moot point, and very little will change in animal production practices. However, rarely is there a "silver bullet" available, and food irradiation should be viewed as just another management tool in the course of food safety. Only time will tell whether food irradiation will have a significant effect on enteropathogens in the pork food chain.

Conclusions

The job of protecting the pork chain from contamination with enteropathogens such as *Salmonella, E. coli*, and *Campylobacter* is daunting, but not an impossible task. In this chapter we have discussed preharvest food safety and some of the past and present control procedures for enteropathogens. However, our main focus has been to look forward and to anticipate some new intervention strategies and production practices that may aid in control of these microorganisms. Furthermore, we have tried to envision how new procedures might impinge on pork production. Topics discussed included biosecurity, sanitation and disinfection, early weaning, alternatives to antibiotics, pre-slaughter transport and holding, augmentation of immune function, bioterrorism, and food irradiation. It is hoped that such discussions will stimulate readers to think "outside the box" and to enable them to design their own novel intervention strategies for combating enteropathogens in the pork food chain.

References

Aarestrup, F. M., Seyfarth, A. M., Emborg, H. D., Pedersen, K., Hendriksen, R. S. and Bager, F. 2001. Effect of abolishment of the use of antimicrobial agents for growth promotion on occurrence of antimicrobial resistance in fecal enterococci from food animals in Denmark. *Antimicrobial Agents and Chemotherapy* **45**:2054–2059.

Anderson, R. C., Buckley, S. A., Callaway, T. R., Genovese, K. J., Kubena, L. F., Harvey, R. B. and Nisbet, D. J. 2001. Effect of sodium chlorate on *Salmonella typhimurium* concentrations in the weaned pig gut. *Journal of Food Protection* **64**:255–258.

Anderson, R. C., Buckley, S. A., Kubena, L. F., Stanker, L. H., Harvey, R. B. and Nisbet, D. J. 2000. Bactericidal effect of sodium chlorate on *Escherichia coli* O157:H7 and *Salmonella typhimurium* DT104 in rumen contents in vitro. *Journal of Food Protection* **63**:1038–1042.

Anderson, R. C., Stanker, L. H., Young, C. R., Buckley, S. A., Genovese, K. J., Harvey, R. B., DeLoach, J. R., Keith, N. K. and Nisbet, D. J. 1999. Effect of competitive exclusion treatment on colonization of early-weaned pigs by *Salmonella* serovar Choleraesuis. *Swine Health and Production* **7**:155–160.

Berends, B. R., Urlings, H. A. P., Snijders, J. M. A. and Van Knapen, F. 1996. Identification and quantification of risk factors in animal management and transport regarding *Salmonella* spp. in pigs. *International Journal of Food Microbiology* **30**:37–53.

Borie, C., Monreal, Z., Guerrero, P., Sanchez, M. L., Martinez, J., Arellano, C. and Prado, V. 1997. Prevalence and characterization of enterohaemorrhagic *Escherichia coli* isolated from healthy cattle and pigs slaughtered in Santiago, Chile. *Archivos de Medicina Veterinaria (Valdivia)* **29**:205–212.

Christensen, J., Baggesen, D. L., Nielsen, A. and Nielsen, B. 1999. Prevalence of *Salmonella enterica* in pigs before the start of the Danish *Salmonella* Control Program (1993/94) and four years later (1998), pp. 333–335. *In* Proceedings of the 3rd International Symposium on the Epidemiology and Control of *Salmonella* in Pork, P. B. Bahnson (Ed.). Urbana-Champaign, Ill.: University of Illinois.

Collins, J. E. 1997. Impact of changing consumer lifestyles on the emergence/reemergence of foodborne pathogens. *Emerging Infectious Diseases (Special Issue)* **3**:471–479.

Dahl, J. 1999. Success-rate for eradication of *Salmonella* by cleaning and restocking pig herds and the use of antemortem-blood samples in herds after restocking, pp. 336–339. *In* Proceedings of the 3rd International Symposium on the Epidemiology and Control of *Salmonella* in Pork, P. B. Bahnson (Ed.). Urbana-Champaign, Ill.: University of Illinois.

Fedorka-Cray, P. J., Harris, D. L. and Whipp, S. C. 1997. Using isolated weaning to raise Salmonella-free swine. *Veterinary Medicine: Food Animal Practice* **April**:375–382.

Genovese, K. J., Anderson, R. C., Harvey, R. B. and Nisbet, D. J. 2000. Competitive exclusion treatment reduces the mortality and fecal shedding associated with enterotoxigenic *Escherichia coli* infection in nursery-reared neonatal pigs. *Canadian Journal of Veterinary Research* **64**:204–207.

Genovese, K. J., Harvey, R. B., Anderson, R. C. and Nisbet, D. J. 2001. Protection of suckling neonatal pigs against an enterotoxigenic *Escherichia coli* expressing 987p fimbriae by the administration of a bacterial competitive exclusion culture. *Microbial Ecology in Health and Disease* **13**:223–228.

Harvey, R. B., Anderson, R. C., Young, C. R., Swindle, M. M., Genovese, K. J., Hume, M. E., Droleskey, R. E., Farrington, L. A., Ziprin, R. L. and Nisbet, D. J. 2001. Effects of feed withdrawal and transport on cecal environment and *Campylobacter* concentration in a swine surgical model. *Journal of Food Protection* **64**:730–733.

Harvey, R. B., Droleskey, R. E., Hume, M. E., Anderson, R. C., Genovese, K. J., Andrews, K. and Nisbet, D. J. 2002a. In vitro inhibition of *Salmonella enterica* serovars Choleraesuis and Typhimurium, *Escherichia coli* F-18, and *Escherichia coli* O157:H7 by a recombined porcine continuous-flow competitive exclusion culture. *Current Microbiology* **45**:226–229.

Harvey, R. B., Ebert, R. C., Andrews, K., Genovese, K. J., Anderson, R. C. and Nisbet, D. J. 2003. Competitive exclusion culture reduces mortality from F-18 strain *E. coli* in nursery pigs—field trial results, pp. 485–486. *In* Proceedings of the Annual Meeting of the American Association of Swine Veterinarians. Des Moines, Iowa: American Association of Swine Veterinarians.

Harvey, R. B., Ebert, R. C., Genovese, K. J., Andrews, K., Anderson, R. C. and Nisbet, D. J. 2002b. Control of enterotoxigenic *E. coli* in a commercial swine operation by use of a competitive exclusion culture, p. 13. *In* Proceedings of the Allen D. Leman Swine Conference, Vol. 29 (Suppl), W. C. Scruton, S. Claas (Eds.). St. Paul, Minn.: University of Minnesota.

Harvey, R. B., Young C. R., Anderson, R. C., Droleskey, R. E., Genovese, K. J., Egan, L. F. and Nisbet, D. J. 2000. Diminution of *Campylobacter* colonization in neonatal pigs reared off-sow. *Journal of Food Protection* **63**:1430–1432.

Harvey, R. B., Young, C. R., Ziprin, R. L., Hume, M. E., Genovese, K. J., Anderson, R. C., Droleskey, R. E., Stanker, L. H. and Nisbet, D. J. 1999. Prevalence of *Campylobacter* species isolated from the intestinal tract of pigs raised in an integrated swine production system. *Journal of the American Veterinary Medical Association* **215**:1601–1604.

Hurd, H. S., McKean, J. D., Griffith, R. W., Wesley, I. V. and Rostagno, M. H. 2002. *Salmonella enterica* infections in market swine with and without transport and holding. *Applied and Environmental Microbiology* **68**:2376–2381.

Isaacson, R. E., Firkins, L. D., Weigel, R. M., Zuckermann, R. A. and DiPietro, J. A. 1999. Effects of transportation and feed withdrawal on shedding of *Salmonella typhimurium* among experimentally infected pigs. *American Journal of Veterinary Research* **60**:1155–1158.

Lloyd, A. B., Cumming R. B. and Kent, R. D. 1977. Prevention of *Salmonella typhimurium* infection in poultry by pretreatment of chickens and poults with intestinal extracts. *Australian Veterinary Journal* **53**:82–87.

Morris, J. G. and Potter, M. 1997. Emergence of new pathogens as a function of changes in host susceptibility. *Emerging Infectious Diseases (Special Issue)* **3**:435–441.

National Animal Health Monitoring System Report. 2000. Swine 2000 Part II: Reference of Swine Health and Health Management in the United States. Centers for Epidemiology and Animal Health, U.S. Department of Agriculture Veterinary Services, Ft. Collins, CO.

Nattress, F. M. and Murray, A. C. 2000. Effect of antemortem feeding regimes on bacterial numbers in the stomachs and ceca of pigs. *Journal of Food Protection* **63**:1253–1257.

Nielsen, E. M., Enberg, J. and Madsen, M. 1997. Distribution of serotypes of *Campylobacter jejuni* and *C. coli* from Danish patients, poultry, cattle, and swine. *FEMS Immunology and Medical Microbiology* **19**:47–56.

Nurmi, E. and Rantala, M. 1973. New aspects in *Salmonella* infections in broiler production. *Nature* **241**:210–211.

Rios, M., Prado, V., Trucksis, M., Arellano, C., Borie, C., Alexandre, M., Fica, A. and Levine, M. M. 1999. Clonal diversity of Chilean isolates of enterohemorrhagic *Escherichia coli* from patients with hemolytic-uremic syndrome, asymptomatic subjects, animal reservoirs, and food products. *Journal of Clinical Microbiology* **37**:778–781.

Schaedler, R. W. 1973. The relationship between the host and its intestinal microflora. *Proceedings of the Nutrition Society* **32**:41–47.

Schwartz, K. 1991. Salmonellosis in swine. *Compendium of Continuing Education for Practicing Veterinarians* **13**:139–147.

Smith, H. W. and Crabb, W. 1961. The faecal bacterial flora of animals and man: its development in the young. *Journal of Pathological Bacteriology* **82**:53–66.

Tauxe, R. V. 1992. Epidemiology of *Campylobacter jejuni* infections in the United States and other industrialized nations, pp. 9–19. *In Campylobacter Jejuni*: Current Status and Future Trends, I. Nachamkin, M. J. Balser, L. S. Tompkins (Eds.). Washington, D.C.: American Society of Microbiology.

Tauxe, R. V. 1997. Emerging foodborne diseases: an evolving public health challenge. *Emerging Infectious Diseases (Special Issue)* **3**:425–434.

Weijtens, M. J. B. M., Bijker, P. G. H., Van der Plas, J., Urlings, H. A. P. and Bieshevel, M. H. 1993. Prevalence of *Campylobacter* in pigs during fattening: an epidemiological study. *Veterinary Quarterly* **15**:138–143.

Williams, L. P. and Newell, K. W. 1970. *Salmonella* excretion in joy-riding pigs. *American Journal of Public Health* **60**:926–929.

13 Environmental Reservoirs and Transmission of Foodborne Pathogens

Scot E. Dowd, Jeanette A. Thurston-Enriquez, and Mindy Brashears

Introduction

The interaction of livestock with their environment is a primary factor modulating the health of the animal. Common sense dictates that a sanitary environment should promote health and productivity. Thus, the importance of proper environmental control to reduce the occurrence of pathogens as part of good production practices (GPPs) cannot be overstated. An environment containing uncontrolled pathogen reservoirs promotes disease among food animals and crops. Ultimately, environmental contamination can cause loss of productivity and higher production costs and, more important, may result in contamination of the final food product. Research directed toward the development of GPP will be the foundation and future of a good on-farm food safety system and will ultimately facilitate the development of farm-specific Hazard Analysis Critical Control Point (HACCP) plans (see Chapter 20). The development of production systems that use effective environmental controls such as providing proper drainage, on-farm water treatment, sanitation, pathogen monitoring, effective pest control, and appropriate animal and waste management will be paramount in preventing the accumulation, survival, and transmission of pathogens to and from livestock.

The environment of a farm can be broken down into four primary components: air, water, soil, and fomites. The impact of the environment on the occurrence, survival, and fate of pathogens depends on biotic and abiotic factors such as temperature, competition, moisture, and so forth. The environments of individual farms can vary tremendously, ultimately affecting the occurrence, survival, and fate of pathogens. For instance, airborne transmission of certain potential foodborne pathogens in a warm humid climate may be more significant than that experienced in a hot, dry climate.

Air

The transport of pathogen-containing aerosols via air currents to other animals is a process that can rapidly infect an entire farm. Airborne transmission can be one of the primary modes of pathogen dispersal among livestock (Proux et al. 2001, Wathes et al. 1988, Holt et al. 1998), and yet air is one of the most hostile environments from a microorganism's perspective (Dowd 2002).

Mention of a trade name, proprietary product, or specific equipment does not constitute a guarantee or warranty by the U.S. Department of Agriculture and does not imply its approval to the exclusion of other products that may be suitable.

The atmosphere is generally inhospitable to microorganisms because of environmental stress factors. This results in a limited time frame in which microbes can remain biologically active when airborne. Many environmental factors have been shown to influence the ability of microorganisms to survive. The most important of these factors are relative humidity and temperature. Other factors such as oxygen content, specific ions, various pollutants, and exposure to ultraviolet radiation are also associated with loss of biological activity.

There are various mechanisms that can be used to control bioaerosols, including ventilation, filtration, ultraviolet treatment, biocidal agents, and physical isolation. Ventilation, for instance, creates a flow of air throughout areas in which pathogen contamination or accumulation occurs, resulting in a dilution-type effect. Ventilation can simply mean opening a window and allowing outside air to circulate into and out of a room. Compared to filtration, strict ventilation is considered one of the least effective, but still most important, methods for controlling airborne pathogens and reducing their effect on health (Dowd and Maier 1999). The effectiveness as noted is an outcome of the influx of less contaminated air, resulting in dilution of air contaminants within the occupied area. The effectiveness of ventilation depends on factors such as the quality of the input air and the turnover rate of air within the occupied area in relation to the rate of contaminant production. However, ventilation used as the sole method of control can result in the input of contaminated air to an occupied environment as easily as it can act to dilute contaminants produced within this environment.

Ventilation combined with filtration is an effective method for control of airborne contamination. Some filters, for example, high-efficiency particulate air filters, are reported to remove virtually all infectious particles. Thus, outdoor air can be filtered as it passes into an area, preventing contamination of the occupants with pathogens from other areas or buildings. Air that is removed from quarantine areas can be filtered as it is exhausted to prevent contamination of other areas.

Biocidal control represents an added treatment that can be used in conjunction with filtration to reduce or eliminate airborne microorganisms. Many methods of biocidal control are available; for example, superheating, superdehydration, ozonization, and ultraviolet irradiation. These systems, along with filtration, are used to treat air coming into or being exhausted from indoor environments.

Control of the indoor airborne environment can be achieved through the use of positive or negative pressurized air pressure gradients and airtight seals (Dowd and Maier 1999). Air from negative-pressure units is forced out of a room through a single port and exhausted into the atmosphere after passing through a high-efficiency particulate air filter and possibly a biocidal unit. Thus, contamination does not escape these areas as long as the filtration and airflow are maintained. This type of system is designed to protect the rest of the farm from pathogens generated in the isolated area. Positive-pressure isolation chambers work on the opposite principle by forcing filtered sterile air into the room through a single port, thus protecting the animals in the area from outside contamination. Quarantine buildings should be under negative pressure, and the rest of the buildings should be under positive pressure to prevent spread of contamination.

Water

On-farm water environments range from freshwater distribution systems to waste-water lagoons. They also include incoming sources such as rainwater, including runoff, groundwater, rivers, and streams. The control of waterborne diseases depends on a multiple-barrier approach that takes into account source-water protection and physical (filtration, sedimentation) and chemical (chemical and physical disinfection) treatment. In addition, wastewater disposal and treatment, protection of distribution systems, flood and other types of runoff control, and many other factors relating to pathogen transport and dissemination within water are also vital.

Waste and wastewater control has arguably received more attention in the literature, yet freshwater distribution systems are of obvious importance in maintaining livestock health. Cattle, as an example, can require between 15 and 21 gallons of water per head per day. A feedlot with 10,000 animals would require a well with continuous pumping capacity of hundreds of gallons every minute. With this much demand for water from a single source, it becomes vital to protect freshwater distribution from an abundance of potential sources of contamination that may even originate in the farm itself. For example, drinking-water wells should be located upstream of the farm so that the flow of groundwater passing under the farm will not carry potential contaminants to the well. Similarly, it is important that water runoff from a farm or surrounding area does not carry contamination to a well head.

Typically, source water would be pumped and stored in a water tower and then distributed throughout the farm. The farm's freshwater distribution system itself must also be maintained to prevent contamination, accumulation of biofilms, decay, or accidental breach and subsequent contamination by wastewater. The water tower as a source or freshwater should be chlorinated before distribution, just like domestic water towers.

Surface waters such as ponds, lakes, rivers, and streams are also used as sources of freshwater for many farms. It is important to prevent unlimited access, organic accumulation, microbial disturbances, and fecal contamination of surface waters. Surface water, especially nonflowing surface water such as lakes, water troughs, or ponds, can be a prime reservoir for accumulation of pathogens. Even if this type of surface water is a flowing source such as a stream or river, contamination from one farm can carry downstream to other farms and communities. Geographical, hydrological, and physical isolation of source waters and the use of watering and pump systems that carry surface water to disinfection and storage areas is ultimately the best way to reduce the risk of disease transmission or fecal contamination of freshwater supplies.

Water safety and quality is determined by analyses of water samples. A bacterial analysis of a water source will determine whether the supply water contains microorganisms that may be harmful. Water safety and quality tests should be conducted and documented at least once per year, but possibly more often as demonstrated by previous history, water source, and experience. For instance, simple coliform bacterial counts of over 1 per 100 mL are correlated with the occurrence of scours in calves, and a total bacterial count of over 20 per 100 mL is correlated with the occurrence of diarrhea in cows and with cows going off feed (Rodenburg 1985).

Soil

The third environment that is significant from a microbiological standpoint is soil. Soil is physically the most complex environment. Soil contains solid, liquid, and gas components interacting to affect the fate of pathogens. Furthermore, once contaminated with certain pathogens, soil is very difficult to disinfect compared to air and water. During rain events, soils can become saturated, and runoff from these events can carry contamination to wells, water supplies, or other environmental reservoirs. Another concern is the accumulation of water (e.g., stagnant water ponds produced by flood events), as amplification of the pathogens themselves and insect propagation can occur and increase the risk of pathogen transmission to livestock and humans.

Fomites

Fomites, for the purposes of this discussion, include any inanimate (non-soil or water) surface found on a farm. Fomites include anything from ropes, gates, floors, walls, and pens to the clothing of livestock workers. The wide variety of material and surface textures that make up farm fomites make this a complex environmental topic. Thus, we must speak in generalities and broadly painted examples.

The relative ability of fomites as an environment to sustain microorganisms is affected by the material they are made of, the texture of this material, and its porosity. For example, metal surfaces are typically very inhospitable for microorganisms. Metal is typically smooth with no porosity, presenting no niches suitable for the survival of a microorganism. Plastics used for livestock flooring, similar to metals, are smooth and non-porous and can even be impregnated with antimicrobial substances. Wood and clothing, however, have more texture and porosity.

In most cases, fomite environments are easily disinfected and pathogens are controlled naturally through desiccation, temperature, sunlight, and oxygen-related stress. For instance, the metal bars that make up holding pens or the metal panels of transportation vehicles provide an inhospitable climate, effectively limiting the survival and amplification of enteric pathogens. Microbial survival, however, is always a relative concept. Pathogens that are associated with large deposits of organic matter—feces, for example—can remain viable for many months and can be effectively "hidden" from disinfectants and natural inactivation processes.

Environmental Occurrence, Survival, and Transport of Pathogens

The types and presence of pathogen carriers or hosts, characteristics of the host to propagate and spread enteric pathogens, the ability of a particular pathogen to withstand conditions outside of a particular host's body, and characteristics of the pathogen and the outside environment that allow pathogen transport are all important in maintaining pathogen occurrence and persistence within, and outside of, animal-feeding operations. Important factors relating to the occurrence, survival, and transport of manure-borne pathogens associated with the preharvest environment are listed in Table 13.1.

Environmental Fecal Contamination

Because zoonotic enteric pathogens are shed in feces, they can inevitably contaminate environmental sources, including soil, water, air, and fomites, by many different routes. These manure-borne pathogens can infect a variety of hosts, including wildlife, livestock, and humans, and remain viable outside of their host for long periods. Feces and the pathogens harbored in it can be widely disseminated in the environment as a result of human, animal, and insect activity.

Several studies have shown that *E. coli* O157:H7 can survive in manure for several weeks under a wide variety of conditions (Wang et al. 1996, Kudva et al. 1998, Fenlon and

Table 13.1 Factors that affect the environmental occurrence and fate of manure-borne pathogens.

Occurrence	Survival/persistence	Transport
Age, type, and numbers of host	Individual pathogen traits	Individual pathogen traits: size, shape, charge, motility
Host's immunological status	Physical and chemical factors of the environment: pH, temperature, water, oxygen, sunlight, antimicrobial substances	Soil transport factors: clay content, pore size, water content, cations, plant roots, worms
Infectious dose		
Severity of illness		
Pathogen concentrations shed in feces	Presence of predatory organisms	Air transport factors: wind speed, humidity, sunlight exposure
Animal management	Ability of pathogen to form environmentally resistant forms	Water transport factors: flow rate, land topography, rainfall rate, water volume
Proximity to and control of potential contamination reservoirs	Aggregation and adsorption of pathogen to particulate matter	
	Pathogen amplification	
	Pathogen survival in other organisms (algae, insects, etc.)	Fomite transmission (shoes, equipment)
	Disinfection practices (water, pens, etc.)	Vectors including wildlife or insects

Wilson 2000). *Campylobacter* has been observed to survive up to 3 weeks in feces. In cattle slurry, *E. coli* survived less than 10 days (Rasmussen and Casey 2001), and a 3.5 log reduction within 12 weeks (McGee et al. 2001) was reported by two separate studies. In the solid fraction of swine slurry, *Salmonella* Typhimurium survived for 26 and 85 days during summer and winter conditions, respectively (Placha et al. 2001).

Factors that can affect microbial survival in slurry or other livestock waste are concentrations of toxic chemicals such as ammonia, the presence of predatory or competitive microorganisms, temperature, pH, aeration, and pathogen multiplication or regrowth. Extremes in pH can be deleterious to pathogenic microorganisms. A large drop in pH, from 8.08 to 5.91, in solid livestock waste significantly decreased *Salmonella* concentrations (Placha et al. 2001). Differences in pathogen reduction may be the result of the amount of dry matter or solids (Jones 1980, Strauch 1991). Stanley et al. (1998) observed that bacterial pathogens are reduced more quickly in an aerated compared with a nonaerated dairy farm slurry when held in storage tanks throughout the year.

Pathogens in Livestock Feed and Water

Feed, water, facilities, and equipment have tested positive for *E. coli* O157:H7 and may contribute to the spread of this organism through herds (Zhao et al. 1995, Hancock et al. 1998). Shere et al. (1998) reported that 6% of feed samples from a dairy farm in Wisconsin tested positive for *E. coli* O157:H7. *E. coli* O157:H7 was demonstrated to have the ability to increase by 1,000- to 10,000-fold in feed (Hancock et al. 1998, Rasmussen and Casey 2001). It has been suggested that fecal contamination of forage followed by inadequate ensiling may contribute to an increased incidence of *E. coli* O157:H7 in cattle (Fenlon and Wilson 2000).

Several scientists have reported that *E. coli* can survive in sediments in water troughs for up to 6 months (Hancock et al. 1998). Some evidence indicates that drinking-water troughs are contaminated via an oral rather than fecal route, as covered tanks that were unlikely to be contaminated with manure contained *E. coli* O157:H7 (Shere et al. 1998). Growth and survival of *E. coli* O157:H7 has also been discovered in trough water and detected after 4 months in trough water sediments (Hancock et al. 1998, Rasmussen and Casey 2001).

Environmental Vectors

Hancock et al. (1998) have suggested that cattle may not be the primary animal reservoir for *E. coli* O157:H7 but, rather, a vehicle of transmission of the organism. A "reservoir" is typically defined as a place in which you can always find the organism. This definition is not supported by the current *E. coli* O157:H7 occurrence data in cattle. Some cattle carry this organism, and others do not. Those that do carry it may carry it only for certain amounts of time. Other animals, insects, and birds carry the organism, and it is also found in a variety of environmental sources that suggest that even though cattle are a major carrier of this pathogen, some nonbovine reservoirs may exist. Transmission of other important foodborne pathogens is also likely to occur by animals, birds, or insects that are not defined as reservoirs.

It has been demonstrated that fruit flies can carry *E. coli* O157:H7 to the surfaces of apples (Janisiewicz et al. 1999), and inoculated houseflies have also been shown to carry and excrete *E. coli* O157 (Sasaki et al. 2000). Because flies are typically ubiquitous in feedlots, they may transfer the organism from one animal to another or from environmental sources to the animals. Wild animals also have access to feedlots, and the presence of wild deer has been shown to increase the risk of *E. coli* O157:H7 transmission to livestock (Sargeant et al. 1999). Birds have also tested positive for *E. coli* O157:H7 (Wallace et al. 1997, Hancock et al. 1998), and Wesley et al. (2000) identified accessibility of birds to animal feed as a risk factor for *Campylobacter* infection in cattle. *Listeria* Monocytogenes has also been isolated from insects and birds (Pell 1997). Because reservoirs of infection outside of the farm environment can affect the occurrence of fecal pathogens within the farm environment, on-farm control of fecal pathogens needs to account for wildlife inputs.

Seasonal Environmental Factors

Although seasonal occurrence of foodborne pathogens has not been observed in some reports (Stanley et al. 1998), in general there is a peak in occurrence seen during the warmer months of the year in both animals and the environment. Increased insect and animal activity and livestock birthing usually occurs during the warmer months of the year, so it is not surprising that peak incidence of *Campylobacter* and *E. coli* have also been reported during these times. Conversely, the peak prevalence of *Listeria,* which is known for its ability to grow at colder temperatures, occurs during winter months (Pell 1997). Placha et al. (2001) reported increased survival of *Salmonella* Typhimurium and various indicator microorganisms during winter compared to summer months (Natvig et al. 2002). Although host and vector activities can affect pathogen occurrence, factors of the pathogen and the environment in which it resides greatly affect its survival and transport to new hosts.

Environmental Pathogen Survival

Many enteric pathogens are able to survive conditions outside their host for extended periods of time. Genetically acquired traits enable some microbial groups to be more resistant to environmental conditions than others. Some bacteria alter their cellular morphology and metabolic activity outside of the host or in response to adverse conditions such as extreme temperatures and low nutrient conditions. For example, thermophillic *Campylobacter* species possibly enter a dormant stage when exposed to colder temperatures, where their cellular morphology changes from spiral to coccoid and their metabolic activities decrease (Jones et al. 1991). Pathogenic strains of *Salmonella* and *E. coli* can also enter a dormant state when introduced to suboptimal environmental conditions (Reissbrodt et al. 2002). Another bacterial protective mechanism is the formation of endospores that are extremely resistant to high temperatures and chemicals. Common spore formers include *Bacillus* and *Clostridium* species.

As mentioned previously, pathogens may be contained within fecal material, which protects them from sunlight, predation, desiccation, oxygen availability, and the effects of disinfectants, surfactants, or other damaging chemicals. In addition to aggregation, uptake by macro invertebrates (nematodes and amphipods) and protozoa can protect pathogens such as *E. coli* O157:H7. In fact, the survival of *E. coli* O157:H7 in soil protozoa has been suggested by Rasmussen and Casey (2001) to be a potential source of survival and subsequent livestock reinfection. Table 13.2 provides information regarding biotic and abiotic factors that affect pathogen survival in the environment.

Transport in Soil

Soil properties can limit or enhance microbial transport through the soil profile. There are many factors that contribute to the potential for contamination of wellheads and subsequent pollution of groundwater. The main factors are soil texture, depth to bedrock, and depth to groundwater.

The texture of the soil is one of the most important determining factors influencing the rate of wastewater movement through the soil to groundwater. Coarse-textured soils such as sands have large pore spaces between soil particles, allowing water to quickly percolate downward to the groundwater. Conversely, in fine-textured soils such as clays, movement of water and contaminants are typically slower. These fine-textured soils can act as natural filters, leaching out microbial contaminants before they reach ground water.

Open fractures in the bedrock are thought to enhance movement of wastewater and microbial contaminants to groundwater. The depth and texture of soil over fissured bedrock is thus a major issue, especially if this layer is very shallow. Shallow layers provide little opportunity for the soil to filter contaminated water. As a consequence, once the water reaches the bedrock, contamination of groundwater can be very swift through fissures or fractures in the bedrock.

The natural "treatment" filtration of contaminated water primarily takes place in soil above the water table (the unsaturated zone of soil). A high water table results in a short travel time for water and contaminants to move through the unsaturated soil before reaching the groundwater. Water table depths can also fluctuate dramatically depending on the season of the year.

Table 13.2   Biotic and abiotic factors affecting pathogen survival in the environment.

Environmental factors	Pathogen effects
Sunlight	Nucleic acid damage (ultraviolet radiation)
Heat	Desiccation; denaturation of enzymes, nucleic acid and other
Freezing	cellular components
temperatures	Damage to cell wall, cell membrane, and other cellular
Oxygen	components caused by ice crystal formation
Extremes in pH	Presence of O_2: Lethal to anaerobic microorganisms (lack
Salt	enzyme systems for reducing lethal respiratory byproducts);
Organic and other	required by strict aerobic microorganisms
particulate matter	Microaerophillic conditions: Low oxygen and high carbon
Disinfectants	dioxide conditions; required by some pathogens such as
Predators	*Campylobacter*
Other microbes	Enzyme denaturation; hydrolysis of cellular components
Water	Hypotonic conditions can cause cells to burst; hypertonic
	conditions cause cells to dehydrate
	Protects or shields pathogens from predators, disinfectants,
	sunlight, and other threats
	Cell wall and membrane damage (chlorine, alcohols, ozone);
	nucleic acid damage (ultraviolet light)
	Pathogen uptake by other organisms (protozoa, algae, worms,
	insects)
	Competition for nutrients; Production of antimicrobial
	substances
	Saturated conditions produce anoxic conditions (detrimental to
	strict aerobes); low water activity (amount of water available for
	microbial use) is detrimental because it's required for all
	biochemical activities

Surface properties such as surface appendages, charge, cell size, and cell shape can influence microbial transport in soil (Mawdsley et al. 1995). Bitton and Gerba (1984) provide reviews of data related to physical factors influencing the mobility of microorganisms in soil. Knowledge of the factors that affect pathogen occurrence, survival, and transport within an animal-feeding operation can enable producers to develop production practices that limit transport within—or to—animals and other food products.

The Future of On-Farm Environmental Control Research

Good On-Farm HACCP Production Practice

Over the last several years, terms such as "farm to table" and "gate to plate" have cropped up to describe a comprehensive process used to control food safety hazards. These concepts support the idea that every segment of the food chain, from the producer to the processor to the supplier and distributor, from the grocery chain and restaurant all the way to the consumer, plays a key role in keeping the food supply safe. One management system used to control hazards is termed HACCP. HACCP has been implemented in food processing and food service environments, but the use in preharvest or on-farm environment has been limited because of a number of factors. Immediate and future research is needed to address these factors.

HACCP Principles 1 and 2—Conduct a Hazard Analysis and Identify Critical Control Points

Microbial hazards pose a unique challenge when it comes to preharvest HACCP. U.S. Department of Agriculture data, as well as that of several regional studies, report the prevalence of various pathogens in the production environment. However, the exact source or "step" in the production process that contributes to the pathogen prevalence has not yet been pinpointed, and research into this issue is vital to progress of on-farm systems for hazard control. Valid contaminant tracking tools such as molecular methodologies will be vital to this effort in the future.

Although we know that various pathogens originate in the production environment, we do not necessarily know the source of the pathogens. The animal itself may carry the pathogen in the intestinal tract or on the hide. However, various pathogens exist in the environmental areas surrounding the animal. The water, soil, flies, and various other environmental sources can all be potential sources of pathogens. This part of the process would make a hazard analysis very difficult. Some important research questions that must be addressed in the future are the following: What are the primary sources or reservoirs of pathogens? Should the primary focus of HACCP be on the animal or the environment? At what step in the process can we control the pathogen? When do pathogen loads really increase in the animal (e.g., following dietary changes, mixing of groups, during shipping, or on arrival at processing)?

During hazard analysis, a particular point or step in the process should be identified so that a control can be applied at this step to prevent, eliminate, or reduce the likelihood of having the pathogen enter the environment. In addition, control measures are an essential part of a hazard analysis. Once the problem has been identified, a control must be applied. Although research in the area of preharvest interventions is growing, the application of these methods is still limited. We currently do not have enough information to identify particular points in the process to apply interventions to reduce the pathogen load in a preharvest environment. Research to identify these points and appropriate interventions will be vital to improving GPPs. Some interventions (potential control measures) and potential CCPs are summarized in Table 13.3.

HACCP Principles 3, 4, and 5—Identify Critical Limits, Monitoring, and Corrective Actions

Critical limits could be applied to the production environment. For example, a beef processor may stipulate that no more than 20% (or any other number) of the animals shipped from

Table 13.3 Potential control measures and critical control points for on-farm Hazard Analysis and Critical Control Point System.

Preharvest interventions	Critical control points
Diet/prebiotics	Animal receiving
Probiotics/direct-fed	Animal shipping
Microbials	Sick-animal care
Vaccination	Water sources
Bacteriophage therapy	Feed supplies
Antimicrobials	Wild animals and insects

his feedlot test positive for *E. coli* O157:H7. The drawback to this is that the animals must be tested and held before shipping before we know whether the critical limit is met. Rapid determinations such as visible hide contamination could be a more rapid assessment if these factors are proven to cause subsequent contamination of the product.

Corrective actions must be taken if the animals exceed critical limits. Again, the following questions are valid, and research to answer them is vital to improving the sustainability of agriculture: Will the animals be held and tested? How will the producer prevent this from happening again?

HACCP Principle 6—Verification

Verification activities will be required in the preharvest environment. Activities required will vary based on the CCP and the monitoring methods associated with the CCP. Verification activities will be based on currently available and future research that details how to verify that a pathogen is controlled in the processing environment.

HACCP Principle 7—Recordkeeping

As with implementation of HACCP programs in a food processing environment, the amount of recordkeeping required for HACCP implementation of the farms will be significant. A management system must be in place to ensure that all critical limits are met and corrective actions are taken to make sure that the product is safe. The development of such recordkeeping systems will require development and continual improvement as HACCP programs are implemented during production.

Conclusions

The future of on-farm research will certainly be directed at identification of the reservoirs, sources, and fate of pathogens and at development and validation of methods to rapidly identify contamination, track its sources, and control pathogen reservoirs. Identification of all potential zoonotic vectors will also be prime research objectives. Ultimately, research into the microbial ecology of farms, the development of HACCP plans, and implementation of management practices to accomplish these tasks will be vital to the survival and sustainability of the livestock industry. The answers provided by this research will lead us to new safer and cost-effective on-farm production systems. Without this type of research, however, implementation of on-farm HACCP will be disastrous and detrimental to the producers whom we are trying to protect.

References

Bitton, G. and Gerba, C. P. 1984. Groundwater Pollution Microbiology. New York: Wiley.

Dowd, S. E. 2002. Wastewater and Biosolids as Sources of Airborne Microorganisms, pp. 3320–3330. *In* Encyclopedia of Environmental Microbiology, G. Bitton (Ed.). New York: Wiley.

Dowd, S. E. and Maier, R. M. 1999. Aeromicrobiology, pp. 91–122. *In* Environmental Microbiology, R. M. Maier, I. L. Pepper, C. P. Gerba (Eds.). San Diego, Calif.: Academic Press.

Fenlon, D. R. and Wilson, J. 2000. Growth of *Escherichia coli* O157 in poorly fermented laboratory silage: a possible environmental dimension in the epidemiology of *E. coli* O157. *Letters in Applied Microbiology* **30**:118–121.

Hancock, D. D., Besser, T. E. and Rice, D. H. 1998. Ecology of *Escherichia coli* O157:H7 in cattle and impact of management practices, pp. 85–91. *In Escherichia coli* O157:H7 and Other Shiga Toxin-Producing *E. coli* Strains, J. B. Kaper, A. D. O'Brien (eds.). Washington, D.C.: American Society for Microbiology.

Holt, P. S., Mitchell, B. W. and Gast, R. K. 1998. Airborne horizontal transmission of *Salmonella enteritidis* in molted laying chickens. *Avian Diseases* **42**:45–52.

Janisiewicz, W. J., Conway, W. S., Brown, M. W., Sapers, G. M., Fratamico, P. and Buchanan, R. L. 1999. Fate of *Escherichia coli* O157:H7 on fresh-cut apple tissue and its potential for transmission by fruit flies. *Applied and Environmental Microbiology* **65**:1–5.

Jones, D. M., Sutcliffe, E. M. and Curry, A. 1991. Recovery of viable but non-culturable *Campylobacter jejuni*. *Journal of General Microbiology* **137**:2477–2482.

Jones, P. W. 1980. Health hazards associated with the handling of animal wastes. *Veterinary Record* **106**:4.

Kudva, I. T., Blanch, K. and Hovde, C. J. 1998. Analysis of *Escherichia coli* O157:H7 survival in ovine or bovine manure and manure slurry. *Applied Environmental Microbiology* **64**:3166–3174.

Mawdsley, J. L., Bardgett, R. D., Merry, R. J., Pain, B. F. and Theodorou, M. K. 1995. Pathogens in livestock waste, their potential for movement through soil and environmental pollution. *Applied Soil Ecology* **2**:1–15.

McGee, P., Bolton, D. J., Sheridan, J. J., Earley, B. and Leonard, N. 2001. The survival of *Escherichia coli* O157:H7 in slurry from cattle fed different diets. *Letters in Applied Microbiology* **32**:152–155.

Natvig, E., Ingham, S., Ingham, B., Cooperband, L. and Roper, T. 2002. *Salmonella enterica* serovar Typhimurium and *Escherichia coli* contamination of root and leaf vegetables grown in soils with incorporated bovine manure. *Applied and Environmental Microbiology* **68**:2737–2744.

Pell, A. N. 1997. Manure and microbes: public and animal health problem? *Journal of Dairy Science* **80**:2673–2681.

Placha, I., Venglovsky, J., Sasakova, N. and Svoboda, I. F. 2001. The effect of summer and winter seasons on the survival of *Salmonella typhimurium* and indicator micro-organisms during the storage of solid fraction of pig slurry. *Journal of Applied Microbiology* **91**:1036–1043.

Proux, K., Cariolet, R., Fravalo, P., Houdayer, C., Keranflech, A. and Madec, F. 2001. Contamination of pigs by nose-to-nose contact or airborne transmission of *Salmonella Typhimurium*. *Veterinary Research* **32**:591–600.

Rasmussen, M. A. and Casey, T. A. 2001. Environmental and food safety aspects of *Escherichia coli* O157:H7 infections in cattle. *Critical Reviews in Microbiology* **27**:57–73.

Reissbrodt, R., Rienaecker, I., Romanova, J. A., Freestone, P. P. E., Haigh, R. D., Lyte, A., Tschape, H. and Williams, P. H. 2002. Resuscitation of *Salmonella enterica* serovar *Typhimurium* and enterohemorrhagic *Escherichia coli* from the viable but nonculturable state by heat-stable enterobacterial autoinducer. *Applied and Environmental Microbiology* **68**:4788–4794.

Rodenburg, J. 1985. Water quality can affect production and health. *Ontario Milk Producer* **62**:12–14.

Sargeant, J. M., Hafer, D. J., Gillespie, J. R., Oberst, R. D. and Flood, S. J. A. 1999. Prevalence of *Escherichia coli* O157:H7 in white-tailed deer sharing rangeland with cattle. *Journal of the American Veterinary Medical Association* **215**:792–794.

Sasaki, T., Kobayashi, M. and Agui, N. 2000. Epidemiological potential of excretion and regurgitation by *Musca domestica* (Diptera: Muscidae) in the dissemination of *Escherichia coli* O157:H7 to food. *Journal of Medical Entomology* **37**:945–949.

Shere, J. A., Bartlett, K. J. and Kaspar, C. W. 1998. Longitudinal study of *Escherichia coli* O157:H7 dissemination on four dairy farms in Wisconsin. *Applied and Environmental Microbiology* **64**:1390–1399.

Stanley, K. N., Wallace, J. S. and Jones, K. 1998. Note: Thermophilic campylobacters in dairy slurries on Lancashire farms: seasonal effects of storage and land applications. *Journal of Applied Microbiology* **85**:405-409.

Strauch, D. 1991. Survival of pathogenic micro-organisms and parasites in excreta, manure and sewage sludge. *Reviews in Science and Technology* **10**:813.

Wallace, J. S., Cheasty, T. and Jones, K. 1997. Isolation of vero cytotoxin-producing *Escherichia coli* O157 from wild birds. *Journal of Applied Microbiology* **82**:399–404.

Wang, G., Zhao, T. and Doyle, M. P. 1996. Fate of enterohemorrhagic *Escherichia coli* O157:H7 in bovine feces. *Applied and Environmental Microbiology* **62**:2567–2570.

Wathes, C. M., Zaidan, W. A., Pearson, G. R., Hinton, M. and Todd, N. 1988. Aerosol infection of calves and mice with *Salmonella typhimurium*. *Veterinary Record* **123**:590–594.

Wesley, I. V., Wells, S. J., Harmon, K. M., Green, A., Schroeder-Tucker, L., Glover, M. and Siddique, I. 2000. Fecal shedding of *Campylobacter* and *Arcobacter* spp. in dairy cattle. *Applied and Environmental Microbiology* **66**:1994–2000.

Zhao, T., Doyle, M. P., Shere, J. A. and Garber, L. 1995. Prevalence of enterohemorrhagic *Escherichia coli* O157:H7 in a survey of dairy herds. *Applied and Environmental Microbiology* **61**:1290–1293.

14 Do Animal Transmissible Spongiform Encephalopathies Pose a Risk for Human Health?

Mary Jo Schmerr

Introduction

Transmissible spongiform encephalopathies (TSEs) or prion diseases in humans and in animals are rare fatal neurodegenerative diseases. TSEs have long incubation times with relatively short clinical phases usually lasting for only a few months. The human TSEs include Creutzfeldt Jakob disease (CJD), kuru (Gajdusek et al. 1967), Gerstmann-Sträussler-Scheinker, and familial cases of CJD (Prusiner 2001). The human TSEs belong to a larger group of amyloid diseases that include Alzheimer's, amylotrophic lateral sclerosis (commonly known as Lou Gehrig's disease), and Huntington's disease. Common domestic and wild animal TSEs include scrapie in sheep and goats (Pattison 1990), chronic wasting disease (CWD) in cervids (Williams and Young 1980), bovine spongiform encephalopathy (BSE) in cattle (Wells et al. 1987, Hope et al. 1988), and transmissible mink encephalopathy (Marsh et al. 1976). A list of the common TSEs is shown in Table 14.1. Other TSEs have been transmitted experimentally to other species or by inadvertent contamination of foodstuffs.

Table 14.1 Major natural transmissible spongiform encephalopathies.

Name	Species infected	Place	Date reported	Suspected route of infection
Scrapie	Sheep and goats	United Kingdom	>250 years ago	Oral, vertical, iatrogenic,[a] environmental
Bovine spongiform encephalopathy	Cattle	United Kingdom	1986	Oral
Variant CJD	Humans	United Kingdom	1996	Oral
Creutzfeld Jakob Disease	Humans	Germany	1920s	Oral, iatrogenic
Kuru	Humans	New Guinea Islands	1967	Oral
Chronic wasting disease	Cervids	Colorado	1980	Oral, vertical, environmental
Feline spongiform encephalopathy	Domestic and big cats	United Kingdom	1991	Oral
Transmissible mink encephalopathy	Mink	Wisconsin	1976	Oral

[a]Iatrogenic = medical procedures, vaccines, and transplantation.

Biochemical Characteristics of the Prion Protein

Most of the current evidence points to an altered protein, the prion protein (PrP), as the agent that causes these diseases. Some controversy still remains about the agent and the search for another source of the infection continues (Priola et al. 2003, Manuelidis and Lu 2003). The PrP is a normal host-surface glycoprotein that is changed in conformation so that it aggregates, becomes insoluble, and forms amyloid or amyloid-like fibers in the brains of the affected individuals. The deposition of these fibers formed from the abnormal form of the PrP may be the cause of the neurological stage of the disease.

The normal PrP is required for the abnormal prion protein to establish a TSE infection. This was shown by experiments with knockout mice for the prion gene (Brandner et al. 1996a, Brandner et al. 1996b). These mice could not be infected with a TSE. A dramatic demonstration of this was when brain grafts expressing PrP were placed in brains of the knockout mice and the mice were subsequently injected with abnormal PrP. Abnormal PrP was found only in the grafts, not in the rest of the brain.

There are several theories that attempt to explain how the normal PrP undergoes a conformational change. The protein secondary structure is changed from approximately 30% α-helix and approximately 3% β-sheet to approximately 30% α-helix and approximately 30% β-sheet, resulting in semi-resistance to proteases, aggregation, and insolubility in biological buffer systems. One of these theories suggests that the abnormal prion acts as a seed for crystallization of the normal PrP (Lansbury and Caughey 1995). Another theory suggests that chaperone-like proteins are involved in changing the conformation (Prusiner 2001). Another suggests that the acidic pH conditions provide the environment for the normal prion to become altered (Zou and Cashman 2002). The presence of prion peptides may have an effect on conformational change (Kaneko et al. 1997, Chabry et al. 1998). Recent studies have shown that the dimerization of the normal PrP may prevent the formation of the abnormal prion protein (Meier et al. 2003).

Another feature of the PrP is that it binds Cu (II) in a region of octa-repeats near the amino terminus of both the normal and abnormal form. Substitution of Cu (II) with manganese causes normal PrP to become semi-resistant to proteases (Brown et al. 2000). This suggests that Cu (II) may play a role in the course of the disease and may be required for the function of the normal PrP (Brown 2001, Rachidi et al. 2003). Because large quantities of protein are required to perform protein characterization, only recombinant derived prion protein has been characterized (Billeter et al. 1997), and full elucidation of the three-dimensional structure of the abnormal PrP has not been attained (Wuthrich and Riek 2001).

Paramithiotos et al. (2003) recently demonstrated that the abnormal prion protein has a repeating motif of three amino acids, tyrosine-tyrosine-arginine, that are exposed and are accessible to interactions with antibodies. In the normal prion protein these amino acids are sequestered within the structure and are not available for binding. This is a major breakthrough and provides the promise of a potential vaccine for these diseases. Other researchers have shown that certain antibodies to the prion protein also delay the formation of the abnormal form (White et al. 2003). These observations and information regarding this mechanism will be helpful in establishing methods or potential drugs to prevent transmission of TSEs (Demaimay et al. 2000).

Strain Typing

There are primarily two biological systems used for strain typing. One, developed in Edinburgh, defines the length of incubation time in a strain of mouse, and the lesion profiles in the brains of these mice are dependent on the prion strain that is used to inoculate the mouse (Fraser and Dickinson 1973). In this model, the variations in incubation times and in lesion profiles of the mice strains are used to determine the TSE strain. BSE has a definite pattern, both for the incubation times and for the lesion profiles (Bruce et al. 1994). All of the different species of animals inoculated with BSE showed the same profile in this strain-typing scheme. The same profile was obtained for variant Creutzfeldt Jakob disease (vCJD) as well, indicating that BSE caused the human disease (Bruce et al. 1997).

Another biological strain-typing method employs transgenic mice in which the mouse prion gene is replaced with a gene for the prion protein of another species. When these mice had the prion gene replaced with bovine or human genes, both the bovinized and humanized mice had similar incubation times whether they were injected with BSE or vCJD. This is strong evidence that BSE and vCJD have the same TSE characteristics (Hill et al. 1997, Scott et al. 1999). Similar experiments have been done for CWD (Safar et al. 2002). Recently, it has been reported (Asante et al. 2002) that when transgenic mice expressing human PrP are derived from different mouse strains, different forms of CJD are prevalent in each mouse. This suggests that the genetic background of the mouse may also play a role in defining the strains of TSEs. Although these experiments are useful, they are very expensive and take several years to complete. It is clear that a much more rapid method is required to resolve problems related to emerging TSEs.

Some biochemical methods for strain typing have been developed. The first breakthrough occurred when it was demonstrated (Caughey et al. 1995, Caughey et al. 1997) that the abnormal prion protein could convert the normal prion protein into a protease K resistant form. Subsequently, it was demonstrated that conversion by the abnormal form is most efficient when it converts normal prion from the same species and the same genotype (Bossers et al. 1997).

Another strain-typing method is based on the amount of each isoform of the abnormal prion protein and its migration patterns on Western blot (Somerville and Ritchie 1990, Collinge et al. 1996, Somerville et al. 1997). The abnormal prion protein has three isoforms that include di-glycosylated, mono-glycosylated, and non-glycosylated forms. The amounts and the migration times vary for different TSEs. Collinge et al. (1996) showed that the patterns are similar for BSE and vCJD but quite different for most strains of sheep scrapie. CWD also displays a different pattern. One of the problems with this method is that the patterns vary from one tissue to another and the resultant bands are quite large, suggesting that more than one species of the protein is present within each band (Race et al. 2002b).

Other methods measure the conformational state of the prion protein (Caughey et al. 1998). A conformational dependent immunoassay (Safar et al. 1998, Bellon et al. 2003) capitalizes on the differences in conformation of different strains of the abnormal prion protein. In this method, the amount of folding of the abnormal prion protein in its native state is compared to the completely unfolded denatured states. The difference in the reactivity of specific antibody with the native and denatured state is indicative of the strain. The explanation is that some strains are more unfolded, exposing sites not available in other strains in the native state (Safar et al. 1998). Folding of the native abnormal prion protein also affects the protease sensitivity of the protein. BSE and vCJD had similar reactivity in the conformational dependent assay, indicating that these two TSEs have very similar structures. All of the data—strain typing in mice, incubation times of transgenic mice, and biochemical characterization—have clearly shown that BSE and variant CJD are the same TSE strain.

Because sheep in the British Isles and in Europe were given the same feed as the cattle, it is possible that sheep were also infected with BSE. Although it is possible to distinguish between BSE and scrapie in sheep by using distinct patterns of staining with a specific antibody by Western blot (J. Langeveld, unpublished data) and by histological methods (Jeffrey et al. 2001a, Gonzalez et al. 2003), more rapid tests that use samples available from live sheep need to be developed (Schreuder and Somerville 2003). The European Union has mandated that efficient strategies be developed to differentiate between BSE and scrapie in sheep because of the concern of possible transmission of vCJD to humans through BSE-infected sheep. As a result, new rapid methods that can determine the signature for each TSE are needed. The challenge for all of the technologies will be to develop a very sensitive method in which a body fluid such as blood could be used as the sample.

Distribution in the Tissues

The abnormal PrP, the putative cause of TSEs, is distributed primarily in the central nervous system in later stages of the disease when clinical disease is apparent. However, it is also found in lymphoid and other tissues, including the eye (Head et al. 2003), in lesser amounts earlier in the disease before it is found in the central nervous system (Hadlow et al. 1982, van Keulen et al. 1996, van Keulen et al. 2000, Jeffrey et al. 2001b, Jeffrey et al. 2002). Recent reports have established that the abnormal prion protein is present in the olfactory epithelium of CJD patients (Zanusso et al. 2003b) and is present in the skeletal muscle and the tongue of hamsters (Bartz et al. 2003, Thomzig et al. 2003). In earlier studies (Pattison 1990), muscle from a scrapie-infected goat was found to be infectious to mice. Deposits of abnormal prion protein have been recently described in mouse skeletal muscle as well (Bosque et al. 2002). It has been established that blood of scrapie-infected sheep and sheep infected with BSE is infectious (Houston et al. 2000, Hunter et al. 2002). The donor sheep in this study were approximately halfway through the incubation period when their blood was transfused to another naïve sheep. The abnormal prion protein has also been found in the blood using a new extraction and testing protocol by Schmerr et al. (1999). Urine contains the abnormal prion protein, but it has not been found to be infectious (Shaked et al. 2001).

Experimental Transmission

Experimental transmission of sheep scrapie to other species has been accomplished in laboratory rodents, cattle (Gibbs et al. 1990, Cutlip et al. 1994, Robinson et al. 1995), elk (Hamir et al. 2001), mink (Marsh and Hadlow 1992), and primates (Gibbs and Gajdusek 1972). Most of these transmissions have been done by intracranial injection, which is the most efficient route of infection. Veterinary vaccines contaminated with scrapie-infected brain can also be a source of infection (Caramelli et al. 2001, Zanusso et al. 2003a). Oral transmission of scrapie is much less efficient and has been successful with only a few species (Cutlip et al. 2001).

BSE has been transmitted to laboratory rodents (Bruce et al. 1994), pigs (Wells et al. 2003), sheep, cats (Pearson et al. 1991), mink (Robinson et al. 1994), and lower primates by intracranial injection (Bons et al. 1999, Lasmezas et al. 2001). BSE also will infect the same species listed above, with the exception of pigs, by oral feeding. It also has inadvertently infected additional species including several zoo ungulates, members of the cat family, and humans (Ironside et al. 1996).

In addition to members of the cervid family, CWD has been experimentally transmitted by intracranial injection to laboratory rodents and ferrets (Bartz et al. 1998), cattle (Hamir et al. 2003), squirrel monkeys, and sheep (Williams 2002). Oral infection with CWD occurs in the cervids and ferrets. To date, transmission to cattle and sheep from other species has not occurred during the time that they are intermixed on the same grazing range with mule deer and elk in the endemic area in Colorado and Wyoming.

Although intracranial injection of infectious material is the most efficient method of transmitting TSEs, it is definitely not a natural route of infection. Transmission studies by natural routes such as orally or by infusion are more significant in assessing the possible risks for inter- and intra-species infections. It is important to document unambiguously the tissues that contain sufficient abnormal prion protein to transmit the infection. This is especially true for muscle tissue of domestic animals, as this is the major part of animals used for human consumption. Several other materials used from animals such as internal organs, collagen, and gelatin have been under intense scrutiny for potential contamination of products intended for humans, such as cosmetics, biomedical devices (Doerr et al. 2003), vaccines, and medicines (Robinson 1996). Although much attention is paid to these experimental transmission experiments, the information, especially if the studies are repeated studies, still does not provide the answers to the questions that are crucial for TSEs; that is, how are these diseases transmitted, and how can we test for them early in the infection?

Natural Transmission

The horizontal route for natural transmission of TSEs appears to be mostly oral but can occur from infection in cuts (Bartz et al. 2003) or infusion in the eye. The mechanism of how the abnormal prion protein travels from the digestive system or peripheral tissues to the central nervous system is not completely understood. Temporal studies in sheep have shown that the abnormal prion protein is first deposited in the lymphoid tissue, followed by deposition in the enteric nervous system, then in the central nervous system, and finally in brain, which usually marks the onset of clinical disease (van Keulen et al. 2000). This pattern of deposition of the abnormal prion protein is similar in variant CJD (Wadsworth et al. 2001) and in CWD in cervids (Sigurdson et al. 2002). However, in BSE and sporadic CJD, the deposition of the abnormal prion protein has not generally been detected in the lymphoid tissues. Little or no deposition of the abnormal prion protein in lymphoid tissues is also found in sheep that are heterozygotes for alleles of the prion protein that determines susceptibility or resistance to sheep scrapie (Bossers et al. 1997).

In sheep scrapie, a large percentage of scrapie infection occurs through vertical transmission from the ewe to the lamb. Again, as in horizontal transmission, the route is not completely elucidated. The placenta and the genotype of the fetus play a role in the transmission of the infection (Tuo et al. 2002), and colostrum may also play a role in transmission.

Many of the characteristics of CWD are similar to sheep scrapie, including deposition of the abnormal prion in lymphoid tissues in a pattern similar to that in sheep scrapie. Vertical as well as horizontal transmission probably plays a role in CWD transmission. In both cases—sheep scrapie and CWD—the role of environmental contamination needs further definition.

The origins of BSE are strongly linked to the consumption of TSE-contaminated rendered meat and bone meal by cattle and by the subsequent recycling of dead animals (Wilesmith 1988, Wilesmith et al. 1991). This contamination was attributed to a change in the rendering practices (Taylor and Woodgate 2003). During the early 1980s, the organic solvent trichloroethylene was removed from the rendering process because of its carcinogenic properties. This solvent inactivates the infectious activity of TSEs. Extensive studies by Taylor and Woodgate (2003) have demonstrated that BSE may remain infectious after heating the material on glass surfaces, which under ordinary circumstances, would inactivate most pathogens.

Transmission to Humans

Although no formal studies have demonstrated that sheep scrapie has been transmitted to humans, and there is a long history indicating that it has not, development of new methods that can differentiate strains may be useful in producing scientific data that confirm these anecdotal observations. Because there is still no known cause for sporadic CJD, it will be informative to clearly define whether there is involvement of sheep scrapie with this disease.

Following the outbreak of BSE in the United Kingdom in the late 1980s and early 1990s, there was a vigilant surveillance undertaken to determine if this "new" TSE would cross the species barrier and infect humans. Previous experience with sheep scrapie strongly indicated that infection of humans was highly unlikely. In early 1996, an announcement was made regarding the discovery of a new variant of CJD (Will et al. 1996, Collinge and Rossor 1996). Consumption of contaminated beef was considered to be the source of vCJD in the United Kingdom. In 1997, it was clearly demonstrated by strain typing that the same agent caused both BSE and the new variant of CJD (Bruce et al. 1997). A few cases of vCJD have been reported in Europe (Deslys et al. 1997, Streichenberger et al. 2000), Asia, and North America. Most, but not all, of these cases have occurred in individuals who resided in the British Isles for extensive periods of time.

Because of the spread of CWD from the endemic area in Wyoming and Colorado to other areas in the United States (Joly et al. 2003) and Canada, there is considerable concern about the potential transmission of CWD to humans. Although there have been several anecdotal accounts of persons consuming venison or meat from wild cervids and becoming ill with CJD, there are no scientific experiments that would confirm this speculation. Although Raymond et al. (2000) has demonstrated *in vitro* that the normal prion protein of cattle and humans can be converted to the abnormal prion protein in the presence of abnormal prion protein from CWD, it was considered a highly unfavorable reaction compared with a same-species conversion. Scientific experiments have shown that it is very difficult to transmit CWD to species of animals other than cervids by natural routes. Unlike other species, ferrets appear to be susceptible to CWD by oral infection. Surveillance in the United States for a new variant of CJD caused by CWD is ongoing, and at this time, no case of a new variant of CJD has been observed. Because CJD is probably somewhat under-reported in the United States, the surveillance is not as rigorous as it could be. Both increased education of neurologists and medical staff to better help identify cases of CJD and increased reporting may improve the reliability of the surveillance.

Efforts to prevent infection of animals with TSEs have resulted in several regulatory actions throughout a number of countries, especially in the British Isles, Europe, Asia, and North America (Bradley 2001). One of the primary actions is to ban the practice of feeding rendered ruminant materials back to ruminants. This action was taken in the United Kingdom in the late 1980s and then in North America in the late 1990s. In European countries, feed bans are now seriously enforced after the discovery of BSE in native cattle in 2000. Although generally speaking, these bans are effective, it does take some time to enforce them (Gizzi et al. 2003). Once they are effectively enforced, the incidence of BSE appears to be considerably reduced. Another action is to prohibit the importation of animals and materials of animal origin from countries with BSE.

Many risk analyses in the British Isles (Bradley 2001) Europe (Morley et al. 2003), Asia (Ozawa 2003), and North America (Cohen et al. 2001, Kellar and Lees 2003) have been done to estimate the risk of BSE appearing in a country. One of the major factors used to assess the risk is an examination of the regulations, especially those regarding feed bans. In spite of regulations and a very low risk factor of one per million head of cattle, cases of BSE still can occur, as demonstrated by the identification of a native cow with BSE in Canada in April 2003. The risk assessments are still basically a combination of assumptions and estimates, and they are dependent on complete compliance with regulations. This demonstrates the necessity of a statistically significant testing program to ensure public confidence in the safety of the food supply and to prevent the economic loss (ca. $11 million USD/day in Canada) and hardship that ensues in the agricultural sector. This experience happened first in the United Kingdom (Matthews 2003), then in Europe and Japan, and now in Canada. It was only after an effective testing program was in place that public confidence was restored in these countries. In several of these countries, the number of animals tested exceeds one million per year. This includes cattle over the age of 30 months as well as high-risk animals and cohorts of a BSE-infected animal. In Japan, every animal that is sent to slaughter is tested for BSE. Even though the United States has a much larger population of cattle (>100 million), only 19,990 animals in the high-risk category were tested in the year 2002. High-risk animals are animals that show signs of neurological disease or that are not ambulatory.

Although there has been considerable attention from the popular media about the transmission of BSE to the human population, the number of documented transmissions is very rare. Approximately 135 cases of vCJD have been documented over a period of 8 years in a population of 58.8 million people. The infection rate is potentially higher in the human population, but at this time, it is difficult to determine the rate with accuracy. Transmission from one species to another through natural routes is a rather rare event. A more likely event is a transmission that induces a subclinical infection similar to that which has been observed in rodents (Race et al. 2001, Race et al. 2002a, Hill and Collinge 2002, Houston et al. 2003). In these cases, the lifetime of the animal is shorter than the length of time needed to develop clinical disease. The inherent danger in this situation is the possible intra-species transmission through organ transplant, blood transfusion, or blood products from a preclinical individual. Contaminated products could be prevented by the use of a preclinical test to establish the TSE status of source animals or use of a test to analyze animal byproducts for TSE contamination.

Data from experiments (Asante et al. 2002) demonstrate that abnormal prion protein from BSE may contain more than one strain. This raises issues regarding the transmission of sheep scrapie to humans, especially with regard to sporadic CJD, as its origins are unknown. At this time, sheep scrapie is not considered to transmit to humans. This fact was primarily the basis for the original statements of the British government that the new TSE, BSE, would not infect humans. Nonetheless, there are some possible links to sheep scrapie as a source of infection for familial forms of CJD. Sheep scrapie was shown to infect primates (Gibbs and Gajdusek 1972) through an oral route. To establish that sheep scrapie is transmitted to the human population, better methods will be needed to distinguish strains of TSEs. Until this is accomplished, the link of sheep scrapie to sporadic CJD remains hypothetical.

Vaccines present a potential hazard for transmission of TSEs, as was found in sheep vaccines (Zanusso et al. 2003a). It has been determined that BSE was not transmitted by using rabies vaccines. Because of the extensive use of animal products in vaccine production, there is some concern that an animal TSE could be transmitted to humans by this route. (Arya 1996, Cashman 2001). At this point, no TSE transmission from a vaccine has been demonstrated.

Tests

One of the major issues with prion diseases is the lack of a validated practical preclinical test to identify infected animals and humans. In Table 14.2, there is a list of tests for prion diseases. Most of these tests use postmortem samples, but some of the prion tests under development make use of tissues and bodily fluids easily obtainable from live animals.

Table 14.2   Published tests for the prion diseases.

Test	Sample used	Time of testing	Commercially available (tissue specified)
Histology	Brain	Postmortem	no
Immunohistochemistry Haritani et al. 1994	Brain, lymphoid tissue, tonsil	Postmortem, preclinical	no
Western blot Oesch et al. 2000, Shaked et al. 2001	Brain, lymphoid tissue, urine	Postmortem	Prionics: For central nervous system tissue
ELISA Safar et al. 1998, Grassi 2003	Brain, lymphoid tissue	Postmortem	EnferBio-Rad: For central nervous system tissue
Fluorescence immunoassay Schmerr et al. 1999	Blood, cerebrospinal fluid, brain, lymphoid tissue	Postmortem, preclinical	No
SIFT[a] Bieschke et al. 2000	CSF	Clinical signs	No
FTIR[b] Schmitt et al. 2002	Serum	Clinical Signs	No
14,3,3-Protein Hsieh et al. 1996, Zerr and Poser 2002	CSF	Clinical Signs	No

[a]SIFT 5 scanning for intensely fluorescent targets.
[b]FTIR = fourier transform infrared spectroscopy.

In the tissue-based tests (Haritani et al. 1994), cellular changes are observed in slide mounted tissue. In immunohistochemistry (IHC), specific antibodies are used to detect the prion protein in the tissue. The IHC test is considered to be the most sensitive test for brain material. It can also be used for lymphoid tissue (Schreuder et al. 1998, O'Rourke et al. 2000). However, there are two problems with the IHC test. It is not rapid, and it generates considerable hazardous waste.

There are three commercially available validated tests for prion diseases (See Table 14.2). These tests are based on either Western blot (Oesch et al. 2000) or enzyme-linked immunoassay technology (Deslys et al. 2001, Grassi et al. 2001, Grassi 2003). Homogenates of the tissues are used, and specific antibodies for the prion protein are also used. For these tests, the amount of abnormal prion protein that can be detected is in the nanogram per milliliter range. These tests are rapid, with a 24-hour turnaround time, and are very sensitive.

New approaches using microtechniques for detecting the abnormal prion protein are under development. These tests use fluorescent dyes and cutting-edge instrumentation to detect the abnormal prion protein. The tests are SIFT (scanning for intensely fluorescent targets; Bieschke et al. 2000) and a fluorescence immunoassay (Schmerr et al. 1999). The sensitivity of these tests is in the picogram per milliliter range and makes it possible to detect the low amounts of abnormal prion protein found in blood and cerebral spinal fluid. Other tests measure metabolic changes that occur as a result of a TSE infection. These tests include Fourier transform infrared spectroscopy (Schmitt et al. 2002) that can detect the difference between normal serum and serum from TSE-infected hamsters. In cerebral spinal fluid of TSE-infected individuals, metabolites are found that are identified as 14-3-3 proteins (Hsieh et al. 1996, Zerr and Poser 2002).

The tests that analyze bodily fluids by analytical methods have the most potential for automation and have the sensitivity to detect the abnormal prion in very low amounts. The development and validation of a preclinical test are a top priority and are essential to prevent the spread of the TSEs. Such tests may encourage drug companies to produce effective drugs for treating these diseases. Because of the nature of analytical tests for TSEs, they can be used to test both processed and manufactured materials intended for human use. Full development of analytical tests will make it possible to eliminate TSEs from the domestic animal population and, potentially, the wild animal population. This in turn will result in considerably lowering the risk of transmission of these diseases to the human population.

Conclusions

To avoid the now-familiar scenarios in which it is announced that a case of BSE has been found in a country, the subsequent quarantines and bans by other countries, assurances from the government of the country in which BSE was found that beef is safe to eat, extensive media coverage, public hysteria with regard to eating meat, and large economic losses to the agricultural sector, it is necessary that statistically significant testing programs are put into place as soon as possible. It has been clearly shown in Europe and in Japan that public confidence is restored only after an extensive testing program using rapid validated tests is established. Although an analytical blood assay has been developed, it needs further improvements and validation before routine application (Schmerr et al. 1999). A validated preclinical test that analyzes blood is the next step needed to ensure that animal products destined for human use are free of TSEs.

References

Arya, S. C. 1996. Blood donated after vaccination with rabies vaccine derived from sheep brain cells might transmit CJD. *British Medical Journal* **313**:1405.

Asante, E. A., Linehan, J. M., Desbruslais, M., Joiner, S., Gowland, I., Wood, A. L., Welch, J., Hill, A. F., Lloyd, S. E., Wadsworth, J. D. and Collinge, J. 2002. BSE prions propagate as either variant CJD-like or sporadic CJD-like prion strains in transgenic mice expressing human prion protein. *European Molecular Biology Organization Journal* **21**:6358–6366.

Bartz, J. C., Kincaid, A. E. and Bessen, R. A. 2003. Rapid prion neuroinvasion following tongue infection. *Journal of Virology* **77**:583–591.

Bartz, J. C., Marsh, R. F., McKenzie, D. I. and Aiken, J. M. 1998. The host range of chronic wasting disease is altered on passage in ferrets. *Virology* **251**:297–301.

Bellon, A., Seyfert-Brandt, W., Lang, W., Baron, H., Groner, A. and Vey, M. 2003. Improved conformation-dependent immunoassay: suitability for human prion detection with enhanced sensitivity. *Journal of General Virology* **84**:1921–1925.

Bieschke, J., Giese, A., Schulz-Schaeffer, W., Zerr, I., Poser, S., Eigen, M. and Kretzschmar, H. 2000. Ultrasensitive detection of pathological prion protein aggregates by dual-color scanning for intensely fluorescent targets. *Proceedings of the National Academy of Sciences (USA)* **97**:5468–5473.

Billeter, M., Riek, R., Wider, G., Hornemann, S., Glockshuber, R. and Wuthrich, K. 1997. Prion protein NMR structure and species barrier for prion diseases. *Proceedings of the National of Academy Sciences (USA)* **94**:7281–7285.

Bons, N., Mestre-Frances, N., Belli, P., Cathala, F., Gajdusek, D. C. and Brown, P. 1999. Natural and experimental oral infection of nonhuman primates by bovine spongiform encephalopathy agents. *Proceedings of the National Academy of Sciences (USA)* **96**:4046–4051.

Bosque, P. J., Ryou, C., Telling, G., Peretz, D., Legname, G., DeArmond, S. J. and Prusiner, S. B. 2002. Prions in skeletal muscle. *Proceedings of the National Academy of Sciences (USA)* **99**:3812–3817.

Bossers, A., Belt, P. B. G. M., Raymond, G. J., Caughey, B., de Vries, R. and Smits, M. A. 1997. Scrapie susceptibility-linked polymorphisms modulate the in vitro conversion of sheep prion protein to protease-resistant forms. *Proceedings of the National Academy of Sciences (USA)* **94**:4931–4936.

Bradley, R. 2001. A brief overview of bovine spongiform encephalopathy and related diseases including a TSE risk analysis of bovine starting materials used during the manufacture of vaccines for use in humans. *Przeglad Epidemiologiczny* **55**:387–405.

Brandner, S., Isenmann, S., Raeber, A., Fischer, M., Sailer, A., Kobayashi, Y., Marino, S., Weissmann, C. and Aguzzi, A. 1996a. Normal host prion protein necessary for scrapie-induced neurotoxicity. *Nature* **379**:339–343.

Brandner, S., Raeber, A., Sailer, A., Blattler, T., Fischer, M., Weissmann, C. and Aguzzi, A. 1996b. Normal host prion protein (PrPC) is required for scrapie spread within the central nervous system. *Proceedings of the National Academy of Sciences (USA)* **93**:13148–13151.

Brown, D. R. 2001. Copper and prion disease. *Brain Research Bulletin* **55**:165–173.

Brown, D. R., Hafiz, F., Glasssmith, L. L., Wong, B. S., Jones, I. M., Clive, C. and Haswell, S. J. 2000. Consequences of manganese replacement of copper for prion protein function and proteinase resistance. *European Molecular Biology Organization Journal* **19**:1180–1186.

Bruce, M., Chree, A., McConnell, I., Foster, J., Pearson, G. and Fraser, H. 1994. Transmission of bovine spongiform encephalopathy and scrapie to mice: strain variation and the species barrier. *Philosophical Transactions of the Royal Society of London B Biological Sciences* **343**:405–411.

Bruce, M. E., Will, R. G., Ironside, J. W., McConnell, I., Drummond, D., Suttie, A., McCardle, L., Chree, A., Hope, J., Birkett, C., Cousens, S., Fraser, H. and Bostock, C. J. 1997. Transmissions to mice indicate that "new variant" CJD is caused by the BSE agent. *Nature* **389**:498–501.

Caramelli, M., Ru, G., Casalone, C., Bozzetta, E., Acutis, P. L., Calella, A. and Forloni, G. 2001. Evidence for the transmission of scrapie to sheep and goats from a vaccine against *Mycoplasma* agalactiae. *Veterinary Record* **148**:531–536.

Cashman, N. R. 2001. Transmissible spongiform encephalopathies: vaccine issues. *Developmental Biology* **106**:455–459.

Caughey, B., Kocisko, D. A., Raymond, G. J. and Lansbury Jr., P. T. 1995. Aggregates of scrapie-associated prion protein induce the cell-free conversion of protease-sensitive prion protein to the protease-resistant state. *Chemistry and Biology* **2**:807–817.

Caughey, B., Raymond, G. J. and Bessen, R. A. 1998. Strain-dependent differences in beta-sheet conformations of abnormal prion protein. *Journal of Biological Chemistry* **273**:32230–32235.

Caughey, B., Raymond, G. J., Kocisko, D. A. and Lansbury Jr., P. T. 1997. Scrapie infectivity correlates with converting activity, protease resistance, and aggregation of scrapie-associated prion protein in guanidine denaturation studies. *Journal of Virology* **71**:4107–4110.

Chabry, J., Caughey, B. and Chesebro, B. 1998. Specific inhibition of in vitro formation of protease-resistant prion protein by synthetic peptides. *Journal of Biological Chemistry* **273**:13203–13207.

Cohen, J. T., Duggar, K., Gray, G. M., Kreindel, S., Abdelrahman, H., Habtemariam, T., Oryang, D. and Tameru, B. 2001. Evaluation of the Potential for Bovine Spongiform Encephalopathy in the United States. Boston: Harvard Center for Risk Analysis and Harvard School of Public Health.

Collinge, J. and Rossor, M. 1996. A new variant of prion disease. *Lancet* **347**:916–917.

Collinge, J., Sidle, K. C., Meads, J., Ironside, J. and Hill, A. F. 1996. Molecular analysis of prion strain variation and the aetiology of "new variant" CJD. *Nature* **383**:685–690.

Cutlip, R. C., Miller, J. M., Hamir, A. N., Peters, J., Robinson, M. M., Jenny, A. L., Lehmkuhl, H. D., Taylor, W. D. and Bisplinghoff, F. D. 2001. Resistance of cattle to scrapie by the oral route. *Canadian Journal Veterinary Research* **65**:131–132.

Cutlip, R. C., Miller, J. M., Race, R. E., Jenny, A. L., Katz, J. B., Lehmkuhl, H. D., DeBey, B. M. and Robinson, M. M. 1994. Intracerebral transmission of scrapie to cattle. *Journal of Infectious Diseases* **169**:814–820.

Demaimay, R., Chesebro, B. and Caughey, B. 2000. Inhibition of formation of protease-resistant prion protein by Trypan Blue, Sirius Red and other Congo Red analogs. *Archives of Virology Supplement* **16**:277–283.

Deslys, J. P., Lasmezas, C. I., Comoy, E. and Domont, D. 2001. Diagnosis of bovine spongiform encephalopathy. *Veterinary Journal* **161**:1–3.

Deslys, J. P., Lasmezas, C. I., Streichenberger, N., Hill, A., Collinge, J., Dormont, D. and Kopp, N. 1997. New variant Creutzfeldt-Jakob disease in France. *Lancet* **349**:30–31.

Doerr, H. W., Cinatl, J., Sturmer, M. and Rabenau, H. F. 2003. Prions and orthopedic surgery. *Infection* **31**:163–171.

Fraser, H. and Dickinson, A. G. 1973. Scrapie in mice. Agent-strain differences in the distribution and intensity of grey matter vacuolation. *Journal of Comparative Pathology* **83**:29–40.

Gajdusek, C., Gibbs, C. J. and Alpers, M. 1967. Slow-acting virus implicated in kuru. *Journal of the American Medical Association* **199**:34.

Gibbs Jr., C. J. and Gajdusek, D. C. 1972. Transmission of scrapie to the cynomolgus monkey (*Macaca fascicularis*). *Nature* **236**:73–74.

Gibbs Jr., C. J. Safar, J., Ceroni, M., Di Martino, A., Clark, W. W. and Hourrigan, J. L. 1990. Experimental transmission of scrapie to cattle. *Lancet* **335**:1275.

Gizzi, G., Van Raamsdonk, L. W. D., Baeten, V., Murray, I., Berben, G., Brambilla, G. and Von Holst, C. 2003. An overview of tests for animal tissues in feeds applied in response to public health concerns regarding bovine spongiform encephalopathy. *Review of Science and Technology* **22**:311–331.

Gonzalez, L., Martin, S. and Jeffrey, M. 2003. Distinct profiles of PrP(d) immunoreactivity in the brain of scrapie- and BSE-infected sheep: implications for differential cell targeting and PrP processing. *Journal of General Virology* **84**:1339–1350.

Grassi, J. 2003. Pre-clinical diagnosis of transmissible spongiform encephalopathies using rapid tests. *Transfusion Clinique et Biologique* **10**:19–22.

Grassi, J., Comoy, E., Simon, S., Creminon, C., Frobert, Y., Trapmann, S., Schimmel, H., Hawkins, S. A., Moynagh, J., Deslys, J. P. and Wells, G. A. 2001. Rapid test for the pre-clinical postmortem diagnosis of BSE in central nervous system tissue. *Veterinary Record* **149**:577–582.

Hadlow, W. J., Kennedy, R. C. and Race, R. E. 1982. Natural infection of Suffolk sheep with scrapie virus. *Journal of Infectious Diseases* **146**:657–664.

Hamir, A. N., Cutlip, R. C., Miller, J. M., Williams, E. S., Stack, M. J., Miller, M. W., O'Rourke, K. I. and Chaplin, M. J. 2001. Preliminary findings on the experimental transmission of chronic wasting disease agent of mule deer to cattle. *Journal of Veterinary Diagnostic Investigation* **13**:91–96.

Hamir, A. N., Miller, J. M., Cutlip, R. C., Stack, M. J., Chaplin, M. J. and Jenny, A. L. 2003. Preliminary observations on the experimental transmission of scrapie to elk (*Cervus elaphus nelsoni*) by intracerebral inoculation. *Veterinary Pathology* **40**:81–85.

Haritani, M., Spencer, Y. I. and Wells, G. A. 1994. Hydrated autoclave pretreatment enhancement of prion protein immunoreactivity in formalin-fixed bovine spongiform encephalopathy-affected brain. *Acta Neuropathologica* **87**:86–90.

Head, M. W., Northcott, V., Rennison, K., Ritchie, D., McCardle, L., Bunn, T. J., McLennan, N. F., Ironside, J. W., Tullo, A. B. and Bonshek, R. E. 2003. Prion protein accumulation in eyes of patients with sporadic and variant Creutzfeldt-Jakob disease. *Investigative Ophthalmology and Visual Science* **44**:342–346.

Hill, A. F. and Collinge, J. 2002. Species-barrier-independent prion replication in apparently resistant species. *Acta Pathologia, Microbiologica et Immunologica Scandinavica* **110**:44–53.

Hill, A. F., Desbruslais, M., Joiner, S., Sidle, K. C., Gowland, I., Collinge, J., Doey, L. J. and Lantos, P. 1997. The same prion strain causes vCJD and BSE. *Nature* **389**:448–450.

Hope, J., Reekie, L. J., Hunter, N., Multhaup, G., Beyreuther, K., White, H., Scott, A. C., Stack, M. J., Dawson, M. and Wells, G. A. 1988. Fibrils from brains of cows with new cattle disease contain scrapie-associated protein. *Nature* **336**:390–392.

Houston, F., Foster, J. D., Chong, A., Hunter, N. and Bostock, C. J. 2000. Transmission of BSE by blood transfusion in sheep. *Lancet* **356**:999–1000.

Houston, F., Goldmann, W., Chong, A., Jeffrey, M., Gonzalez, L., Foster, J., Parnham, D. and Hunter, N. 2003. Prion diseases: BSE in sheep bred for resistance to infection. *Nature* **423**:498.

Hsieh, G., Kenney, K., Gibbs, C. J., Lee, K. H. and Harrington, M. G. 1996. The 14-3-3 brain protein in cerebrospinal fluid as a marker for transmissible spongiform encephalopathies. *New England Journal of Medicine* **335**:924–930.

Hunter, N., Foster, J., Chong, A., McCutcheon, S., Parnham, D., Eaton, S., MacKenzie, C. and Houston, F. 2002. Transmission of prion diseases by blood transfusion. *Journal of General Virology* **83**:2897–2905.

Ironside, J. W., Sutherland, K., Bell, J. E., McCardle, L., Barrie, C., Estebeiro, K., Zeidler, M. and Will, R. G. 1996. A new variant of Creutzfeldt-Jakob disease: neuropathological and clinical features. *Cold Spring Harbor Symposia on Quantitative Biology* **61**:523–530.

Jeffrey, M., Begara-McGorum, I., Clark, S., Martin, S., Clark, J., Chaplin, M. and González, L. 2002. Occurrence and distribution of infection-specific PrP in tissues of clinical scrapie cases and cull sheep from scrapie-affected farms in Shetland. *Journal of Comparative Pathology* **127**:264–273.

Jeffrey, M., Martin, S., González, L., Ryder, S. J., Bellworthy, S. J. and Jackman, R. 2001a. Differential diagnosis of infections with the bovine spongiform encephalopathy (BSE) and scrapie agents in sheep. *Journal of Comparative Pathology* **125**:271–284.

Jeffrey, M., Martin, S., Thomson, J. R., Dingwall, W. S., Begara-McGorum, I. and González, L. 2001b. Onset and distribution of tissue PrP accumulation in scrapie-affected Suffolk sheep as demonstrated by sequential necropsies and tonsillar biopsies. *Journal of Comparative Pathology* **125**:48–57.

Joly, D. O., Ribic, C. A., Langenberg, J. A., Beheler, K., Batha, C. A., Dhuey, B. J., Rolley, R. E., Bartelt, G., Van Deelen, T. R. and Samuel, M. D. 2003. Chronic wasting disease in free-ranging Wisconsin white-tailed deer. *Emerging Infectious Diseases* **9**:599–601.

Kaneko, K., Wille, H., Mehlhorn, I., Zhang, H., Ball, H., Cohen, F. E., Baldwin, M. A. and Prusiner, S. B. 1997. Molecular properties of complexes formed between the prion protein and synthetic peptides. *Journal of Molecular Biology* **270**:574–586.

Kellar, J. A. and Lees, V. W. 2003. Risk management of the transmissible spongiform encephalopathies in North America. *Review of Science and Technology* **22**:201–225.

Lansbury Jr., P. T. and Caughey, B. 1995. The chemistry of scrapie infection: implications of the "ice 9" metaphor. *Chemistry and Biology* **2**:1–5.

Lasmezas, C. I., Fournier, J. G., Nouvel, V., Boe, H., Marce, D., Lamoury, F., Kopp, N., Hauw, J. J., Ironside, J., Bruce, M., Dormont, D. and Deslys, J. P. 2001. Adaptation of the bovine spongiform encephalopathy agent to primates and comparison with Creutzfeldt-Jakob disease: implications for human health. *Proceedings of the National Academy of Sciences (USA)* **98**:4142–4147.

Manuelidis, L. and Lu, Z. Y. 2003. Virus-like interference in the latency and prevention of Creutzfeldt-Jakob disease. *Proceedings of the National Academy of Sciences (USA)* **100**:5360–5365.

Marsh, R. F. and Hadlow, W. J. 1992. Transmissible mink encephalopathy. *Review of Science and Technology* **11**:539–550.

Marsh, R. F., Sipe, J. C., Morse, S. S. and Hanson, R. P. 1976. Transmissible mink encephalopathy. Reduced spongiform degeneration in aged mink of the Chediak-Higashi genotype. *Laboratory Investigation* **34**:381–386.

Matthews, D. 2003. BSE: a global update. *Journal of Applied Microbiology* **94**:120S–125S.

Meier, P., Genoud, N., Prinz, M., Maissen, M., Rulicke, T., Zurbriggen, A., Raeber, A. J. and Aguzzi, A. 2003. Soluble dimeric prion protein binds PrPSc in vivo and antagonizes prion disease. *Cell* **113**:49–60.

Morley, R. S., Chen, S. and Rheault, N. 2003. Assessment of the risk factors related to bovine spongiform encephalopathy. *Review of Science and Technology* **22**:157–178.

Oesch, B., Doherr, M., Heim, D., Fischer, K., Egli, S., Bolliger, S., Biffiger, K., Schaller, O., Vandevelde, M. and Moser, M. 2000. Application of Prionics Western blotting procedure to screen for BSE in cattle regularly slaughtered at Swiss abattoirs. *Archives of Virology Supplement* **16**:189–195.

O'Rourke, K. I., Baszler, T. V., Besser, T. E., Miller, J. M., Cutlip, R. C., Wells, G. A., Ryder, S. J., Parish, S. M., Hamir, A. N., Cockett, N. E., Jenny, A. and Knowles, D. P. 2000. Preclinical diagnosis of scrapie by immunohistochemistry of third eyelid lymphoid tissue. *Journal of Clinical Microbiology* **38**:3254–3259.

Ozawa, Y. 2003. Risk management of transmissible spongiform encephalopathies in Asia. *Review of Science and Technology* **22**:237–249.

Paramithiotis, E., Pinard, M., Lawton, T., LaBoissiere, S., Leathers, V. L., Zou, W. Q., Estey, L. A., Lamontagne, J., Lehto, M. T., Kondejewski, L. H., Francoeur, G. P., Papadopoulos, M., Haghighat, A., Spatz, S. J., Head, M., Will, R., Ironside, J., O'Rourke, K., Tonelli, Q., Ledebur, H. C., Chakrabartty, A. and Cashman, N. R. 2003. A prion protein epitope selective for the pathologically misfolded conformation. *Nature Medicine* **9**:893–899.

Pattison, I. 1990. Scrapie agent in muscle. *Veterinary Record* **126**:68.

Pearson, G. R., Gruffydd-Jones, T. J., Wyatt, J. M., Hope, J., Chong, A., Scott, A. C., Dawson, M. and Wells, G. A. 1991. Feline spongiform encephalopathy. *Veterinary Record* **128**:532.

Priola, S. A., Chesebro, B. and Caughey, B. 2003. A view from the top—prion diseases from 10,000 feet. *Science* **300**:917–919.

Prusiner, S. B. 2001. Shattuck lecture—neurodegenerative diseases and prions. *New England Journal of Medicine* **344**:1516–1526.

Race, R., Meade-White, K., Raines, A., Raymond, G. J., Caughey, B. and Chesebro, B. 2002a. Subclinical scrapie infection in a resistant species: persistence, replication, and adaptation of infectivity during four passages. *Journal of Infectious Diseases* **186**:S166–S170.

Race, R., Raines, A., Raymond, G. J., Caughey, B. and Chesebro, B. 2001. Long-term subclinical carrier state precedes scrapie replication and adaptation in a resistant species: analogies to bovine spongiform encephalopathy and variant Creutzfeldt-Jakob disease in humans. *Journal of Virology* **75**:10106–10112.

Race, R. E., Raines, A., Baron, T. G., Miller, M. W., Jenny, A. and Williams, E. S. 2002b. Comparison of abnormal prion protein glycoform patterns from transmissible spongiform encephalopathy agent-infected deer, elk, sheep, and cattle. *Journal of Virology* **76**:12365–12368.

Rachidi, W., Mange, A., Senator, A., Guiraud, P., Riondel, J., Benboubetra, M., Favier, A. and Lehmann, S. 2003. Prion infection impairs copper binding of cultured cells. *Journal of Biological Chemistry* **278**:14595–14598.

Raymond, G. J., Bossers, A., Raymond, L. D., O'Rourke, K. I., McHolland, L. E., Bryant 3rd, P. K., Miller, M. W., Williams, E. S., Smits, M. and Caughey, B. 2000. Evidence of a molecular barrier limiting susceptibility of humans, cattle and sheep to chronic wasting disease. *European Molecular Biology Organization Journal* **9**:4425–4430.

Robinson, M. M. 1996. Transmissible encephalopathies and biopharmaceutical production. *Development of Biological Standards* **88**:237–241.

Robinson, M. M., Hadlow, W. J., Huff, T. P., Wells, G. A., Dawson, M., Marsh, R. F. and Gorham, J. R. 1994. Experimental infection of mink with bovine spongiform encephalopathy. *Journal of General Virology* **75**:2151–2155.

Robinson, M. M., Hadlow, W. J., Knowles, D. P., Huff, T. P., Lacy, P. A., Marsh, R. F. and Gorham, J. R. 1995. Experimental infection of cattle with the agents of transmissible mink encephalopathy and scrapie. *Journal of Comparative Pathology* **113**:241–251.

Safar, J., Wille, H., Itri, V., Groth, D., Serban, H., Torchia, M., Cohen, F. E. and Prusiner, S. B. 1998. Eight prion strains have PrP(Sc) molecules with different conformations. *Nature Medicine* **4**:1157–1165.

Safar, J. G., Scott, M., Monaghan, J., Deering, C., Didorenko, S., Vergara, J., Ball, H., Legname, G., Leclerc, E., Solforosi, L., Serban, H., Groth, D., Burton, D. R., Prusiner, S. B. and Williamson, R. A. 2002. Measuring prions causing bovine spongiform encephalopathy or chronic wasting disease by immunoassays and transgenic mice. *Nature Biotechnology* **20**:1147–1150.

Schmerr, M. J., Jenny, A. L., Bulgin, M. S., Miller, J. M., Hamir, A. N., Cutlip, R. C. and Goodwin, K. R. 1999. Use of capillary electrophoresis and fluorescent labeled peptides to detect the abnormal prion protein in the blood of animals that are infected with a transmissible spongiform encephalopathy. *Journal of Chromatography A* **853**:207–214.

Schmitt, J., Beekes, M., Brauer, A., Udelhoven, T., Lasch, P. and Naumann, D. 2002. Identification of scrapie infection from blood serum by Fourier transform infrared spectroscopy. *Analytical Chemistry* **74**:3865–3868.

Schreuder, B. E., van Keulen, L. J., Vromans, M. E., Langeveld, J. P. and Smits, M. A. 1998. Tonsillar biopsy and PrPSc detection in the preclinical diagnosis of scrapie. *Veterinary Record* **142**:564–568.

Schreuder, B. E. C. and Somerville, R. A. 2003. Bovine spongiform encephalopathy in sheep? *Review of Science and Technology* **22**:103–120.

Scott, M. R., Will, R., Ironside, J., Nguyen, H. O., Tremblay, P., DeArmond, S. J. and Prusiner, S. B. 1999. Compelling transgenetic evidence for transmission of bovine spongiform encephalopathy prions to humans. *Proceedings of the National Academy of Sciences (USA)* **96**:15137–15142.

Shaked, G. M., Shaked, Y., Kariv-Inbal, Z., Halimi, M., Avraham, I. and Gabizon, R. 2001. A protease-resistant prion protein isoform is present in urine of animals and humans affected with prion diseases. *Journal of Biological Chemistry* **276**:31479–31482.

Sigurdson, C. J., Barillas-Mury, C., Miller, M. W., Oesch, B., van Keulen, L. J., Langeveld, J. P. and Hoover, E. A. 2002. PrP(CWD) lymphoid cell targets in early and advanced chronic wasting disease of mule deer. *Journal of General Virology* **83**:2617–2628.

Somerville, R. A., Chong, A., Mulqueen, O. U., Birkett, C. R., Wood, S. C. and Hope, J. 1997. Biochemical typing of scrapie strains. *Nature* **386**:564.

Somerville, R. A. and Ritchie, L. A. 1990. Differential glycosylation of the protein (PrP) forming scrapie-associated fibrils. *Journal of General Virology* **71**:833–839.

Streichenberger, N., Jordan, D., Verejan, I., Souchier, C., Philippeau, F., Gros, E., Mottolese, C., Ostrowsky, K., Perret-Liaudet, A., Laplanche, J. L., Hermier, M., Deslys, J. P., Chazot, G. and Kopp, N. 2000. The first case of new variant Creutzfeldt-Jakob disease in France: clinical data and neuropathological findings. *Acta Neuropathologica (Berlin)* **99**:704–708.

Taylor, D. M. and Woodgate, S. L. 2003. Rendering practices and inactivation of transmissible spongiform encephalopathy agents. *Review of Science and Technology* **22**:297–310.

Thomzig, A., Kratzel, C., Lenz, G., Kruger, D. and Beekes, M. 2003. Widespread PrPSc accumulation in muscles of hamsters orally infected with scrapie. *European Molecular Biology Organization Reports* **4**:530–533.

Tuo, W., O'Rourke, K. I., Zhuang, D., Cheevers, W. P., Spraker, T. R. and Knowles, D. P. 2002. Pregnancy status and fetal prion genetics determine PrPSc accumulation in placentomes of scrapie-infected sheep. *Proceedings of the National Academy of Sciences (USA)* **99**:6310–6315.

van Keulen, L. J., Schreuder, B. E., Meloen, R. H., Mooij-Harkes, G., Vromans, M. E. and Langeveld, J. P. 1996. Immunohistochemical detection of prion protein in lymphoid tissues of sheep with natural scrapie. *Journal of Clinical Microbiology* **34**:1228–1231.

van Keulen, L. J., Schreuder, B. E., Vromans, M. E., Langeveld, J. P. and Smits, M. A. 2000. Pathogenesis of natural scrapie in sheep. *Archives of Virology Supplement* **16**:57–71.

Wadsworth, J. D., Joiner, S., Hill, A. F., Campbell, T. A., Desbruslais, M., Luthert, P. J. and Collinge, J. 2001. Tissue distribution of protease resistant prion protein in variant Creutzfeldt-Jakob disease using a highly sensitive immunoblotting assay. *Lancet* **358**:171–180.

Wells, G. A., Scott, A. C., Johnson, C. T., Gunning, R. F., Hancock, R. D., Jeffrey, M., Dawson, M. and Bradley, R. 1987. A novel progressive spongiform encephalopathy in cattle. *Veterinary Record* **121**:419–420.

Wells, G. A. H., Hawkins, S. A. C., Austin, A. R., Ryder, S. J., Done, S. H., Green, R. B., Dexter, I., Dawson, M. and Kimberlin, R. H. 2003. Studies of the transmissibility of the agent of bovine spongiform encephalopathy to pigs. *Journal of General Virology* **84**:1021–1031.

White, A. R., Enever, P., Tayebi, M., Mushens, R., Linehan, J., Brandner, S., Anstee, D., Collinge, J. and Hawke, S. 2003. Monoclonal antibodies inhibit prion replication and delay the development of prion disease. *Nature* **422**:80–83.

Wilesmith, J. W. 1988. Bovine spongiform encephalopathy. *Veterinary Record* **122**:614.

Wilesmith, J. W., Ryan, J. B. and Atkinson, M. J. 1991. Bovine spongiform encephalopathy: epidemiological studies on the origin. *Veterinary Record* **128**:199–203.

Will, R. G., Ironside, J. W., Zeidler, M., Cousens, S. N., Estibeiro, K., Alperovitch, A., Poser, S., Pocchiari, M., Hofman, A. and Smith, P. G. 1996. A new variant of Creutzfeldt-Jakob disease in the UK. *Lancet* **347**:921–925.

Williams, E. S. and Young, S. 1980. Chronic wasting disease of captive mule deer: a spongiform encephalopathy. *Journal of Wildlife Diseases* **16**:89–98.

Williams, E. S. and Young, S. 1992. Spongiform encephalopathies in Cervidae. *Review of Science and Technology* **11**:551–567.

Williams, E. S. 2002. The transmissible spongiform encephalopathies: disease risks for North America. *Veterinary Clinics of North America Food Animal Practice* **18**:461–473.

Wuthrich, K. and Riek, R. 2001. Three-dimensional structures of prion proteins. *Advances in Protein Chemistry* **57**:55–82.

Zanusso, G., Casalone, C., Acutis, P., Bozzetta, E., Farinazzo, A., Gelati, M., Fiorini, M., Forloni, G., Sy, M. S., Monaco, S. and Caramelli, M. 2003a. Molecular Analysis of Iatrogenic Scrapie in Italy. *Journal of General Virology* **84**:1047–1052.

Zanusso, G., Ferrari, S., Cardone, F., Zampieri, P., Gelati, M., Fiorini, M., Farinazzo, A., Gardiman, M., Cavallaro, T., Bentivoglio, M., Righetti, P. G., Pocchiari, M., Nicola Rizzuto, N. and Monaco, S. 2003b. Detection of pathologic prion protein in the olfactory epithelium in sporadic Creutzfeldt–Jakob disease. *New England Journal of Medicine* **348**:711–719.

Zerr, I. and Poser, S. 2002. Clinical diagnosis and differential diagnosis of CJD and vCJD with special emphasis on laboratory tests. *Acta Pathologica, Microbiologica et Immunologica Scandinavia* **110**:88–98.

Zou, W. Q. and Cashman, N. R. 2002. Acidic pH and detergents enhance in vitro conversion of human brain PrPC to a PrPSc-like form. *Journal of Biological Chemistry* **277**:43942–43947.

Part III

Antimicrobial Resistance

15 Antimicrobial Susceptibility Testing

Patrick F. McDermott, David G. White,
Shaohua Zhao, Shabbir Simjee, and
Robert D. Walker

Introduction

Since the introduction of penicillin during the 1940s, numerous antibacterial agents have been developed and marketed for clinical use. There are well over 100 antimicrobial agents currently available in human and veterinary medicine. This large and growing number of agents, along with the facility with which bacteria develop resistance to them, makes the selection of an appropriate agent for therapeutic use an increasingly more difficult task. This highlights the importance of *in vitro* antimicrobial susceptibility testing to help guide the decision-making process in selecting a drug for the treatment of many bacterial infections.

In vitro antimicrobial susceptibility testing serves multiple purposes. In the research setting, susceptibility testing provides a means for comparing the contribution of different determinants (genes) to drug resistance in the cell and for developing new anti-infective compounds. In clinical medicine, susceptibility testing can provide insight as to which agents might be effective against a specific pathogen. When performed over time, the results of susceptibility testing can also be used to help establish guidelines for empirical therapy within a local hospital or community. When used in conjunction with well-designed surveillance programs, the results from properly conducted susceptibility testing can serve as a basis for documenting the emergence and dissemination of antimicrobial resistance on a national or global basis. It is perhaps in the arena of food animal and retail food microbiological surveillance that routine susceptibility testing is most relevant to issues of food safety. It is in part by monitoring trends over time that the potential effect of food animal antimicrobial use on human and animal health can be evaluated. In contrast, clinicians infrequently request susceptibility testing of diarrhea pathogens and rely instead on empiric treatment. As with other bacterial pathogens, this is changing as a result of the emergence of some resistance phenotypes, particularly fluoroquinolone resistance in *Campylobacter*.

In vitro antimicrobial susceptibility testing methods were originally developed in clinical microbiology laboratories within individual hospitals to help guide therapy. Because the concept of susceptibility testing is relatively straightforward, initially many laboratories developed their own testing methods. With an increasing number of laboratories generating susceptibility data on an increasing number of antimicrobial agents, along with the evolution of bacterial resistance to these agents, it became necessary to establish standardized susceptibility testing methods to facilitate communication and to ensure the quality of results between and within laboratories and over time. In the United States the National Committee for Clinical Laboratory Standards (NCCLS) has become the organization responsible for standardizing *in vitro* antimicrobial susceptibility testing methods and for communicating the criteria for interpreting the results of such tests. The NCCLS uses a consensus-based approach to develop standardized testing methods, quality-control (QC) ranges, and interpretive criteria that will ensure intra- and interlaboratory reproducibility of data generated by antimicrobial susceptibility testing.

Standardization of *in vitro* antimicrobial susceptibility testing requires the use of specific testing methods, including media, incubation conditions and times, and identification of appropriate QC organisms along with their specific ranges. Although the majority of bacteria of clinical interest may be tested using the same *in vitro* testing conditions, bacteria that require increased incubation times, different incubation temperatures or atmospheric conditions, or a different growth medium may require the identification of specific testing conditions. For some organisms, such as fastidious genera, it may be necessary to validate specific QC organisms and their QC ranges. For example, although *Campylobacter jejuni* may inhabit the same *in vivo* environment as *Escherichia coli*, its *in vitro* cultivation requires a different medium and different atmospheric conditions. As such, the *in vitro* antimicrobial susceptibility testing methods for these two organisms are different.

Numerous articles dealing with the susceptibility of bacterial isolates to antibacterial agents have been published over the years. Unfortunately, not all of these publications have presented results generated from methods that have been standardized for the bacterium–antimicrobial agent combination being studied. Results generated from non standardized testing methods can be misleading, especially when trying to compare results between laboratories or over time. The objective of this chapter is to describe the standardization of *in vitro* antimicrobial susceptibility testing methods as defined by the NCCLS, including how QC data are generated and how interpretive criteria are determined. It is through adherence to standardized testing methods, such as those described by the NCCLS, that our understanding will grow of the contribution that the use of antimicrobial agents in humans and animals has made to the selection for and dissemination of resistant bacterial pathogens and commensal organisms.

Susceptibility Testing Methods

The selection of antimicrobial testing methods is dependent in large part on the data needs of the testing laboratory and their clients. When performing *in vitro* antimicrobial susceptibility tests, the results may be reported qualitatively or quantitatively. Qualitative results are reported as susceptible, intermediate, or resistant, whereas quantitative results are reported as minimal inhibitory concentrations (MICs) in micrograms per milliliter or milligrams per liter. Results generated for most clinical applications are reported qualitatively, although some clinical laboratories may also provide MIC values. For the surveillance purpose of monitoring changes in susceptibility, methods that generate MIC values are required. When surveillance tracks trends in the prevalence of resistance, only qualitative results may be needed.

The most common methods that have been used for performing *in vitro* susceptibility testing are disk diffusion, agar dilution, broth macrodilution, broth microdilution, and a concentration gradient test (Etest, AB BIODISK, Piscataway, NJ). Of these, the disk diffusion and broth microdilution methods, and to a lesser extent the Etest, are currently those methods most commonly used for clinical purposes. Research laboratories and surveillance programs may also use these methods. However, some organisms may only be amenable to testing by specific methods. For example, *Helicobacter pylori* can only be tested by agar dilution. Anaerobic bacteria may be tested by agar dilution or broth microdilution, but this is organism dependent. At present the only NCCLS-standardized method for testing *Campylobacter* is an agar dilution method. In the clinical laboratory, aside from the specific requirements of the organism to be tested, the decision on which test method to use is based on cost, ease of use, and flexibility as it relates to the needs of the laboratory's clientele. However, for surveillance and research purposes, the testing method used is frequently determined by the experience of the laboratory personnel.

Diffusion Susceptibility Tests

The disk diffusion test has been most widely used because of its simplicity and relatively low cost. In addition, it is the most flexible in terms of the types and number of drugs that can be tested on a daily basis. The disk diffusion test is based on the diffusion of an antimicrobial agent from a paper disk into the surrounding agar medium. Commercially prepared disks are available for all antimicrobial classes; however, they may not be available for every agent within a class. For example, ampicillin is the class representative for the aminobenzyl penicillins, so there are disks commercially available for ampicillin but not for amoxicillin or hetacillin. In performing the test, an antimicrobial disk of appropriate potency is placed on an agar surface of standardized growth medium that has been seeded with approximately 1.0×10^8 colony-forming units (CFU) per milliliter of a pure culture of the bacterium to be tested. The bacterial culture may be obtained from either a broth culture or by using a direct colony suspension method, which is the preferred method for many slow-growing organisms. When the disk is applied to the seeded agar surface, a race is initiated between the growth of the bacterium on the agar surface and the diffusion of the drug through the medium. The diffusion of the drug results in a drug concentration gradient that decreases with the distance the drug has diffused from the disk. When the concentration of the antimicrobial agent becomes too dilute to inhibit the growth of the bacterium, a zone of inhibition is formed. Following incubation, agar plates are examined and the diameters of the zones of inhibition surrounding the disk are measured. These zones of inhibition correlate inversely with the MIC of the test organism. In other words, the larger the zone of inhibition, the smaller the concentration of drug required to inhibit the pathogen. Because of the potential in variability in diffusion rates of the different drugs being tested, the zone diameter for each drug is compared with established ranges for that antimicrobial agent to predict whether the agent is likely to be effective *in vivo*.

Because the interpretive criteria for the disk diffusion test have been established under standardized conditions, including inoculation density and composition and depth of the agar, it is critical that the appropriate testing conditions be used. The use of too many bacteria in the suspension used to inoculate the agar surface may result in smaller zones of inhibition, whereas the use of too few organisms, or inoculation with an organism that grows slowly on the test medium, results in larger zones of inhibition and false resistant or susceptible readings, respectively. Plates should also be inoculated within 15 minutes of preparing the bacterial suspension to prevent an increase in the number of organisms from cell division. The depth of the agar matrix is also critical. Variation in agar depth will influence the lateral diffusion of the drug and, thus, the size of the zone of inhibition of the test isolate. For example, agar depths that are less than the recommended 4 mm will result in increased lateral diffusion and a corresponding increase in the size of the zone of inhibition, whereas an agar matrix that is greater than the recommended depth will result in a downward diffusion of the drug and a smaller zone of inhibition. The use of a medium other than that established in developing the reference method for that particular bacterium–antimicrobial agent combination may result in increased or decreased diffusion of the antimicrobial agent through the medium and, thus, altered zones of inhibition.

As indicated above, the disk diffusion test is often preferred because of cost, simplicity, and flexibility. The disadvantage of the disk diffusion test is that the results are reported as qualitative. This makes disk diffusion less useful for surveillance purposes where changes in degrees of susceptibility are monitored for trend analysis. It would be challenging to ensure data quality in a surveillance system based on reporting changes in zone sizes over time. This is because the tolerated variability in zone sizes for QC purposes is relatively large (9–12 mm) for many drugs assayed by disk diffusion. The lack of quantitative interpretation also makes it difficult for an attending clinician to adjust a dose for an organism that is uniquely susceptible (to reduce costs) or to increase the dose to maximize clinical efficacy for those isolates that are less susceptible. To compensate for this deficiency, two variations of the disk diffusion test have been developed. The first is a computerized caliper or camera-based system (Korgenski and Daly 1998), and the second is the epsilometer test (Etest). Both systems were developed for use in clinical medicine but could be applied to research or surveillance programs if appropriate QC data are generated and if the testing methods are properly calibrated. The computerized method for generating quantitative results from a qualitative testing method has been found to be useful when testing human clinical isolates but has not been validated for many of the bacterial pathogens isolated from animals (Hubert et al. 1998).

The Etest is a modified diffusion test that generates quantitative results. The test relies on the diffusion of an antimicrobial agent from a plastic strip that has been impregnated on one side with a continuous concentration gradient of antimicrobial agent. It is placed onto the surface of a seeded agar medium and incubated for a predetermined time. On the top surface of the strip is imprinted an MIC interpretive scale. After incubation, the MIC is determined by reading the value where the zone of inhibition intersects the strip. Because the Etest is a proprietary technology provided by a single company, it has not received formal endorsement from the NCCLS. However, it has been adapted for use with a variety of organisms and is simple to perform. The major drawback of the Etest is cost per isolate, especially when multiple antimicrobial agents are tested.

Dilution Susceptibility Tests

Dilution susceptibility tests generate quantitative data based on twofold serial dilutions of antimicrobials. Dilution testing may be performed using agar dilution, broth macrodilution, or broth microdilution. Among these, agar dilution is considered by many to be the "gold standard" and is at present the only standardized testing method for some fastidious organisms such as *Helicobacter, Campylobacter,* and certain species of anaerobes (NCCLS 2001, NCCLS 2002c). The broth macrodilution method typically uses broth volumes of 1–2 mL. For most uses, this method has been supplanted by semi-automated microdilution technology, which typically requires only volumes of 50–100 μL. The microdilution methods permit higher sample throughput with fewer laboratory supplies and less space.

The broth microdilution test is performed in microtiter plates, usually of the 96-well format, with round or truncated V-shaped well bottoms. The plate layouts are designed to contain antimicrobials of known potency in progressive twofold dilutions that encompass concentrations similar to those obtained in serum and tissue at recommended doses. It is important that the dilution range for each agent also encompass the QC dilution range of the QC organism that will be used. The microtiter trays may be prepared using a variety of formats, but each tray usually contains several antimicrobial agents that are tested against a single isolate. The microtiter trays may be prepared in house or obtained from a number of commercial sources. Those obtained commercially may be purchased as dehydrated trays or as frozen trays. Dehydrated trays generally have a shelf life of 1–2 years and may be stored at room temperature, whereas frozen trays have a shelf life of 6 months and must be stored at $-10°C$ or $-70°C$, depending on the antibacterial agents contained in the tray.

Microdilution tests are more expensive to conduct than disk diffusion tests, and they lack the day-to-day flexibility of disk diffusion in terms of changing antimicrobial agents to be tested. This lack of flexibility may be compounded by buying trays in bulk to reduce cost and the need to finish using a 2-year supply of plates before modifying the panel format. To reduce the cost associated with microtiter dilution trays, some laboratories use breakpoint formats. Breakpoint formats usually consist of the antimicrobial agent being tested over a two or three dilution range. There is usually one dilution below the susceptible breakpoint, the susceptible breakpoint itself, and sometimes one dilution above the susceptible breakpoint. The use of this format allows for more compounds to be included in a single microtiter tray. Alternatively, if few antimicrobial agents are used, then more organisms can be tested in a single microtiter tray. A major limitation of the breakpoint approach is that it does not allow for proper QC in that the dilution ranges used do not encompass those needed for testing the QC organism. Results generated by the use of breakpoint panels are also qualitative and thus similar (although more expensive) to those generated by disk diffusion. For research and surveillance purposes, breakpoint panels are of limited value, as they do not generate data that fully reflect changes in susceptibility and do not provide for proper QC.

To perform broth microdilution, a bacterial suspension is prepared to a density equivalent to that of a 0.5 McFarland turbidity standard (approximately 1.0×10^8 CFU/mL). This can be made from an overnight culture grown on an agar surface or broth culture in logarithmic phase of growth. This suspension is further diluted in sterile water, saline, or broth so that the final concentration of bacteria is approximately 5×10^5 CFU/mL or 5×10^4 CFU/well. Once inoculated, the trays are stacked two to four trays high, placed in an incubator with the appropriate incubation atmosphere and temperature, and incubated for the standard time interval (usually 16–20 hours). As it is with agar dilution, the MIC is recorded as the lowest concentration of antimicrobial agent that completely inhibits the growth of the organism, as determined by the unaided eye (NCCLS 1999, NCCLS 2001, NCCLS 2002c).

Interpretation of Susceptibility Tests

Bacteria subjected to susceptibility testing are classified, where interpretive criteria are available (see following), as susceptible, intermediate, or resistant. Susceptible implies that when the tested drug is administered at the recommended dose, the serum or tissue concentration will be high enough to inhibit the bacterium's growth at the sight of the infection. Intermediate breakpoints are those zones of inhibition or MICs that fall between the susceptible and resistant breakpoints. Traditionally these have represented a "buffer zone" that prevented resistant organisms from being categorized as susceptible, or resistant, because of uncontrolled technical problems in the laboratory. However, this category may also represent susceptible organisms for antimicrobial agents with wide pharmacotoxicity margins, which allows the drug to be administered at the upper end of the dosage range. This is also the case for agents that are concentrated at the site of infection, such as β-lactam agents and some fluoroquinolones when used to treat urinary tract infections.

Numerous definitions have been applied to the term "resistant." For bacteriologists engaged in surveillance studies, resistance may be loosely used to describe an increase in MICs over time of wild-type strains above what is seen with control strains of the same species. In some cases, resistant breakpoints used to describe the activity of an antimicrobial agent against one isolate may be used as interpretive criteria for another bacterium. For the molecular geneticist, resistance is often explained in terms of acquired resistance determinants and whether they are present in certain isolates of a given species. The microbial physiologist studies resistance from the standpoint of metabolic mechanisms that influence the activity of antimicrobials on cells at the site of action. The pharmacologist considers resistance in term of dose and achievable serum or tissue concentrations relative to toxicity. For the clinician, resistance is important in terms of patient outcome and the therapeutic effect of specific antimicrobials used to combat infection. The NCCLS defines resistance as follows:

> Resistant strains are not inhibited by the usually achievable systemic concentrations of the agent with normal dosage schedules and/or fall in the range where specific microbial resistance mechanisms are likely (e.g., β-lactamases) and clinical efficacy has not been reliable in treatment studies. (NCCLS 2000, p. 14)

This definition, although encompassing the different perspectives mentioned above, presupposes a standard process by which interpretive criteria are established.

Developing Standardized Methods and Establishing Interpretive Criteria

There are a number of standards and guidelines currently available for antimicrobial susceptibility testing and interpretive criteria. These include standards and guidelines published by NCCLS, the British Society for Antimicrobial Chemotherapy (BSAC), the Japan Society for Chemotherapy (JSC), the Swedish Reference Group for Antibiotics (SIR), the Deutshes Institute für Normung (DIN), the Comite de L'Antibiogramme de la Societe Francaise de Microbiologie (CASFM), and the Werkgroep richtlijnen gevoeligheidsbepalingen (WRG System, NL), among others. The variations in diffusion and dilution antimicrobial susceptibility testing methods (e.g., choice of agar medium, inoculum size, and growth conditions) and the differing interpretive criteria (including breakpoints) in different countries makes it difficult to compare susceptibility data from one system with another. Several member countries in the European Union are contemplating coordination of AST methods to facilitate resistance surveillance (Cotter and Adley 2001, White et al. 2001).

The NCCLS approach to developing interpretive criteria (i.e., classifying isolates as susceptible, intermediate, or resistant) is based on four factors. The first is identifying the appropriate testing method for the antimicrobial agent in question when tested against the target pathogen, including the identification of the appropriate QC organism or organisms and corresponding QC ranges. A standardized method is a prerequisite to the next step, which is to determine the *in vitro* activity of the antimicrobial agent against several hundred clinical isolates against which the drug will be used in a clinical setting. This is usually expressed as a scattergram, which plots the MICs versus the zones of inhibition for the same bacterial population. Third is the generation of pharmacokinetic and pharmacodynamic parameters (absorption distribution, elimination, and concentration) of the antimicrobial agent in the target animal species, at the site of infection and at the proposed label dose or doses. The forth component in developing interpretive criteria is the results of clinical trials using the drug at label doses. Ideally, the first objective is met early in the drug development process and may be submitted to the NCCLS long before the other data are generated. The remaining information is usually submitted to the NCCLS shortly after the drug has received Food and Drug Administration approval for clinical trials.

Standardization of In Vitro *Testing Methods*

It is important to keep in mind that the use of NCCLS interpretive criteria is only valid if susceptibility testing is conducted using appropriate NCCLS standardized methods for the bacterium/antimicrobial in question. These include using a pure culture grown under the appropriate conditions and inoculated at the correct cell density in conjunction with the use of the appropriate QC organism or organisms, relevant dilution schemes or disk masses, and the growth medium identified in the reference method. Identifying and controlling for these various parameters is the goal of standardization. Even when these essential requirements are met, several factors may alter the results. These include, among other things, the growth rate of the bacterium being tested and the lot-to-lot variations in the growth medium used to conduct the assay. The pH, oxygen tension, temperature, time of incubation, preparation of the antimicrobial compounds, and storage of the antimicrobial compounds are all critical variables that must be standardized to ensure repeatable and reliable results. The use of standardized testing conditions along with the regular use of appropriate QC strains should provide protection against variations in the testing procedures. Current editions of NCCLS guidelines should be consulted for recommendations on how to correct testing procedures when faced with aberrant QC results (NCCLS 2000, NCCLS 2002c).

The most important element of routine antimicrobial susceptibility testing is the use of appropriate QC organisms. This critical factor is often overlooked in reports within the scientific literature. As a rule, QC organisms are selected on the basis of their genetic stability and reproducible MIC values within and between laboratories over time. QC strains should be from a common source, such as the American Type Culture Collection to ensure easy access. In identifying the QC organisms and ranges, the antimicrobial agent is tested against the QC organism(s) under various growth conditions to determine the effect of different variables (pH, atmospheric condition, etc.) on the test results. In addition, broth microdilution test results are compared with agar dilution test results to confirm correlation of the methods. Once these basic parameters have been evaluated, expected QC ranges are generated for the antimicrobial agent for each applicable QC organism. When identifying the QC organism and ranges for dilution testing the QC organism should be susceptible at drug concentrations (QC range) in the middle of the anticipated range encompassed by the susceptible clinical isolates. For disk diffusion the appropriate drug concentration in the disk must be determined before QC zone ranges are ascertained. The ideal drug concentration in the disk should be that which consistently results in zones of inhibition of >15 mm but <45 mm for susceptible strains and small or no detectable zones for resistant strains.

Once the aforementioned parameters have been determined, the antimicrobial agent and the proposed QC organisms are tested in a minimum of seven different laboratories over a period of 10 days on three different lots of medium. When possible, the NCCLS recommends that a control drug, in the same class as the test agent that already has had QC guidelines established, be tested concurrently with the test agent. Obviously it is essential that QC ranges are established for a new antimicrobial agent before testing can be performed on clinical isolates for the purpose of developing interpretive criteria.

Population Distribution

Once a standardized susceptibility testing method is available, the second phase of testing can be performed. This involves an examination of MICs and of zones of inhibition distributions among numerous representative isolates (NCCLS 2002a, NCCLS 2002b). To determine population distributions, disk diffusion and dilution tests are performed on 300–600 recent clinical isolates made up of geographically diverse isolates representing all species of bacteria that are to be targeted by the antimicrobial agent in question. The susceptibility of these isolates may be determined in a single laboratory. The population of bacteria tested should include susceptibility as well as those that exhibit decreased susceptibility to the drug if they are available. Once generated, the susceptibility data, zones of inhibition and MICs, are plotted against one another to generate a scattergram or population distribution. Tentative breakpoints are then determined on the basis of error rate bounding, pharmacokinetics, and verification of the breakpoints by clinical and bacteriological response rates.

Pharmacokinetic/Pharmacodynamic Parameters

Pharmacokinetic parameters are important for assessing the bioavailability of the drug in the host species. Pharmacokinetic data include, but are not limited to, serum absorption, distribution, and elimination. Serum kinetic values should be expressed relative to the free, biologically active drug concentration, or the percentage protein binding should be measured. Closely associated with the pharmacokinetic parameters are the drug's pharmacodynamic parameters. Pharmacodynamic parameters are measurements of the biological and toxic effect of the antimicrobial agent in the host and at the site of infection. These parameters include such considerations as bacteriostatic versus bactericidal effects, first exposure effect, a concentration-dependent killing effect, a post-antibiotic effect (PAE), a sub-MIC effect (referred to as sub-PAE) and a post-antibiotic leukocyte enhancement effect (PALE). These factors, along with the pharmacokinetic parameters in host species, provide insight as to the dosing regimen that can maximize clinical efficacy of the drug. For example, with the β-lactams it is the length of time the serum concentration exceeds the MIC that determines the efficacy of the dosing regimen. In contrast, for the aminoglycosides, the peak serum concentration-to-MIC ratio is the relevant predictor of efficacy. For most other antimicrobial agents it is the area under the serum–drug concentration curve-to-MIC ratio that is used to assess potential efficacy.

Clinical Trials

The final piece of information that is evaluated in determining NCCLS interpretive criteria is the results from clinical trials. Information evaluated from clinical trials includes, but may not be limited to, a description of the patient population studied, the dosage and duration of antimicrobial therapy, microbiological results including the identification and antibiograms of bacteria isolated pre- and posttreatment, and clinical assessments. QC data that are generated when testing clinical isolates are also evaluated to ensure the accuracy of the susceptibility data of the clinical isolates.

For numerous pathogenic bacteria, not all of these data are available for all potentially effective therapeutic agents. As resistance to antimicrobials increases, practitioners are more frequently faced with selecting therapeutic agents for which the NCCLS process has not been attempted or completed. This is especially true in veterinary medicine. In cases in which no guidance is available from the NCCLS, the selection of antimicrobial therapy is driven by clinical judgment and inferences from related drugs and similar pathogens. In some cases, tentative interpretive criteria are proposed, often by adopting those established for other organisms. This is the case for *Campylobacter* infections in humans, where the distribution of MICs in the population, along with what is known about the pharmacodynamic properties of the antimicrobials, has been used by some to set tentative breakpoints for specific antimicrobials (e.g., ciprofloxacin and erythromycin; Danish Integrated Antimicrobial Resistance Monitoring and Research Programme 2002, King 2001). These breakpoints are sometimes used in surveillance studies as arbitrary threshold MICs for reporting purposes. However, they should be evaluated in this context and viewed with caution as a guide to therapy, pending clinical efficacy studies.

Antimicrobial Susceptibility Testing of Animal Pathogens

Because many animal pathogens are common or similar to those of humans (e.g., *Staphylococcus*, *Salmonella*, and *E. coli*), NCCLS M2 and M7 standards have been applied to isolates of animal origin. To address testing issues unique to animal pathogens, the NCCLS established the Subcommittee on Veterinary Antimicrobial Susceptibility Testing in 1992. The subcommittee developed the M31 (NCCLS 1999) and M37 (NCCLS 2002a) consensus documents. The M31 document contains QC information and interpretive criteria for bacterium–antimicrobial agent combinations that have been adopted from the standards developed for human isolates and that are specific for animal pathogens and veterinary-specific antimicrobial agents. The M37 document is a guideline for drug sponsors to follow in the development of QC information and interpretive criteria for veterinary-specific drugs. The M31 document is updated on an annual basis as new information becomes available (as of 2003, it is version M31-A2). Material in the M31 document that is of particular interest to the veterinary diagnostic laboratories includes the testing procedure for fastidious organisms such as *Haemophilus somnus* and *Actinobacillus pleuropneumoniae,* as well as veterinary-specific antimicrobial agents such as ceftiofur, danofloxacin, difloxacin, enrofloxacin, florfenicol, marbofloxacin, orbifloxacin, and tilmicosin.

Conclusion

Although a variety of methods exist, the goal of *in vitro* antimicrobial susceptibility testing is the same: to provide a reliable predictor of how an organism is likely to respond to antimicrobial therapy in the infected host. This type of information aids the clinician in selecting the appropriate antimicrobial agent, provides data for surveillance, and aids in developing antimicrobial use policy. *In vitro* antimicrobial susceptibility testing can be performed using a variety of formats. Because of the simplicity associated with incubating an antimicrobial agent with a bacterium, numerous laboratories have participated in generating antimicrobial susceptibility (resistance) data. As the science of antimicrobial susceptibility testing has progressed, a greater understanding of the multiple factors that could affect the overall outcome of susceptibility testing has become clearer. Unfortunately, not all laboratories follow standardized methods for generating the values that are used to determine the interpretive criteria. In addition, there are many instances in which interpretive criteria are arbitrarily determined. Therefore, it is essential that testing methods provide reproducible results in day-to-day laboratory use and that the data be comparable with those results obtained by an acknowledged "gold standard" reference method. In the absence of standardized methods or reference procedures, susceptibility results from different laboratories cannot be reliably compared.

References

Cotter, G. and Adley, C. C. 2001. Comparison and evaluation of antimicrobial susceptibility testing of enterococci performed in accordance with six national committee standardized disk diffusion procedures. *Journal of Clinical Microbiology* **39**:3753–3756.

Danish Integrated Antimicrobial Resistance Monitoring and Research Programme. 2002. 2002—Consumption of antimicrobial agents and occurrence of antimicrobial resistance in bacteria from food animals, foods, and humans in Denmark. Available at: http://www.vetinst.dk/high_uk.asp?page_id=179.

Hubert, S. K., Nguyen, P. D. and Walker, R. D. 1998. Evaluation of a computerized antimicrobial susceptibility system with bacteria isolated from animals. *Journal of Veterinary Diagnostic Investigation* **10**:164–168.

King, A. 2001. Recommendations for susceptibility tests on fastidious organisms and those requiring special handling. *Journal of Antimicrobial Chemotherapy* **48(Suppl. 1)**:77–80.

Korgenski, E. K. and Daly, J. A. 1998. Evaluation of the BIOMIC video reader system for determining interpretive categories of isolates on the basis of disk diffusion susceptibility results. *Journal of Clinical Microbiology* **36**:302–304.

National Committee for Clinical Laboratory Standards. 1999. Performance standards for antimicrobial disk and dilution susceptibility tests for bacteria isolated from animals; Approved standard. NCCLS document M31-A2. Wayne, PA: National Committee for Clinical Laboratory Standards.

National Committee for Clinical Laboratory Standards. 2000. Performance standards for antimicrobial disk susceptibility tests; approved standard-seventh edition. NCCLS document M2-A7. Wayne, PA: National Committee for Clinical Laboratory Standards.

National Committee for Clinical Laboratory Standards. 2001. Methods for antimicrobial susceptibility testing of anaerobic bacteria; approved standard-fifth edition. NCCLS document M11-A5. Wayne, Penn.: National Committee for Clinical Laboratory Standards.

National Committee for Clinical Laboratory Standards. 2002a. Development of *in vitro* susceptibility testing criteria and quality control parameters for veterinary antimicrobial agents; approved guideline—second edition. NCCLS document M37-A2. Wayne, Penn.: National Committee for Clinical Laboratory Standards.

National Committee for Clinical Laboratory Standards. 2002b. Development of *in vitro* susceptibility testing criteria and quality control parameters; approved guideline—fifth edition. NCCLS document M23-A5. Wayne, Penn.: National Committee for Clinical Laboratory Standards.

National Committee for Clinical Laboratory Standards. 2002c. Methods for dilution antimicrobial susceptibility tests for bacteria that grow aerobically; approved standard—fifth edition. NCCLS document M7-A5. Wayne, Penn.: National Committee for Clinical Laboratory Standards.

White, D. G., Acar, J., Anthony, F., Franklin, A., Gupta, R., Nicholls, T., Tamura, Y., Thompson, S., Threlfall, E. J., Vose, D., van Vuuren, M., Wegener, H. C. and Costarrica, M. L. 2001. Antimicrobial resistance: standardisation and harmonisation of laboratory methodologies for the detection and quantification of antimicrobial resistance. *Revue Scientifique et Technique* **20**:849–858.

16 Antimicrobial Resistance in Food Animals

Kenneth M. Bischoff, Toni L. Poole, and Ross C. Beier

Introduction

Antimicrobials are used in animal feed to manage disease, to maintain herd health, and to promote growth. When individual animals display clinical signs, drugs are often administered to the entire herd to treat asymptomatic animals and to prevent the spread of disease (Gustafson and Bowen 1997). As a result, the incidence of antimicrobial-resistant bacteria in animals is increasing, and with the increase, there is heightened concern that drug-resistant enteric bacteria may enter the food chain via meat products (Coates and Hoopes 1980, Libal and Gates 1982, Aalbaek et al. 1991, Wray et al. 1993, Fairbrother 1999, Gebreyes et al. 2000). This chapter will discuss the general mechanisms employed by bacteria to resist the action of antimicrobial drugs, as well as the mechanisms responsible for the dissemination of resistance genes. Intervention strategies that are being developed to combat the emergence and spread of antimicrobial resistance genes among enteric bacteria of food animals will also be presented.

Biochemical Mechanisms of Antimicrobial Resistance

Antimicrobial drugs generally target one of four essential cellular processes: protein synthesis, cell wall synthesis, nucleic acid replication and synthesis, or folate metabolism (Russell and Chopra 1996, Hooper 2001). Several classes of antibiotics including the aminoglycosides, tetracyclines, macrolides, and chloramphenicol bind to one or both subunits of the ribosome to cause misreading or termination of translation, rendering protein synthesis one of the most common targets for broad-spectrum antibiotics. Inhibition of cell wall biosynthesis also is a common target for some broad-spectrum antibiotics. Members of the β-lactam family (penicillins, cephalosporins) bind to enzymes called transpeptidases and inhibit the formation of the essential cell wall component peptidoglycan. Vancomycin inhibits formation of peptidoglycan by binding to peptidoglycan precursor molecules. The quinolones and fluoroquinolones target DNA replication and synthesis by inhibiting DNA gyrase or topoisomerase—enzymes involved in folding and supercoiling of DNA. Finally, drugs such as trimethoprim and the sulfonamides target folate metabolism by competitively inhibiting the synthesis of folic acid for *de novo* amino acid biosynthesis.

Bacteria have adapted either by mutation or by acquisition of extraneous genes to circumvent these drugs. The specific mechanisms employed are as varied as the types of antimicrobials and their modes of action. In general, bacteria use three strategies to resist the action of antimicrobials: modification of the drug, alteration of the target, or decreased accessibility of the drug to its target.

Modification of the Drug

Direct enzymatic modification of a drug results in a chemical that no longer functions as an effective antimicrobial agent, allowing the organism to survive in the presence of the drug. Examples of this type of strategy include the mechanisms used for resistance to the β-lactams, the aminoglycosides, and chloramphenicol. β-lactamases are enzymes that hydrolyze the lactam ring of penicillins and cephalosporins, inactivating the drug's antimicrobial effect (Kono et al. 1983, Medeiros 1984). Aminoglycosides such as gentamicin, kanamycin, neomycin, and streptomycin may be subject to several chemical modifications that inactivate these drugs. These include *O*-phosphorylation catalyzed by an adenosine triphosphate (ATP)-dependent phosphotransferase, *O*-adenylation catalyzed by an ATP-dependent nucleotidyltransferase, and *N*-acetylation catalyzed by an acetyl-CoA-dependent acetyltransferase (Davies and Wright 1997). Chloramphenicol may be modified at the 3′-hydroxyl position by chloramphenicol acetyltransferase, the enzyme responsible for most enzymatic resistance to chloramphenicol (Shaw 1983, Russell and Chopra 1996).

Alteration of the Target

A genetic or post-translational modification that alters the structure of an antimicrobial's target also can decrease the therapeutic effectiveness of a drug. Despite the high fidelity of DNA replication under optimal growth conditions, point mutations still occur at a frequency of about one single base substitution for each 10^6 nucleotides (Kunkel et al. 1987). Under times of stress, such as that induced by antimicrobial pressure, mutations may occur at an even higher frequency (Boe 1992, Humayun 1998, Ren et al. 1999).

Cells with altered target proteins will have a selective advantage for clonal expansion under antimicrobial selection pressure. Single–amino acid substitutions at position Gly81, Ser83, or Asp87 of DNA gyrase reduce drug binding and consequently confer resistance to the fluoroquinolone antibiotics (Willmott and Maxwell 1993, Weigel et al. 1998). Substitution of glycine for tryptophan at position 30 in *Escherichia coli* dihydrofolate reductase results in a threefold increase in the inhibitory concentration for trimethoprim (Flensburg and Skold 1987, Huovinen et al. 1995). Similarly, a phenylalanine to isoleucine replacement at position 28 of dihydropteroate synthase confers resistance to sulfonamides (Huovinen et al. 1995). Resistance to penicillins also can arise from mutations in penicillin-binding proteins that decrease the affinity with which these proteins bind the drug (Hackbarth et al. 1995).

Alternatively, the target may be post-translationally modified through enzymatic action. The *erm* family of erythromycin-resistance genes encode structurally similar methyltransferases that catalyze methylation of 23S ribosomal RNA (Weisblum 1995). These same modifications may confer resistance to the lincosamides and the streptogramins—classes of antimicrobials that are chemically distinct but functionally similar to the macrolides. The resulting cross resistance has food safety implications, owing to the widespread use of the streptogramin antibiotic virginiamycin and its capacity to select for resistance to the streptogramin B component of Synercid, a drug used to treat vancomycin-resistant enterococcal infections in humans (Bozdogan and Leclercq 1999).

Decreased Accessibility of the Drug to Its Target

Resistance may be conferred by decreasing the accessibility of the antimicrobial to its target. Although changes in membrane structure may sometimes impede diffusion into the cell, the primary mechanism for limiting access to intracellular antimicrobial targets is through the expression of drug-efflux proteins (Nikaido 1996, Paulsen et al. 1996, Williams 1996). These integral-membrane proteins use energy derived from proton motive force or ATP hydrolysis to actively pump the drugs out of the cytoplasm. Such drug efflux systems have been reported in bacteria that are resistant to a broad spectrum of antibiotics including tetracyclines, β-lactams, aminoglycosides, and phenicols.

Eighteen of 29 genes in the *tet* family of tetracycline-resistance genes code for efflux pumps (Chopra and Roberts 2001). The Tet proteins are homologous in primary and secondary structure, share 41% to 78% amino acid sequence identity, two active domains, and from 12 to 14 predicted transmembrane-helices (Rubin and Levy 1991, Chopra and Roberts 2001). Another example is the *cmlA* gene that confers non-enzymatic resistance to chloramphenicol. Although its mechanism has yet to be characterized, similarity in primary structure to bacterial transport proteins suggests that the CmlA protein functions as a drug efflux pump (Bissonnette et al. 1991). Likewise, the *flo* gene, whose product shares 57% amino acid sequence identity to *cmlA*, also encodes a putative efflux pump that confers resistance to both chloramphenicol and florfenicol (Kim and Aoki 1996, Bolton et al. 1999).

Some efflux systems can handle a wide range of structurally dissimilar substrates and consequently may confer multidrug resistance (Paulsen et al. 1996). For example, the AcrAB-TolC system found in *E. coli* encodes an efflux transporter complex that confers resistance to multiple antimicrobial agents including tetracycline, chloramphenicol, fluoroquinolones, β-lactams, erythromycin, ethidium bromide, and sodium dodecylsulfate (Ma et al. 1995, Elkins and Nikaido 2002). This system has three structural components: the AcrB pump, which is localized in the cytosolic membrane; the TolC outer-membrane channel; and the AcrA protein that links AcrB to TolC (Zgurskaya and Nikaido 1999). Other examples of multidrug transporters include MexAB and EmrB in *E. coli* and QacA and NorA in *S. aureus* (Neyfakh et al. 1993, Paulsen et al. 1996). Multidrug transporters present a particularly serious clinical threat to public health because acquisition of a single resistance determinant can confer resistance to a broad range of antimicrobial reagents.

Dissemination of Resistance Genes

Regardless of whether antibiotic resistance in a particular microorganism is intrinsic or has emerged through chromosomal mutations, of particular concern is the spread of resistance genes encoded on mobile DNA elements such as plasmids, transposons, and phage. Plasmids are covalently closed pieces of DNA that are self-replicating and independent of the chromosome. They vary in size from a few thousand base pairs to over 100 kilobase pairs. Genes for resistance to most classes of antibiotics have been found on plasmids, and larger plasmids are capable of carrying multiple resistance genes. Some plasmids are self-transmissible to other cells through a process called conjugation. Although originally thought to be restricted in host range, bacterial conjugation is now known to occur frequently and easily across species (Brisson-Noel et al. 1988, Bertram et al. 1991, Andrup and Andersen 1999). In *E. coli*, genes for virulence factors such as enterotoxins and fimbrae are often found on plasmids, which may compound the resistance problem through the potential co-selection of plasmids encoding genes for both resistance and virulence (Franklin and Mollby 1983, Gonzalez and Blanco 1985, Harnett and Gyles 1985, Blanco et al. 1991, Mainil et al. 1998).

A second type of mobile DNA element is the conjugative transposon (Salyers et al. 1995, Rice 1998). Transposons are discrete segments of the genome that are characterized by two insertion-sequence elements that flank an intervening stretch of DNA (the "transposon"). Transposons can integrate into non-homologous sequences of DNA and move from one plasmid to another or between a plasmid and the chromosome. Similar to plasmids, transposons have the capacity to carry multiple resistance genes. Transposition permits the spread of resistance genes to an even wider range of bacteria, even between genera, than can be achieved by plasmid vectors alone. Transposons permit integration of resistance genes into the chromosome, where they are more stable than plasmids, and thus more difficult to cure from the bacterium.

Transduction via bacteriophage is another vehicle for dissemination of resistance genes (Davison 1999). When a phage enters its lytic cycle, small segments of the host chromosome are packaged with the virus' genetic material. When the viral particle infects a new cell and integrates into the new chromosome, transfer of chromosomal material from the original host to the new recipient occurs. This mechanism has been considered less important in the dissemination of resistance genes because the host range is generally narrow and the size of transferred DNA is relatively small.

Integrons are genetic structures consisting of two conserved segments that flank a central region where cassettes of resistance genes may accumulate (Hall and Collis 1995, Fluitt and Schmitz 1999). The 5′ segment consists of a site-specific integrase gene whose product may facilitate the insertion of genes into the cassette. The 5′ segment also has a promoter for expression of genes in the cassette (Collis and Hall 1995). The 3′ segment generally consists of the *sulI* gene, which confers resistance to sulfonamides. Although they are not mobile in and of themselves, integrons frequently reside within transposable elements or on plasmids and have recently been implicated in the dissemination and persistence of multiple drug resistance (Bass et al. 1999, Adrian et al. 2000).

Intervention Strategies to Combat Antimicrobial Resistance in Food-Producing Animals

In 1999, an interagency task force co-chaired by the Centers for Disease Control, the U.S. Food and Drug Administration, the National Institutes of Health, and other federal agencies including the Department of Agriculture developed the *Public Health Action Plan to Combat Antimicrobial Resistance* (Centers for Disease Control and Prevention 1999). This plan identified four focus areas to serve as a blueprint for federal actions needed to address the antimicrobial resistance problem: surveillance, prevention and control, research, and product development.

Surveillance

The National Antimicrobial Resistance Monitoring System (NARMS) was established in 1996 to monitor changes in antimicrobial susceptibility patterns of enteric bacteria isolated from human and animal diagnostic specimens, healthy farm animals, and carcasses of food animals at slaughter (Tollefson et al. 1998). Data are currently collected on six bacterial organisms: non-Typhi *Salmonella*, *Salmonella* Typhi, *Campylobacter*, *E. coli* O157:H7, *Shigella*, and *Enterococcus*. The broad goal of the surveillance system is to provide useful information about patterns of emerging drug resistance, which in turn can guide mitigation efforts.

Prevention and Control

Prevention and control measures seek to promote the appropriate use of antimicrobial drugs through educational and regulatory actions. National laws and regulations pertaining to licensure and compliance may effectively limit the availability of antimicrobials, but regulatory action may not be sufficient to reduce the prevalence of resistance. In the mid 1980s, concerns over toxicity resulting from potential trace exposure to chloramphenicol led the U.S. Food and Drug Administration to ban its use in food animals. Yet a recent study of antimicrobial resistance among enterotoxigenic *E. coli* from swine reported that 53% were resistant to chloramphenicol (Bischoff et al. 2002). Interestingly, a statistically high level of significance for co-selection was observed for kanamycin, sulfamethoxazole, and tetracycline; agents currently used in swine production. The use of these agents may serve to maintain plasmids on which chloramphenicol resistance resides with other resistance genes. In Denmark, the growth-promoting drugs avoparcin and virginiamycin were banned in 1995 and 1998, respectively, and the food animal industries voluntarily agreed to stop all use of antimicrobial growth promoters in 1999. In 2000, a study on the prevalence of resistance among enterococci from food animals reported a decrease in avilamycin, erythromycin, vancomycin, and virginiamycin resistance from 1995 to 2000 (Aarestrup et al. 2001). The authors of this study also report, however, that in cases in which vancomycin resistance was linked to erythromycin resistance, the prevalence of resistance to either drug did not decrease until the use of avoparcin and tylosin, agents chemically related to vancomycin and erythromycin, respectively, was limited. Thus, simultaneous reductions in the selection pressures of all co-selecting agents may be required to reverse the emergence, spread, and persistence of antimicrobial resistance in the animal production environment.

Research Activities

The fundamental factors that govern the acquisition and spread of antimicrobial resistance are the subject of continuing investigation. The scientific literature is replete with studies on the ability of antibiotics to aid in the dissemination of resistance through clonal expansion and through induction of conjugative transfer (Al-Masaudi et al. 1991, Doucet-Populaire et al. 1991, Nijsten et al. 1995, Igimi et al. 1996, Heinemann 1999, Witte 2000, Carman and Woodburn 2001). In contrast, the factors that inhibit the dissemination of plasmids have not been scrutinized to the same degree. Several studies have suggested that flavomycin, a growth-promoting antimicrobial agent approved for use in poultry and swine, inhibits the growth of bacteria harboring resistance plasmids. Administration of this agent has been observed to suppress the dissemination of multiresistant *E. coli* in the intestines of pigs (Brana et al. 1973, George and Fagerberg 1984, van den Bogaard et al. 2002). Other reports suggest that ascorbic acid can induce plasmid loss in *Staphylococcus aureus* (Amabile-Cuevas 1988, Amabile-Cuevas et al. 1991). In addition, antimutagenic agents, such as green-tea catechins, can reduce the prevalence of resistant bacteria either by preventing point mutations that confer resistance or by potentiating the actions of other antibiotics (Shiota et al. 1999, Pillai et al. 2001). A more thorough investigation of the mechanisms by which these agents inhibit the transfer of genetic material will facilitate the development of new interventions that reduce the dissemination of resistance.

Product Development

Agents that inhibit mechanisms of resistance may be used clinically to combat antimicrobial resistance. Clavulanic acid irreversibly inactivates many β-lactamases by forming a covalent complex that resists hydrolysis (Livermore 1993). Therapy that combines the β-lactam antibiotic amoxicillin with clavulanic acid has been used successfully to overcome β-lactamase-mediated resistance (Heinze-Krause et al. 1998, Chaibi et al. 1999). Other inhibitors currently under development target drug-efflux pumps. Recent studies report the use of inhibitors of the membrane-associated efflux protein NorA to restore the activity of fluoroquinolones against drug-resistant strains of *S. aureus* (Aeschlimann et al. 1999). Other reports suggest that synthetic efflux pump inhibitors can increase the efficacy of plant-derived antimicrobials against Gram-negative bacteria (Tegos et al. 2002).

Although combination therapy is promising for human clinical medicine, its application in food animal agriculture would likely suffer from the same criticisms lodged against current agricultural antimicrobial practices (e.g., resistant bacteria will enter the food chain and cause untreatable human illness). After decades of use in human medicine, inhibitor-resistant β-lactamases are already being encountered. It would therefore appear that for agriculture, particularly food animal production, the primary intervention strategy to combat antimicrobial resistance will be to reduce the selection pressure that gives rise to and sustains antibiotic resistance.

Because the maintenance of animal health and productivity is paramount to the producer, the development of products for use in lieu of antimicrobials is essential. Such products include probiotics (competitive exclusion), alternative feed additives, immune modulators, and bacteriophage therapies that would replace antimicrobials for disease prevention and control, thus reducing the reliance on antibiotics.

Competitive exclusion is a method of controlling gut colonization by *Salmonella* and other enteric pathogens in chickens (Nisbet 1998, Nisbet et al. 1998). The treatment of newly hatched chicks with suspensions of cecal bacteria obtained from healthy adult birds has been shown to protect against *Salmonella* cecal colonization (Corrier et al. 1994). The mechanism is thought to involve the blocking of potential attachment sites, production of bacteriocins, maintenance of gut pH by volatile fatty acids, and competition for nutrients. Defined competitive exclusion cultures are being developed for use in chickens (Nisbet et al. 1998) and in swine (Genovese et al. 1999, Genovese et al. 2000).

Recent investigations of alternatives to antibiotics for the control of *E. coli* O157:H7 and *Salmonella* Typhimurium in the rumen of cattle have focused on the exploitation of facultative metabolism. Facultative anaerobic enterobacteria possess a respiratory nitrate reductase that allows the coupling of anaerobic nitrate reduction to oxidative phosphorylation (Stewart 1988). This enzyme also is capable of reducing chlorate, an analog of nitrate, to the cytotoxic product chlorite. *In vitro*, chlorate has been shown to have a bactericidal effect on *E. coli* O157:H7 and *S.* Typhimurium DT104 (Anderson et al. 2000, Callaway et al. 2001a). Using experimentally infected animals, sodium chlorate supplementation has been demonstrated to reduce *E. coli* O157:H7 populations in cattle and to reduce both *E. coli* O157:H7 or *S.* Typhimurium populations in pigs (Anderson et al. 2001a, Anderson et al. 2001b, Callaway et al. 2001b). The use of chlorate is currently being developed as a preharvest intervention measure to reduce intestinal *E. coli* O157:H7 and *S.* Typhimurium levels in livestock immediately before harvest.

Stimulation of the immune system is another approach used to protect animals from infectious disease. Vaccines stimulate the animal's acquired immune defenses against specific pathogens, but "preventive activation" of the non-specific innate immune system may be an effective alternative, particularly in neonates whose acquired system has yet to mature (Toth et al. 1988). In chickens, one of the main effectors of innate resistance is the heterophil, a polymorphonuclear phagocytic cell that migrates to the site of infection, engulfs the offending microbes, and kills them through a variety of antimicrobial mechanisms. T-cells from adult chickens previously immunized against *Salmonella* Enteritidis secrete factors collectively called immune lymphokines (ILKs) that modulate the functional activities of heterophils in day-old chicks (Kogut et al. 1998). In addition, the activation of heterophils in ILK-treated chicks is strongly associated with protection against organ invasion by *Salmonella* and with reduction of *Salmonella*-induced mortality (Tellez et al. 1993, Kogut et al. 1996). Immune lymphokines derived from the T-cells of *S.* Enteritidis–immunized pigs have been shown to protect weaned piglets from infection by *Salmonella choleraesuis* as well as to enhance growth performance in the presence of an infection (Genovese et al. 1999). Further research is necessary to fully define the ILK preparations, as to date, only one active component of the chicken lymphokine has been identified and characterized (Bischoff et al. 2001, Crippen et al. 2003).

The use of lytic bacteriophage for the prevention and treatment of bacterial infections offers another attractive alternative to antimicrobials. In experimentally infected animals, bacteriophage have been shown to prevent and treat *E. coli*–induced diarrhea in calves, piglets, and lambs and to prevent *E. coli* respiratory infections in broiler chickens (Smith and Huggins 1983, Smith et al. 1987, Huff et al. 2002).

Conclusions

This chapter has presented some of the intervention strategies that are being developed for managing the spread of antimicrobial resistance among enteric bacteria of food animals. Surveillance is a cornerstone of the research activities, but investigations that focus on the prevalence of resistance phenotypes need to be expanded to include characterization of resistance genotypes. These studies should focus on understanding the factors that govern dissemination and persistence of antimicrobial resistance by characterizing the mobility and genetic linkage of resistance genes. Integration of these data will help to develop programs that educate producers and veterinarians on how to maximize the therapeutic effectiveness of drugs, while minimizing the selective pressure favoring resistance, through their appropriate use.

Despite the best efforts to control and treat infectious disease with antimicrobials, bacteria will continue to adapt and survive. The development of alternative products or treatments for use in lieu of antimicrobials is therefore a top research priority. Promising areas of research include the development of probiotics, alternative feed additives, immune modulators, and bacteriophage therapies. Ultimately the application of these products will decrease the need for antimicrobials and will likely have the greatest effect on reduction of antimicrobial resistance among enteric bacteria in food animals.

References

Aalbaek, B., Rasmussen, J., Nielsen, B. and J. E. Olsen. 1991. Prevalence of antibiotic-resistant *Escherichia coli* in Danish pigs and cattle. *Acta Pathologica, Microbiologica et Immunologica Scandinavica* **99**:1103–1110.

Aarestrup, F. M., Seyfarth, A. M., Emborg, H. D., Pedersen, K., Hendriksen, R. S. and Bager, F. 2001. Effect of abolishment of the use of antimicrobial agents for growth promotion on occurrence of antimicrobial resistance in fecal enterococci from food animals in Denmark. *Antimicrobial Agents and Chemotherapy* **45**:2054–2059.

Adrian, P. V., Thomson, C. J., Klugman, K. P. and Amyes, S. G. B. 2000. New gene cassettes for trimethoprim resistance, *dfr13*, and streptomycin-spectinomycin resistance, *aadA4*, inserted on a class 1 integron. *Antimicrobial Agents and Chemotherapy* **44**:355–361.

Aeschlimann, J. R., Dresser, L. D., Kaatz, G. W. and Rybak, M. J. 1999. Effects of NorA inhibitors on *in vitro* antibacterial activities and postantibiotic effects of levofloxacin, ciprofloxacin, and norfloxacin in genetically related strains of *Staphylococcus aureus*. *Antimicrobial Agents and Chemotherapy* **43**:335–340.

Al-Masaudi, S. B., Day, M. J. and Russell, A. D. 1991. Effect of some antibiotics and biocides on plasmid transfer in *Staphylococcus aureus*. *Journal of Applied Bacteriology* **71**:239–243.

Amabile-Cuevas, C. F. 1988. Loss of penicillinase plasmids of *Staphylococcus aureus* after treatment with L-ascorbic acid. *Mutation Research* **207**:107–109.

Amabile-Cuevas, C. F., Pina-Zentella, R. M. and Wah-Laborde, M. E. 1991. Decreased resistance to antibiotics and plasmid loss in plasmid-carrying strains of *Staphylococcus aureus* treated with ascorbic acid. *Mutation Research* **264**:119–125.

Anderson, R. C., Buckley, S. A., Callaway, T. R., Genovese, K. J., Kubena, L. F., Harvey, R. B. and Nisbet, D. J. 2001a. Effect of sodium chlorate on *Salmonella* Typhimurium concentrations in the weaned pig gut. *Journal of Food Protection* **64**:255–258.

Anderson, R. C., Buckley, S. A., Kubena, L. F., Stanker, L. H., Harvey, R. B. and Nisbet, D. J. 2000. Bactericidal effect of sodium chlorate on *Escherichia coli* O157:H7 and *Salmonella typhimurium* DT104 in rumen contents in vitro. *Journal of Food Protection* **63**:1038–1042.

Anderson, R. C., Callaway, T. R., Buckley, S. A., Anderson, T. J., Genovese, K. J., Sheffield, C. L. and Nisbet, D. J. 2001b. Effect of oral sodium chlorate administration on *Escherichia coli* O157:H7 in the gut of experimentally infected pigs. *International Journal of Food Microbiology* **71**:125–130.

Andrup, L. and Andersen, K. 1999. A comparison of the kinetics of plasmid transfer in the conjugation systems encoded by the F plasmid from *Escherichia coli* and plasmid pCF10 from *Enterococcus faecalis*. *Microbiology* **145**:2001–2009

Bass, L., Liebert, C. A., Lee, M. D., Summers, A. O., White, D. G., Thayer, S. G. and Maurer, J. J. 1999. Incidence and characterization of integrons, genetic elements mediating multiple-drug resistance in avian *Escherichia coli*. *Antimicrobial Agents and Chemotherapy* **43**:2925–2929.

Bertram, J., Stratz, M. and Durre, P. 1991. Natural transfer of conjugative transposon Tn916 between Gram-positive and Gram-negative bacteria. *Journal of Bacteriology* **173**:443–448.

Bischoff, K. M., Pishko, E. J., Genovese, K. J., Crippen, T. L., Holtzapple, C. K., Stanker, L. H., Nisbet, D. J. and Kogut, M. H. 2001. Chicken mim-1 protein, P33, is a heterophil chemotactic factor present in *Salmonella enteritidis* immune lymphokine. *Journal of Food Protection* **64**:1503–1509.

Bischoff, K. M., White, D. G., McDermott, P. F., Zhao, S., Gaines, S., Maurer, J. J. and Nisbet, D. J. 2002. Characterization of chloramphenicol resistance in beta-hemolytic *Escherichia coli* associated with diarrhea in neonatal swine. *Journal of Clinical Microbiology* **40**:389–394.

Bissonnette, L., Champetier, S., Buisson, J. and Roy, P. H. 1991. Characterization of the nonenzymatic chloramphenicol resistance (*cmlA*) gene of the In4 integron of Tn1696: similarity of the product to transmembrane transport proteins. *Journal of Bacteriology* **173**:4493–4502.

Blanco, J., Blanco, M., Garabal, J. I. and Gonzalez, E. A. 1991. Enterotoxins, colonization factors and serotypes of enterotoxigenic *Escherichia coli* from humans and animals. *Microbiologia* **7**:57–72.

Boe, L. 1992. Translational errors as the cause of mutations in *Escherichia coli*. *Molecular and General Genetics* **231**:469–471.

Bolton, L. F., Kelley, L. C., Lee, M. D., Fedorka-Cray, P. J. and Mauer, J. J. 1999. Detection of multidrug-resistant *Salmonella enterica* Serotype typhimurium DT104 based on a gene which confers cross-resistance to florfenicol and chloramphenicol. *Journal of Clinical Microbiology* **37**:1348–1351.

Bozdogan, B. and Leclercq, R. 1999. Effects of genes encoding resistance to streptogramins A and B on the activity of quinupristin-dalfopristin against *Enterococcus faecium*. *Antimicrobial Agents and Chemotherapy* **43**:2720–2725.

Brana, H., Hubacek, J. and König, J. 1973. The effect of actinomycin D and flavomycin on *Escherichia coli* R$^+$ strains. *Folia Microbiologica* **18**:257–262.

Brisson-Noel, A., Arthur, M. and Couvalin, P. 1988. Evidence for natural gene transfer from Gram-positive cocci to *Escherichia coli. Journal of Bacteriology* **170**:1739–1745.

Callaway, T. R., Anderson, R. C., Anderson, T. J., Poole, T. L., Bischoff, K. M., Kubena, L. F. and Nisbet D. J. 2001a. *Escherichia coli* O157:H7 becomes resistant to sodium chlorate in pure culture, but not in mixed culture or in vivo. *Journal of Applied Microbiology* **91**:427–434.

Callaway, T. R., Anderson, R. C., Genovese, K. J., Poole, T. L., Anderson, T. J., Byrd, J. A., Kubena, L. F. and Nisbet D. J. 2001b. Sodium chlorate supplementation reduces *E. coli* O157:H7 populations in cattle. *Journal of Animal Science* **80**:1683–1689.

Carman, R. J. and Woodburn, M. A. 2001. Effects of low levels of ciprofloxacin on a chemostat model of human colonic microflora. *Regulatory Toxicology and Pharmacology* **33**:276–284.

Centers for Disease Control and Prevention. 1999. A Public Health Action Plan to Combat Antimicrobial Resistance. Centers for Disease Control and Prevention, Atlanta, GA Available at: http://www.cdc.gov/drugresistance/actionplan/aractionplan.pdf.

Chaibi, E. B., Sirot, D., Paul, G. and Labia, R. 1999. Inhibitor-resistant TEM beta-lactamases: phenotypic, genetic and biochemical characteristics. *Journal of Antimicrobial Chemotherapy* **43**:447–458.

Chopra, I. and Roberts, M. 2001. Tetracycline antibiotics: mode of action, applications, molecular biology, and epidemiology of bacterial resistance. *Microbiology and Molecular Biology Reviews* **65**:232–260.

Coates, S. R. and Hoopes, K. H. 1980. Sensitivities of *Escherichia coli* isolated from bovine and porcine enteric infections to antimicrobial antibiotics. *American Journal of Veterinary Research* **41**:1882–1883.

Collis, C. M. and Hall, C. M. 1995. Expression of antibiotic resistance genes in the integrated cassettes of integrons. *Antimicrobial Agents and Chemotherapy* **39**:155–162.

Corrier, D. E., Hollister, A. G., Nisbet, D. J., Scanlan, C. M., Beier, R. C. and DeLoach, J. R. 1994. Competitive exclusion of *Salmonella enteritidis* in leghorn chicks: comparison of treatment by crop gavage, drinking water, spray, or lyophilized alginate beads. *Avian Diseases* **38**:297–303.

Crippen, T. L., Bischoff, K. M., Lowry, V. K. and Kogut, M. H. 2003. rP33 activates bacterial killing by chicken peripheral blood heterophils. *Journal of Food Protection* **66**:787–792.

Davies, J. and Wright, G. D. 1997. Bacterial resistance to aminoglycoside antibiotics. *Trends in Microbiology* **5**:234–240.

Davison, J. 1999. Genetic exchange between bacteria in the environment. *Plasmid* **42**:73–91.

Doucet-Populaire, F., Trieu-Cuot, P., Dosbaa, I., Andremont, A. and Courvalin, P. 1991. Inducible transfer of conjugative transposon Tn1545 from *Enterococcus faecalis* to *Listeria monocytogenes* in the digestive tracts of gnotobiotic mice. *Antimicrobial Agents and Chemotherapy* **35**:185–187.

Elkins, C. A. and Nikaido, H. 2002. Substrate specificity of the RND-type multidrug efflux pumps AcrB and AcrD of *Escherichia coli* is determined predominantly by two large periplasmic loops. *Journal of Bacteriology* **184**:6490–6498.

Fairbrother, J. M. 1999. Neonatal *Escherichia coli* diarrhea, pp. 433–441. *In* Diseases of Swine, 8th ed. Ames, Iowa: Iowa State University Press.

Flensburg, J. and Skold, O. 1987. Massive overproduction of dihydrofolate reductase in bacteria as a response to the use of trimethoprim. *European Journal of Biochemistry* **162**:473–476.

Fluitt, A. C. and Schmitz, F. J. 1999. Class 1 integrons, gene cassettes, mobility, and epidemiology. *European Journal of Clinical Microbiology and Infectious Diseases* **18**:761–770.

Franklin, A. and Mollby, R. 1983. Concurrent transfer and recombination between plasmids encoding for heat-stable enterotoxin and drug resistance in porcine enterotoxigenic *Escherichia coli. Medical Microbiology and Immunology* **172**:137–147.

Gebreyes, W. A., Davies, P. R., Morgan Morrow, W. E., Funk, J. A. and Altier, C. 2000. Antimicrobial resistance of *Salmonella* isolates from swine. *Journal of Clinical Microbiology* **38**:4633–4636.

Genovese, K. J., Anderson, R. C., Harvey, R. B. and Nisbet, D. J. 2000. Competitive exclusion treatment reduces the mortality and fecal shedding associated with enterotoxigenic *Escherichia coli* infection in nursery-raised neonatal pigs. *Canadian Journal of Veterinary Research* **64**:204–207.

Genovese K. J., Anderson, R. C., Nisbet, D. E., Harvey, R. B., Lowry, V. K., Buckley, S., Stanker, L. H. and Kogut, M. H. 1999. Prophylactic administration of immune lymphokine derived from T cells of *Salmonella enteritidis*-immune pigs. Protection against *Salmonella choleraesuis* organ invasion and cecal colonization in weaned pigs. *Advances in Experimental Medicine and Biology* **473**:299–307.

George, B. A. and Fagerberg, D. J. 1984. Effect of bambermycins, in vitro, on plasmid-mediated antimicrobial resistance. *American Journal of Veterinary Research* **45**:2336–2341.

Gonzalez, E. A. and Blanco, J. 1985. Relation between antibiotic resistance and number of plasmids in enterotoxigenic and non-enterotoxigenic *Escherichia coli* strain. *Medical Microbiology and Immunology* **174**:257–265.

Gustafson, R. H. and Bowen, R. E. 1997. Antibiotic use in animal agriculture. *Journal of Applied Microbiology* **83**:531–541.

Hackbarth, C. J., Kocagoz, T., Kocagoz, S. and Chambers, H. F. 1995. Point mutations in *Staphylococcus aureus* PBP 2 gene affect penicillin-binding kinetics and are associated with resistance. *Antimicrobial Agents and Chemotherapy* **39**:103–106.

Hall, R. M. and Collis, C. M. 1995. Mobile gene cassettes and integrons: capture and spread of genes by site-specific recombination. *Molecular Microbiology* **15**:593–600.

Harnett, N. M. and Gyles, C. L. 1985. Linkage of genes for heat-stable enterotoxin, drug resistance, K99 antigen, and colicin in bovine and porcine strains of enterotoxigenic *Escherichia coli*. *American Journal of Veterinary Research* **46**:428–433.

Heinemann, J. A. 1999. How antibiotics cause antibiotic resistance. *Drug Discovery Today* **4**:72–79.

Heinze-Krauss, I., Angehrn, P., Charnas, R. L., Gubernator, K., Gutknecht, E. M., Hubschwerlen, C., Kania, M., Oefner, C., Page, M. G., Sogabe, S., Specklin, J. L. and Winkler, F. 1998. Structure-based design of beta-lactamase inhibitors. 1. Synthesis and evaluation of bridged monobactams. *Journal of Medicinal Chemistry* **41**:3961–3971.

Hooper, D. C. 2001. Mechanisms of action of antimicrobials: focus on fluoroquinolones. *Clinical Infectious Diseases* **32**:S9–S15.

Huff, W. E., Huff, G. R., Rath, N. C., Balog, J. M., Xie, H., Moore Jr., P. A. and Donoghue, A. M. 2002. Prevention of *Escherichia coli* respiratory infection in broiler chickens with bacteriophage (SPR02). *Poultry Science* **81**:437–441.

Humayun, M. Z. 1998. SOS and Mayday: multiple inducible mutagenic pathways in *Escherichia coli*. *Molecular Microbiology* **30**:905–910.

Huovinen, P., Sundstrom, L., Swedberg, G. and Skold, O. 1995. Trimethoprim and sulfonamide resistance. *Antimicrobial Agents and Chemotherapy* **39**:279–289.

Igimi, S., Ryu, C. H., Park, S. H., Sasaki, Y., Sasaki, T. and Kumagai, S. 1996. Transfer of conjugative plasmid pAM beta 1 from *Lactococcus lactis* to mouse intestinal bacteria. *Letters in Applied Microbiology* **23**:31–35.

Kim, E. and Aoki, T. 1996. Sequence analysis of the florfenicol resistance gene encoded in the transferable T-plasmid of a fish pathogen, *Pasteruella piscicida*. *Microbiology and Immunology* **40**:665–669.

Kogut, M. H., Lowry, V. K., Moyes, R. B., Bowden, L. L., Bowden, R., Genovese, K. and DeLoach, J. R. 1998. Lymphokine-augmented activation of avian heterophils. *Poultry Science* **77**:964–971.

Kogut, M. H., Tellez, G., McGruder, E. D., Wong, R. A., Isibasi, A., Ortiz, V. N., Hargis, B. M. and DeLoach, J. R. 1996. Evaluation of *Salmonella enteritidis*-immune lymphokines on host resistance to *Salmonella enterica* ser. *gallinarum* infection in broiler chicks. *Avian Pathology* **25**:737–749.

Kono, M., Sasatsu, M., O'Hara, K., Shiomi, Y. and Hayasaka, T. 1983. Mechanism of resistance to some cephalosporins in *Staphylococcus aureus*. *Antimicrobial Agents and Chemotherapy* **23**:938–940.

Kunkel, T. A., Sabatino, R. D. and Bambara, R. A. 1987. Exonucleolytic proofreading by calf thymus DNA polymerase delta. *Proceedings of the National Academy of Sciences (USA)* **84**:4865–4869.

Libal, M. C. and Gates, C. E. 1982. Antimicrobial resistance in *Escherichia coli* strains isolated from pigs with diarrhea. *Journal of the American Veterinary Medical Association* **180**:908–909.

Livermore, D. M. 1993. Determinants of the activity of beta-lactamase inhibitor combinations. *Journal of Antimicrobial Chemotherapy* **31(Suppl. A)**:9–21.

Ma, D., Cook, D. N., Alberti, M., Pon, N. G., Nikaido, H. and Hearst, J. E. 1995. Genes *acrA* and *acrB* encode a stress-induced efflux system of *Escherichia coli*. *Molecular Microbiology* **16**:45–55.

Mainil, J. G., Daube, G. E., Jacquemin, E., Pohl, P. and Kaeckenbeeck, A. 1998. Virulence plasmids of enterotoxigenic *Escherichia coli* isolates from piglets. *Veterinary Microbiology* **62**:291–301.

Medeiros, A. A. 1984. Beta-lactamases. *British Medical Bulletin* **40**:18–27.

Neyfakh, A. A., Borsch, C. M. and Kaatz, G. W. 1993. Fluoroquinolone resistance protein NorA of *Staphylococcus aureus* is a multidrug efflux transporter. *Antimicrobial Agents and Chemotherapy* **37**:128–129.

Nijsten, R., London, N., Van den Bogaard, A. and Stobberingh, E. 1995. In-vivo transfer of resistance plasmids in rat, human or pig-derived intestinal flora using a rat model. *Journal of Antimicrobial Chemotherapy* **36**:975–985.

Nikaido, H. 1996. Multidrug efflux pumps of Gram-negative bacteria. *Journal of Bacteriology* **178**:5853–5859.

Nisbet, D. J. 1998. Use of competitive exclusion in food animals. *Journal of the American Veterinary Medical Association* **213**:1744–1746.

Nisbet, D. J., Tellez, G. I., Lowry, V. K., Anderson, R. C., Garcia, G., Nava, G., Kogut, M. H., Corrier, D. E. and Stanker, L. H. 1998. Effect of a commercial competitive exclusion culture (Preempt) on mortality and horizontal transmission of *Salmonella gallinarum* in broiler chickens. *Avian Diseases* **42**:651–656.

Paulsen, I. T., Brown, M. H. and Skurray, R. A. 1996. Proton-dependent multdrug efflux systems. *Microbiological Reviews* **60**:575–608.

Pillai, S. P., Pillai, C. A., Shankel, D. M. and Mitscher, L. A. 2001. The ability of certain antimutagenic agents to prevent development of antibiotic resistance. *Mutation Research* **496**:61–73.

Ren, L., Rahman, M. S. and Humayun, M. Z. 1999. *Escherichia coli* cells exposed to streptomycin display a mutator phenotype. *Journal of Bacteriology* **181**:1043–1044.

Rice, L. B. 1998. Tn916 family conjugative transposons and dissemination of antimicrobial resistance determinants. *Antimicrobial Agents and Chemotherapy* **42**:1871–1877.

Rubin, R. A. and Levy, S. B. 1991. Tet protein domains interact productively to mediate tetracycline resistance when present on separate polypeptides. *Journal of Bacteriology* **173**:4503–4509.

Russell, A. D. and Chopra, I. 1996. Understanding Antibacterial Action and Resistance, 2nd ed. Hemel Hempstead: Ellis Horwood Ltd.

Salyers, A. A., Shoemaker, N. B., Stevens, N. A. and Li, L. Y. 1995. Conjugative transposons: an unusual and diverse set of integrated gene transfer elements. *Microbiological Reviews* **59**:579–590.

Shaw, W. V. 1983. Chloramphenicol acetyltransferase: enzymology and molecular biology. *CRC Critical Reviews in Biochemistry* **14**:1–46.

Shiota, S., Shimizu, M., Mizushima, T., Ito, H., Hatano, T., Yoshida, T. and Tsuchiya, T. 1999. Marked reduction in the minimum inhibitory concentration (MIC) of β-lactams in methicillin-resistant *Staphylococcus aureus* produced by epicatechin gallate, an ingredient of green tea (*Camellia sinensis*). *Biological and Pharmaceutical Bulletin* **22**:1388–1390.

Smith, H. W. and Huggins, M. B. 1983. Effectiveness of phages in treating experimental *Escherichia coli* diarrhoea in calves, piglets and lambs. *Journal of General Microbiology* **129**:2659–2675.

Smith, H. W., Huggins, M. B. and Shaw, K. M. 1987. The control of experimental *Escherichia coli* diarrhoea in calves by means of bacteriophages. *Journal of General Microbiology* **133**:1111–1126.

Stewart, V. 1988. Nitrate respiration in relation to facultative metabolism in enterobacteria. *Microbiological Reviews* **52**:190–232.

Tegos, G., Stermitz, F. R., Lomovskaya, O. and Lewis, K. 2002. Multidrug pump inhibitors uncover remarkable activity of plant antimicrobials. *Antimicrobial Agents and Chemotherapy* **46**:3133–3141.

Tellez, G. T., Kogut, M. H. and Hargis, B. M. 1993. Immunoprophylaxis of *Salmonella enteritidis* infection by lymphokines in leghorn chicks. *Avian Diseases* **37**:1062–1070.

Tollefson, L., Angulo, F. J. and Fedorka-Cray, P. J. 1998. National surveillance for antibiotic resistance in zoonotic enteric pathogens. *The Veterinary Clinics of North America: Food Animal Practice* **14**:141–150

Toth, T. E., Veit, H., Gross, W. B. and Siegel, P. B. 1988. Cellular defense of the avian respiratory system: protection against *Escherichia coli* airsacculitis by *Pasteurella multocida*-activated respiratory phagocytes. *Avian Diseases* **32**:681–687.

van den Bogaard, A. E., Hazen, M., Hoyer, M., Oostenbach, P. and Stobberingh, E. E. 2002. Effects of flavophospholipol on resistance in fecal *Escherichia coli* and enterococci of fattening pigs. *Antimicrobial Agents and Chemotherapy* **46**:110–118.

Weigel, L. M., Steward, C. D. and Tenover, F. C. 1998. gyrA mutations associated with fluoroquinolone resistance in eight species of Enterobacteriaceae. *Antimicrobial Agents and Chemotherapy* **42**:2661–2667.

Weisblum, B. 1995. Erythromycin resistance by ribosome modification. *Antimicrobial Agents and Chemotherapy* **39**:577–585.

Williams, J. B. 1996. Drug efflux as a mechanism of resistance. *British Journal of Biomedical Science* **53**:290–293.

Willmott, C. J. and Maxwell, A. 1993. A single point mutation in the DNA gyrase a protein greatly reduces binding of fluoroquinolones to the gyrase-DNA complex. *Antimicrobial Agents and Chemotherapy* **37**:126–127.

Witte, W. 2000. Selective pressure by antibiotic use in livestock. *International Journal of Antimicrobial Agents* **16**:S19–S24.

Wray, C., McLaren, I. M. and Carroll, P. J. 1993. *Escherichia coli* isolated from farm animals in England and Wales between 1986 and 1991. *Veterinary Record* **133**:439–442.

Zgurskaya, H. I. and Nikaido, H. 1999. Bypassing the periplasm: reconstitution of the AcrAB multidrug efflux pump of *Escherichia coli*. *Proceedings of the National Academy of Sciences (USA)* **96**:7190–7195.

17 Antimicrobial Resistance and the Microflora of the Gastrointestinal Tract

Toni L. Poole, Kenneth J. Genovese, Ross C. Beier, Todd R. Callaway, and Kenneth M. Bischoff

Introduction

To maintain a safe food supply, preharvest and postharvest management practices are implemented to reduce bacterial contamination of animal products. This often includes the use of antimicrobials. Preharvest food safety management starts on the farm and focuses on practices that improve animal health and performance, and it benefits the producer by minimizing economic losses resulting from disease. Unfortunately, the emergence of antimicrobial resistance in enteric pathogens of food animals is a by-product of these practices.

The fundamental tenet with regard to the emergence of antimicrobial resistant bacteria is very simple: antimicrobial use selects for resistant bacteria. From that point on it becomes a very complex field of study.

Antimicrobial Use in Food-Animal Production

Antimicrobials are used in three ways for food-animal production: clinical treatment, disease prophylaxis, and growth enhancement. There are broad arrays of protocols for use of antimicrobials in poultry and livestock production (Lechtenberg et al. 1998, Anonymous 2002). Disease prophylaxis and growth promotion practices that use antimicrobials commonly used in human medicine are a source of controversy (Witte 2000, Teuber 2001). In both instances, antimicrobials are applied in subtherapeutic concentrations and are given to large numbers of animals regardless of their health status (Gustafson and Bowen 1997). This type of exposure to antimicrobials has become a public health issue in recent years because of the widespread emergence of multidrug-resistant pathogenic bacteria. Low-level exposure to antibiotics over extended time periods is believed to increase selection for antibiotic-resistant bacteria (Baquero et al. 1997, Zhao and Drlica 2002). Second, the treatment of large herds or flocks with antimicrobials selects for resistance in thousands of healthy animals that may not need treatment. Approximately 90% of the antibiotics used in food-animal production are given as prophylactic agents or as growth promotants in feed (Khachatourians 1998). It has become apparent that antimicrobial consumption needs to be substantially reduced in all venues. However, decreasing antimicrobial use has health and economic implications.

Mention of a trade name, proprietary product, or specific equipment does not constitute a guarantee or warranty by the U.S. Department of Agriculture and does not imply its approval to the exclusion of other products that may be suitable.

The use of antimicrobials as nutritional supplements for food-animal production began after the discovery that antibiotics promoted rapid weight gain in animals (Taylor 1957). In addition, the incidence of bacterial diseases that result in large economic losses was reduced. Antimicrobials provide a dual benefit to producers by improving animal health and bringing animals to market weight earlier. The exact mechanism by which antimicrobials enhance animal growth is not understood and may differ for each antibiotic. It is believed by many that antimicrobials exert their effect as a result of their effect on the gastrointestinal microflora (Taylor 1957, Gustafson and Bowen 1997). Early growth-promotion studies addressed this issue in two ways: studies in germ-free animals devoid of gastrointestinal bacteria, and studies using inactivated antibiotics administered by various means (Taylor 1957, Coates et al. 1972). The results from many of these early studies were conflicting but suggested that some antimicrobials may also promote growth by mechanisms apart from their antibacterial properties. Tylosin, for example, is known to enhance immune function in the host and to inhibit bacterial virulence (Shryock et al. 1998). Whether these attributes function separately from its antimicrobial activity is unknown.

Most oral antimicrobials used to promote growth are not well absorbed in the gut and act primarily on Gram-positive microflora. Experimental evidence suggests that inhibition of *Clostridium perfringens* significantly contributes to enhanced growth in poultry (Stutz et al. 1983, Stutz and Lawton 1984, Engberg et al. 1999, Knarreborg et al. 2002). Studies by Engberg et al. (1999) and Knarreborg et al. (2002) suggest that impaired broiler growth may result from bacterial production of enzymes that deconjugate bile salts. It is believed that deconjugation of bile salts reduces fat emulsification and lipid absorption. On the basis of these data, ionophore anticoccidials may be effective in suppressing necrotic enteritis without selecting for resistance against important human antimicrobials. The ionophores, monensin, lasalocid, laidlomycin, salinomycin and narasin, are commonly fed to ruminant animals to improve feed efficiency but are not used in human medicine and are not known to confer cross-resistance to antimicrobials currently used in human medicine. Although ionophores pose a low risk of adding new resistance determinants to the environment, they have been shown to increase the carriage of Gram-negative pathogens by ruminants (Chen and Wolin 1979, Newbold et al. 1993).

Alternatives to antibiotics for maintaining animal health are necessary because very few antibiotics are in development and only oxazolidinones and cationic peptides represent new classes of antibiotics (Copra 1998). Most new antibiotics are derivatives of established antibiotic classes. Because the mechanism of action is the same for most members of an antibiotic class, resistant isolates present in the environment will often be cross-resistant to new drugs. Expanding molecular knowledge of resistance mechanisms and mechanisms of growth promotion may provide new targets for antimicrobial drug development and growth-promoting agents. At present there is pressure on the Food and Drug Administration (FDA) to cease approval of new antibiotics for growth promotion if a chemically similar antibiotic is used in human medicine. Growth-promoting antibiotics have not been banned in the United States; however, stricter guidelines have been imposed for the approval of new antibacterial drugs for use in animal feeds (FDA 2003). Under these guidelines most of the antimicrobials currently on the market would not be approved for use.

Antimicrobial Action

To understand the manifestations of antimicrobial resistance, it is necessary to understand the complex means by which antimicrobials exert their effect. Antibacterials differ in the variety and number of bacterial species they inhibit, as well as in their mode of action (Yao and Moellering 1999). The mechanism by which antimicrobials bind or interact with their biochemical targets and the effect elicited depends on the drug class (see Chapter 16).

The spectrum of activity against bacterial species also differs between and within antimicrobial classes. For example, the cephalosporins are categorized in "generations" based on spectrum of activity (Yao and Moellering 1999). First-generation cephalosporins are narrow-spectrum drugs that inhibit a narrow range of bacterial species. Second-generation cephalosporins are expanded-spectrum drugs, third-generation cephalosporins are broad-spectrum drugs, and fourth-generation cephalosporins are extended-spectrum drugs. When the bacterial species causing a disease is unknown, treatment with broad-spectrum drugs is beneficial. However, the lack of inhibitory specificity contributes to the emergence of resistance by disrupting the natural ecological balance of microbial communities and selecting for resistance among non-pathogenic and opportunistic pathogens.

Antibiotics are also categorized according to their mode of inhibition. Bacteriostatic antibiotics act by temporarily inhibiting growth of a bacterial cell. Depending on the length of time growth is impaired and the ecological factors present, a cell may be able to resume growth when the antibiotic is removed (National Committee for Clinical Laboratory Standards 2001). In an active mixed bacterial ecosystem, static cells may be competitively excluded from the niche by actively dividing cells (Poole et al. 2003). An animal's immune system may remove static cells before they can resume growth (de la Cruz and Garcia-Lobo 2002).

Bactericidal antibiotics kill susceptible cells. If the drug treatment kills 100% of the susceptible cells, growth will not resume. If a few resistant mutants were present in the population at the time of drug exposure, they may remain viable; these cells have been termed "persisters" (Gunnison et al. 1963), and under selective pressure they may predominate in an ecological niche. The spectrum of activity and the mode of bacterial inhibition depend on the drug's mechanism of action, chemical structure, and the bacterial species. Vancomycin inhibits cell-wall formation in Gram-positive bacteria (Yao and Moellering 1999). This mechanism of inhibition is bactericidal for most Gram-positive bacteria; however, enterococci are an exception in which case vancomycin is bacteriostatic (Storch and Krogstad 1981). Gram-negative bacteria are unaffected by vancomycin because the size of the drug prevents it from penetrating the outer cell membrane. Gram-positive bacteria are also more susceptible to chemical biocides and ionophores because they lack an outer cell membrane (White and McDermott 2001, Callaway et al. 2003). Aminoglycosides are bactericidal: They inhibit protein synthesis by irreversibly binding to the bacterial ribosome. Macrolide antibiotics also inhibit protein synthesis but are bacteriostatic; they are sometimes bactericidal at high concentrations (Yao and Moellering 1999). It is clear that appropriate dose, duration, and type of antibiotic are important clinical decisions. These decisions are best made on a per case or per animal basis by clinicians who are trained in pharmacology. Herein lies one of the problems in food-animal production: It is simply not feasible or practical to identify and treat individual animals. In addition, animals are infectious during the incubation period before clinical symptoms. By the time an outbreak occurs, the entire flock or herd may have been exposed and will be in need of treatment.

Ecology of Antimicrobial Resistance

In nature, bacteria reside in complex ecological habitats. It is estimated that only 1% of the bacterial species on earth are culturable. This factor alone has greatly facilitated the emergence of resistance and hampered scientists' ability to fully understand the ecology of resistance. The vast array of bacterial species in the environment and the corresponding genomic diversity provide an infinite reservoir for the evolution of resistance. The emergence of multidrug resistance exemplifies the genetic fluidity that exists within microbial communities. Most experiments designed to study antimicrobial resistance are done with pure cultures of bacteria removed from their natural habitats. There is a need for experimental systems that model natural microbial niches so that we can gain a greater knowledge of the ecology of antimicrobial resistance.

The emergence of microbial resistance has an ecological progression that can be described in three stages. The first occurs at the molecular level and is genetic in nature. This can be intrinsic to the cell or a newly acquired genetic element or mutation. During the second stage, the population of resistant clones expands under selective pressure. The third stage is dissemination of resistant bacteria. A great deal of focus has been applied to dissemination of resistance genes (see Chapter 16); however, this would be a mute point if the bacteria that carried these genes were present in dead-end niches and could not be disseminated to new hosts. Dissemination of resistant bacteria from the niche provides the greatest opportunity for the cycle to continue and for resistance determinants to accumulate on mobile genetic elements. Fecal shedding, from both human and animal populations, is a significant source of bacterial dissemination. If we are to curtail the global emergence of multidrug resistance—while still using antibiotics—all three stages must be targeted for control.

The Genetic Reservoir

The reservoir for resistance is the bacterial genome. All mechanisms of antimicrobial resistance have a genetic basis, including insensitivity to antibiotics caused by biochemical or structural factors. Intrinsic resistance refers to the biochemical or structural composition of the cell that naturally provides resistance to an antimicrobial. The genes that confer intrinsic resistance are located on the bacterial chromosome and are vertically transferred as part of cell division. For this reason intrinsic resistance is a characteristic of an entire species or, in some cases, entire genera. Chromosomal genes are not transferred horizontally to other bacteria via conjugation unless they are translocated to a mobile genetic element, such as a plasmid or transposon. Integration of DNA segments into chromosomal or plasmid DNA occurs by recombination and involves cutting and splicing of DNA. Recombination occurs by three different mechanisms: recA-dependent recombination involves gene exchange between two molecules of DNA and requires homology between both DNA strands; transposition involves excision and integration of a transposable element into heterologous DNA, either plasmid or chromosomal; and site-specific recombination involves DNA that contains the specific sequences that allow acceptance of heterologous DNA (Bennett 1999). The third mechanism has been characterized in bacteria in the form of genetic elements termed integrons that can accept foreign genes. Integrons cannot excise themselves from DNA, but they have become localized on plasmids and transposons that are transferable to other bacteria via conjugation. Localization of resistance genes on mobile genetic elements facilitates horizontal transfer of DNA between bacteria, providing a rapid means of dissemination at the molecular level. Moreover, many mobile genetic elements contain clusters of genes that confer resistance. *Salmonella enterica* serotype Typhimurium DT104 represents a bacterial strain that possesses an integron containing metabolically unrelated genes (Briggs and Fratamico 1999). This type of cassette confers resistance to multiple antibiotics, providing a common ecological advantage. Genetic elements of this nature are believed to have resulted from successive recombination events and may be selected because of consecutive exposure to antibiotics. Cycling of antibiotics has been suggested as a means to curtail emergence of resistance; however, this concept may actually be the most effective means to generate multidrug resistance.

A gene cassette may consist of different genes whose action is related to one metabolic pathway subsequently conferring resistance to one antibiotic. This is exemplified by certain types of vancomycin resistance. For example, VanA resistance results from a multigene cassette that encodes seven proteins (Arthur et al. 1993). These proteins have coordinated functions that provide an effective alternative pathway for Gram-positive cell-wall synthesis. The origin of the VanA gene cassette is unknown. It is hypothesized that it may have originated from vancomycin-producing or vancomycin-resistant bacteria commonly isolated from soil (Rippere et al. 1998). Genes with homology to VanX and VanA have been identified in two Gram-negative species, *Salmonella* and *Escherichia coli* (Daub et al. 1988, Zawadzke et al. 1991, Hilbert et al. 1999). These genes have a chromosomal location and may be involved in peptidoglycan synthesis (Hilbert et al. 1999). Experiments in our laboratory demonstrated that other genes with homology to vancomycin resistance were present in field isolates of *Salmonella,* but these genes were not adjacent to one another on the bacterial chromosome, as seen on the VanA resistance gene cassette (T. Poole, unpublished data). Because Gram-negative microorganisms are intrinsically resistant to glycopeptide antibiotics as a result of their outer cell membrane, they would not need additional genes to provide vancomycin resistance. It is generally accepted that bacteria do not retain extraneous genetic material because it incurs a fitness cost (Andersson and Levin 1999, Spratt 1996). However, Sander et al. (2002) found that chromosomal drug resistance incurs a very small fitness cost. In addition, gene products that confer antibiotic resistance are present in the bacterial gene pool because their primary function is essential to cell metabolism, although enzymes may have only one catalytic function that does not limit them from performing multiple metabolic tasks. Furthermore, bacterial species that are insensitive to antibiotics should not be discounted as a source of resistance genes. They may serve as a source of entire operons that provide alternative biochemical pathways for other bacteria. This demonstrates the stealthy nature of antibiotic resistance.

Point mutations in chromosomal genes can also confer resistance. Because these genes are not present on mobile genetic elements, they do not represent a great risk for horizontal gene transmission to new bacterial species. Instead, the risk of dissemination is from clonal expansion and dissemination of resistant clones.

Competition and Selection in the Gastrointestinal Reservoir

The gastrointestinal tract is an open system that supports a high density of diverse microflora; it also represents an important reservoir for the emergence of resistance, but only under selective pressure. In the absence of selection pressure, microbial diversity provides competition that prevents overpopulation of individual species. When antibiotics are used to eradicate a pathogenic infection, the entire body is not sterilized, and the concentration of antibiotic may vary in different compartments of the body, resulting in selection of confined niches (Baquero et al. 1997). Selection changes the profile of normal microflora in compartmentalized niches and reduces species diversity (Jensen 1998, Baggesen et al. 1999). Some species may be eliminated, thus allowing the clonal expansion of resistant or tolerant species.

As a bacterial niche, the gastrointestinal tract should be considered from an ecological perspective. An ecological niche can support growth of a finite number of individuals, and in the case of a mass extinction, as would occur with broad-spectrum antimicrobial treatment, an evolutionary burst from surviving species would occur to refill the niche. The population that refills the gastrointestinal niche is the population that will subsequently be shed and disseminated throughout the environment. This is significant when you consider the extensive number of animals treated during food-animal production. The use of antimicrobials allows the clonal expansion of many intrinsically resistant microorganisms that are also human enteric pathogens.

Our knowledge regarding the effect of antibiotic growth promoters on the ecology of the gastrointestinal population is incomplete because many of the species present in the gut are difficult to culture. Modern culture-independent molecular methods used to identify bacterial strains are useful in expanding our knowledge of species diversity but are not as useful for understanding ecological relationships among microflora. The diversity of gastrointestinal microflora differs among animal species, among individual animals, and along the length of the gastrointestinal tract itself (Savage 1977, Brook 1999, Hooper et al. 1999, Apajalahti et al. 2001, Gong et al. 2002, Knarreborg et al. 2002). Synergistic and antagonistic interactions between the host and the bacteria entering the gut shortly after birth lead to a succession of bacterial species until a stable consortium results (Savage 1977, Brook 1999, Hooper et al. 1999). The bacteria that persist in the gut must tolerate or adapt to the conditions present in the niche during the initial colonization. Many bacterial species produce substances that are inhibitory to other bacterial species. These substances include bacteriocins, organic acids, and hydrogen peroxide (Hinton and Hume 1995, Nisbet et al. 1996, Brook 1999). Adaptation to such an environment may involve acquisition of specific genes that encode immunity proteins against inhibitory substances. These adaptive processes are the same as previously discussed for antimicrobial resistance. Immunity proteins confer resistance against antagonistic factors present in the gastrointestinal niche. Production of inhibitory substances maintains niche-specific selection pressure on the resistant clones (Reeves 1992, Brook 1999, Lan and Reeves 2000). Coordination of nutrient use, attachment site specificity, and host-specific factors also contribute to the selection process that ultimately determines the species profile of the gut microflora (Hooper et al. 1999). Autochthonous bacteria are those species considered to have evolved with the host and are considered normal flora to the host (Savage 1977, Klaenhammer 2001). Allochthonous species are those bacteria non-indigenous to a particular host species; however, they are ubiquitous in the environment and can establish whether the competitive stringency of the niche is disrupted (Savage 1977, Klaenhammer 2001). This is an important distinction when considering issues of food safety and antibiotic resistance. The use of antimicrobials not only reshapes the profile of established bacteria already present in the gut as discussed above but also disrupts the inhibitory stringency of the niche, allowing allochthonous species to persist that would normally be transient (Poole et al. 2001a, Poole et al. 2001b, Poole et al. 2003). Disruption of the normal flora leads to an increase in susceptibility to intestinal disease by microorganisms that are pathogenic to food animals (Collins and Carter 1978, Klaenhammer 2001). Extended residence also increases opportunity for allochthonous species to acquire resistance to antibiotics and niche-selective factors.

Although certain members of the gastrointestinal microflora may be responsible for inhibiting rapid growth of the host animal, many provide a benefit by preventing colonization by enteric pathogens (Nurmi and Rantala 1973, Snoeyenbos et al. 1978, Barnes et al. 1979). The mechanisms that provide protection are thought to be the same as those that are involved in the initial selection of the normal flora. The most efficient protection in chickens is provided when healthy flora colonize the gastrointestinal tract immediately after birth (Corrier et al. 1998). Modern food-animal production practices often prevent normal colonization of the gut by bacteria in the environment (Jensen 1998, Gustafson and Bowen 1997). In addition to the use of antimicrobials, newborn animals may not have contact with their mothers or natural environmental factors such as soil that contribute to the establishment of healthy flora (Klaenhammer 2001). Nurmi and Rantala (1973) were the first to treat day-old chicks with cecal contents from adult chickens in an attempt to provide the chicks with normal flora. This technique provided effective protection against colonization by enteric pathogens and was termed "competitive exclusion."

Nisbet et al. (1993) further developed the technology of competitive exclusion by establishing an anaerobic continuous-flow culture of chicken caecal bacteria (CF3). Because the bacteria in CF3 were obtained from the contents of an adult chicken cecum, the bacteria were niche-adapted to the chicken intestinal tract before *in vitro* culture (Corrier et al. 1995). This may have contributed to their ability to maintain complex species diversity as well as stable cell concentrations within the continuous-flow system. Attempts in our laboratory to generate stable, diverse continuous-flow cultures by selecting bacterial isolates that originated from different animals of the same or different species have failed. Defined cultures of selected isolates have only been successful in continuous-flow culture if all of the isolates originated from one animal. We have also noted that successful continuous-flow cultures derived from gut contents of individual chickens or pigs differ in their inhibitory stringency against enteric pathogens (Nisbet et al. 1996, Genovese et al. 2000).

A previously described derivative of the CF3 culture (CCF) was found to inhibit exogenous *Enterococcus* spp. (Poole et al. 2001a). This was surprising because four enterococcal species, *E. faecium*, *E. faecalis*, *E. avium*, and *E. gallinarum*, were naturally present in CCF. Because the enterococci predominate in the CCF culture, it had been assumed that CCF would easily support enterococci added exogenously. This was not the case, as all enterococci added to the CCF culture were rapidly eliminated. Moreover, when 100 mL of CCF was added to a 1,050-mL monoculture of vancomycin-resistant *E. faecium* (VRE), at 10^8 colony-forming units per milliliter, the CCF microflora displaced and completely eliminated the vancomycin-resistant *E. faecium* (Poole et al. 2001a). The question has been put forth, If exogenous enterococci cannot survive in CCF, how do the enterococci that are present in CCF survive? We believe, in part, that CCF enterococci coadapted to inhibitory factors in the chicken ceca during the initial microbial colonization or niche-adaptation process. It is likely that multiple-resistance determinants are involved and that exogenous enterococci are unable to acquire all of these resistance determinants before their demise in the CCF culture.

The initial CCF studies that used exogenous enterococci may have elucidated a significant role for niche adaptation in the establishment and stability of gastrointestinal microbial ecology (Poole et al. 2001a, Poole et al. 2001b, Poole et al. 2003). This may have relevance to the ecology of antibiotic resistance. Instead of cycling antibiotics and allowing bacteria to consecutively adapt, it may be preferable to hit them with multifaceted inhibitory cocktails, as occurs during niche selection. The multiple-attack approach is analogous to treatment of HIV. This approach is believed to decrease the generation of resistant mutants by inhibiting more than one of the viral enzymes necessary for viral replication (Kaufmann and Cooper 2000).

Niche selection pressure may explain why most probiotic preparations fed to animals are not maintained after oral administration is discontinued (Netherwood et al. 1999). Wagner et al. (2002) developed an assay on short-term mixed-batch cultures to assess the efficacy of competitive exclusion cultures. They compared PREEMPT, the FDA product derived from CF3, to a similar batch culture derived by selecting similar species from American Type Culture Collection (ATCC). Use of a short-term culture of selected ATCC isolates provides no data on the survivability of these isolates in a long-term culture. Our experience suggests that 29 ATCC isolates would be reduced to two or three isolates in a week's time in a continuous-flow culture. Although selected probiotic cultures may be strongly inhibitory to enteropathogens *in vitro*, they are of little value if they cannot become stable inhabitants of the gut. The health benefits of probiotic cultures may be greatly increased if cultures are derived from niche-adapted mixed cultures of bacteria.

Dissemination

No single source of bacterial contamination is to blame for the resistance problem; however, fecal shedding from man and animals should be considered a major facilitator of bacterial dissemination. Control of fecal shedding into the environment represents a significant control point for bacterial dissemination. Disposal of human and animal fecal waste has been a long-standing problem for society. Sewage sludge from human sanitation plants and fecal waste from animal production facilities are routinely spread on crops as fertilizer. The practice of spreading human sewage sludge has been an environmental concern from its inception because heavy metals and other pollutants collect in sludge and are transferred through the food chain via uptake in plants (Atlas and Bartha 1981). Such sources of fertilization also pollute many surface water and groundwater sources in this country (Colwell 1980, Hagedorn et al. 1999, Chee-Sanford et al. 2001). Similar concerns have arisen with regard to dissemination of antibiotic-resistant bacteria. Antibiotic resistance profiles of fecal coliforms and fecal streptococci are used to monitor fecal contamination of water sources (Hagedorn et al. 1999).

Bacteria are ubiquitous and mobile in the environment: As long as antibiotics are in use, bacteria will adapt, and resistant species will pervade all of their natural habitats. A greater risk comes not from food products we cook but from the viable flora in the environment to which humans and animals are exposed on a daily basis. The ability of bacteria to disseminate throughout the environment as a result of fecal shedding is exemplified by a 1996 *E. coli* O157:H7 outbreak. The epidemiological investigation suggested the source of contamination was apples, collected from the ground, that had been contaminated with *E. coli:* O157:H7-infected deer feces (Parish 1997). The fact that wildlife carries pathogens such as *E. coli* O157:H7 illustrates the pervasive nature of bacteria throughout the environment.

Public health officials often overlook the benefit gained by the food safety measures already in place to reduce dissemination of foodborne pathogens into the food supply. Many postharvest food safety measures are in place to prevent fecal contamination of meat products post-slaughter. As these measures continue to improve, the risk of resistant bacterial isolates entering the food chain will also be reduced. Public awareness as to the health risks of undercooked meat has increased significantly as a result of well-publicized outbreaks of *E. coli* O157:H7.

Reduction of Antibacterial Consumption

Many public forums have been convened to discuss the problem of antibiotic resistance. Prudent use of antimicrobials has been called for to decrease consumption of antimicrobials; this implies antibiotics be used only for individuals clinically diagnosed with bacterial infections. It excludes antibiotic treatment for viral infections, disease prophylaxis, and growth promotion (van den Bogaard and Stobberingh 1999). Although most agree that unnecessary use of antimicrobials should be eliminated, few agree on what constitutes unnecessary use. Both modern medicine and modern food-animal production practices have contributed to the current problem, and more than cessation of antimicrobial use for prophylaxis and growth promotion is necessary to reduce the incidence of multidrug-resistant pathogens in hospitals and the environment.

A significant proportion of both human and animal infections that require treatment are caused by opportunistic pathogens—organisms that cause disease only when an underlying condition predisposes the host to infection and pathogenesis. In both human and animal populations, immune status and viral diseases predispose the host to secondary bacterial infections. Demographic studies of human populations have shown an increase in groups with compromised immunity over the last 50 years (Altekruse et al. 1998). This corresponds with increased use of antimicrobials. Alternative treatments are needed for the immunocompromised human and animal populations if we are to substantially reduce consumption of antimicrobials. This will require technological advances that will reduce the incidence of infectious disease. Two approaches already in use include vaccines and immunomodulators. There are over 2,000 vaccines approved for use in domestic animals; they include modified-live, killed, DNA, and subunit vaccines (Roth and Henderson 2001). Immunomodulators include cytokines, pharmaceuticals, probiotics, nutraceuticals, and medicinal plant products (Blecha 2001). It is important that products in both of these categories elicit an appropriate immune response without inducing immune-mediated tissue damage. In addition to clinical efficacy, technologies for food-animal producers must provide an application protocol suitable for large numbers of animals without overdue stress on the individual animals. Many new technologies that have been developed require changes in management practices that introduce significant additional expense to the producer and are therefore not considered practical.

Newer, more controversial technologies include genetic modification and *in vitro* technologies. In human medicine, new cancer treatments that specifically target tumor cells without injuring other host cells or organs are promising. These types of treatments dramatically reduce or eliminate damage to the patient's immune system. The ability to reproduce an individual's own organs for transplant would eliminate waiting lists and the postoperative need for immunosuppressive drugs. Crops and food animals that are bred or genetically modified to resist disease could have a huge effect on the need for antibiotics. Although at present some technologies are controversial, many technologies that are now in everyday use were at one time also controversial.

Advances in molecular biology have allowed us to elucidate many specific genetic mechanisms that confer resistance to antimicrobials. From this we have a better understanding of the metabolic processes that provide resistance. This may allow the development of pathogen-specific inhibitors, as have been achieved with a few antivirals that target viral proteins without inhibiting essential host cell proteins. It is more difficult in bacterial systems because pathogens are so closely related to important normal flora. Ideally, drug design would target only certain pathogenic strains. The technologies mentioned above should not affect the normal microflora of the human and animal gastrointestinal tract and, therefore, would not select for resistance in the gastrointestinal reservoir.

Conclusions

We cannot eliminate the genetic reservoir that provides antimicrobial resistance to bacteria, but we may be able to make vast improvements over the current situation. To curtail the emergence of multidrug-resistant bacteria, several approaches need to be pursued. New management strategies must be used that reduce clonal expansion and dissemination of resistant bacterial isolates into the environment. These may include nutritional supplements that promote growth without affecting the bacterial flora in the gut. Such supplements should not select for resistance, nor should they increase fecal shedding of enteric pathogens. Combination therapy should also be considered for clinical therapeutics. This approach has been successful in reducing the emergence of resistant HIV mutants (Kaufmann and Cooper 2000) and appears to be similar to the mechanism used in nature to competitively exclude enteric pathogens from the gut. Strategies that maintain competition in the gut may reduce the ability of pathogens to disseminate. Finally, improved health care treatments that reduce the susceptibility of humans and animals to opportunistic pathogens are necessary.

References

Altekruse, S. F., Swerdlow, D. B. L. and Wells, S. J. 1998. Factors in the emergence of foodborne diseases. *Veterinary Clinics of North America: Food Animal Practice* **14**:1–15.

Andersson, D. I. and Levin, B. R. 1999. The biological cost of antibiotic resistance. *Current Opinion in Microbiology* **2**:489–493.

Anonymous. 2002. Feed Additive Compendium; Minnetonka, Minn: The Miller Publishing Company.

Apajalahti, J. H. A., Kettunen, A., Bedford, M. R. and Holben, W. E. 2001. Percent G+C profiling accurately reveals diet related difference in the gastrointestinal microbial community of broiler chickens. *Applied and Environmental Microbiology* **67**:5656–5667.

Arthur, M., Molinas, C., Depardieu, F. and Courvalin, P. 1993. Characterization of Tn*1546*, a TN3-related transposon conferring glycopeptide resistance by synthesis of depsipeptide peptidoglycan precursors in *Enterococcus faecium* BM4147. *Journal of Bacteriology* **175**:117–127.

Atlas, R. M. and Bartha, R. 1981. Ecological aspects of biodeterioration control: soil water and waste management, pp. 382–414. *In* Microbial Ecology Fundamentals and Applications, R. M. Atlas, R. Bartha (Eds.). Reading, MA: Addison-Wesley.

Baggesen, D. L., Wingstran, A., Carstensen, B., Nielsen, B. and Aarestrup, F. M. 1999. Effects of the antimicrobial growth promoter tylosin on subclinical infection of pigs with *Salmonella enterica* serotype Typhimurium. *American Journal of Veterinary Research* **60**:1201–1206.

Baquero, F., Negri, C., Morosini, M. I. and Blazquez, J. 1997. The antibiotic selective process: concentration-specific amplification of low-level resistant populations, pp. 93–111. *In* Antibiotic Resistance: Origins, Evolution, Selection and Spread. Ciba Foundation Symposium 207. Chichester: Wiley.

Barnes, E. M., Impey, C. S. and Stevens, B. J. H. 1979. Factors affecting the incidence and anti-salmonella activity of the anaerobic caecal flora of the young chick. *Journal of Hygiene-Cambridge* **82**:263–283.

Bennett, P. M. 1999. Integrons and cassettes: a genetic construction kit for bacteria. *Journal of Antimicrobial Chemotherapy* **43**:1–4.

Blecha, F. 2001. Immunomodulators for prevention and treatment of infectious diseases in food-producing animals. *Veterinary Clinics of North America: Food Animal Practice* **17**:621–633.

Briggs, C. E. and Fratamico, P. M. 1999. Molecular Characterization of an antibiotic gene cluster of *Salmonella typhimurium* DT104. *Antimicrobial Agents and Chemotherapy* **43**:846–849.

Brook, I. 1999. Bacterial interference. *Critical Reviews in Microbiology* **25**:155–172.

Callaway, T. R., Edrington, T. S., Rychlik, J. L., Genovese, K. J., Poole, T. L., Elder, R. O., Bischoff, K. M., Anderson, R. C., Russell, J. B. and Nisbet, D. J. 2003. Ionophores: their use as ruminant growth promotants and impact on food safety. *Current Issues in Intestinal Microbiology* **4**:43–51.

Chee-Sanford, J. C., Aminov, R. I., Krapac, I. J., Garrigues-Jeanjean, N. and Mackie, R. I. 2001. Occurrence and diversity of tetracycline resistance genes in lagoons and groundwater underlying two swine production facilities. *Applied and Environmental Microbiology* **67**:1494–1502.

Chen, M. and Wolin, M. J. 1979. Effect of monensin and lasalocid-sodium on the growth of methanogenic and rumen saccharolytic bacteria. *Applied and Environmental Microbiology* **38**:72–77.

Coates, M. E., Mead, G. C., Barnum, D. A. and Harry, M. G. 1972. A comparison of growth of chicks in the Gustafsson germ free apparatus and in a conventional environment, with and without dietary supplements of penicillin. *British Journal of Nutrition* **17**:141–150.

Collins, F. M. and Carter, P. B. 1978. Growth of salmonellae in orally infected germ-free mice. *Infection and Immunity* **21**:41–47.

Colwell, R. R. 1980. Human Pathogens in the Aquatic Environment, pp. 377–379. *In* Microbiology—1980, D. Schlessinger (Ed.). Washington, D.C.: American Society of Microbiology Press.

Copra, I. 1998. Research and development of antibacterial agents. *Current Opinion Microbiology* **1**:495–501.

Corrier, D. E., Byrd II, J. A., Hume, M. E., Nisbet, D. J. and Stanker, L. H. 1998. Effect of simultaneous or delayed competitive exclusion treatment on the spread of *Salmonella* in chicks. *Applied Poultry Science* **7**:132–137.

Corrier, D. E., Nisbet, D. J., Scanlan, C. M., Hollister, A. G. and DeLoach, J. R. 1995. Control of *Salmonella typhimurium* colonization in broiler chicks with a continuous-flow culture. *Poultry Science* **74**:916–924.

de la Cruz, F. and Garcia-Lobo, J. M. 2002. Antibiotic resistance: how bacterial populations respond to a simple evolutionary force, pp. 19–36. *In* Bacterial Resistance to Antimicrobials, K. Lewis, A. A. Salyers, H. W. Taber, R. G. Wax (Eds.). New York: Marcel Dekker.

Daub, E., Zawadzke, L. E., Botstein, D. and Walsh, C. T. 1988. Isolation, cloning, and sequencing of the *Salmonella typhimurium ddlA* gene with purification and characterization of its product, D-alanine:D-alanine ligase (ADP forming). *Biochemistry* **27**:3701–3708.

Engberg, R. M., Hedemann, M. S., Leser, T. D. and Jensen, B. B. 1999. Effect of zinc bacitracin and salinomycin on intestinal microflora and performance of broilers. *Poultry Science* **79**:1311–1319.

Food and Drug Administration. 2003. Guidance for industry: evaluation of the human health impact of the microbial effects of antimicrobial new animal drugs intended for use in food-producing animals. Available at: www.fda.gov/cvm/guidance/guideline18.html.

Gong, J., Forster, R. J., Yu, H., Chambers, J. R., Sabour, P. M., Wheatcroft, R. and Chen, D. 2002. Diversity and phylogenetic analysis of bacteria in the mucosa of chicken ceca and comparison with bacteria in the cecal lumen. *FEMS Microbiology Letters* **208**:1–7.

Genovese, K. J., Anderson, R. C., Harvey, R. B. and Nisbet, D. J. 2000. Competitive exclusion treatment reduces the mortality and fecal shedding associated with enterotoxigenic *Escherichia coli* infection in nursery-raised neonatal pigs. *Canadian Journal of Veterinary Research* **64**:204–207.

Gunnison, J. B., Fraher, M. A. and Jawetz, E. 1963. Persistence of *Staphylococcus aureus* in penicillin *in vitro*. *Journal of General Microbiology* **34**:335–349.

Gustafson, R. H. and Bowen, R. E. 1997. Antibiotic use in animal agriculture. *Journal of Applied Microbiology* **83**:531–541.

Hagedorn, C., Robinson, S. L., Filtz, J. R., Grubbs, S. M., Angier, T. A. and Reneau Jr., R. B. 1999. Determining sources of fecal pollution in a rural Virginia watershed with antibiotic resistance patterns in fecal streptococci. *Applied and Environmental Microbiology* **65**:5522–5531.

Hilbert, F., del Portillo, F. G. and Groisman, E. A. 1999. A periplasmic D-Alanyl-D-Alanine dipeptidase in the Gram-negative bacterium *Salmonella enterica*. *Journal of Bacteriology* **181**:2158–2165.

Hinton, A. and Hume, M. E. 1995. Antibacterial activity of the metabolic by-products of a *Veillonella* sp. and *Bacteroides fragilis*. *Anaerobe* **1**:121–127.

Hooper, L. V., Xu, J., Falk, P. G., Midtvedt, T. and Gordon, J. I. 1999. A molecular sensor that allows a gut commensal to control its nutrient foundation in a competitive ecosystem. *Proceedings of the National Academy of Sciences (USA)* **96**:9833–9838.

Jensen, B. B. 1998. The impact of feed additives on the microbial ecology of the gut in young pigs. *Journal of Animal and Feed Sciences* **7**:45–64.

Kaufmann, G. R. and Cooper, D. A. 2000. Antiretroviral therapy of HIV-1 infection: established treatment strategies and new therapeutic options. *Current Opinion in Microbiology* **3**:508–514.

Khachatourians, G. G. 1998. Agricultural use of antibiotics and the evolution and transfer of antibiotic-resistant bacteria. *Canadian Medical Association Journal* **159**:1129–1136.

Klaenhammer, T. R. 2001. Probiotics and prebiotics, pp. 797–811. *In* Food Microbiology: Fundamentals and Frontiers, 2nd ed., M. P. Doyle (Ed.). Washington, D.C.: American Society for Microbiology Press.

Knarreborg, A., Simon, M. A., Enberg, R. M., Jensen, B. B. and Tannock, G. W. 2002. Effects of dietary fat source and subtherapeutic levels of antibiotic on bacterial community in the ileum of broiler chickens at various ages. *Applied and Environmental Microbiology* **68**:5918–5924.

Lan, R. and Reeves, P. R. 2000. Intraspecies variation in bacterial genomes: the need for a species genome concept. *Trends in Microbiology* **8**:396–401.

Lechtenberg, K. F., Smith, R. A. and Stokka, G. L. 1998. Feedlot Health and Management. *Veterinary Clinics of North America: Food Animal Practice* **14**:177–197.

National Committee for Clinical Laboratory Standards (M26-A). 2001. Methods for determining bactericidal activity of antimicrobial agent; approved guideline. Approved Standard, National Committee for Clinical Laboratory Standards, Villanova, Penn.

Newbold, C. J., Wallace, R. J. and Walker, N. D. 1993. The effect of tetronasin and monensin on fermentation, microbial numbers and the development of ionophore-resistant bacteria in the rumen. *Journal of Applied Microbiology* **75**:129–134.

Netherwood, T., Bowden, R., Harrison, P., O'Donnell, A. G., Parker, D. S. and Gilbert, H. J. 1999. Gene transfer in the gastrointestinal tract. *Applied and Environmental Microbiology* **65**:5139–5141.

Nisbet, D. J., Corrier, D. E. and DeLoach, J. R. 1993. Effect of a defined continuous-flow derived bacterial culture and dietary lactose on *Salmonella* colonization in broiler chicks. *Avian Diseases* **37**:1017–1025.

Nisbet, D. J., Corrier, D. E., Ricke, S. C., Hume, M. E., Byrd II, J. A. and DeLoach, J. R. 1996. Caecal propionic acid as a biological indicator of the early establishment of a microbial ecosystem inhibitory to *Salmonella* in the chicks. *Anaerobe* **2**:345–350.

Nurmi, E. and Rantala, M. 1973. New aspects of salmonella infection in broiler production. *Nature* **241**:210–211.

Parish, M. E. 1997. Public health and nonpasteurized fruit juices. *Critical Reviews in Microbiology* **23**:109–119.

Poole, T. L., Genovese, K. J., Anderson, T. J., Sheffield, C. L., Bischoff, K. M., Callaway, T. R. and Nisbet D. J. 2001a. Inhibition of vancomycin-resistant enterococci by an anaerobic continuous flow culture of chicken microflora. *Microbial Ecology in Health and Disease* **13**:246–253.

Poole, T. L., Genovese, K. J., Knape, K. D., Callaway, T. R., Bischoff, K. M. and Nisbet, D. J. 2003. Effect of subtherapeutic concentrations of tylosin on the inhibitory stringency of a mixed anaerobe continuous-flow culture of chicken microflora against *Escherichia coli* O157:H7. *Journal of Applied Microbiology* **94**:73–79.

Poole, T. L., Hume, M. E., Genovese, K. J., Anderson, T. J., Sheffield, C. L., Bischoff, K. M. and Nisbet, D. J. 2001b. Persistence of a vancomycin-resistant *Enterococcus faecium* in an anaerobic continuous-flow culture of porcine microflora in the presence of subtherapeutic concentrations of vancomycin. *Microbial Drug Resistance* **7**:343–348.

Reeves, P. R. 1992. Variation in O-antigens, niche-specific selection and bacterial populations. *FEMS Microbiology Letters* **100**:509–516.

Rippere, K., Patel, R., Ull, J. R., Piper, K. E., Steckelberg, J. M., Kline, B. C., Cockerill III, F. R. and Yousten, A. A. 1998. DNA sequence resembling vanA and vanB in the vancomycin-resistant biopesticide *Bacillus popilliae*. *Journal of Infectious Diseases* **178**:584–588.

Roth, J. A. and Henderson, L. M. 2001. New technology for improved vaccine safety and efficacy. *Veterinary Clinics of North America Food Animal Practice*. **17**:585–597.

Sander, P., Springer, B., Prammananan, T., Sturmfels, A., Kappler, M., Pletschett, M. and Bottger, E. C. 2002. Fitness cost of chromosomal drug resistance-conferring mutations. *Antimicrobial Agents and Chemotherapy* **46**:1204–1211.

Savage, D. C. 1977. Microbial ecology of the gastrointestinal tract. *Annual Review of Microbiology* **31**:107–133.

Shryock, T. R., Mortensen, J. E. and Baumholtz, M. 1998. The effects of macrolides on the expression of bacterial virulence mechanisms. *Journal of Antimicrobial Chemotherapy* **41**:505–512.

Snoeyenbos, G. H., Weinak, O. M. and Smyser, C. F. 1978. Protecting chicks and poultry from salmonellae by oral administration of "normal" gut microflora. *Avian Diseases* **22**:273–287.

Spratt, B. G. 1996. Antibiotic resistance: counting the cost. *Current Biology* **6**:1219–1221.

Storch, G. A. and Krogstad, D. J. 1981. Antibiotic-induced lysis of enterococci. *Journal of Clinical Investigation* **68**:639–645.

Stutz, M. W., Johnson, S. L., Judith, F. R. and Muir, L. A. 1983. Effect of the antibiotic thiopeptin on Clostridium perfringens, and growth and feed efficiency of broiler chicks. *Poultry Science* **62**:1633–1638.

Stutz, M. W. and Lawton, G. C. 1984. Effects of diet and antimicrobials on growth, feed efficiency, intestinal Clostridium perfringens, and ileal weight of broiler chicks. *Poultry Science* **63**:2036–2042.

Taylor, J. H. 1957. The mode of action of antibiotics in promoting animal growth. *The Veterinary Record* **69**:278–288.

Teuber, M. 2001. Veterinary use and antibiotic resistance. *Current Opinion in Microbiology* **4**:493–499.

van den Bogaard, A. E and Stobberingh, E. E. 1999. Antibiotic usage in animals. *Drugs* **58**:589–607.

Wagner, D. R., Holland, M. and Cerniglia, C. E. 2002. An *in vitro* assay to evaluate competitive exclusion products for poultry. *Journal of Food Protection* **65**:746–751.

White, D. G. and McDermott, P. F. 2001. Biocides, drug resistance and microbial evolution. *Current Opinion in Microbiology* **4**:313–317.

Witte, W. 2000. Selective pressure by antibiotic use in livestock. *International Journal of Antimicrobial Agents* **16**:S19–S24.

Yao, J. D. C. and Moellering, R. C. 1999. Antibacterial Agents, pp. 1474–1504. *In* Manual of Clinical Microbiology, P. R. Murray, E. J. Baron, M. A. Pfaller, F. C. Tenover, R. H. Yolker (Eds.). Washington, D.C.: American Society for Microbiology Press.

Zawadszke, L. E., Bugg, T. D. H. and Walsh, C. T. 1991. Existence of two D-Alanine:D-Alanine Ligases in *Escherichia coli*: cloning and sequencing of the *ddlA* gene and purification and characterization of the DdlA and DdlB enzymes. *Biochemistry* **30**:1673–1682.

Zhao, X. and Drlica, K. 2002. Restricting the selection of antibiotic-resistant mutant bacteria: measurement and potential use of the mutant selection window. *Journal of Infectious Diseases* **185**:561–565.

18 Disinfectants (Biocides) Used in Animal Production: Antimicrobial Resistance Considerations

Ross C. Beier, Kenneth M. Bischoff, and Toni L. Poole

Introduction

There is ongoing and increasing concern over antibiotic and biocide resistance. Does cross-resistance between antibiotics and biocides occur (Lambert et al. 2001)?

Antibiotics are chemicals produced by microorganisms and other living systems that are capable of inhibiting growth or killing other microorganisms. There is also a group of chemicals that are referred to as antibacterial agents, but these are not strictly antibiotics. These antibacterial agents are produced by synthesis and not by a living system. Some of these are nalidixic acid, quinolone carboxylic acids, nitrofurans, and sulfonamides (Grayson 1982). Biocides are chemicals that usually inhibit or kill a broad spectrum of microorganisms (White and McDermott 2001). Biocides also are not produced by living systems and are used as disinfectants and antiseptics. An antiseptic prevents or arrests the growth or action of microorganisms on living tissue. A disinfectant frees from infection by destroying disease-producing microorganisms on inanimate objects (Boothe 1998).

Much work is needed to determine the extent and significance of the cause-and-effect relationships between biocide use and antibiotic resistance. This chapter attempts to bring together and put in perspective the information on this issue that is available at this time, to identify research gaps, and to propose useful research directions for the future.

Antibiotic Resistance

Since the middle to late 1990s, a widespread concern has developed over the possibility that the use of antimicrobial agents in food animal production might select antibiotic- resistant bacteria that could in turn compromise antimicrobial therapy in humans. In 1997, a group from Denmark published an exhaustive study on this issue (Seyfarth et al. 1997). The animal- and human-derived bacterial strains evaluated were tested against 22 antimicrobial agents used in both veterinary and human medicine. They were phage-typed, all resistant strains were evaluated for plasmid content, and ribotyping was performed. The investigators concluded that although *Salmonella* Typhimurium isolated from animals and humans showed antimicrobial resistance, multiple resistance was most often acquired outside of Denmark (Seyfarth et al. 1997).

Mention of a trade name, proprietary product, or specific equipment does not constitute a guarantee or warranty by the U.S. Department of Agriculture and does not imply its approval to the exclusion of other products that may be suitable.

The zoonotic nature of *Salmonella* (the transmission of *Salmonella* infections under normal conditions from animals to humans; Acha and Szyfres 1987) is well established. It is clear that with increased antibiotic use in animals there was a similar increase in the number of isolated antibiotic-resistant bacterial strains (Levy et al. 1987). An increase in human cases of antimicrobial-resistant salmonellosis in Denmark was associated with increasing levels of infection in pigs and poultry in Danish animal husbandry (Baggesen and Wegener 1994, Wegener et al. 1994). Workers in other countries have also demonstrated high levels of *S.* Typhimurium resistance to antimicrobial agents in animal and human isolates (van Leeuwen et al. 1979, Threlfall 1992, Threlfall et al. 1993, Espinasse 1993), and antimicrobial-resistant *S.* Typhimurium was demonstrated to be associated with the use of antimicrobial agents in animal herds (van Leeuwen et al. 1979, Threlfall 1992).

In the foodborne zoonoses, the gastrointestinal tract of an animal serves not only as a reservoir of bacterial propagation but also as the primary spot for exchange of genetic information (van den Bogaard 1997) and amplification of organisms within the animal population (Teale 2002). The dependence on large amounts of antibiotics for disease control in animal production provides a favorable environment for the development and spread of antimicrobial-resistant bacteria. Aarestrup and Wegener (1999) stated in their review that "There is an urgent need to implement strategies for prudent use of antibiotics in food animal production to prevent further increases in the occurrence of antimicrobial resistance in foodborne human pathogenic bacteria such as *Campylobacter* and *E. coli*" (p. 639) and "the stage is set for the occurrence of noncurable zoonoses" (p. 641). Before Aarestrup and Wegener's review was published, their concerns became reality. A variant of *S.* Typhimurium strain DT104, resistant to quinolones, was responsible for the death of two people in 1998 (Ferber 2000). Researchers traced the strain of *Salmonella* from the outbreak, which included 25 culture-confirmed cases and nine hospitalized patients in addition to the two deaths, back to a single Danish swine herd that had been treated with quinolones (Mølbak et al. 1999). The investigations carried out in the DT104 outbreak clearly documented the spread of zoonotic bacteria from animals to humans (Mølbak et al. 1999).

Biocide Resistance

In addition to the use of antibiotics, strategies to control infection include the use of biocides in the form of antiseptics and disinfectants. The Royal Pharmaceutical Society of Great Britain (1997) proposed a link between antibiotic and biocide resistance. Since the first use of a chlorinated lime hand wash by Semmelweiss 150 years ago, these agents have become indispensable in infection control and biosecurity programs (White and McDermott 2001). Biocides do play an important role in infection control and appear to be effective, but there is a growing concern about the increasing use of biocides and potential resistance development, including cross-resistance to clinically important antibiotics (Heath et al. 1998, McMurry et al. 1998a, Russell et al. 1998, Heath et al. 1999, Levy et al. 1999, Tattawasart et al. 1999, Levy 2000, Levy 2001). The development of antibiotic resistance may ultimately influence the efficacy of antimicrobial therapy in humans (Donnelly et al. 1996, Piddock 1996, Das et al. 1997, Wegener 1998, Aarestrup and Wegener 1999, Mølbak et al. 1999).

Some researchers have focused on the possibility that the use of biocides will select for resistant bacteria, which may also have antibiotic resistance. In some cases, this is already known to occur. Bacterial resistance to biocides was described in the 1950s and appears to be increasing (Russell 2002). The cationic agents, quaternary amine compounds (QACs), chlorhexidine, diamidines, and acridines along with the antimicrobial agent triclosan have been implicated as possible causes for selection of bacterial strains with low-level antibiotic resistance (Russell 2002).

Triclosan

Triclosan is an antiseptic with broad-spectrum antibacterial and antifungal activity (Furia and Schenkel 1968, Bhargava and Leonard 1996). It has been used in underarm deodorants and deodorant soaps since the 1960s, it was first used during 1972 in surgical scrubs, and it was used in a toothpaste in Europe during 1985 (Jones et al. 2000). Triclosan is used today in a large number of products including deodorants, deodorant soaps, hand creams, hand lotions, hand soaps, hand washes, mouthwashes, shower gels, surgical scrubs, toothpastes (Bhargava and Leonard 1996, DeSalva et al. 1989), body washes, dish soaps, and facial cleaners. A review by Jones et al. (2000) presents an overview of the effectiveness and safety of triclosan in health care settings. The review states that triclosan has demonstrated broad-spectrum antimicrobial effectiveness and utility in the health care setting (Jones et al. 2000). However, the use of triclosan in soap decreases triclosan's effectiveness against *E. coli* (Levy 2001).

Triclosan was long thought to be a nonspecific biocide that disrupted cell membranes, leaving bacteria unable to assimilate nutrients and proliferate (Regös and Hitz 1974, Vischer and Regös 1974, Regös et al. 1979). The view of triclosan as a nonspecific biocide is what propelled its use in consumer products, with the understanding that the emergence of resistant bacterial strains was unlikely (Heath et al. 1998). Despite its widespread use, triclosan was previously known to readily cause *E. coli* isolates to become resistant (Furia and Schenkel 1968). Recent research demonstrates that triclosan acts at specific targets: the *fabI* gene (Heath et al. 1998, McMurry et al. 1998a, Heath et al. 1999, Hoang and Schweizer 1999, McMurry et al. 1999, Heath et al. 2000) and the *fabK* gene (Heath and Rock 2000). These studies reveal that resistant *E. coli* strains do arise from missense mutations. The basis of triclosan activity has been shown to be as an inhibitor of fatty acid biosynthesis (Heath et al. 1998, Heath et al. 1999, Levy et al. 1999).

Triclosan is a substrate of a multidrug efflux pump in *Pseudomonas aeruginosa* (Chuanchuen et al. 2001) and in laboratory and clinical strains of *E. coli* (McMurry et al. 1998b). Efflux is one of the common mechanisms resulting in antibiotic resistance. Efflux systems are responsible for bacterial resistance to macrolides (Ma et al. 1993, Sanchez et al. 1997, Aires et al. 1999, Moore et al. 1999, Zarantonelli et al. 1999, Li et al. 2000, Masuda et al. 2000a), β-lactams (Masuda et al. 1999, Srikumar et al. 1999, Masuda et al. 2000b, Mazzariol et al. 2000), aminoglycosides (Aires et al. 1999, Moore et al. 1999, Rosenberg et al. 2000), tetracycline (Nikaido 1998, Masuda et al. 2000a) and fluoroquinolones (Poole 2000a, Poole 2000b). Susceptibility studies on 186 isolates of methicillin-resistant (MRSA) and methicillin-sensitive (MSSA) *Staphylococcus aureus* determined that 14 isolates (7.5%) had an elevated minimal inhibitory concentration (MIC) for triclosan (Bamber and Neal 1999). It was suggested that susceptibility testing of the *Staphylococci* against triclosan might be indicated, especially in the assessment of treatment failure (Bamber and Neal 1999). Triclosan will select for resistance to itself, but it also will select for multidrug-resistant bacteria (Chuanchuen et al. 2001).

Chlorhexidine

Chlorhexidine, 1,1'-hexamethylenebis[5-(4-chlorophenyl)biguanide], was first described by Davies et al. (1954). Chlorhexidine is a biguanide that is widely used in medical and veterinary applications as a topical antiseptic and disinfectant (Nicoletii et al. 1993). Antimicrobial activity of chlorhexidine and its mechanism of action have been reviewed (Hugo 1992, Russell 1986, Carlotti and Maffart 1996). Briefly, chlorhexidine interferes with cellular membrane function by introducing alterations in bacterial cell membranes (Boothe 1998); this seems to occur primarily on the cytoplasmic membrane. The binding of biguanide groups to membrane phospholipids induces structural modifications with concomitant leakage of intracellular components (Carlotti and Maffart 1996).

A hand-washing study on the effects of pathogenicity of *S.* Typhimurium showed that the amount of pathogen reduction by the use of chlorhexidine compared with soap or alcohol was negligible (Rotter et al. 1988). This result leads one to look further in an attempt to evaluate the effectiveness of biocides and how they affect the resistance characteristics of various bacteria. Cytological observations on chlorhexidine-sensitive and chlorhexidine-resistant *Pseudomonas stutzeri* by various microscopy analyses suggest that changes in the outer membrane are involved when *P. stutzeri* becomes resistant to chlorhexidine (Tattawasart et al. 2000).

Researchers also have evaluated the resistance of other clinical isolates to chlorhexidine. Unstable resistance to chlorhexidine was induced in *Proteus mirabilis* (Prince et al. 1978). During attempts to induce resistance in *E. coli* and *S. aureus* isolates to chlorhexidine, a few resistant colonies were isolated, but the resistance was not stable and the cells quickly reverted to their natural sensitivity (Fitzgerald et al. 1992). However, stable resistance to chlorhexidine was developed in *Serratia marcescens* and *Pseudomonas cepacia* (Prince et al. 1978), in *P. stutzeri* (Russell et al. 1998, Tattawasart et al. 1999), and in *P. aeruginosa* and *S. marcescens* isolates (Nicoletti et al. 1993). Both *P. aeruginosa* and *S. marcescens* chlorhexidine-resistant strains were also resistant to cetyl trimethyl ammonium bromide. The *P. stutzeri* strains that developed stable resistance to chlorhexidine also showed reduced sensitivity to triclosan (Tattawasart et al. 1999). With the ability to develop stable resistance among different bacteria to chlorhexidine, and with some of these same resistant isolates showing reduced sensitivity to other biocides, it would be beneficial to compare results obtained from evaluating clinical isolates and animal-derived isolates for chlorhexidine resistance.

Clinical Isolates and Chlorhexidine

Clinical isolates of *P. mirabilis*, *S. marcescens*, and *P. cepacia* were shown to display increased resistance to chlorhexidine gluconate (Prince et al. 1978). Prince et al. found that the development of drug resistance may be an important factor in the choice of a skin antiseptic. During studies of 30 strains of *S. aureus*, resistant to both methicillin and gentamicin (MGRSA), Brumfitt et al. (1985) demonstrated that resistance to chlorhexidine was increased in the isolates of *S. aureus* resistant to MGRSA. In another study of hospital-derived clinical isolates of *S. aureus* that were MSSA and MRSA, the MRSA isolates exhibited low-level resistance to chlorhexidine with MICs of up to threefold over controls (Suller and Russell 1999). Suller and Russell tried to select for increased resistance with staphylococcal strains against chlorhexidine but observed only unstable resistance. In studies using patient-unique clinical isolates, however, chlorhexidine was significantly less active against MRSA isolates compared with MSSA isolates (Block et al. 2000). Mengistu et al. (1999) showed that a significant number of clinical Gram-negative bacterial isolates were not growth inhibited by chlorhexidine at concentrations used for disinfection of wounds or instruments.

Since the early 1990s, there has been concern that antiseptic resistance in MRSA may be an emerging problem (Cookson 1994). There was a quantifiable reduction in killing of MRSA isolates dried on stainless steel surfaces by chlorhexidine compared with that observed for MSSA isolates (Block et al. 2000). Various European approaches for testing surface disinfectants have been reviewed (van Klingeren 1995). The AOAC (Association of Analytical Chemists) hard-surface carrier test (Hamilton et al. 1995) is widely accepted and used in the United States to evaluate surface contamination. A hybrid of these techniques was used by Block and coworkers. It was shown that surface-dried *S. aureus* is not fully inactivated even after prolonged exposure to chlorhexidine (Block et al. 2000).

Studies on antibiotic and chlorhexidine susceptibility of 70 clinical isolates showed that Gram-negative bacteria resistant to antibiotics also had increased chlorhexidine resistance (Kõljalg et al. 2002). This suggests a common resistance mechanism operating on these two groups of chemicals. Continuing surveillance is needed to detect chlorhexidine-resistant bacteria and to compare antibiotic resistance in these bacteria.

Animal-Derived Isolates and Chlorhexidine

There is very limited information concerning chlorhexidine and animal-derived isolates. However, a link between the development of resistance to antiseptics and antibiotics has been shown in animal-derived isolates (Maris 1991). The *in vitro* efficacy of chlorhexidine was higher against American Type Culture Collection (ATTC) strains of *S. aureus* and *P. aeruginosa* than it was against isolates obtained from cases of canine and feline soft-tissue diseases (Odore et al. 2000). That same trend was observed between ATCC strains and the human clinical isolates discussed in the section above. We need to examine any chlorhexidine-resistance similarities between animal and human isolates and antibiotic resistance patterns.

In our laboratory, we have evaluated 89 β-hemolytic enterotoxigenic *E. coli* strains isolated from neonatal pigs with diarrhea obtained from five farms in Oklahoma during 1998 (Bischoff et al. 2002). Of the 89 *E. coli* isolates tested, 43.8% showed low and stable resistance to chlorhexidine digluconate in comparison with the laboratory control strain (R. Beier, K. Bischoff, and D. Nisbet, unpublished data). These chlorhexidine-resistant isolates also have multiple antibiotic resistance, many are highly correlated to one ribogroup, and they correlate to multiple virulence factors (R. Beier, K. Bischoff, and D. Nisbet, unpublished data). These data are further evidence that the trend seen in human clinical isolates is also seen in animal-derived isolates. Based on these results, further surveillance is needed to detect bacteria that demonstrate decreased susceptibilities to chlorhexidine in animal production. Emphasis also is needed on the comparison of other biocide resistant organisms and their susceptibilities to antibiotics used in human medicine.

QACs

QACs have been used as biocides since the 1930s (Russell 2002). A survey of the disinfectants used in the U.K. food industry acknowledged the persistence of *Listeria monocytogenes* and

E. coli throughout the industry (Holah et al. 2002) and documented that QACs were the most popular of all the disinfectants used in the United Kingdom. However, these workers felt that the observed *L. monocytogenes* and *E. coli* persistence was not the result of disinfectant use.

The observation that certain strains of *S. aureus* possess a plasmid that carries genes for resistance to both gentamicin and antiseptics of the QAC class was very important (Brumfitt et al. 1985). It was postulated that strains possessing the *qac* genes may have enhanced survival in the clinical environment (Russell et al. 1998). Ninety-seven epidemiologically unrelated strains of *L. monocytogenes* were investigated for sensitivities to QACs (Mereghetti et al. 2000). These isolates were obtained from both the environment and foods. None of the isolates was obtained from humans or animals. Seven strains were QAC-resistant. The intrinsic resistance seen here was not plasmid associated and was not caused by the *qacA* and *smr* genes (Mereghetti et al. 2000). We can see that the use of these QAC compounds may be just as important in developing enhanced survival of microorganisms in the food industry setting as they are in the clinical or animal production environments.

Clinical Isolates and QACs

Clinical isolates of *P. cepacia* were resistant to QAC and benzalkonium chloride (BC). Clinical isolates of *P. mirabilis* and *S. marcescens* developed resistance to chlorhexidine gluconate and also show cross-resistance to BC (Prince et al. 1978). Prince et al. (1978) confirmed the previous work of others and suggested that the significance of their findings should not be underestimated. Work on MGRSA clinical isolates also confirmed earlier work showing that *S. aureus* possesses a plasmid that carries genes for resistance to both gentamicin and antiseptics of the QAC class (Brumfitt et al. 1985). *Pseudomonas stutzeri* strains showed stable resistance to the QAC, cetyl pyridinium chloride, and *P. aeruginosa* NCIMB 8626 showed a high level of resistance to cetyl pyridinium chloride (Tattawasart et al. 1999).

Clinical isolates of *S. aureus* ($n = 61$) and coagulase-negative staphylococcal strains ($n = 177$) were obtained from blood and skin of both cancer and human immunodeficiency virus–positive patients (Sidhu et al. 2002). MIC analyses demonstrated that 50% of these isolates were resistant to the QAC BC. The work pointed out that *qac* resistance genes are common and that cross-resistance to disinfectants and penicillin is frequent in Norway clinical isolates. It was also noted that the presence of either resistance determinant (i.e., antibiotic or biocide resistance) selects for the other based on the frequency of antibiotic resistance among the BC-resistant strains (Sidhu et al. 2002).

Animal-Derived Isolates and QACs

A *S. aureus* strain, designated I. 24, was isolated from broiler flocks and characterized by pulsed-field gel electrophoresis (McCullagh et al. 1998). This strain was the predominant type recovered from broilers showing disease signs in Northern Ireland (McCullagh et al. 1998). I. 24 and other *S. aureus* strains obtained from the poultry industry were used to evaluate the efficacy of 18 commercial disinfectants (Rodgers et al. 2001). In the presence of hatchery organic matter, one of the six commercial QACs tested was not effective against I. 24 when tested at the manufacturer's recommended concentration (Rodgers et al. 2001). The results emphasized the importance of proper usage and application areas.

Recently, we evaluated the QAC susceptibility of β-hemolytic enterotoxigenic *E. coli* strains isolated from neonatal pigs with diarrhea (R. Beier, unpublished data). The commercial QACs tested comprised a number of different individual QAC components. Of the 89 *E. coli* isolates tested, approximately 29% showed low resistance to the individual QAC components, and approximately 8% showed low resistance to the commercial QACs tested. The resistance of these isolates to the QAC components and commercial QACs was not as stable as seen with chlorhexidine. At times high resistance to the QACs was observed, but it was unstable, and on regrowth, the isolates reverted back to their natural sensitivity.

Disinfectant Resistance Mechanisms

Mechanisms of bacterial resistance to antibiotics have been widely studied, but mechanisms of resistance to biocides are less well understood (Tattawasart et al. 1999). Resistance to biocides had been thought to be less common compared with antibiotic resistance and, therefore, have been studied less. Researchers today are finding that biocide resistance is more common. For instance, the work by Sidhu et al. (2002) shows that 50% of the clinical isolates (*S. aureus* and coagulase-negative staphylococcal strains) they studied were resistant to QACs. Molecular and genetic studies demonstrated that *qac* resistance genes were common in their isolates. Mechanisms of resistance typically involve one or more of the following: alteration of the target in the bacterial cell, enzymatic modification or destruction of the chemical, or limitation of the chemical accumulation as a result of exclusion or efflux (Poole 2002). Resistance to biocides is considered to be mostly caused by intrinsic cellular mechanisms rather than acquired resistance (Russell 1998, White and McDermott 2001). The intrinsic resistance of *P. aeruginosa* to inhibitors of fatty acid biosynthesis is caused by efflux (Schweizer 1998). However, acquired resistance mechanisms are important in some cases (Russell 1998), as plasmid-carrying genes have been shown to be important in some biocide resistance (Brumfitt et al. 1985, Russell et al. 1998, Sidhu et al. 2002). Certainly, acquired resistance plays a larger role than expected and is important in resistance to QACs and cross-resistance to antibiotics in clinical isolates from Norway (Sidhu et al. 2002).

Resistance caused by efflux is very common with many antibiotics and biocides (Levy 2002, Poole 2002), but target alteration and impermeability are also important mechanisms in bacterial resistance (Poole 2002). Some clinical isolates with QAC resistance possess cross-resistance to certain antibiotics (Sidhu et al. 2002), and it is reported that triclosan will select for multidrug-resistant bacteria (Chuanchuen et al. 2001). There appears to be a link between the development of resistance to antiseptics and antibiotics in animal-derived isolates (Maris 1991). Sidhu et al. (2002) indicated that the frequency of antibiotic resistance among benzalkonium chloride-resistant *S. aureus* isolates suggests that the presence of either resistance determinant selects for the other.

Summary and Conclusions

The investigations carried out in the *S.* Typhimurium DT104 outbreak in Denmark well document the spread of zoonotic bacteria from animals to humans. Clearly this shows that bacteria can develop antibiotic resistance in food animals and can be a problem for humans. There is documented concern about the increasing use of biocides and potential resistance development, as well as cross-resistance to clinically important antibiotics. It was shown that the biocide triclosan will select for resistance to itself, as well as selecting for multidrug resistance in bacteria. This suggests common resistance mechanisms between the biocides and antibiotics. A common mechanism for resistance is efflux, and bacteria have been shown to use efflux for resistance to both antibiotics and biocides, thereby opening the door to cross-resistance.

Since the early 1990s there has been concern that antiseptic resistance in MRSA may be an emerging problem. A number of clinical Gram-negative bacterial isolates are not inhibited by chlorhexidine at concentrations used for disinfection of wounds or instruments. What is also troublesome is that surface-dried *S. aureus* is not fully inactivated after prolonged exposure to chlorhexidine. In studies using patient-unique clinical isolates, chlorhexidine was significantly less active against MRSA isolates when compared with MSSA isolates. In addition, *P. stutzeri* strains that develop stable resistance to chlorhexidine show reduced sensitivity to triclosan. With the ability to develop stable resistance among different bacteria to chlorhexidine, and with some of these same resistant isolates showing reduced sensitivity to other biocides, it would be prudent to look at a wide array of isolates for biocide resistance, using appropriate surveillance methods.

The *in vitro* efficacy of chlorhexidine has proven to be higher against ATCC strains of *S. aureus* and *P. aeruginosa* than it is against isolates obtained from cases of canine and feline soft-tissue diseases. A link between the development of resistance to antiseptics and antibiotics has been shown in animal-derived isolates, as is seen with clinical isolates. A high percentage of β-hemolytic enterotoxigenic *E. coli* strains isolated from neonatal pigs shows low but stable resistance to chlorhexidine. This suggests that further surveillance is needed to detect bacteria that demonstrate decreased susceptibilities to chlorhexidine in animal production.

The use of QAC compounds may be just as important in developing enhanced bacterial survival in the food industry as they are in clinical or animal production environments. In Norwegian clinical isolates, *qac* resistance genes are common, and cross-resistance to disinfectants and penicillin is frequent. In addition, the presence of either resistance determinant (antibiotic or biocide resistance) selects for the other based on the frequency of antibiotic resistance among benzalkonium chloride-resistant strains.

We need to develop much more data derived from animal isolates to follow the apparent trend of increased resistance to chlorhexidine and the QACs. Continuing and improving surveillance is needed to detect biocide-resistant bacteria. Research on these isolated resistant bacteria should be carried out to evaluate and understand any cross-resistance they may have to antibiotics.

We also should conduct biocide-resistance surveillance on opportunistic pathogens that commonly evolve in animal production. There should be a strong program for evaluating *E. coli* O157:H7 and *S.* Typhimurium DT104 in animal production environments for resistance to biocides and for evaluating their cross-resistance to antibiotics. This is an important issue that is currently not being adequately investigated. The biocide resistance issue may be almost as important as the antibiotic resistance issue because today we are using large amounts of a wide array of biocides in animal production, in the clinical environment, and in the food industry. Many of the biocides used are the same or are chemically very similar throughout the spectrum of different applications, whether they are used in animal production, clinical applications, or as sanitizers in food processing plants. We already see cross-resistance between biocides and antibiotics. Significant outbreaks of one or more multidrug resistant pathogens, aided by concomitant biocide resistance, would appear to be possible if not likely in the near future. Therefore, it is of prime importance to understand biocide resistance patterns to predict problems in food animals and humans. To this end there is a need for future research to be directed at surveillance, epidemiology, resistance mechanisms, cross-resistance patterns, and improving the knowledge base of biocides implicated in resistance.

References

Aarestrup, F. M. and Wegener, H. C. 1999. The effects of antibiotic usage in food animals on the development of antimicrobial resistance of importance for humans in *Campylobacter* and *Escherichia coli*. *Microbes and Infection* **1**:639–644.

Acha, P. N. and Szyfres, B. 1987. Zoonoses and Communicable Diseases Common to Man and Animals, 2nd ed. Scientific Publication No. 503. Washington, D.C.: Pan American Health Organization, Pan American Sanitary Bureau, Regional Office of the World Health Organization.

Aires, J. R., Kohler, T., Nikaido, H. and Plesiat, P. 1999. Involvement of an active efflux system in the natural resistance of *Pseudomonas aeruginosa* to aminoglycosides. *Antimicrobial Agents and Chemotherapy* **43**:2624–2628.

Baggesen, D. L. and Wegener, H. C. 1994. Phage types of *Salmonella enterica* spp. *enterica* serovar Typhimurium isolated from production animals and humans in Denmark. *Acta Veterinaria Scandinavica* **35**:349–354.

Bamber, A. I. and Neal, T. J. 1999. An assessment of triclosan susceptibility in methicillin-resistant and methicillin-sensitive *Staphylococcus aureus*. *Journal of Hospital Infection* **41**:107–109.

Bhargava, H. N. and Leonard, P. A. 1996. Triclosan: applications and safety. *American Journal of Infection Control* **24**:209–218.

Bischoff, K. M., White, D. G., McDermott, P. F., Zhao, S., Gaines, S., Maurer, J. J. and Nisbet, D. J. 2002. Characterization of chloramphenicol resistance in beta-hemolytic *Escherichia coli* associated with diarrhea in neonatal swine. *Journal of Clinical Microbiology* **40**:389–394.

Block, C., Robenshtok, E., Simhon, A. and Shapiro, M. 2000. Evaluation of chlorhexidine and povidone iodine activity against methicillin-resistant *Staphylococcus aureus* and vancomycin-resistant *Enterococcus faecalis* using a surface test. *Journal of Hospital Infection* **46**:147–152.

Brumfitt, W., Dixson, S. and Hamilton-Miller, J. M. 1985. Resistance to antiseptics in methicillin and gentamicin resistant *Staphylococcus aureus*. *Lancet* **1**:1442–1443.

Boothe, H. W. 1998. Antiseptics and disinfectants. *Veterinary Clinics of North America Small Animal Practice* **28**:233–248.

Carlotti, D. N. and Maffart, P. 1996. La chlorhexidine, revue bibliographique. *Pratique Médicale et Chirurgicale de l'Animal de Compagnie* **31**:553–563.

Chuanchuen, R., Beinlich, K., Hoang, T. T., Becher, A., Karkhoff-Schweizer, R. R. and Schweizer, H. P. 2001. Cross-resistance between Triclosan and antibiotics in *Pseudomonas aeruginosa* is mediated by multidrug efflux pumps: exposure of a susceptible mutant strain to triclosan selects *nfxB* mutants overexpressing *MexCD-OprJ*. *Antimicrobial Agents and Chemotherapy* **45**:428–432.

Cookson, B. D. 1994. Antiseptic resistance in methicillin resistant *Staphylococcus aureus*: an emerging problem? *In* Staphylococci and Staphylococcal Infections, Proceedings of the 7th International Symposium, Stockhom, Sweden, June 29–July 3, 1992. New York: G. Fischer, pp. 227–234.

Das, I., Fraise, A. and Wise, R. 1997. Are glycopeptide-resistant enterococci in animals a threat to human beings? *Lancet* **349**:997–998.

Davies, G. E., Francis, J., Martin, A. R., Rose, F. L. and Swain, G. 1954. 1:6-Di-4′-Chlorophenyldiguanidohexane ("hibitane"). Laboratory investigation of a new antibacterial agent of high potency. *British Journal of Pharmacology* **9**:192–196.

DeSalva, S. J., Kong, B. M. and Lin, Y. J. 1989. Triclosan: a safety profile. *American Journal of Dentistry* **2**:185–196.

Donnelly, J. P., Voss, A., Witte, W. and Murray, B. E. 1996. Does the use in animals of antimicrobial agents, including glycopeptide antibiotics, influence the efficacy of antimicrobial therapy in humans? *Journal of Antimicrobial Chemotherapy* **37**:389–397.

Espinasse, J. 1993. Responsible use of antimicrobials in veterinary medicine: perspectives in France. *Veterinary Microbiology* **35**:289–301.

Ferber, D. 2000. Superbugs on the hoof? *Science* **288**:792–794.

Fitzgerald, K. A., Davies, A. and Russell, A. D. 1992. Sensitivity and resistance of *Escherichia coli* and *Staphylococcus aureus* to chlorhexidine. *Letters in Applied Microbiology* **14**:22–36.

Furia, T. E. and Schenkel, A. G. 1968. New, broad spectrum bacteriostat. *Soap/Cosmetics/Chemical Specialties* **44**:47–50, 116–122.

Grayson, M. (Ed.). 1982. Antibiotics, Chemotherapeutics, and Antibacterial Agents for Disease Control. New York: Wiley.

Hamilton, M. A., DeVries, T. A. and Rubino, J. R. 1995. Hard surface carrier test as a quantitative test of disinfection: a collaborative study. *Journal of AOAC International* **78**:1102–1109.

Heath, R. J. and Rock, C. O. 2000. A triclosan-resistant bacterial enzyme. *Nature* **406**:145–146.

Heath, R. J., Li, J., Roland, G. E. and Rock, C. O. 2000. Inhibition of the *Staphylococcus aureus* NADPH-dependent enoyl-acyl carrier protein reductase by triclosan and hexachlorophene. *Journal of Biological Chemistry* **275**:4654–4659.

Heath, R. J., Rubin, J. R., Holland, D. R., Zhang, E., Snow, M. E. and Rock, C. O. 1999. Mechanism of triclosan inhibition of bacterial fatty acid synthesis. *Journal of Biological Chemistry* **274**:11110–11114.

Heath, R. J., Yu, Y.-T., Shapiro, M. A., Olson, E. and Rock, C. O. 1998. Broad spectrum antimicrobial biocides target the FabI component of fatty acid synthesis. *Journal of Biological Chemistry* **273**:30316–30320.

Hoang, T. T. and Schweizer, H. P. 1999. Characterization of *Pseudomonas aeruginosa* enoyl-acyl carrier protein reductase (FabI): a target for the antimicrobial triclosan and its role in acylated homoserine lactone synthesis. *Journal of Bacteriology* **181**:5489–5497.

Holah, J. T., Taylor, J. H., Dawson, D. J. and Hall, K. E. 2002. Biocide use in the food industry and the disinfectant resistance of persistent strains of *Listeria monocytogenes* and *Escherichia coli*. *Journal of Applied Microbiology* **92(Suppl.)**:111S–120S.

Hugo, W. B. 1992. Disinfection mechanisms, pp. 187–210. *In* Principles and Practice of Disinfection, Preservation and Sterilization, A. D. Russell, W. B. Hugo, G. A. J. Ayliffe (Eds.). Oxford: Blackwell Scientific.

Jones, R. D., Jampani, H. B., Newman, J. L. and Lee, A. S. 2000. Triclosan: a review of effectiveness and safety in health care settings. *American Journal of Infection Control* **28**:184–196.

Kõljalg, S., Naaber, P. and Mikelsaar, M. 2002. Antibiotic resistance as an indicator of bacterial chlorhexidine susceptibility. *Journal of Hospital Infection* **51**:106–113.

Lambert, R. J. W., Joynson, J. and Forbes, B. 2001. The relationships and susceptibilities of some industrial, laboratory and clinical isolates of *Pseudomonas aeruginosa* to some antibiotics and biocides. *Journal of Applied Microbiology* **91**:972–984.

Levy, C. W., Roujeinikova, A., Sedelnikova, S., Baker, P. J., Stuitje, A. R., Slabas, A. R., Rice, D. W. and Rafferty, J. B. 1999. Molecular basis of triclosan activity. *Nature* **398**:383–384.

Levy, S. B. 2000. Antibiotic and antiseptic resistance: impact on public health. *Pediatric Infectious Disease Journal* **19**:S120–S122.

Levy, S. B. 2001. Antibacterial household products: cause for concern. *Emerging Infectious Diseases* **7**:512–515.

Levy, S. B. 2002. Active efflux, a common mechanism for biocide and antibiotic resistance. *Journal of Applied Microbiology* **92(Suppl.)**:65S–71S.

Levy, S. B., Burke, J. P. and Wallace, C. K. 1987. Antibiotic use and antibiotic resistance worldwide. *Reviews of Infectious Diseases* **9(Suppl. 3)**:S231–S316.

Li, X.-Z., Zhang, L. and Poole, K. 2000. Interplay between the *MexA-MexB-OprM* multidrug efflux system and the outer membrane barrier in the multiple antibiotic resistance of *Pseudomonas aeruginosa*. *Journal of Antimicrobial Chemotherapy* **45**:433–436.

Ma, D., Cook, D. N., Alberti, M., Pon, N. G., Nikaido, H. and Hearst, J. E. 1993. Molecular cloning and characterization of *acrA* and *acrE* genes of *Escherichia coli*. *Journal of Bacteriology* **175**:6299–6313.

Maris, P. 1991. Resistance of 700 Gram-negative bacterial strains to antiseptics and antibiotics. *Annuales de Recherches Veterinaires* **22**:11–23.

Masuda, N., Gotoh, N., Ishii, C., Sakagawa, E., Ohya, S. and Nishino, T. 1999. Interplay between chromosomal β-lactamase and the *MexAB-OprM* efflux system in intrinsic resistance to β-lactams in *Pseudomonas aeruginosa*. *Antimicrobial Agents and Chemotherapy* **43**:400–402.

Masuda, N., Sakagawa, E., Ohya, S., Gotoh, N., Tsujimoto, H. and Nishino, T. 2000a. Contribution of the *MexX-MexY-OprM* efflux system to intrinsic resistance in *Pseudomonas aeruginosa*. *Antimicrobial Agents and Chemotherapy* **44**:2242–2246.

Masuda, N., Sakagawa, E., Ohya, S., Gotoh, N., Tsujimoto, H. and Nishino, T. 2000b. Substrate specificities of *MexAB-OprM*, *MexCD-OprJ*, and *MexXY-OprM* efflux pumps in *Pseudomonas aeruginosa*. *Antimicrobial Agents and Chemotherapy* **44**:3322–3327.

Mazzariol, A., Cornaglia, G. and Nikaido, H. 2000. Contributions of the *AmpC* β-lactamase and the *AcrAB* multidrug efflux system in intrinsic resistance of *Escherichia coli* K-12 to β-lactams. *Antimicrobial Agents and Chemotherapy* **44**:1387–1390.

McCullagh, J. J., McNamee, P. T., Smyth, J. A. and Ball, H. J. 1998. The use of pulsed-field gel electrophoresis to investigate the epidemiology of *Staphylococcus aureus* infection in commercial broiler flocks. *Veterinary Microbiology* **63**:275–281.

McMurry, L. M., McDermott, P. F. and Levy, S. B. 1999. Genetic evidence that InhA of *Mycobacterium smegmatis* is a target for triclosan. *Antimicrobial Agents and Chemotherapy* **43**:711–713.

McMurry, L. M., Oethinger, M. and Levy, S. B. 1998a. Triclosan targets lipid synthesis. *Nature* **394**:531–532.

McMurry, L. M., Oethinger, M. and Levy, S. B. 1998b. Overexpression of *marA*, *soxS*, or *acrAB* produces resistance to triclosan in laboratory and clinical strains of *Escherichia coli*. *FEMS Microbiology Letters* **166**:305–309.

Mengistu, Y., Erge, W. and Bellete, B. 1999. *In vitro* susceptibility of Gram-negative bacterial isolates to chlorhexidine gluconate. *East African Medical Journal* **76**:243–246.

Mereghetti, L., Quentin, R., Marquet-Van der Mee, N. and Audurier, A. 2000. Low sensitivity of *Listeria monocytogenes* to quaternary ammonium compounds. *Applied Environmental Microbiology* **66**:5083–5086.

Mølbak, K., Baggesen, D. L., Aarestrup, F. M., Ebbesen, J. M., Engberg, J., Frydendahl, K., Gerner-Smidt, P., Petersen, A. M. and Wegener, H. C. 1999. An outbreak of multidrug-resistant, quinolone-resistant *Salmonella enterica* serotype Typhimurium DT104. *The New England Journal of Medicine* **341**:1420–1425.

Moore, R. A., DeShazer, D., Reckseidler, S., Weissman, A. and Woods, D. E. 1999. Efflux-mediated aminoglycoside and macrolide resistance in *Burkholderia pseudomallei*. *Antimicrobial Agents and Chemotherapy* **43**:465–470.

Nicoletti, G., Boghossian, V., Gurevitch, F., Borland, R. and Morgenroth, P. 1993. The antimicrobial activity *in vitro* of chlorhexidine, a mixture of isothiazolinones ('Kathon' CG) and cetyl trimethyl ammonium bromide (CTAB). *Journal of Hospital Infection* **23**:87–111.

Nikaido, H. 1998. Antibiotic resistance caused by Gram-negative multidrug efflux pumps. *Clinical Infectious Diseases* **27(Suppl. 1)**:S32–S41.

Odore, R., Valle, V. C. and Re, G. 2000. Efficacy of chlorhexidine against some strains of cultured and clinically isolated microorganisms. *Veterinary Research Communications* **24**:229–238.

Piddock, L. J. V. 1996. Does the use of antimicrobial agents in veterinary medicine and animal husbandry select antibiotic-resistant bacteria that infect man and compromise antimicrobial chemotherapy? *Journal of Antimicrobial Chemotherapy* **38**:1–3.

Poole, K. 2000a. Efflux-mediated resistance to fluoroquinolones in Gram-negative bacteria. *Antimicrobial Agents and Chemotherapy* **44**:2233–2241.

Poole, K. 2000b. Efflux-mediated resistance to fluoroquinolones in Gram-positive bacteria and the mycobacteria. *Antimicrobial Agents and Chemotherapy* **44**:2595–2599.

Poole, K. 2002. Mechanisms of bacterial biocide and antibiotic resistance. *Journal of Applied Microbiology* **92(Symp. Suppl.)**:55S–64S.

Prince, H. N., Nonemaker, W. S., Norgard, R. C. and Prince, D. L. 1978. Drug resistance studies with topical antiseptics. *Journal of Pharmaceutical Sciences* **67**:1629–1631.

Regös, J. and Hitz, H. R. 1974. Investigations on the mode of action of Triclosan, a broad spectrum antimicrobial agent. *Zentralblatt für Bakteriologie, Parasitenkunde, Infektionskrankheiten und Hygiene—Erste Alteilung Originale—Reihe A: Medizinische Mikrobiologie und Parasitologie* **226**:390–401.

Regös, J., Zak, O., Solf, R., Vischer, W. A. and Weirich, E. G. 1979. Antimicrobial spectrum of Triclosan, a broad-spectrum antimicrobial agent for topical application. II. Comparison with some other antimicrobial agents. *Dermatologica* **158**:72–79.

Rodgers, J. D., McCullagh, J. J., McNamee, P. T., Smyth, J. A. and Ball, H. J. 2001. An investigation into the efficacy of hatchery disinfectants against strains of *Staphylococcus aureus* associated with the poultry industry. *Veterinary Microbiology* **82**:131–140.

Rosenberg, E. Y., Ma, D. and Nikaido, H. 2000. AcrD of *Escherichia coli* is an aminoglycoside efflux pump. *Journal of Bacteriology* **182**:1754–1756.

Rotter, M. L., Hirschl, A. M. and Koller, W. 1988. Effect of chlorhexidine-containing detergent, non-medicated soap or isopropanol and the influence of neutralizer on bacterial pathogenicity. *Journal of Hospital Infection* **11**:220–225.

Royal Pharmaceutical Society of Great Britain. 1997. Resistance to antimicrobial agents: submission to House of Lords subcommittee. *Pharmaceutical Journal* **259**:919–921.

Russell, A. D. 1986. Chlorhexidine: antibacterial action and bacterial resistance. *Infection* **14**:212–215.

Russell, A. D. 1998. Bacterial resistance to disinfectants: present knowledge and future problems. *Journal of Hospital Infection* **43(Suppl.)**:S57–S68.

Russell, A. D. 2002. Introduction of biocides into clinical practice and the impact on antibiotic-resistant bacteria. *Journal of Applied Microbiology* **92(Suppl.)**:121S–135S.

Russell, A. D., Tattawasart, U., Maillard, J.-Y. and Furr, J. R. 1998. Possible link between bacterial resistance and use of antibiotics and biocides. *Antimicrobial Agents and Chemotherapy* **42**:2151.

Sanchez, L., Pan, W., Vinas, M. and Nikaido, H. 1997. The *acrAB* homolog of *Haemophilus influenzae* codes for a functional multidrug efflux pump. *Journal of Bacteriology* **179**:6855–6857.

Schweizer, H. P. 1998. Intrinsic resistance to inhibitors of fatty acid biosynthesis in *Pseudomonas aeruginosa* is due to efflux: application of a novel technique for generation of unmarked chromosomal mutations for the study of efflux systems. *Antimicrobial Agents and Chemotherapy* **42**:394–398.

Seyfarth, A. M., Wegener, H. C. and Frimodt-Møller, N. 1997. Antimicrobial resistance in *Salmonella enterica* subsp. *enterica* serovar Typhimurium from humans and production animals. *Journal of Antimicrobial Chemotherapy* **40**:67–75.

Sidhu, M. S., Heir, E., Leegaard, T., Wiger, K. and Holck, A. 2002. Frequency of disinfectant resistance genes and genetic linkage with β-lactamase transposon Tn*552* among clinical Staphylococci. *Antimicrobial Agents and Chemotherapy* **46**:2797–2803.

Srikumar, R., Tsang, E. and Poole, K. 1999. Contribution of the *MexAB-OprM* multidrug efflux system to the β-lactam resistance of penicillin-binding protein and β-lactamase-derepressed mutants of *Pseudomonas aeruginosa*. *Journal of Antimicrobial Chemotherapy* **44**:537–540.

Suller, M. T. E. and Russell, A. D. 1999. Antibiotic and biocide resistance in methicillin-resistant *Staphylococcus aureus* and vancomycin-resistant enterococcus. *Journal of Hospital Infection* **43**:281–291.

Tattawasart, U., Hann, A. C., Maillard, J.-Y., Furr, J. R. and Russell, A. D. 2000. Cytological changes in chlorhexidine-resistant isolates of *Pseudomonas stutzeri*. *Journal of Antimicrobial Chemotherapy* **24**:145–152.

Tattawasart, U., Maillard, J.-Y., Furr, J. R. and Russell, A. D. 1999. Development of resistance to chlorhexidine diacetate and cetylpyridinium chloride in *Pseudomonas stutzeri* and changes in antibiotic susceptibility. *Journal of Hospital Infection* **42**:219–229.

Teale, C. J. 2002. Antimicrobial resistance and the food chain. *Journal of Applied Microbiology* **92(Suppl.)**:85S–89S.

Threlfall, E. J. 1992. Antibiotics and the selection of food-borne pathogens. *Society for Applied Bacteriology Symposium Series* **21**:96S–102S.

Threlfall, E. J. Rowe, B. and Ward, L. R. 1993. A comparison of multiple drug resistance in salmonellas from humans and food animals in England and Wales, 1981 and 1990. *Epidemiology and infection* **111**:189–197.

van den Bogaard, A. E. 1997. Antimicrobial resistance—relation to human and animal exposure to antibiotics. *Journal of Antimicrobial Chemotherapy* **40**:453–454.

van Klingeren, B. 1995. Disinfectant testing on surfaces. *Journal of Hospital Infection* **30(Suppl.)**:397–408.

van Leeuwen, W. J., van Embden, J., Guinée, P., Kampelmacher, E. H., Manten, A., van Schothorst, M. and Voogd, C. E. 1979. Decrease of drug resistance in *Salmonella* in the Netherlands. *Antimicrobial Agents and Chemotherapy* **16**:237–239.

Vischer, W. A. and Regös, J. 1974. Antimicrobial spectrum of Triclosan, a broad-spectrum antimicrobial agent for topical application. *Zentralblatt für Bakteriologie, Parasitenkunde, Infektionskrankheiten und Hygiene—Erste Abteilung Originale—Reihe A: Medizinische Mikrobiologie und Parasitologie* **226**:376–389.

Wegener, H. C. 1998. Historical yearly usage of glycopeptides for animals and humans: the American-European paradox revisited. *Antimicrobial Agents and Chemotherapy* **42**:3049.

Wegener, H. C., Baggesen, D. L. and Gaarslev, K. 1994. *Salmonella typhimurium* phage types from human salmonellosis in Denmark 1988–1993. *APMIS. Acta Pathologica, Microbiologia, et Immunologica Scandinavica* **102**:521–525.

White, D. G. and McDermott, P. F. 2001. Biocides, drug resistance and microbial evolution. *Current Opinion in Microbiology* **4**:313–317.

Zarantonelli, L., Borthagaray, G., Lee, E. H. and Shafer, W. M. 1999. Decreased azithromycin susceptibility of *Neisseria gonorrhoeae* due to *mtrR* mutations. *Antimicrobial Agents and Chemotherapy* **43**:2468–2472.

19 Prevalence of Antimicrobial-Resistant Bacteria in Retail Foods

David G. White, Shaohua Zhao, Shabbir Simjee, Jianghong Meng, Robert D. Walker, and Patrick F. McDermott

Introduction

Foodborne microbial illnesses are an important public health issue worldwide. Each year in the United States, it is estimated that there are 76 million foodborne illnesses, resulting in economic losses of approximately $6.5–$34.9 billion (1995 US$) (Buzby and Roberts 1997, Mead et al. 1999). Although these illnesses are usually a mild to moderate self-limiting gastroenteritis, invasive diseases and complications may occur. For example, *Campylobacter* infection has been identified as the predominant cause of Guillain-Barré syndrome, whereas *Listeria* can frequently invade the central nervous system, leading to meningoencephalitis, cerebritis, and brain abscesses (Blaser 1997, Lorber 1997). Systemic salmonellae infections can be life threatening, and Shiga toxin–producing *Escherichia coli* (STEC), particularly *E. coli* O157:H7, can cause hemorrhagic colitis (HC) and hemolytic uremic syndrome (HUS; Schroeder et al. 2002).

Campylobacter, *Salmonella*, and *E. coli* (pathogenic and commensal varieties) all colonize the gastrointestinal tracts of a wide range of wild and domestic animals, especially animals raised for human consumption (Meng and Doyle 1997). *Listeria monocytogenes* is an opportunistic pathogen widely distributed in the environment, and its transmission via food has been associated with poultry, dairy products, prepared or processed meats, seafood, and fresh vegetables (Kathariou 2002). Food contamination with these pathogens can occur at multiple steps along the food chain, including production, processing, distribution, and preparation. An additional concern is the growing incidence of antimicrobial-resistant foodborne bacterial pathogens. This chapter will focus on the prevalence of antimicrobial-resistant phenotypes among four of the most relevant foodborne bacterial pathogens: *Salmonella*, *Campylobacter*, *E. coli*, and *Listeria*.

Salmonella

The genus *Salmonella* currently includes more than 2,400 different serotypes. *Salmonella* serotypes are ubiquitous in the environment and can colonize and cause disease in a variety of animals. Non-typhoidal *Salmonella* gastroenteritis in humans usually manifests as a self-limiting diarrhea that does not require antimicrobial therapy. The length of time between ingestion of the pathogen and onset of clinical disease depends on the infecting dose of the bacterium. Symptoms of nausea, vomiting, cramping, and diarrhea typically begin 12–48 hours after ingestion and usually resolve within 2–7 days. There are occasions, however, when salmonellosis can lead to life-threatening systemic infections (enteric fever) that may be fatal if antimicrobials are not promptly administered (Lee et al. 1994). The drugs of choice are fluoroquinolones (e.g., ciprofloxacin) in adults and extended-spectrum cephalosporins (e.g., ceftriaxone) in children (Hohmann 2001).

In the United States it is estimated that approximately 1.4 million cases of salmonellosis occur each year, 95% of which are attributed to foodborne transmission (Mead et al. 1999). The most common animal reservoirs are chickens, turkeys, pigs, and cows, although a number of wild and domestic animals can also harbor the organisms. *S.* Typhimurium, one of the most prevalent serotypes in animals, is also the most common etiology of human infections in the United States, Canada, and Europe (Glynn et al. 1998, Poppe et al. 1998).

Antimicrobial resistance among *Salmonella* isolates is increasing on a global scale (Threlfall et al. 1993, Lee et al. 1994, Yang et al. 1998, Yildirmak et al. 1998). A recent 7-year study in Spain revealed that ampicillin resistance in *Salmonella* species has increased from 8% to 44%, tetracycline resistance from 1% to 42%, chloramphenicol resistance from 1.7% to 26%, and nalidixic acid resistance from 0.1% to 11% (Prats et al. 2000). Similarly, in Great Britain, the reported rates of antimicrobial resistance for *S.* Typhimurium more than doubled between 1981 and 1989 (Threlfall et al. 1993). In the United States, resistance to tetracycline has increased from 9% in 1980 to 24% in 1990, and ampicillin resistance increased from 10% to 14% (Lee et al. 1994). Fluoroquinolones and expanded-spectrum cephalosporins, the primary antimicrobials used for treating human infections caused by multidrug-resistant strains (Cherubin and Eng 1991, Fey et al. 2000), are also showing decreased activity against *Salmonella* isolates (Mølbak et al. 1999, Fey et al. 2000, Winokur et al. 2000).

An important factor in this increase in resistance over the last 10 years has been the epidemic spread through food animals and humans of multidrug-resistant *S.* Typhimurium DT104 (Threlfall et al. 1993, Glynn et al. 1998). DT104 is characterized by resistance to five agents: ampicillin, chloramphenicol or florfenicol, streptomycin, sulfonamides, and tetracycline (ACSSuT). This strain type was first isolated from sea gulls, then cattle, and later from humans in England. Since then, it has been isolated from multiple animal species including poultry, cattle, pigs, sheep, and nondomestic birds (Besser et al. 2000, Hudson et al. 2000). Some strains in Great Britain have acquired additional resistance to trimethoprim and aminoglycosides, as well as decreased susceptibility to fluoroquinolones (Threlfall et al. 1996, Low et al. 1997). The majority of resistant DT104 isolates has a unique multidrug-resistance chromosomal gene cluster encoding the complete ACSSuT phenotype (Threlfall et al. 1996, Ridley and Threlfall 1998, Arcangioli et al. 1999, Briggs and Fratamico 1999). This gene cluster typically consists of a chromosomal locus 12.5 kb in size with flanking integrons (Arcangioli et al. 1999, Briggs and Fratamico 1999). The first integron possesses the *aadA2* gene, conferring resistance to streptomycin and spectinomycin; the second contains the β-lactamase gene bla_{PSE-1} encoding resistance to ampicillin. A gene encoding sulfonamide resistance (*sul-1*) is present within conserved sequences of both integrons. Flanked by these two integron structures are the *flo* efflux gene, conferring cross-resistance to chloramphenicol and florfenicol, and the tetracycline-resistance genes *tetR* and *tetA* (Arcangioli et al. 1999, Briggs and Fratamico 1999). The DT104-like resistance gene cluster has also been described in poultry strains of *S.* Agona (Cloeckaert et al. 2000), indicating that it is transmittable between serotypes. It has been shown experimentally that the DT104 MDR cluster can be efficiently transduced by P22-like phage (Schmieger and Schicklmaier 1999). Upstream of the first integron in the MDR locus is a gene encoding a putative resolvase enzyme, which demonstrates greater than 50% identity with the Tn*3* resolvase family (Arcangioli et al. 1999). This indicates that the MDR antibiotic-resistance gene cluster could be part of a much larger transposon or pathogenicity island.

The presence of resistant *Salmonella* in retail meats has been assessed in numerous studies. In a pilot survey by White et al. (2001), 200 ground meat samples (51 chicken, 50 beef, 50 turkey, and 49 pork) were purchased from retail stores representing three different supermarket chains in the greater Washington, D.C., area between June and September of 1998. Products were fresh, prepackaged meats coming from four poultry and one pork processing plants, and store-ground and packaged beef. Salmonellae were recovered from 41 of 200 (21%) ground meat samples. *Salmonella* was isolated more frequently from poultry (33% of chicken and 24% of turkey samples) than red meats (18% of pork and 6% of beef samples). All salmonellae were susceptible to amikacin, apramycin, ciprofloxacin, and nalidixic acid. Eighty-four percent (38 of 45) of the isolates displayed resistance to at least one antibiotic. The most common resistance observed was to tetracycline (80%), streptomycin (73%), sulfamethoxazole (69%), and to a lesser extent, ampicillin (27%). In addition, 16% displayed resistance to amoxicillin/clavulanic acid, cephalothin, ceftiofur, and ceftriaxone. Ceftiofur- and ceftriaxone-resistant *Salmonella* were isolated from ground turkey, chicken, and beef.

Antimicrobial-resistant *Salmonella* have also been reported from imported foods. Zhao et al. (2003) reported the antimicrobial susceptibilities of 187 *Salmonella* isolates, representing 80 serotypes, recovered from imported foods by Food and Drug Administration field laboratories in 2000. Fifteen (8%) were resistant to at least one antimicrobial, and five (2.7%) were resistant to three or more antimicrobials. Nine isolates exhibited resistance to tetracycline. Four isolates also demonstrated resistance to nalidixic acid—all were isolated from imported catfish or tilapia from Taiwan or Thailand. All four nalidixic acid–resistant *Salmonella* isolates possessed amino acid substitutions at the Ser83 or Asp87 position in DNA gyrase. One *S.* Derby isolated from frozen anchovies imported from Cambodia was resistant to six antimicrobials including ampicillin, amoxicillin/clavulanic acid, chloramphenicol, trimethoprim/sulfamethoxazole, sulfamethoxazole, and tetracycline.

Studies conducted outside the United States have also examined foods for the presence of antimicrobial-resistant salmonellae. One study from Spain determined the extent of antimicrobial resistance among 112 *Salmonella* isolates recovered from 691 samples of frozen and fresh chicken meat (Hernandez et al. 2002). Almost half of the isolates tested (46%) were susceptible to all antimicrobials tested. However, resistance was commonly seen to chloramphenicol (45%), ampicillin (35%), and tetracycline (34%). Resistance to multiple antimicrobials was observed in 44% of the isolates. *S.* Typhimurium isolates tended to be more resistant than other serotypes tested. Another study from Europe, authored by Mammina et al. (2002), investigated the distribution of serotypes and patterns of drug resistance of 206 strains of *Salmonella* isolated in southern Italy. Salmonellae were obtained between 1998 and 2000 from 172 samples of raw foods of animal origin, 22 fecal samples from food animals, and 12 animal feed samples (Mammina et al. 2002). Salmonellae resistant to three or more antimicrobials were considered multi-resistant. Among non-Typhimurium isolates tested, 46 of 122 (38%) strains were categorized as multidrug resistant. *S.* Typhimurium was the predominant serotype recovered, with 35 of 67 (52%) isolates displaying multidrug resistance. The characteristic DT104 antimicrobial resistance phenotype of ACSSuT was identified in 17 of these isolates.

Recently there has been a national emergence of strains of *S.* Newport in the United States, known as Newport-MDRAmpC, which are resistant to at least nine antimicrobials, including extended-spectrum cephalosporins. In the United States, the prevalence of Newport-MDRAmpC among *S.* Newport isolates from humans increased from 0% during

1996–1997 to 26% in 2001. At least 26 states have isolated these strains from humans, cattle, or ground beef. Similar to DT104, these strains were resistant to ampicillin, chloramphenicol, streptomycin, sulfamethoxazole, and tetracycline. In addition, Newport-MDRAmpC isolates were resistant to amoxicillin/clavulanic acid, cephalothin, cefoxitin, and ceftiofur and exhibited decreased susceptibility to ceftriaxone (minimal inhibitory concentration [MIC] $\geq$ 16 μg/mL). In 2001, Newport-MDRAmpC strains were responsible for 2.6% of the more than 1 million estimated *Salmonella* infections in 2001 (Centers for Disease Control 2002b).

Prevention and control of Newport-MDRAmpC infection in humans and cattle requires an understanding of how this strain is introduced onto farms and disseminated among cattle. The management practices that promote its spread are unknown, so it is essential that further research in this area continues. It has been suggested that the use of tetracycline-containing milk replacer as well as other antimicrobials commonly used to treat diseases in dairy cattle may contribute to the dissemination of Newport-MDRAmpC (Quigley et al. 1997, Hornish and Kotarski 2002). Non-antimicrobial management practices also may contribute to the spread of resistant strains. For example, dairy cattle are frequently moved between farms and transported over long distances to abattoirs in other states. This movement of cattle may provide ample opportunity for dissemination of Newport-MDRAmpC among animals. The emergence of Newport-MDRAmpC, and increasing resistance among other foodborne *Salmonella* strains, is an unfavorable development with considerable implications for the future treatment and prevention of *Salmonella* infections in both animals and humans. Although much scientific information is available on this subject, many aspects of the development and dissemination of antimicrobial resistant *Salmonella* remain undecided. As more is learned about the ecology of this important foodborne pathogen, interventions can be developed to limit the transmission of resistant salmonella through the food supply.

Campylobacter

Campylobacter was first recognized as an important enteric pathogen in the 1970s, when improvements in isolation methods linked this bacterium to diarrheal disease. Today it is recognized as one of the leading causes of acute bacterial diarrhea worldwide (Mead et al. 1999). It affects people of all ages, but is most common in young adults. The majority of *Campylobacter* infections in humans are caused by *C. jejuni* (90%–95%), followed by *C. coli* (5%–10%). Other species such as *C. lari*, *C. upsaliensis*, and *C. consisus* also have been recovered from humans. The relevance of these species in human disease is unclear.

Although large outbreaks do occur, campylobacteriosis is usually sporadic and is most often associated with ingestion of contaminated food (Friedman et al. 2000). Among foodborne sources in developing countries, raw poultry has been identified as the largest potential source of human infection (Friedman et al. 2000). Travel to certain countries is another known important risk factor (Smith et al. 1999). Intestinal illness typically manifests within 24–72 hours after ingestion of contaminated food and is characterized by fever, abdominal pain, and diarrhea with or without blood (Blaser 2000). The diarrhea usually resolves spontaneously within a few days. However, in some patients, diarrhea may last more than 1 week, whereas others (5%–10%) may suffer relapsing illness (Blaser 2000). In patients with prolonged, severe, or relapsing disease, antimicrobial therapy may be required. It is also recommended in cases of extra-intestinal disease (e.g., bacteremia, meningitis, and endocarditis) and for infections in pregnant women, the immunocompromised, and those at the extremes of age.

In those relatively few cases of gastroenteritis in which the etiologic agent is known to be *Campylobacter*, a macrolide such as erythromycin is the drug of choice. Because stool cultures are rarely done in cases of gastroenteritis, a fluoroquinolone such as ciprofloxacin is often used as empiric therapy to manage diarrhea of unknown etiology, including traveler's diarrhea (Adachi et al. 2000). There are indications that fluoroquinolone resistance in *Campylobacter* has risen dramatically, making these agents less attractive for therapy (Nachamkin et al. 2002).

Campylobacter has been recovered from the intestinal tracts of both wild and domestic animals. In food-producing animals such as cattle, poultry, and swine, fecal *C. jejuni and C. coli* can be regarded as a commensal organism. As a result of fecal contact during processing, enteric bacteria frequently contaminate foods derived from animals. This is the case with *Campylobacter* (*C. jejuni*), which are present in high numbers in the poultry intestine (Corry and Atabay 2001). As a consequence, *Campylobacter* can be readily cultured from most retail chicken carcasses.

Several surveys have addressed the prevalence of *Campylobacter* in retail meats without regard to their susceptibility profiles. One of the earliest such studies was conducted in 1986 by the Washington State Department of Health and showed that 57% of poultry processing plant samples and 23% of retail chickens carried *C. jejuni* (Harris et al. 1986).

The U.S. Department of Agriculture (USDA) Food Safety Inspection Service (FSIS) periodically analyzes samples of raw meat taken from slaughter plants for the prevalence of certain bacteria, including *Campylobacter*. A 1996 FSIS report approximated that 88% of chicken broiler carcasses ($n = 1,297$) were contaminated with *Campylobacter* species (USDA 1996a). FSIS conducted a similar study of young turkey carcasses processed from August 1996 to July 1997 ($n = 1,221$) and recovered *C. jejuni* and *C. coli* from 90% of analyzed samples (USDA 1998). In 1996, FSIS published The Nationwide Ground Raw Chicken Microbiological Survey (USDA 1996b) and The Nationwide Raw Ground Turkey Microbiological Survey (USDA 1996c). In the survey of raw ground chicken samples ($n = 285$), FSIS estimated the national prevalence of *C. jejuni* and *C. coli* on raw ground chicken to be around 60%. The Nationwide Raw Ground Turkey Survey estimated approximately 25% of the ground turkey samples to be contaminated with *Campylobacter*.

In a Minnesota survey, 91 chicken meat products were purchased in the Minneapolis–St. Paul area from September to November 1997 (Smith et al. 1999). These products came from 15 poultry-processing plants in nine states. All chicken meat products were tested for *Campylobacter*, and recovered isolates were tested for susceptibility to nalidixic acid and ciprofloxacin. *Campylobacter* was isolated from 88% of the meat samples tested; *C. jejuni* was isolated from 74% of chicken samples, and *C. coli* was recovered from 21% of samples tested. Ciprofloxacin-resistant *Campylobacter* (MIC > 32 μg/mL) were recovered from 20% of tested chicken samples ($n = 18$). Using molecular subtyping methods (*fla*-RFLP), the researchers further identified six of seven subtypes among *Campylobacter* isolates from both retail chicken products and quinolone-resistant *C. jejuni* isolates from humans. These data demonstrated that domestic chicken obtained from retail markets had high rates of contamination with ciprofloxacin-resistant *C. jejuni*. Furthermore, the authors identified an association between molecular subtypes of quinolone-resistant *C. jejuni* strains that were acquired domestically in humans and those found in retail chicken products (Dudley and Ambrose 2000).

Another study of retail chicken meats was conducted by the Centers for Disease Control and Prevention in 1999 (Rossiter et al. 2000). Between January and June 1999, public health laboratories in Georgia, Maryland, and Minnesota each tested 10 retail samples per month. One-hundred eighty retail meats, making up 23 domestic brand names, were purchased from 25 grocery stores. Eighty (44%) retail chicken meat samples were found to carry *Campylobacter*. Among the 80 *Campylobacter* isolates tested, 19 (24%) were resistant to ciprofloxacin. Overall, 19 of 180 (11%) retail chicken meats yielded ciprofloxacin-resistant *Campylobacter*.

Zhao et al. (2001a) examined 719 retail raw meat samples (chicken, turkey, pork and beef) obtained from 59 stores of four supermarket chains during 107 sampling visits in the Greater Washington, D.C., area, from June 1999 to July 2000. *Campylobacter* was detected in 71% of 184 chicken samples, and a large percentage of the stores visited (91%) had *Campylobacter*-contaminated chicken. Approximately 14% of the 172 turkey samples were culture-positive, whereas few pork (1.7%) and beef (0.5%) samples were positive. Approximately one-half (53.6%) of the isolates were identified as *C. jejuni*, 41.3% were identified as *C. coli*, and 5.1% were identified as other species. *Campylobacter coli* were recovered more often from retail turkey samples than were *C. jejuni*. The most common resistance observed among the *Campylobacter* poultry isolates was to tetracycline (82%), followed by erythromycin (54%), nalidixic acid (41%), and ciprofloxacin (35%) (Ge et al. 2003). *C. coli* isolates displayed higher resistance rates to ciprofloxacin and erythromycin than did *C. jejuni*. Turkey isolates, regardless of species, showed elevated resistance rates to ciprofloxacin and erythromycin compared with *Campylobacter* isolates recovered from retail chickens. Eighty-seven percent of samples were contaminated with *Campylobacter* resistant to at least two antimicrobial agents, and 22% to at least five antimicrobials. Co-resistance to ciprofloxacin and erythromycin was found in *Campylobacter* from 26% of the meat samples. Multidrug resistance, including co-resistance to fluoroquinolones and erythromycin, has been reported previously in *Campylobacter* isolated from retail meat products (Sáenz et al. 2000) and from humans (Engberg et al. 2001). The finding of co-resistance to fluoroquinolones and erythromycin in *Campylobacter* is undesirable, as these two antimicrobials are generally advocated as first-line drugs for treatment of human campylobacteriosis.

In 2001, the Food and Drug Administration Center for Veterinary Medicine began a study to assess the prevalence of resistant foodborne pathogens in retail raw meats (Carter et al. 2002). Chicken breast, ground turkey, ground beef, and pork chops were sampled from March 2001 to March 2002 from 300 supermarket grocery locations throughout the state of Iowa. The objective was to assess the prevalence of *Salmonella*, *Enterococcus*, and *Campylobacter* and to determine their antimicrobial susceptibility patterns. Preliminary data revealed that 20% of 654 retail samples tested for *Campylobacter* were positive, with chicken accounting for most of these (73%). *Campylobacter* was less frequently recovered from turkey (6%) and pork (1%) and was not recovered at all from ground beef. Overall, 27% of *C. jejuni* isolates, 27% of *C. coli* isolates, and 21% of nonspeciated *Campylobacter* isolates exhibited resistance to ciprofloxacin (MIC ≥ 4 μg/mL).

In 2002, The National Antimicrobial Resistance Monitoring System (NARMS) expanded into surveillance of retail meats to determine the prevalence of antimicrobial resistance among *Salmonella*, *Campylobacter*, *E. coli*, and *Enterococcus*. As of January 2003, nine FoodNet sites were participating (California, Colorado, Connecticut, Georgia, Minnesota, New York, Tennessee, Maryland, and Oregon). Each FoodNet site purchases a total of 40 food samples per month, consisting of 10 samples each of chicken breast, ground turkey, ground beef, and pork chops. Preliminary data show that 58% of 356 chicken breasts, 8% of 372 ground turkey, 3% of 373 ground beef samples, and 2% of 343 pork chop samples cultured positive for *Campylobacter* (D. White, unpublished data). This surveillance is ongoing and being expanded to include additional state labs.

Campylobacter have been recovered from many foods of animal origin. Numerous surveys of raw agricultural products provide epidemiologic evidence that implicates poultry, meat, and raw milk as sources of human infection (Altekruse et al. 1999). It is therefore necessary to readdress hygienic practices at every step in the food production continuum, from farm to consumer, in preventing and reducing the incidence of *Campylobacter* infections.

Escherichia coli

Escherichia coli are widely distributed in the intestine of humans and warm-blooded animals and are the predominant facultative anaerobe in the intestinal tract (Conway 1995). However, there are many pathogenic *E. coli* strains that can cause a variety of diseases in animals and humans. Pathogenic types of *E. coli* differ from those that predominate as the normal enteric flora of healthy animals in that they are more likely to express certain virulence factors (Donnenberg and Whittam 2001). Pathogenic types of *E. coli* are classified by their specific virulence mechanisms (e.g., toxins, adhesins, and invasiveness), serotypes, and pathogenesis. At present, pathogenic *E. coli* are divided into at least six distinct classes: enteropathogenic *E. coli* (EPEC), enterotoxigenic *E. coli* (ETEC), enteroinvasive *E. coli* (EIEC), diffuse-adhering *E. coli* (DAEC), enteroaggregative *E. coli* (EAEC), and enterohemorrhagic *E. coli* (EHEC; Nataro and Kaper 1998, Wasteson 2001).

Consumption of contaminated foods is a documented route by which pathogenic *E. coli* infect humans. This is most likely because *E. coli* is present throughout the food production continuum, where they may contaminate meats through various routes, including post-evisceration processing, handling, and packaging of finished products (Jackson et al. 2001). The spectrum of *E. coli* types colonizing animals and humans has not been fully explored. However, there are numerous reports in the literature indicating that O-serogroups commonly isolated from food animals and humans are similar (Orskov and Orskov 1992, Aarestrup and Wegener 1999). In addition, specific antimicrobial-resistant *E. coli* have been traced from the gut contents of pigs, calves, and chickens to carcasses at slaughter and ultimately shown to colonize the gut of a human volunteer handling and eating the meat (Linton 1986).

Antimicrobial-resistant *E. coli* have been recovered from a variety of foods, including minced meat (Osterblad et al. 1999a), vegetables (Osterblad et al. 1999b), cakes and confectionary (Pinegar and Cooke 1985), custards and desserts (Persson et al. 1980), and milk and milk products (Chang 1975, Johnston et al. 1983). However, most data describing antimicrobial-resistant phenotypes in foodborne *E. coli* have come from isolates recovered from retail meats (Meng et al. 1998, Zhao et al. 2001b, Schroeder et al. 2003).

A number of studies outside the United States from the early 1980s looked for the presence in foods of antimicrobial-resistant Gram-negative bacteria, including *E. coli* (Bensink et al. 1981, Khor et al. 1982, Wood et al. 1983). One of the first studies was described by Persson et al. (1980). *E. coli* was the predominant bacterium identified and was most often isolated from raw meat and egg products. All isolates were considered susceptible to streptomycin, neomycin, and trimethoprim-sulfamethoxazole. Fifty-four percent (166 of 308) of bacteria tested, however, were resistant to at least one of the nine antibiotics tested, with 40% displaying multidrug resistance. Thirty percent of the bacteria were resistant to nitrofurantoin, and 25% were resistant to sulphaisodimidine. The lowest numbers of resistant strains identified were recovered from custards and desserts (42%) and from raw meat products (46%), and the highest from ice-cream (89%). Multidrug-resistant strains were recorded most frequently from pasteurized milk products, custards, and desserts.

Bensink et al. (1981) examined Australian beef and pig carcasses, meat products, and frozen chickens for the presence of antibiotic resistant coliforms. *E. coli* was isolated from 18 of 50 beef carcasses, and only resistance to tetracycline was detected. However, the situation was different with *E. coli* recovered from pig carcasses, meat products, and chickens, where numerous resistance phenotypes were observed, including multidrug-resistant strains. A later study by the same researchers investigated antibiotic resistance in coliforms isolated from poultry carcasses immediately after slaughter and at retail (Bensink and Botham 1983). Approximately 85% of a total of 13,858 isolates examined were found to be resistant to at least one antibiotic. Highly significant differences were found in the levels of antibiotic resistance from the two sources: Ampicillin, chloramphenicol, and sulphonamide resistances were found more frequently in isolates from poultry at retail, whereas resistance to streptomycin and neomycin occurred more frequently in isolates from poultry at slaughter.

Wood et al. (1983) investigated the prevalence of antibiotic-resistant, Gram-negative bacteria from foods in Mexico. Of 235 Gram-negative organisms recovered from frozen food samples, 34 isolates were resistant to trimethoprim, 15 were resistant to trimethoprim-sulfamethoxazole, and 33 demonstrated multidrug-resistant phenotypes.

Several more recent studies have investigated specifically the prevalence of antimicrobial resistant *E. coli* in retail foods. Meng et al. (1998) determined antimicrobial susceptibilities of 118 *E. coli* O157:H7 and seven O157:NM isolates from animals, foods, and humans in the United States. Among the 125 isolates tested, 30 (24%) were resistant to at least one antibiotic and 24 (19%) were resistant to three or more antibiotics. Cattle strains were more often resistant than other isolates tested (34%). The seven resistant food *E. coli* isolates were all recovered from ground beef. Two *E. coli* O157:NM isolates from cattle were resistant to six antibiotics: ampicillin, kanamycin, sulfisoxazole, streptomycin, tetracycline, and ticarcillin. Streptomycin was the most common antibiotic to which *E. coli* O157:H7 and O157:NM were resistant (29 of 30 isolates), followed by tetracycline (26 isolates). Overall, the most frequent resistance phenotype observed was to streptomycin-sulfisoxazole-tetracycline, which accounted for over 70% of the resistant strains.

Sáenz et al. (2001) investigated the prevalence of antimicrobial-resistant *E. coli* from animals, foods, and humans in La Rioja, Spain. Food products of animal origin sampled for the presence of antimicrobial-resistant *E. coli* included hamburger, sausage, chicken, and turkey. Using disk diffusion, they reported that among 47 *E. coli* isolates recovered from foods, 53% were resistant to nalidixic acid, 47% to ampicillin, and 40% to kanamycin. Resistance was also observed, but to a lesser extent, to gentamicin (17%), ciprofloxacin (13%), and amoxicillin-clavulanic acid (Sáenz et al. 2001).

Schroeder et al. (2003) reported antimicrobial susceptibilities among 472 generic *E. coli* isolates recovered from ground and whole retail beef, chicken, pork, and turkey obtained from greater Washington, D.C., during the years 1998 and 2000. Isolates displayed resistance to tetracycline (59%), sulfamethoxazole (45%), streptomycin (44%), cephalothin (38%), and ampicillin (35%). Lower resistance rates were observed for gentamicin (12%), nalidixic acid (8%), chloramphenicol (6%), ceftiofur (4%), and ceftriaxone (1%). Sixteen percent of the isolates displayed resistance to one antimicrobial, followed by 23% to two, 23% to three, 12% to four, 7% to five, 3% to six, 2% to seven, and 2% to eight.

Other studies demonstrate that *E. coli* recovered from foods are not necessarily resistant to multiple antimicrobials. Zhao et al. (2002) tested 404 fresh ground beef samples obtained at retail stores from New York, San Francisco, Philadelphia, Denver, Atlanta, Houston, and Chicago for the presence of *Salmonella* and *E. coli*. Among the 102 generic *E. coli* isolates obtained, only three were resistant to multiple antibiotics.

There is still debate about the public health implications of the presence of antimicrobial-resistant *E. coli* in foods. Nevertheless, the numerous reports that have documented the presence of antimicrobial-resistant *E. coli* in retail foods highlights the public health value of continuing efforts to educate consumers in proper food handling and preparation practices.

Listeria monocytogenes

Listeria monocytogenes is an opportunistic pathogen that was recognized as an important foodborne pathogen in the early 1980s. The organism is widely distributed in nature and can be found in soils, animal waste, sewage, and water. *Listeria* is frequently present in the gut of humans and animals including cattle, poultry, and pigs (Meng and Doyle 1997). *Listeria monocytogenes* grows well in cold temperatures with minimal nutrients and is able to survive and multiply in the environment. Its ubiquity in the environment allows it ample opportunity to reach food products during various phases of processing and distribution. *Listeria monocytogenes* has been found in a variety of foods, including fresh fruits and vegetables, fresh and frozen and processed meats, dairy products, eggs, and seafood (Meng and Doyle 1997, Kathariou 2002). The prevalence of *L. monocytogenes* contamination of raw and processed meat products ranges from <1% to 70%. It has been estimated that approximately 25% of raw and ready-to-eat seafood and fish products are contaminated with *Listeria* (Meng and Doyle 1997). Outbreaks largely have been attributed to ready-to-eat foods, which are typically consumed without prior cooking or reheating (Schuchat et al. 1991). Examples of high-risk ready-to-eat foods are coleslaw, pasteurized milk, soft cheeses, pate, pork tongue in jelly, shrimp, smoked mussels, hot dogs, and salami (Meng and Doyle 1997).

The global incidence of listeriosis has increased over the last two decades. A number of large foodborne outbreaks have been reported in different countries, including England, Germany, Sweden, New Zealand, Switzerland, Australia, France, and the United States (Rocourt and Cossart 1997). Listeriosis is one of the most severe foodborne infections (causing meningitis, septicemia, and abortion), with low morbidity (annual incidence rate ranging from two to 10 cases per million population) but high lethality (30%) and a high predilection for people with impaired T-cell immunity (pregnant women and neonates, immunocompromised patients, and the elderly; Rocourt et al. 2003). Although surveillance studies from the Centers for Disease Control have shown that the incidence of *Listeria* infection in the United States has decreased, several large U.S. outbreaks have occurred in recent years. One outbreak associated with hot dogs involved more than 50 cases in 11 states (Centers for Disease Control 1999). Six adults died and two pregnant women had spontaneous abortions. Two multi-state outbreaks of listeriosis, one in 2000 and another in 2002, were both linked to turkey deli meat (Centers for Disease Control 2000, Centers for Disease Control 2002a). The 2002 outbreak consisted of 46 confirmed cases, seven deaths, and three miscarriages in eight states.

The spectrum of listeriosis is broad and includes an asymptomatic carrier state, cutaneous lesions, flu-like illness, miscarriage, stillbirth, septicemia, encephalitis, and meningitis. Foodborne infection with *L. monocytogenes* also can cause febrile illness with gastroenteritis in immunocompetent persons (Aureli et al. 2000). Although listeriosis can occur in otherwise healthy adults and children, immunocompromised individuals, including the immunosuppressed, the elderly, newborns, pregnant women, and persons suffering a ranging of underlying diseases, are at higher risk.

The degree and prevalence of antibiotic-resistant *L. monocytogenes* in food is not clear. Few studies report the frequency of antimicrobial resistance in *L. monocytogenes* isolated from foods. In addition, there is no standardized *in vitro* susceptibility testing method for this organism, making the comparison of results difficult. Ampicillin, penicillin, or rifampin, plus gentamicin is the treatment of choice for most *Listeria* infections. Cotrimoxazole is considered to be a second-choice therapy (Jones and MacGowan 1995, Charpentier and Courvalin 1999). Vancomycin and erythromycin, respectively, are used to treat *Listeria* bacteremia and infections in pregnant women.

Regardless of the source of isolates, *Listeria* is generally susceptible *in vitro* to the antimicrobials active against Gram-positive pathogens. Poulsen et al. (1988) reported that 156 strains of *L. monocytogenes* from human cerebral spinal fluid and blood between 1958 and 1985 were susceptible to 12 antimicrobials throughout the 25-year sampling period. However, several subsequent studies indicate that multidrug-resistant *L. monocytogenes* have emerged. The first *L. monocytogenes* strains resistant to antimicrobials were reported in 1988 (Charpentier and Courvalin 1999). Since then, *Listeria* spp. isolated from foods and the environment, or in sporadic cases of human illness, have displayed resistance to one or several antimicrobials.

In the United States, Prazak et al. (2002) determined antimicrobial susceptibilities via disk diffusion of 21 *L. monocytogenes* isolates recovered from cabbage farms and packing sheds in Texas. Ninety-five percent (20 of 21) of the isolates tested were resistant to two or more antimicrobial agents, with eighty-five percent (17 of 20) of the multi-resistant strains showing resistance to penicillin. Safdar and Armstrong (2003) reported antimicrobial susceptibilities of 84 clinical *L. monocytogenes* isolates recovered from 1955 to 1997. During this time frame, susceptibilities to penicillin (97.6%), ampicillin (90.7%), erythromycin (98.8%), tetracycline (96.9%), and gentamicin (98.0%) remained virtually unchanged; however, only 4% of isolates were considered susceptible to clindamycin (Safdar and Armstrong 2003). All isolates were susceptible to amikacin, ciprofloxacin, imipenem, rifampin, trimethoprim-sulfamethoxazole, and vancomycin.

A survey for listeriae in retail poultry in Porto, Portugal, identified *L. monocytogenes* in 41% of 63 samples, with a high percentage (73%) resistant to one or more antimicrobial agents of different classes. Using disk diffusion, the most prevalent resistances were to clindamycin (54%) and enrofloxacin (43%), followed by tetracycline (15%), streptomycin (7%), erythromycin (2%), and ofloxacin (2%). This study suggested a high incidence of *L. monocytogenes* on Portuguese poultry products at retail and that poultry could be a vehicle for resistant foodborne *Listeria* infections.

Charpentier et al. (1995) screened 1,100 *Listeria* spp. (60 from human patients and 1,040 from food and the environment) collected worldwide and found many of them were resistant to antimicrobials. Of the 61 tetracycline- and minocycline-resistant strains (37 *L. monocytogenes*), 57 harbored the tetracycline-resistance determinant encoded by tet(M), and four non-*L. monocytogenes* isolates contained tet(S). Three clinical isolates of *L. mono-*

cytogenes were resistant to low levels of streptomycin, and one strain of *L. monocytogenes* was trimethoprim resistant.

Study of the trimethoprim-resistant strain showed that the trimethoprim-resistance gene *dfr*D was present on a 3.7-kb plasmid (pIP823) of staphylococcal origin (Charpentier et al. 1999). It was found that pIP823 had a broad host range, including *L. monocytogenes, Enterococcus faecalis, S. aureus, Bacillus subtilis,* and *E. coli.* The emergence of trimethoprim resistance in *L. monocytogenes* is noteworthy because the trimethoprim–sulfamethoxazole combination is a second-line drug for treating human listeriosis.

Erythromycin-resistant strains of *L. monocytogenes* isolated from food have also been reported (Roberts et al. 1996). The resistance gene *ermC* was identified and was transferable by conjugation to recipient strains, of *L. monocytogenes, Listeria invanovii,* and *E. faecalis* but did not appear to be associated with conjugative plasmids. In addition, vancomycin resistance (*vanA*) has been successfully transferred from *Enterococci* to *L. monocytogenes* and other *Listeria* spp. *in vitro* (Biavasco et al. 1996).

Resistance to other antibiotics in *L. monocytogenes* has also been reported, primarily from human clinical strains. A strain resistant to gentamicin, streptomycin, chloramphenicol and clindamycin was isolated from a neonate who developed meningitis in Greece (Abrahim et al. 1998). Analysis of 98 *L. monocytogenes* isolates in Italy revealed that two strains were resistant to streptomycin, sulfamethoxazole, and kanamycin: one was resistant to streptomycin, sulfamethoxazole, kanamycin, and rifampin, and one was resistant to the latter four drugs plus erythromycin and chloramphenicol. Resistance to ciprofloxacin has also been reported (Oethinger et al. 2000) and has recently been attributed to expression of an efflux pump termed *lde* (Godreuil et al. 2003).

With the exception of tetracycline resistance, antibiotic resistance among *Listeria* spp. from clinical, food, or environmental sources remains low. Nevertheless, antibiotic-resistant *L. monocytogenes* variants have emerged. This pathogen is capable of acquiring antibiotic-resistance genes from other bacterial genera, both Gram-positive and Gram-negative (Aureli et al. 2000). Emergence and dissemination of drug resistance in *L. monocytogenes,* such as resistance to trimethoprim, sulfamethoxazole, gentamicin, and erythromycin, may influence the selection of antimicrobial agents for treating listeriosis in the future.

Conclusions

Microbial food safety is an important public health concern worldwide. In response to an increasing number of food safety problems and rising consumer concerns, industry and government agencies from numerous nations are intensifying their efforts to improve food safety (Tollefson et al. 1998, Torrence 2001, Bull et al. 2002, Rose et al. 2002). Although the sanitary standards of meat production are quite high in most developed countries, they cannot completely prevent fecal contamination of animal-derived food products. Antimicrobial-resistant bacteria from animals, both commensal and pathogenic variants, can reach the general public via exposure to contaminated food products of animal origin if they are improperly cooked or otherwise mishandled (Witte 2000, Sáenz et al. 2001). It has been theorized that these resistant bacteria have the potential to colonize humans and to transfer their resistance determinants to resident constituents of the human microflora, including pathogens. Therefore, it is necessary to continue work to advance food processing technologies while emphasizing hygienic food handling at all stages of food production. There is also a continuing need for education programs aimed at improving food-safety behaviors of consumers of all ages (see Chapter 30). Finally, surveillance programs designed to detect emerging antimicrobial resistance phenotypes among foodborne pathogens in retail foods will continue to be a foundational tool for helping to ensure a safe food supply.

References

Aarestrup, F. M. and Wegener, H. C. 1999. The effects of antibiotic usage in food animals on the development of antimicrobial resistance of importance for humans in *Campylobacter* and *Escherichia coli*. *Microbes and Infection* **1**:639–644.

Abrahim, A., Papa, A., Soultos, N., Ambrosiadis, I. and Antoniadis, A. 1998. Antibiotic resistance of *Salmonella* spp. and *Listeria* spp. isolates from traditionally made fresh sausages in Greece. *Journal of Food Protection* **61**:1378–1380.

Adachi, J. A., Ostrosky-Zeichner, L., DuPont, H. L. and Ericsson, C. D. 2000. Empirical antimicrobial therapy for traveler's diarrhea. *Clinical Infectious Diseases* **31**:1079–1083.

Altekruse, S. F., Stern, N. J., Fields, P. I. and Swerdlow, D. L. 1999. *Campylobacter jejuni*—an emerging foodborne pathogen. *Emerging Infectious Diseases* **5**:28–35.

Arcangioli, M. A., Leroy-Setrin, S., Martel, J. L. and Chaslus-Dancla, E. 1999. A new chloramphenicol and florfenicol resistance gene flanked by two integron structures in *Salmonella typhimurium* DT104. *FEMS Microbiology Letters* **174**:327–332.

Aureli, P., Fiorucci, G. C., Caroli, D., Marchiaro, G., Novara, O., Leone, L. and Salmaso, S. 2000. An outbreak of febrile gastroenteritis associated with corn contaminated by *Listeria monocytogenes*. *New England Journal of Medicine* **342**:1236–1241.

Bensink, J. C. and Botham, F. P. 1983. Antibiotic resistant coliform bacilli, isolated from freshly slaughtered poultry and from chilled poultry at retail outlets. *Australian Veterinary Journal* **60**:80–83.

Bensink, J. C., Frost, A. J., Mathers, W., Mutimer, M. D., Rankin, G. and Woolcock, J. B. 1981. The isolation of antibiotic resistant coliforms from meat and sewage. *Australian Veterinary Journal* **57**:12–16, 19.

Besser, T. E., Goldoft, M., Pritchett, L. C., Khakhria, R., Hancock, D. D., Rice, D. H., Gay, J. M., Johnson, W. and Gay, C. C. 2000. Multiresistant *Salmonella* Typhimurium DT104 infections of humans and domestic animals in the Pacific Northwest of the United States. *Epidemiology and Infection* **124**:193–200.

Biavasco, F., Giovanetti, E., Miele, A., Vignaroli, C., Facinelli, B. and Varaldo, P. E. 1996. *In vitro* conjugative transfer of VanA vancomycin resistance between enterococci and *Listeriae* of different species. *European Journal of Clinical Microbiology and Infectious Diseases* **15**:50–59.

Blaser, M. J. 1997. Epidemiologic and clinical features of *Campylobacter jejuni* infections. *Journal of Infectious Diseases* **176(Suppl. 2)**:S103–S105.

Blaser, M. J. 2000. *Campylobacter jejuni* and Related Species, pp. 2276–2285. *In* Mandell, Douglas, and Bennett's Principles and Practice of Infectious Diseases, 5th edn., G. L. Mandell, J. E. Bennett, R. Dolin (Eds.). Philadelphia, Penn.: Churchill Livingstone.

Briggs, C. E. and Fratamico, P. M. 1999. Molecular characterization of an antibiotic resistance gene cluster of *Salmonella typhimurium* DT104. *Antimicrobial Agents and Chemotherapy* **43**:846–849.

Bull, A. L., Crerar, S. K. and Beers, M. Y. 2002. Australia's Imported Food Program—a valuable source of information on micro-organisms in foods. *Communicable Diseases Intelligence* **26**:28–32.

Buzby, J. C. and Roberts, T. 1997. Economic costs and trade impacts of microbial foodborne illness. *World Health Statistics Quarterly* **50**:57–66.

Carter, P. J., English, L., Cook, B., Proescholdt, T. and White, D. G. 2002. Prevalence and antimicrobial susceptibility profiles of *Salmonella* and *Campylobacter* isolated from retail meats. Abstract No. P-84, Abstracts of the 102nd General Meeting of the American Society for Microbiology, May 19–23, Salt Lake City, Utah. Washington, D.C.: American Society for Microbiology Press.

Centers for Disease Control. 1999. Update: multistate outbreak of listeriosis—United States, 1998–1999. *Morbidity and Mortality Weekly Report* **47**:1117–1118.

Centers for Disease Control. 2000. Multistate outbreak of listeriosis—United States, 2000. *Morbidity and Mortality Weekly Report* **49**:1129–1130.

Centers for Disease Control. 2002a. Outbreak of listeriosis—northeastern United States, 2002. *Morbidity and Mortality Weekly Report* **51**:950–951.

Centers for Disease Control. 2002b. Outbreak of multidrug-resistant *Salmonella Newport*—United States, January–April 2002. *Morbidity and Mortality Weekly Report* **51**:545–548.

Chang, S. H. 1975. *Escherichia coli* isolation from raw milk in central and southern Taiwan and their susceptibility to drugs. *Zhonghua Min Guo Wei Sheng Wu Xue Za Zhi* **8**:142–145.

Charpentier, E. and Courvalin, P. 1999. Antibiotic resistance in *Listeria* spp. *Antimicrobial Agents and Chemotherapy* **43**:2103–2108.

Charpentier, E., Gerbaud, G. and Courvalin, P. 1999. Conjugative mobilization of the rolling-circle plasmid pIP823 from *Listeria monocytogenes* BM4293 among Gram-positive and Gram-negative bacteria. *Journal of Bacteriology* **181**:3368–3374.

Charpentier, E., Gerbaud, G., Jacquet, C., Rocourt, J. and Courvalin, P. 1995. Incidence of antibiotic resistance in *Listeria* species. *Journal of Infectious Disease* **172**:277–281.

Cherubin, C. E. and Eng, R. H. 1991. Quinolones for the treatment of infections due to *Salmonella*. *Reviews of Infectious Diseases* **13**:343–344.

Cloeckaert, A., Boumedine, K. S., Flaujac, G., Imberechts, H., D'Hooghe, I. and Chaslus-Dancla, E. 2000. Occurrence of a *Salmonella enterica* serovar Typhimurium DT104-like antibiotic resistance gene cluster including the *floR* gene in *S. enterica* serovar Agona. *Antimicrobial Agents and Chemotherapy* **44**:1359–1361.

Conway, P. L. 1995. Microbial ecology of the human large intestine, pp. 1–24. *In* Human Colonic Bacteria: Role in Nutrition, Physiology, and Pathology, G. R. Gibson, G. T. Macfarlane (Eds.). Boca Raton, Fla.: CRC Press.

Corry, J. E. and Atabay, H. I. 2001. Poultry as a source of *Campylobacter* and related organisms. *Symposium Series of the Society for Applied Microbiology* **30**:96S–114S.

Donnenberg, M. S. and Whittam, T. S. 2001. Pathogenesis and evolution of virulence in enteropathogenic and enterohemorrhagic *Escherichia coli*. *Journal of Clinical Investigation* **107**:539–548.

Dudley, M. N. and Ambrose, P. G. 2000. Pharmacodynamics in the study of drug resistance and establishing *in vitro* susceptibility breakpoints: ready for prime time. *Current Opinion in Microbiology* **3**:515–521.

Engberg, J., Aarestrup, F. M., Taylor, D. E., Gerner-Smidt, P. and Nachamkin, I. 2001. Quinolone and macrolide resistance in *Campylobacter jejuni* and *C. coli*: resistance mechanisms and trends in human isolates. *Emerging Infectious Diseases* **7**:24–34.

Fey, P. D., Safranek, T. J., Rupp, M. E., Dunne, E. F., Ribot, E., Iwen, P. C., Bradford, P. A., Angulo, F. J. and Hinrichs, S. H. 2000. Ceftriaxone-resistant *Salmonella* infection acquired by a child from cattle. *New England Journal of Medicine* **342**:1242–1249.

Friedman, C. R., Neimann, J., Wegener, H. C. and Tauxe, R. V. 2000. Epidemiology of *Campylobacter jejuni* infections in the United States and other industrialized nations, pp. 121–138. *In* Campylobacter, 2nd ed., I. Nachamkin, M. J. Blaser (Eds.). Washington, D.C.: American Society for Microbiology Press.

Ge, B., White, D. G., McDermott, P. F., Girard, W., Zhao, S., Hubert, S. and Meng, J. 2003. Antimicrobial resistant *Campylobacter* species from retail raw meats. *Applied and Environmental Microbiology* **69**:3005–3007.

Glynn, M. K., Bopp, C., Dewitt, W., Dabney, P., Mokhtar, M. and Angulo, F. J. 1998. Emergence of multidrug-resistant *Salmonella enterica* serotype Typhimurium DT104 infections in the United States. *New England Journal of Medicine* **338**:1333–1338.

Godreuil, S., Galimand, M., Gerbaud, G., Jacquet, C. and Courvalin, P. 2003. Efflux pump Lde is associated with fluoroquinolone resistance in *Listeria monocytogenes*. *Antimicrobial Agents and Chemotherapy* **47**:704–708.

Harris, N. V., Weiss, N. S. and Nolan, C. M. 1986. The role of poultry and meats in the etiology of *Campylobacter jejuni/coli* enteritis. *American Journal of Public Health* **76**:407–411.

Hernandez, T., Rodriguez-Alvarez, C., Arevalo, M. P., Torres, A., Sierra, A. and Arias, A. 2002. Antimicrobial-resistant *Salmonella enterica* serovars isolated from chickens in Spain. *Journal of Chemotherapy* **14**:346–350.

Hohmann, E. L. 2001. Nontyphoidal salmonellosis. *Clinical Infectious Diseases* **32**:263–269.

Hornish, R. E. and Kotarski, S. F. 2002. Cephalosporins in veterinary medicine—ceftiofur use in food animals. *Current Topics in Medicinal Chemistry* **2**:717–731.

Hudson, C. R., Quist, C., Lee, M. D., Keyes, K., Dodson, S. V., Morales, C., Sanchez, S., White, D. G. and Maurer, J. J. 2000. Genetic relatedness of *Salmonella* isolates from nondomestic birds in Southeastern United States. *Journal of Clinical Microbiology* **38**:1860–1865.

Jackson, T. C., Marshall, D. L., Acuff, G. R. and Dickson, J. S. 2001. Meat, poultry, and seafood, pp. 91–109. *In* Food Microbiology: Fundamentals and Frontiers, M. P. Doyle, L. R. Beuchat, T. J. Montville (Eds.). Washington, D.C.: American Society for Microbiology Press.

Johnston, D. W., Bruce, J. and Hill, J. 1983. Incidence of antibiotic-resistant *Escherichia coli* in milk produced in the west of Scotland. *Journal of Applied Bacteriology* **54**:77–83.

Jones, E. M. and MacGowan, A. P. 1995. Antimicrobial chemotherapy of human infection due to *Listeria monocytogenes*. *European Journal of Clinical Microbiology and Infectious Diseases* **14**:165–175.

Kathariou, S. 2002. *Listeria monocytogenes* virulence and pathogenicity, a food safety perspective. *Journal of Food Protection* **65**:1811–1829.

Khor, S. Y., Lim, Y. S. and Jegathesan, M. 1982. Antibiotic resistance and R plasmids in coliforms isolated from some Malaysian cooked foods. *Southeast Asian Journal of Tropical Medicine and Public Health* **13**:270–274.

Lee, L. A., Puhr, N. D., Maloney, E. K., Bean, N. H. and Tauxe, R. V. 1994. Increase in antimicrobial-resistant *Salmonella* infections in the United States, 1989–1990. *Journal of Infectious Diseases* **170**:128–134.

Linton, A. H. 1986. Flow of resistance genes in the environment and from animals to man. *Journal of Antimicrobial Chemotherapy* **18(Suppl. C)**:189–197.

Lorber, B. 1997. Listeriosis. *Clinical Infectious Diseases* **24**:1–9.

Low, J. C., Angus, M., Hopkins, G., Munro, D. and Rankin, S. C. 1997. Antimicrobial resistance of *Salmonella enterica typhimurium* DT104 isolates and investigation of strains with transferable apramycin resistance. *Epidemiology and Infection* **118**:97–103.

Mammina, C., Cannova, L., Massa, S., Goffredo, E. and Nastasi, A. 2002. Drug resistances in *Salmonella* isolates from animal foods, Italy 1998–2000. *Epidemiology and Infection* **129**:155–161.

Mead, P. S., Slutsker, L., Dietz, V., McCaig, L. F., Bresee, J. S., Shapiro, C., Griffin, P. M. and Tauxe, R. V. 1999. Food-related illness and death in the United States. *Emerging Infectious Diseases* **5**:607–625.

Meng, J. and Doyle, M. P. 1997. Emerging issues in microbiological food safety. *Annual Reviews in Nutrition* **17**:255–275.

Meng, J., Zhao, S., Doyle, M. P. and Joseph, S. W. 1998. Antibiotic resistance of *Escherichia coli* O157:H7 and O157:NM isolated from animals, food, and humans. *Journal of Food Protection* **61**:1511–1514.

Mølbak, K., Baggesen, D. L., Aarestrup, F. M., Ebbesen, J. M., Engberg, J., Frydendahl, K., Gerner-Smidt, P., Petersen, A. M. and Wegener, H. C. 1999. An outbreak of multidrug-resistant, quinolone-resistant *Salmonella enterica* serotype Typhimurium DT104. *New England Journal of Medicine* **341**:1420–1425.

Nachamkin, I., Ung, H. and Li, M. 2002. Increasing fluoroquinolone resistance in *Campylobacter jejuni* in Philadelphia, 1982–2001. *Emerging Infectious Diseases* **8**:1501–1503.

Nataro, J. P. and Kaper, J. B. 1998. Diarrheagenic *Escherichia coli*. *Clinical Microbiology Reviews* **11**:142–201.

Oethinger, M., Kern, W. V., Jellen-Ritter, A. S., McMurry, L. M. and Levy, S. B. 2000. Ineffectiveness of topoisomerase mutations in mediating clinically significant fluoroquinolone resistance in *Escherichia coli* in the absence of the AcrAB efflux pump. *Antimicrobial Agents and Chemother*apy **44**:10–13.

Orskov, F. and Orskov, I. 1992. *Escherichia coli* serotyping and disease in man and animals. *Canadian Journal of Microbiology* **38**:699–704.

Osterblad, M., Kilpi, E., Hakanen, A., Palmu, L. and Huovinen, P. 1999a. Antimicrobial resistance levels of enterobacteria isolated from minced meat. *Journal of Antimicrobial Chemotherapy* **44**:298–299.

Osterblad, M., Pensala, O., Peterzens, H. and Huovinen, P. 1999b. Antimicrobial susceptibility of *Enterobacteriaceae* isolated from vegetables. *Journal of Antimicrobial Chemotherapy* **43**:503–509.

Persson, L., Olsson, B. and Franklin, A. 1980. Antibiotic resistance patterns of coliform bacteria isolated from food. *Scandinavian Journal of Infectious Diseases* **12**:289–294.

Pinegar, J. A. and Cooke, E. M. 1985. *Escherichia coli* in retail processed food. *Journal of Hygiene (London)* **95**:39–46.

Poppe, C., Smart, N., Khakhria, R., Johnson, W., Spika, J. and Prescott, J. 1998. *Salmonella typhimurium* DT104: a virulent and drug-resistant pathogen. *Canadian Veterinary Journal* **39**:559–565.

Poulsen, P. N., Carvajal, A., Lester, A. and Andreasen, J. 1988. *In vitro* susceptibility of *Listeria monocytogenes* isolated from human blood and cerebrospinal fluid. A material from the years 1958–1985. *Acta Pathologica, Microbiologica, et Immunologica Scandinavica* **96**:223–228.

Prats, G., Mirelis, B., Llovet, T., Munoz, C., Miro, E. and Navarro, F. 2000. Antibiotic resistance trends in enteropathogenic bacteria isolated in 1985–1987 and 1995–1998 in Barcelona. *Antimicrobial Agents and Chemotherapy* **44**:1140–1145.

Prazak, M. A., Murano, E. A., Mercado, I. and Acuff, G. R. 2002. Antimicrobial resistance of *Listeria monocytogenes* isolated from various cabbage farms and packing sheds in Texas. *Journal of Food Protection* **65**:1796–1799.

Quigley, III, J. D., Drewry, J. J., Murray, L. M. and Ivey, S. J. 1997. Body weight gain, feed efficiency, and fecal scores of dairy calves in response to galactosyl-lactose or antibiotics in milk replacers. *Journal of Dairy Science* **80**:1751–1754.

Ridley, A. and Threlfall, E. J. 1998. Molecular epidemiology of antibiotic resistance genes in multiresistant epidemic *Salmonella typhimurium* DT 104. *Microbial Drug Resistance* **4**:113–118.

Roberts, M. C., Facinelli, B., Giovanetti, E. and Varaldo, P. E. 1996. Transferable erythromycin resistance in *Listeria* spp. isolated from food. *Applied and Environmental Microbiology* **62**:269–270.

Rocourt, J., BenEmbarek, P., Toyofuku, H. and Schlundt, J. 2003. Quantitative risk assessment of *Listeria monocytogenes* in ready-to-eat foods: the FAO/WHO approach. *FEMS Immunology and Medical Microbiology* **35**:263–267.

Rocourt, J. and Cossart, P. 1997. *Listeria monocytogenes*, pp. 337–352. *In* Food Microbiology: Fundamentals and Frontiers, M. P. Doyle, L. R. Beuchat, T. J. Montville (Eds.). Washington, D.C.: American Society for Microbiology Press.

Rose, B. E., Hill, W. E., Umholtz, R., Ransom, G. M. and James, W. O. 2002. Testing for *Salmonella* in raw meat and poultry products collected at federally inspected establishments in the United States, 1998 through 2000. *Journal of Food Protection* **65**:937–947.

Rossiter, S., Joyce, K., Ray, M., Benson, J., Mackinson, C., Gregg, C., Sullivan, M., Vought, K., Leano, F., Besser, J., Marano, N. and Angulo, F. 2000. High prevalence of antimicrobial-resistant, including fluoroquinolone-resistant *Campylobacter* on chicken in U.S. grocery stores. Abstract No. C-296, Abstracts of the 100th Annual Meeting of the American Society for Microbiology, May 21–25, Los Angeles, CA. Washington, D.C.: American Society for Microbiology Press.

Sáenz, Y., Zarazaga, M., Lantero, M., Gastañares, M. J., Baquero, F. and Torres, C. 2000. Antibiotic resistance in *Campylobacter* strains isolated from animals, foods, and humans in Spain in 1997–1998. Antimicrobial Agents and Chemotherapy **44**:267–271.

Sáenz, Y., Zarazaga, M., Briñas, L., Lantero, M., Ruiz-Larrea, F. and Torres, C. 2001. Antibiotic resistance in *Escherichia coli* isolates obtained from animals, foods and humans in Spain. *International Journal of Antimicrobial Agents* **18**:353–358.

Safdar, A. and Armstrong, D. 2003. Antimicrobial activities against 84 *Listeria monocytogenes* isolates from patients with systemic listeriosis at a comprehensive cancer center (1955–1997). *Journal of Clinical Microbiology* **41**:483–485.

Schmieger, H. and Schicklmaier, P. 1999. Transduction of multiple drug resistance of *Salmonella enterica* serovar Typhimurium DT104. *FEMS Microbiology Letters* **170**:251–256.

Schroeder, C. M., White, D. G., Ge, B., Zhang, Y., McDermott, P. F., Ayers, S., Zhao, S. and Meng, J. 2003. Isolation of antimicrobial-resistant *Escherichia coli* from retail meats purchased in greater Washington, DC, USA. *International Journal of Food Microbiology* **85**:197–202.

Schroeder, C. M., Zhao, C., DebRoy, C., Torcolini, J., Zhao, S., White, D. G., Wagner, D. D., McDermott, P. F., Walker, R. D. and Meng, J. 2002. Antimicrobial resistance of *Escherichia coli* O157 isolated from humans, cattle, swine, and food. *Applied and Environmental Microbiology* **68**:576–581.

Schuchat, A., Swaminathan, B. and Broome, C. V. 1991. Epidemiology of human listeriosis. *Clinical Microbiology Reviews* **4**:169–183.

Smith, K. E., Besser, J. M., Hedberg, C. W., Leano, F. T., Bender, J. B., Wicklund, J. H., Johnson, B. P., Moore, K. A. and Osterholm, M. T. 1999. Quinolone-resistant *Campylobacter jejuni* infections in Minnesota, 1992–1998. *New England Journal of Medicine* **340**:1525–1532.

Threlfall, E. J., Frost, J. A., Ward, L. R. and Rowe, B. 1996. Increasing spectrum of resistance in multiresistant *Salmonella typhimurium*. *Lancet* **347**:1053–1054.

Threlfall, E. J., Rowe, B. and Ward, L. R. 1993. A comparison of multiple drug resistance in salmonellas from humans and food animals in England and Wales, 1981 and 1990. *Epidemiology and Infection* **111**:189–197.

Tollefson, L., Angulo, F. J. and Fedorka-Cray, P. J. 1998. National surveillance for antibiotic resistance in zoonotic enteric pathogens. *The Veterinary Clinics of North America. Food Animal Practice* **14**:141–150.

Torrence, M. E. 2001. Activities to address antimicrobial resistance in the United States. *Preventive Veterinary Medicine* **51**:37–49.

U.S. Department of Agriculture. 1996a. United States Department of Agriculture. Nationwide broiler chicken microbiological baseline data collection program (July 1994–June 1995). Washington, D.C.: Food Safety Inspection Service.

U.S. Department of Agriculture. 1996b. United States Department of Agriculture. Nationwide raw ground chicken microbiological survey (May 1996). Washington, D.C.: Food Safety Inspection Service.

U.S. Department of Agriculture. 1996c. United States Department of Agriculture. Nationwide raw ground turkey microbiological survey (May 1996). Washington, D.C.: Food Safety Inspection Service.

U.S. Department of Agriculture. 1998. Nationwide young turkey microbiological baseline data collection program (August 1996–July 1997). Washington, D.C.: Food Safety Inspection Service.

Wasteson, Y. 2001. Zoonotic *Escherichia coli*. *Acta Veterinaria Scandinavica. Supplementum* **95**:79–84.

White, D. G., Zhao, S., Sudler, R., Ayers, S., Friedman, S., Chen, S., McDermott, P. F., McDermott, S., Wagner, D. D. and Meng, J. 2001. The isolation of antibiotic-resistant salmonella from retail ground meats. *New England Journal of Medicine* **345**:1147–1154.

Winokur, P. L., Brueggemann, A., DeSalvo, D. L., Hoffmann, L., Apley, M. D., Uhlenhopp, E. K., Pfaller, M. A. and Doern, G. V. 2000. Animal and human multidrug-resistant, cephalosporin-resistant salmonella isolates expressing a plasmid-mediated CMY-2 AmpC beta-lactamase. *Antimicrobial Agents and Chemotherapy* **44**:2777–2783.

Witte, W. 2000. Ecological impact of antibiotic use in animals on different complex microflora: environment. *International Journal of Antimicrobial Agents* **14**:321–325.

Wood, L. V., Morgan, D. R. and DuPont, H. L. 1983. Antimicrobial resistance of Gram-negative bacteria isolated from foods in Mexico. *Journal of Infectious Diseases* **148**:766.

Yang, Y. J., Liu, C. C., Wang, S. M., Wu, J. J., Huang, A. H. and Cheng, C. P. 1998. High rates of antimicrobial resistance among clinical isolates of nontyphoidal *Salmonella* in Taiwan. *European Journal of Clinical Microbiology and Infectious Diseases* **17**:880–883.

Yildirmak, T., Yazgan, A. and Ozcengiz, G. 1998. Multiple drug resistance patterns and plasmid profiles of nontyphi salmonellae in Turkey. *Epidemiology and Infection* **121**:303–307.

Zhao, C., Ge, B., De Villena, J., Sudler, R., Yeh, E., Zhao, S., White, D. G., Wagner, D. and Meng, J. 2001a. Prevalence of *Campylobacter* spp., *Escherichia coli*, and *Salmonella* serovars in retail chicken, turkey, pork, and beef from the Greater Washington, D.C., area. *Applied and Environmental Microbiology* **67**:5431–5436.

Zhao, S., Data, A. R., Ayers, S., Friedman, S., Walker, R. D. and White, D. G. 2003. Antimicrobial-resistant *Salmonella* serovars isolated from imported food. *International Journal of Food Microbiology* **84**:87–92.

Zhao, S., White, D. G., McDermott, P. F., Friedman, S., English, L., Ayers, S., Meng, J., Maurer, J. J., Holland, R. and Walker, R. D. 2001b. Identification and expression of cephamycinase *bla*(CMY) genes in *Escherichia coli* and *Salmonella* isolates from food animals and ground meat. *Antimicrobial Agents and Chemotherapy* **45**:3647–3650.

Zhao, T., Doyle, M. P., Fedorka-Cray, P. J., Zhao, P. and Ladely, S. 2002. Occurrence of *Salmonella enterica* serotype Typhimurium DT104A in retail ground beef. *Journal of Food Protection* **65**:403–407.

Part IV

Verification Tests

20 The Hazard Analysis and Critical Control Point System and Importance of Verification Procedures

Jimmy T. Keeton and Kerri B. Harris

Introduction

The Hazard Analysis and Critical Control Point (HACCP) system is a preventative food safety management system that is recognized internationally as the most effective way to produce safe food (Stevenson and Bernard 1999, Mortimore and Wallace 2001). This system is a methodical and systematic application of scientific and technological principles designed to control and document the safe production of foods. Its primary objective is to make a product safe to consume and to be able to prove it (Stevenson and Bernard 1999). Its premise is based on the prevention of hazards by identifying food safety problems during processing and thus reducing or eliminating the need for finished-product inspection and testing. Hazards are categorized as microbiological, physical, or chemical, with microbiological hazards posing the greatest threat to public health.

This chapter provides a general overview of the basic principles of HACCP with an emphasis on meeting the current requirements specified by the U.S. Department of Agriculture Food Safety Inspection Service (USDA-FSIS) in Title 9 of the Code of Federal Regulations Part 304 et al. (USDA-FSIS 2002). Application of the HACCP system to the food industry requires that periodic reviews be conducted because of changes in regulatory requirements, development of new and innovative technologies, emergence of new hazards (especially microbial pathogens), or failure in part of the system. Special emphasis will be given to critical verification procedures that provide confirmation that the HACCP system is reliable and effective for reducing, removing, or preventing hazards in our foods.

A Brief History

HACCP concepts were developed in 1959 by the Pillsbury Company in conjunction with the U.S. Army Natick Laboratories and the National Aeronautics and Space Administration (NASA) to meet the safety requirements imposed by NASA primarily to both prevent crumbs and water in a zero-gravity environment that could interfere with space capsule electrical systems and to prevent a food poisoning outbreak in an enclosed space environment. The concepts were based on the engineering system of Failure, Mode, and Effect Analysis (FMEA), which considers what potentially could go wrong at each step of an operation and then places controls at critical points in the operation to prevent hazards. Three basic HACCP principles were first presented at the National Conference on Food Protection in 1971 (U.S. Department of Health, Education and Welfare 1971). In 1985, a Subcommittee of the Food Protection Committee of the National Academy of Sciences issued a report on the establishment of microbiological criteria for foods and recommended that regulators and the food industry adopt the HACCP system as the most effective means of protecting the food supply. In 1988, the National Advisory Committee on Microbiological Criteria for Foods (NACMCF), an expert scientific advisory panel, was established by the Secretaries of Agriculture, Commerce, Defense, and Health and Human Services to encourage the adoption of the HACCP system by the food industry (NACMCF 1997, NACMCF 1998). Seven HACCP principles were identified along with a systematic approach for establishing HACCP programs in the food industry.

The U.S. General Accounting Office, in a series of reports between 1992 and 1994, endorsed HACCP as an effective, scientific, risk-based system for protecting the public from foodborne illness. On December 18, 1995, the U.S. Food and Drug Administration published the final rule requiring the adoption of HACCP systems in seafood processing plants. Other international and governmental bodies such as the United Nations World Health Organization (WHO)/Food and Agriculture Organization (FAO) Codex Alimentarius Commission also advocated adoption of HACCP systems in the food industry and worked in concert with the NACMCF to revise the HACCP principles.

As part of its overall food safety strategy to reduce the risk of foodborne illness associated with pathogens such as *Salmonella*, *Escherichia coli* O157:H7, and *Listeria monocytogenes*, the USDA-FSIS issued a final rule on July 25, 1996 (Federal Register 1996), requiring meat and poultry establishments to adopt a science-based HACCP system and to develop plant- and product-specific HACCP plans. According to Title 9 of the Code of Federal Regulations Part 304 et al. (USDA-FSIS 2002), the regulations require that each establishment develop and implement written sanitation standard operating procedures; that generic *E. coli* testing be performed by slaughter establishments to verify the adequacy of the establishments' process controls for the prevention and removal of fecal contamination and associated bacteria; that slaughter establishments and those producing raw ground products meet pathogen-reduction performance standards for *Salmonella*; and that all meat and poultry establishments develop and implement a system of preventative controls designed to improve the safety of their products. Thus, these regulations require that all meat and poultry plants under federal or state inspection develop, implement, and maintain a HACCP program.

Basic Principles

Seven HACCP principles provide a framework for identification, evaluation, and control of food safety hazards. The seven principles (NACMCF 1998) are applied to the development of an effective HACCP plan and require the HACCP team to conduct a hazard analysis, determine the critical control points (CCPs) in a processing sequence, establish critical limits (CLs) for the CCPs, establish monitoring procedures for the CCPs, establish corrective actions when monitoring indicates a CCP is out of control, establish verification procedures to ensure that the HACCP system is working, and establish effective record keeping and documentation procedures.

Briefly, HACCP system plans are developed by first accomplishing five preliminary tasks (Stevenson and Bernard 1999, NACMCF 1998) before application of the principles. These tasks require assembling a HACCP team (usually from in-plant personnel); describing the food products produced and their distribution; detailing the intended use and potential users of the food products described; developing a flow diagram listing all of the steps in the food production process to include receipt of raw materials, receipt of ingredients, receipt of packaging, handling of rework, handling of shipping/distribution, and handling of product returns; and verifying the flow diagram for accuracy and completeness.

Application of HACCP principles in the development of a HACCP plan can be summarized in the following steps (Stevenson and Bernard 1999, NACMCF 1998, Mayes 2001, Federal Register 1996):

(1) Conduct a hazard analysis—Each step of the flow diagram is evaluated to identify hazards (microbiological, physical, or chemical) that are reasonably and likely to occur. Consideration should also be given to the severity of a potential hazard, its magnitude for harm, and the duration or aftereffect of a disease or injury.

(2) Determine CCPs—A CCP is a point, step, or procedure in the process at which control can be applied and a food safety hazard prevented, eliminated, or reduced to acceptable levels. If a hazard cannot be controlled at a particular step, that processing step cannot be a CCP.

(3) Establish CLs—A CL is a maximum or minimum value to which a hazard must be controlled at a CCP to keep food safe. CLs are usually established on the basis of scientific studies or regulatory requirements/guidelines that reduce the risk of or eliminate a hazard.

(4) Establish monitoring procedures—Monitoring is a planned sequence of observations or measurements to assess whether a CCP is under control, to indicate a trend toward loss of control, and to produce an accurate record or log for future use in verification. Monitoring must accurately and rapidly measure the CL in real time at a specified frequency or interval (continuous would be ideal) and should be performed by a designated individual trained in monitoring procedures. Results should be recorded, signed, and dated by the person monitoring the CL. These records are legal documents and provide evidence that the process is under control or that corrective action is warranted. Examples of monitoring activities are given in Table 20.1.

Table 20.1 Examples of real-time monitoring methods used in a Hazard Analysis and Critical Control Point plan.

Methods for Monitoring Critical Limits		
Physical	Chemical	Biological
Temperature indicators	Water activity (a_w)	Rapid pathogen
Continuous chart recorders	pH	screens/detection
Equipment alarm systems	Nitrite	methods
Metal detectors	Chlorine	
Magnets	Enzyme-linked	
X-ray devices	immunosorbent assay tests	
Sifters/sieves	for allergens	
Bone/object removal devices	Titratable acidity	
Direct observation to remove	Moisture	
visible contamination	Moisture:protein ratio	

(5)   Establish corrective actions—Corrective actions are taken any time a process deviates from its CLs. The HACCP Final Rule requires the following four corrective actions to be taken and documented: determine the cause of a deviation and correct or eliminate the cause, bring the CCP under control after corrective action or actions are taken, document that corrective measures were taken and procedures implemented to prevent recurrence, and ensure that no product that is injurious to health or otherwise adulterated as a result of the deviation enters commerce. A sample worksheet for a HACCP plan summary is shown in Table 20.2.

(6)   Establish verification procedures—Verification procedures are those activities (other than monitoring) that determine the validity of the HACCP plan and verify that the system is operating as intended (Table 20.3). Validation is an important and often overlooked component of this principle. Validation is required to ensure that the process will accomplish what it is designed to do. To fulfill the USDA's HACCP regulation, the ongoing verification procedures include, but are not limited to, the calibration of process-monitoring instruments, direct observations of monitoring activities and corrective actions, and a review of records generated and maintained during processing events.

(7)   Establish record-keeping and documentation procedures—Documentation for each plan should include, first, a written hazard analysis (including all supporting documentation); a listing of the HACCP team and assigned responsibilities; all materials that describe the food, its ingredients, distribution, and intended use; and potential consumers of the product. Second, the written HACCP plan, including decision-making documents associated with the selection and development of CCPs and CLs, documents supporting both the monitoring and verification procedures selected, and the frequency of those procedures. Third, monitoring of CCPS (and their CLs) such as times, temperatures, quantifiable values, corrective actions taken, calibration of monitoring instruments, verification procedures accomplished, and appropriate product, lot number, or production date codes. Entries onto records or into computers should be made in real time (as an event occurs) and dated, signed, or initialed by the person monitoring the process.

Hazard Analysis and Critical Control Points

One of the most difficult parts of developing a HACCP plan is conducting the hazard analysis. Every step of the process must be considered to ensure that food safety hazards are properly identified. It is appropriate to identify a food safety hazard at the step that it is being introduced into the process flow and to have subsequent steps in the process at which it will be controlled. For example, the hazard analysis for a fully cooked product may identify pathogens as reasonably likely to occur as biological food safety hazards at the point of receiving the raw material, and yet, the CCP may be at the subsequent step for fully cooking the product. It is important to remember that the identification of the food safety hazard does not have to be at the same step as the control measure.

After the food safety hazards are identified, it is crucial that the appropriate points of control are established. By definition (NACMCF 1998), CCPs are steps at which control can be applied and are essential to prevent or eliminate a food safety hazard or reduce it to an acceptable level. During the production of raw foods, such as raw, ground products, it is argued that as of today there is no scientifically validated CCPs that can be applied during

Table 20.2 A Hazard Analysis and Critical Control Point plan summary and worksheet designating the critical control point description, critical limits, monitoring procedures, and corrective action or actions.

Critical control point	Hazard	Critical limit or limits	Monitoring procedures	Corrective action or actions
Step to be monitored where a hazard can be prevented, eliminated, or reduced to acceptable levels	List the specific hazard or hazards based on the following categories: Biological, physical, chemical	A maximum or minimum value to be monitored at a critical control point; critical limits are established on the basis of scientific evidence that the critical limit is effective at reducing hazards	What is monitored Measurement methodology (How the critical control point is to be monitored) Frequency of monitoring Responsible person Record observations on the specified form, initial, date, and sign	Records should document corrective actions to: 1) Identify and eliminate cause of deviation 2) Bring the critical control point under control 3) Identify measures taken to prevent recurrence of a deviation 4) Ensure that no product that is injurious to health or adulterated enters commerce Quality Assurance is responsible for ensuring that all corrective actions are taken and documented.
Date:	Signature:		Revision No.	

Adapted from Harris (2002).

Table 20.3 Examples of verification and validation activities.

Critical control point	Verification	Validation
Marination: Marination of tenderized marinated beef (boneless NY strip steaks) Hazard: Pathogen growth may occur if temperature is abused or if temperature exceeds critical limit of 50°F	Quality control (QC): QC personnel or designee will take the internal product temperature from five steaks exiting the tumbler and record all five temperatures on the Product Temperature Log. Temperatures shall not exceed 44°F. Frequency: Performed once every 2 weeks; date and time of day are randomly selected using a random number table. Calibration: Calibration of recording thermometer in cooler is conducted once each month. Calibrate to manufacture's specifications. Calibrate handheld thermocouple thermometer to ±1°F. Calibrate before each verification test using slush ice and a National Institute of Standards and Technology–certified reference thermometer. Recording: All recording charts verified daily by QC manager or designee. Check temperature recorded on Product Temperature Log, Recording Chart, and Calibration Log. Check for correct date and appropriate signatures.	Microbial loads on food and contact surfaces do not increase as long as room temperature is maintained at ≤50°F for <4 hours (Brashears et al. 2002).
Final beef carcass wash at 200°F Hazard: Pathogens (*Salmonella* spp., *Escherichia coli* O157:H7) will be reduced on carcass surfaces by exposure to hot water	Directly observe corrective actions as needed. QC: QC manager or designee directly observes monitoring personnel performing final wash procedure weekly. Findings are recorded on the verification form. Final wash records will be reviewed weekly. Calibration: Calibration of in-line thermometer of Chad beef wash cabinet and chain speed calibrated weekly in indicated delivery time/temperature of 200°F water. Review of corrective action log weekly.	Spraying carcasses with 180°F water results in a significant reduction in aerobic plate count at all levels of inoculation from 2.5 to 6.7 logs (Dorsa et al. 1996). Mean reduction in aerobic plate counts and *E. coli* counts on carcasses were 2 log colony-forming units/cm^2 when Chad spray washers were used. Temperature range of 162–184°F for a wash duration of 11–18 seconds (Reagan et al. 1966).

Table 20.3 Examples of verification and validation activities. *(Cont.)*

Critical control point	Verification	Validation
Mild heating post-fermentation Hazard: Heating to 142°F for ≥18 minutes will destroy pathogens (*Salmonella* spp., *E. coli* O157:H7, *Listeria monocytogenes*, *Staphylococcus aureus*)	QC: QC supervisor or designee will review smokehouse oven circular recording charts, daily heating log, corrective action log, and instrument calibration records once per week. QC supervisor observes oven operator measure internal temperature of product twice per week. Calibration: Temperature recorders on smokehouse oven will be calibrated by engineering department personnel once per week. Digital handheld thermometers will be calibrated daily by QC personnel.	Heating to 142°F for 8 minutes produces a 7 $\log_{10}$ lethality for *Salmonella* (Appendix A, U.S. Department of Agriculture Food Safety Inspection Service Compliance Guidelines for Meeting Lethality Performance Standards for Certain Meat and Poultry Products; http://www.fsis.usda.gov/OA/fr/95033F-a.htm).

the production process that will prevent, eliminate, or reduce to an acceptable level biological food safety hazards, such as *E. coli* O157:H7. (It is noted that irradiation is a process that can be applied to the finished product to reduce microbial contamination.) Producers of raw products often rely on the programs in place by their suppliers to reduce the likelihood of occurrence of pathogenic microorganisms. Therefore, the HACCP in a raw process may focus on minimizing the potential growth of biological food safety hazards rather than on reducing them. Although the control of pathogen growth may not meet the true definition of a CCP because one cell of *E. coli* O157:H7 may cause illness, it is one of the only points of control that can be applied in raw-food processing establishments. It is unrealistic to expect that raw food products, such as ground beef and poultry, will be pathogen free; therefore, optimal food safety must include all sectors of the food production process as well as the end-users and consumers.

To be successful, the HACCP system must clearly identify the food safety hazards that are reasonably likely to occur and the points of control for each. Every establishment must be able to support their decisions in the hazard analysis and in the selection of the CLs for each CCP.

Verification

Verification procedures are ongoing activities performed to ensure that the HACCP plan is being implemented or functioning properly through regularly scheduled activities and reviews or audits of records. These activities are done to confirm the accuracy of the records, to observe that procedures are being performed, to determine whether monitoring activities and corrective actions are effective, and to ensure the reliability of process-monitoring devices by calibration. Verification activities also include auditing the entire HACCP plan or performing an in-depth review and analysis of specific records, logs, or data to ensure compliance with CLs (i.e., review of microbiological/chemical tests of product or ingredient samples, calibration charts, sanitation schedules, or equipment maintenance records, and even evaluation of consumer complaints to understand potential problems with a product; Compliance Manual for Food Quality and Safety 2002).

Records review includes verifying correct dates on forms, having an appropriate signature or initials of the responsible individual, having a description of monitoring results, performing the corrective actions, and ensuring that missing data do not imply deviations from a CL. Direct observation of a person performing a monitoring function on a periodic basis is an additional method for providing evidence that the HACCP plan is effective. Calibration of instruments used to monitor temperature, pressure, flow rate, pH, or water activity at CCPs is another means of process verification.

Verification audits or reviews should be carried out on a routine basis but also may be unannounced. Events that may necessitate verification of the HACCP plan include

- Identification of deficiencies through random sampling
- Deviations between monitoring records and the corrective actions taken
- Emergence of new pathogens
- Adaptation of new technologies in the plant
- Foods similar to those manufactured in the plant being implicated in a foodborne disease outbreak (Center for Disease Control summaries/*Morbidity and Mortality Weekly*) or product recall (USDA-FSIS)

- When foods similar to those manufactured in the plant have been implicated in a food-borne disease outbreak or a product recall, information regarding these foods, other related disease outbreaks, or the status of similar recalls may be found at the following Web sites: http://www.cdc.gov/mmwr//weekcvol.html; http://www.fsis.usda.gov/
- New lines or equipment being installed or ingredients/formulations changed.

Additional events might include

- Adoption of new testing methods
- Changes in pathogen predictive models
- Changes in product distribution or the target consumer
- Regulatory agency alerts and new scientific information
- Changes in training procedures
- Modifications in the existing HACCP plan.

Comprehensive audits of the HACCP plan are typically conducted by personnel internal or external to the food operation. Often these are third-party experts or regulatory personnel that serve as an independent authority (i.e., food safety consultants, customer audits, or USDA-FSIS Consumer Safety Officers).

Verification versus Validation

Once a HACCP plan has been developed, it must be validated before implementation to determine whether all hazards have been identified and will be effectively controlled. An initial validation includes evaluating activities or providing evidence that the plan is functioning as intended. This may include repeated tests of the CCPs, CLs, monitoring and corrective action procedures, and record keeping. Other determinants of plan effectiveness are to provide scientifically valid and technically sound evidence that all hazards have been identified and that they can be controlled. This may include advice from food safety experts, scientific studies, regulatory mandates, in-plant observations, and company-generated data. Examples of validation procedures that might be performed to establish control criteria for a CL include performing a study to determine the "cold spots" in a processing oven that can then be monitored to ensure that products have received sufficient heat to destroy human pathogens; performing a study to show that the combined effect of drying, acidification, and mild heat treatment controls pathogens of concern for a particular food product; or providing a controlled scientific study that proves that a newly approved decontamination agent is effective for reducing, inhibiting, or eliminating *L. monocytogenes* on the surfaces of ready-to-eat (RTE) meat products postcooking.

Validation is required to ensure that the CLs in place will actually control the identified hazards as designed. In one instance, a plant could implement a HACCP system that is not valid, and the identified hazards would still be likely to occur. In another, a plant could have designed a valid HACCP system but failed to properly implement it, and the identified hazards would still be likely to occur. Therefore, proper validation and verification activities are needed to ensure that the HACCP system will control the identified food safety hazards and that the plan is consistently implemented.

Regulatory Verification Methods and Testing

In the United States, the USDA-FSIS issued a directive (FSIS 2002) that established a more intense stratified sampling plan for plants that produce RTE meat products depending on the risk of *L. monocytogenes* contamination (high, medium, and low risk) for a specified product category (not heat treated—shelf stable, heat treated—shelf stable, fully cooked—not shelf stable, and product with secondary inhibitors—not shelf stable) and on the size of the operation. Regulatory actions are described for incidents in which a sample tests positive for *Salmonella* or *E. coli* O157:H7. These verification procedures are in addition to a processor's verification tasks and were established on the basis of product contamination risk and the agency's goal of further reducing microbial hazards.

Further regulatory changes by the USDA-FSIS now mandate that for some categories of products (i.e., RTE) meat and poultry establishments must specifically address the control of *Listeria* through a written program such as their HACCP systems, sanitation standard operating procedures, or other prerequisite programs as well as verifying the effectiveness of these actions through testing, and then share the results with FSIS (Federal Register 2003). Under the rule, processing establishments must choose one of three alternatives to control *L. monocytogenes* and verify the treatment's effectiveness: first, employ both post-lethality treatment (i.e., post-heating pasteurization in a sealed package) and a growth inhibitor (i.e., sodium diacetate, potassium lactate, sodium citrate). Establishments will still be subject to FSIS verification sampling procedures (directed sampling by inspection personnel as described previously) and must validate the post-lethality treatment's effectiveness. Second, employ either a post-lethality treatment or a growth inhibitor for the RTE product; more frequent sampling will occur by FSIS than in the first alternative. Third, employ sanitation measures only. Plants choosing this option will incur the most frequent sampling by FSIS, especially those producing hot dogs and deli meats, as these were identified by FSIS and the U.S. Food and Drug Administration as being high-risk products for listeriosis. The procedures described indicate the regulatory agency's move toward a more risk-based verification testing program especially for large-volume processors. As a consequence of increasing verification methods and testing protocols to reduce the risk of *Listeria* contamination, product manufacturers will now be allowed to identify products with enhanced safety specific to this pathogen.

Improvements in HACCP

The HACCP system, as viewed from either a regulatory or non-regulatory perspective, has some inherent pitfalls and could be improved. The following points are intended to identify some of the critical issues that reduce the effectiveness of the HACCP system, allow for creative changes in the system to enhance the safety of our food supply, and foster innovative research to address the problems identified.

First, HACCP systems could be more effective if HACCP concepts were applied to the entire food chain (producer, processor, distributor, food service, and retailer or commercial end user).

Second, effective HACCP systems and food safety programs must have a high priority and financial support at the corporate level to allow optimum development and implementation of plant-specific plans.

Third, HACCP must be science-based for effective application and should focus on control and prevention of food safety hazards whether applied from the regulatory or non-regulatory perspective. Science-based HACCP initiatives are foundational to systems operations and regulatory monitoring.

Fourth, regulatory HACCP needs greater uniformity of interpretation and enforcement across the meat and poultry industry. Education and retraining of the inspection force are needed to increase uniformity of regulatory enforcement in addition to accountability for the enforcement decisions made.

Fifth, sufficient numbers of scientifically trained inspection personnel are not available to provide comprehensive risk assessment and interpretation of effective intervention procedures. An increased number of scientifically trained personnel in the inspection force is needed.

Sixth, adequate education and training of non-regulatory employees and management can be a weak link in the HACCP system. Food safety education and training of employees must be continuous.

Seventh, co-training of regulatory and non-regulatory personnel could balance HACCP expectations and enhance understanding of food safety requirements.

Eighth, testing programs employed by regulatory and non-regulatory entities that are not science-based give a false sense of security and do not truly reduce the risk of foodborne disease. An independent commission such as one established by the National Academy of Sciences should have a mandate to critically review the scientific basis of regulatory testing programs.

Ninth, application of HACCP may engender unrealistic consumer expectations with regard to food safety risks. Consumer education to prevent foodborne disease must be vigilant (see Chapter 30).

Tenth, scientific guidelines for optimal use of standardized testing methods to validate hazard interventions are neither available nor codified. A standardized sampling and testing manual is needed for regulatory and non-regulatory entities.

Eleventh, HACCP principles were designed to eliminate end-product testing for food safety, and current regulations that require testing for the presence of pathogens actually may be detrimental to or discourage further testing

Twelfth, meat and poultry industries need to aggressively develop and implement new pathogen interventions. Research support is needed for the development of new technologies that prevent and detect pathogen contamination of foods.

Thirteenth, pathogen-detection sensitivity may exceed the limits of scientific control of pathogens in foods. Critical analysis of pathogen outbreaks and a thorough understanding of pathogen incidents and infectivity are needed to assess disease risks and perhaps establish tolerable limits for certain pathogens.

Fourteenth, use of pathogens for validation and verification of process effectiveness may not be scientifically sound, as statistical frequencies of occurrence can limit microbial detection in foods.

Finally, regulatory approval within and between federal agencies of new hazard prevention innovations or pathogen interventions needs to be expedited.

Conclusions

HACCP is a dynamic process control system that can be applied throughout all sectors of the food industry, including the end-user and consumer, to produce safe food and protect public health. Verification procedures performed before the initiation of a HACCP plan in a food processing facility, or on an ongoing basis, are designed to ensure that the food safety system is operating effectively and that appropriate safeguards (identification of hazards, establishment of CLs, monitoring procedures, and corrective actions) are in place to produce a food product safely and to be able to prove it. Ongoing verification consists of a series of audits that include, but are not limited to, periodic review of the HACCP plan, calibration of process-monitoring instruments, random sampling, direct observation of monitoring activities and corrective actions, and review of records generated by the HACCP system. These audits may be conducted internally or externally and often involve third-party experts or regulatory personnel that serve as an independent authority (i.e., food safety consultants, International HACCP Alliance or USDA-FSIS Consumer Safety Officers). Effective verification procedures should reduce or eliminate the need for end-product testing.

Validation provides evidence that the HACCP system and components of the HACCP plan have adequate control measures in place for preventing hazards and that they have a sound scientific basis. Regulatory agencies have increased pathogen sampling of RTE meat and poultry establishments through a risk-based verification testing program to reduce the incidence of listeriosis from high-risk products. This program should be used in addition to a plant's HACCP verification program and provides an external means of evaluating a plant's HACCP program effectiveness.

HACCP development and implementation is an evolving process, and the food industry is constantly striving to use new technologies and scientific information to strengthen the existing food safety programs. Everyone, from the producer and processor to the consumer, must accept responsibility for food safety.

References

Brashears, M. M., Dormedy, E. S., Mann, J. E. and Burson, D. E. 2002. Validation and optimization of chilling and holding temperature parameters as critical control points in raw meat and poultry processing establishments. *Dairy, Food, and Environmental Sanitation* **22**:246–251.

Compliance Manual for Food Quality and Safety. 2002. Hazard Analysis and Critical Control Points, pp. 1–115. *In* Compliance Manual for Food Quality and Safety. Neenah, Wisc.: J. J. Keller and Associates.

Dorsa, W. J., Cutter, C. N., Siragusa, G. R. and Koohmaraie, M. 1996. Microbial decontamination of beef and sheep carcasses by steam, hot water spray washes, and a steam-vacuum sanitizer. *Journal of Food Protection* **59**:127–135.

Federal Register. 1996. Pathogen Reduction; Hazard Analysis and Critical Control Point (HACCP) Systems; Final Rule. Department of Agriculture, Food Safety and Inspection Service. Title 9 CFR Parts 304, 308, 310, 320, 327, 381, 416 and 417. *Federal Register* **61**:38805–38989.

Federal Register. 2003. Control of *Listeria monocytogenes* in Ready-to-Eat Meat and Poultry Products; Interim Final Rule Department of Agriculture, Food Safety and Inspection Service. Title 9 CFR Part 430. *Federal Register* **68**:34207–34254.

Food Safety Inspection Service. 2002. Microbial Sampling of Ready-To-Eat (RTE) Products for the FSIS Verification Testing Program. 10,240.3, December 9, 2002. Washington, D.C.: United States Department of Agriculture, Food Safety Inspection Service.

Harris, K. B. 2002. Example plans. *In* Developing and Implementing HACCP Plans for the Meat Industry—A Short Course Manual. College Station, Tex.: International HACCP Alliance.

Mayes, T. 2001. Introduction, pp. 1–10. *In* Making the Most of HACCP, T. Mayes, S. Mortimore (Eds.). New York: CRC Press.

Mortimore, S. and Wallace, C. 2001. Introduction to HACCP, pp. 1–4, 70–75. *In* Food Industry Briefing Series: HACCP. Osney Mead, Oxford: Blackwell Science.

National Advisory Committee on Microbiological Criteria for Foods. 1997. Hazard Analysis and Critical Control Point Principles and Application Guidelines. National Advisory Committee on Microbiological Criteria for Foods. Washington, D.C.: U.S. Department of Agriculture Food Safety Inspection Service.

National Advisory Committee on Microbiological Criteria for Foods. 1998. Hazard analysis and critical control point principles and application guidelines. National Advisory Committee on Microbiological Criteria for Foods. *Journal of Food Protection* **61**:1246–1259.

Reagan, J. O., Acuff, G. R., Buege, D. R., Buyck, M. J., Dickson, J. S., Kastner, C. L., Marsden, J. L., Morgan, J. B., Nickelson II, R., Smith, G. C. and Sofos, J. N. 1996. Trimming and washing of beef carcasses as a method of improving the microbiological quality of meat. *Journal of Food Protection* **59**:751–756.

Stevenson, K. E. and Bernard, D. T. 1999. Introduction to Hazard Analysis and Critical Control Point Systems, pp. 1–4. *In* HACCP: A Systematic Approach to Food Safety, 3rd ed. Washington, D.C.: The Food Processors Institute.

U.S. Department of Agriculture Food Safety Inspection Service. 2002. Pathogen Reduction; Hazard Analysis and Critical Control Point (HACCP) Systems. Title 9 Code of Federal Regulations, Parts 304, 308, 310, 320, 327, 381, 416 and 417. Washington, D.C.: U.S. Government Printing Office.

U.S. Department of Health, Education, and Welfare. 1971. Proceedings of the 1971 National Conference on Food Protection. U.S. Department of Health, Education and Welfare. Washington, D.C.: Public Health Service.

21 Are They Vibrios? How Do You Know?

Sam W. Joseph

Introduction

Attention is usually drawn to the disease cholera on hearing or seeing the term *"Vibrio,"* which is not surprising, as the prevalence of cholera goes back into antiquity (McNicol and Doetsch 1983, Barua 1992). For our purposes, however, real knowledge of this disease dates back to Pacini and his study of cholera in the mid 1800s (Pacini 1854). In fact, he was the first to view the comma-shaped bacillus microscopically. On the basis of his diverse study of the disease, he concluded that it was the cause of an epidemic, intestinal, mucous-damaging, diarrheal-type disease in human hosts. His work chronologically paralleled that of John Snow, who was performing his pioneering epidemiologic studies in cholera-ravaged London, England. Snow essentially discovered the link between cholera and a source of drinking water in Central London (the infamous Broad Street Pump; Snow 1936).

After Pacini's work was largely forgotten, Robert Koch revived those studies in 1884 while investigating an epidemic of cholera in Cairo, Egypt. He was the first to actually culture and macroscopically view the organism on the surface of an artificial, solid medium and to demonstrate that the cause of the disease was bacterial (Koch 1884). Since those early studies, nations almost globally have been involved with pandemic cholera. The emergence of cholera as a pandemic disease seemed to occur with the advent of travel over trade routes (Lacy 1995). Eventually, spread of the disease was associated with pilgrimages (hajjs) to Mecca from various points of the Islamic world. There have been seven pandemics, with the seventh originating in Indonesia in 1961. According to Pollitzer (1959), the dates of the recognized pandemics are as follows: first pandemic 1817–1823, second pandemic 1829–1851, third pandemic 1852–1859, fourth pandemic 1863–1879, fifth pandemic 1881–1896, and the sixth pandemic 1899–1923. The United States was affected by four of the seven pandemics. There were over 150,000 deaths after two outbreaks during the second pandemic in the mid-1800s and 50,000 more during the fourth pandemic in 1866 (Morris and Black 1985).

The disease was essentially not detected in the United States after 1900. The only reported cases thereafter were either imported or the result of laboratory accidents. There was one case of a longshoreman dying of cholera in 1941 that was unexplainable. The next reported case of native cholera occurred in 1973 in Port Lavaca, Texas, where a shrimp fisherman was diagnosed, and a cholera toxin (CT)-producing *Vibrio cholerae* O1 Biotype El Tor was isolated from his stool (Weissman et al. 1974). Since the report of that case, there continue to be sporadic and erratic occurrences of apparently indigenous cholera mostly occurring around the Gulf Coast area of the United States (Blake 1994).

Until the seventh pandemic, the primary biotype of *V. cholerae*, the actual cause of pandemic cholera, was the so-called "classical" biotype. With the advent of the seventh pandemic, a new biotype, "El Tor," emerged out of Indonesia (Table 21.1).

The particular *V. cholerae* agent (either biotype) that was responsible for causing pandemic cholera is characteristically serogroup O1 and CT positive. There are presently over

Table 21.1   Differentiation of *V. cholerae* biotypes.[a]

Test	Classical	El Tor
Susceptibility to polymyxin B	+[b]	−[c]
Hemolysis (sheep erythrocytes)	−	V[d]
Voges-Proskauer	−	+
Susceptibility to El Tor phage V	−	+
Susceptibility to Classical phage IV	+	−
Hemagglutination (chicken erythrocytes)	−	+

(a) Adapted from Elliott et al. 1992
(b) +, positive.
(c) −, negative.
(d) V, variable.

200 serogroups of *V. cholerae*, some of which rarely produce CT and are sometimes associated with human disease, but none that are associated with pandemic cholera. Neither is O1, non-CT producing *V. cholerae*. These non-O1 *V. cholerae* organisms are generally termed non-Cholera-Vibrios.

In fact, there are presently over 20 species described for the genus *Vibrio*, approximately 10 of which have been associated with diseases of humans (Table 21.2). The species most frequently isolated after *V. cholerae* O1 is *V. parahaemolyticus*. Therefore, the isolation, identification, and verification of *V. cholerae* and *V. parahaemolyticus* organisms will be primarily featured here.

Vibrio cholerae

Cholera is one of the few diseases that commands the attention of people from all walks of life, especially professionals in the medical field, because of its capacity to spread rapidly

Table 21.2   Association of *Vibrio* spp. with different clinical conditions.[a,b]

Species	Clinical condition				
	Gastroenteritis	Wound infection	Ear infection	Primary septicemia	Secondary septicemia
V. cholerae O1	+++	+			
V. cholerae non-O1	+++	++	+	+	+
V. mimicus	++		+		
V. fluvialis	++				
V. parahaemolyticus	+++	+	+		+
V. alginolyticus	(+)	++	++	+	
V. cincinnatiensis			+		
V. hollisae	++			+	
V. vulnificicus	+	++		++	++
V. damsela		++			
V. carchariae		+			

(a) Adapted from Elliott et al. 1992.
(b) +++ = frequent, ++ = less common; + = rare, (+) extremely rare.

and globally. The epidemic in South America in 1991 is an excellent example of the spontaneous and rapid nature of this feared disease. In 1991, cholera began in Peru and quickly spread to other parts of the continent, eventually reaching Central America as far North as Mexico. By 1994, over one million people had contracted the disease, resulting in approximately 10,000 deaths (Centers for Disease Control [CDC] 1995). Cholera is still prevalent in Lima, Peru, and people, in particular those who are forced to live in low sanitary conditions, are mostly at risk.

A sudden increase in cholera occurring first in Madras, India, in 1992 spread soon thereafter, approaching epidemic numbers, to 11 other countries in southern Asia, primarily Bangladesh and Burma (Cholera Working Group 1993). The cause was found to be a presumably new strain of *V. cholerae*, designated "O139 Bengal." Thus, the prevailing notion that only Group O1 *V. cholerae* could cause epidemic cholera appeared to be purely dogma. This epidemic soon led investigators to believe that the world was on the verge of the eighth pandemic of cholera. Fortunately, the epidemic subsided, although the disease caused by this organism continues to persist. Interestingly, it was discovered through molecular analyses that O139 Bengal was essentially the same organism as *V. cholerae* O1 except that it has certain unique features including a different somatic antigen, the capacity to survive and spread in the environment, and a potential for capsule expression (Morris and the Cholera Laboratory Task Force 1994).

The discovery that a non-O1 *V. cholerae*, that is, O139 Bengal, could cause disease of epidemic proportions was almost iconoclastic. In fact, since the discovery of the organism by Koch in 1884, all other vibrios had been essentially regarded as so-called nonagglutinating vibrios (Hugh and Feeley 1972). They were deemed ordinary, unimportant aquatic organisms. In 1970, a large group of O1 and non-O1 *V. cholerae* strains was assembled for a polyphasic study in which the organisms were examined phenotypically for a large number of biochemical characteristics and were also examined genotypically by DNA–DNA hybridization analysis (Citarella and Colwell 1970, Colwell 1970). The majority of the strains were found to be essentially the same organism. These early findings and further confirmatory findings established that the nonagglutinating vibrios were definitively *V. cholerae*. In 1977, *V. cholerae* other than O1 was discovered in the Chesapeake Bay, and although they were considered a very low risk for the inhabitants of the region, the experience in Southern Asia with the O139 Bengal strain suggested that the finding may have been a portend that, in fact, a non-O1 strain might eventually challenge a human population (Colwell et al. 1977).

After the discovery of non-O1 *V. cholerae* in the Chesapeake Bay, other estuarine studies were performed with similar findings (Colwell et al. 1981, Desmarchelier and Reichelt 1981, Davey et al. 1982, Hood and Ness 1982, Lee et al. 1982, Madden et al. 1982, Roberts et al. 1982, Hood et al. 1983, Bradford 1984, Garay et al. 1985, Kaysner et al. 1987, Bravo Farinas 1991, Yamai et al. 1997). Cholera is transmitted through contaminated drinking water and can be isolated from household drinking water containers. The organism has also been isolated from surface-water reservoirs (Grizhebovskii et al. 2001, Ziatdinov 2002).

Microscopic and Culture Characteristics

Vibrio cholerae is the type species of the genus *Vibrio.* Members of the genus *Vibrio* are facultatively anaerobic, asporogenous, motile, curved or straight, Gram-negative rods

1.4–2.5 μm in length. They may require either minimal or larger amounts of salt. Some are enhanced by the addition of larger amounts of NaCl, whereas others are mainly halotolerant. *Vibrio cholerae* can grow optimally with as little as 0.2% Na^+ added to its growth medium (Baumann et al. 1984, Huq et al. 1984). It grows well at a pH of 9.0 and even at pH 10 but is inhibited at pH 6.0 or lower (Pollitzer 1959, Huq et al. 1984).

Sources of *Vibrio cholerae*

Water in the Transmission of Cholera

It is now established that the occurrence of *V. cholerae* in the aquatic environment is not dependent on recontamination via human excrement (Kaper et al. 1979, Colwell et al. 1981, West 1992). *Vibrio cholerae* is a component of the common microbiota in the aquatic environment and is frequently found to be associated with phytoplankton and copepods (Huq et al. 1983). The relative numbers isolated (colony-forming units [CFU]/mL) are a reflection of the optimal combination of water temperature, pH and salinity (A_w; Kaper et al. 1979, Singleton et al. 1982, Huq et al. 1984, Miller et al. 1984, Eyles and Davey 1988, Dalsgaard et al. 1995).

In less-than-optimal conditions, *V. cholerae* may reside in a viable-but-nonculturable stage until the return of favorable circumstances for cell growth and multiplication (Xu et al. 1982, Colwell and Huq 1994, Huq and Colwell 1996, Lipp et al. 2002). This cycle of the lifespan of the organism dictates their detection when aquatic samples are cultured for *V. cholerae* (Lipp et al. 2002). In the absence of direct isolation in culture media, fluorescence microscopy has been used to detect *V. cholerae* (Huq et al. 1990). Although non-detectable, these viable-but-nonculturable *V. cholerae* are still capable of resuscitation in the human intestine and causing disease, as shown by Colwell et al. (1996) in a study of human volunteers.

Cholera is caused primarily by drinking water or food contaminated by *V. cholerae*–containing water. Cholera is not believed to be transmitted from human to human. Water contaminated with *V. cholerae* is usually the primary source of the organism during epidemics and pandemics. Small outbreaks and sporadic cases are usually associated with various types of foods—especially seafoods. In one instance bottled water was responsible for a small outbreak of cholera in Portugal in 1974 (Blake et al. 1977a, Blake et al. 1977b). Migration of cholera as suspected in the South American outbreak has been investigated in several studies. The Latin American strain of *V. cholerae* O1 (El Tor, Inaba) was isolated along the coast of Alabama, leading to suspicion that bilge water dumped as ballast might be a means of importation of particular strains from one region to another (DePaola et al. 1992). Furthermore, analysis of bilge water from ships traveling from South America to the Northern Hemisphere provided additional credence for this hypothesis (McCarthy et al. 1992). In another study, ballast water of vessels arriving to the Chesapeake Bay from foreign ports was examined to estimate the abundance of total bacteria, virus-like particles, and *V. cholerae* (O1 and O139) in water and plankton samples. *Vibrio cholerae* O1 was found in plankton samples taken from the ballast tanks of all ships, and *V. cholerae* O1 and *V. cholerae* O139 were found in 93% of the ships. Furthermore, there were 100× more *V. cholerae* (O1 and O139) in the water samples than were found in the plankton samples (Ruiz et al. 2000).

Foods in the Transmission of Cholera

Seafoods, especially oysters, are high on the list of foods that can harbor *V. cholerae* O1 and transmit cholera, especially after inadequate cooking or recontamination after preparation. Various shellfish have been implicated in outbreaks including mussels (Dutta et al. 1971, Baine et al. 1974), cockles (Blake et al. 1977a), oysters (Weissman et al. 1974, Cameron et al. 1977, Klontz et al. 1987, Pavia et al. 1987, CDC 1989), crabs (Lin et al. 1986, CDC 1990), crab salad (CDC 1992a, CDC 1992b), and shrimp (Joseph et al. 1965). Other foods have been implicated including coconut milk imported from Thailand to Maryland in the United States (CDC 1991), fin fish (Merson et al. 1977, Tamplin and Parodi 1991, Epstein 1992), uncooked beef (Swaddiwudhipong et al. 1992), uncooked pork (Swaddiwudhipong et al. 1990), peanut sauce (St. Louis et al. 1990), and vegetables (Glass et al. 1992). Food contaminated with *V. cholerae* O1 and left unrefrigerated provides improved conditions for surviving organisms to grow and multiply, as described in the report of an outbreak on an oil rig off the coast of Texas. A potable water system was contaminated because of spill-over from a sewage line. Rice prepared with water from this source was left standing after preparation and was consumed several hours later. Apparently there was *V. cholerae* O1 in the sewage, because consumption of the rice led to contraction of cholera by 15 workers on the rig (Johnston et al. 1983). Ironically, this isolated situation in the United States provided a microcosm of a much more frequent occurrence of foodborne cholera in endemic and epidemic areas of Asia, South America, and Africa that harbor *V. cholerae* O1 in large numbers.

Outbreaks have occurred in passengers on flights originating in countries in which cholera is either epidemic or endemic. There were several instances of flights from South America arriving with passengers who, soon after landing, began having cholera-like disease. In one instance, the food served on a flight to the United States from Argentina with a stopover in Peru was suspected of causing 75 cases of cholera and one death. Other cases have involved isolated instances in which passengers have brought food (such as succulent crabs) from the originating country to share with family only to have the diners contract cholera.

Interestingly, *V. cholerae* O1 can survive refrigerated and frozen conditions in foods shipped internationally (CDC 1991). The increasingly cosmopolitan appetites of citizens in developed countries raise the possibility that contaminated foods may enter the marketplace from distant cholera-endemic shores, thus increasing the challenge to food safety regulators and inspectors.

The majority of outbreaks occur during the warmer months of the year, corresponding to the higher prevalence of *V. cholerae* in the water column (Kaper et al. 1979). Additional information on this subject can by found in reviews by DePaola (1981), Rippey (1991) and Lipp et al. (2002).

Non–Group O1 *Vibrio cholerae*

Non-O1 *V. cholerae* are ubiquitous in aquatic environments worldwide. Occasional cases of gastrointestinal infection occur in the United States, most frequently after eating oysters. Intestinal illness is usually accompanied by blood in the stool and fever. Unlike O1 *V. cholerae*, non-O1 *V. cholerae* can also cause skin and ear infections among other extraintestinal infections. It can be invasive following intestinal infection, leading to sepsis, which may result in death, but not because of volemic shock as observed with traditional cholera. Non-O1 *V. cholerae* very rarely produces the cholera toxin of O1 *V. cholerae* (Morris 1990).

Isolation of *Vibrio cholerae*

Procedures for the isolation and identification of vibrios are well described in the Bacteriological Analytical Manual (BAM) of the U.S. Food and Drug Administration (FDA; Elliott et al. 1992). This section will provide an overview of the methods described therein, with supplementary information from other sources. The materials for the methods described and their sources, as well as equipment and reagent formulae, are thoughtfully provided in the BAM (Elliott et al. 1992).

Processing Water Samples for Isolation of Vibrios

Because *V. cholerae* are usually present in very low numbers in water samples, in contrast to other members of the microbiota, they are difficult to culture. As a consequence, inability to detect the organism does not guarantee that the aqueous environment is free of cholera or, for that matter, other vibrios.

In an attempt to optimize the opportunity for growth, conditions are established to take advantage of the rapid growing and multiplication time of the organism and its tolerance to an elevated pH of 9. So the presumptive broth used for initial isolation is alkaline peptone water, which provides an advantage to the vibrios.

Filtration

The following method is used in the author's laboratory. Other methods are described in "Standard Methods for the Examination of Water" (Anonymous 1989). Because of the low numbers usually present in the water column, filtration using large volumes of a water sample (2.0–4.0 L) is passed through a 0.45-μM filter attached to a sterile 1.0-L bottle. The filter is then cut into small pieces using sterile scissors and placed in a 50-mL conical tube containing 25 mL of alkaline peptone broth. The tube is vortexed vigorously for 2–3 minutes, and then the contents are deposited in 250 mL of sterile alkaline peptone broth contained in a 500-mL Erlenmeyer flask. The suspension is incubated for 6–8 hours at 35°–37°C and processed as described later.

Quantification

In some instances it is desirable to enumerate the number of vibrios in the sample. Procedures are described to examine the filtered sample using the "Most Probable Number" (MPN) method for quantification. The procedure is statistically based, using a set of numerical codes based on the number of positive tubes out of five tubes inoculated to estimate the CFU per milliliter (Anonymous 1989).

Moore Swabs

Because of the difficulty in isolating vibrios, some investigators prefer to use the Moore swab, which is simply a bundle of sterile gauze swabs attached to a wire or strong cord that is lowered into a stream, river, sewer, and so forth. The swab is left in the sampling site for a prolonged period of time (at least 24 hours). The working principle here is that organisms will attach to the cotton material, eventually concentrating the organisms from the environment. Although this is not a quantitative method, it has proven effective over the years (Moore 1948).

Processing Food Samples for Isolation of Vibrios

Dice 25 g of a food sample and place it into 225 mL of alkaline peptone broth in a 500-mL Stomacher bag or blender jar. Process and then transfer the blended contents to a sterile 500-mL Erlenmeyer flask. For samples suspected of containing large numbers of other organisms, dilution (1:100 and 1:1,000) of the blended suspension is recommended.

Incubate the flask 6–8 hours at 45°–37°C except when processing oysters, which should be incubated at 42°C (Elliott et al. 1992). After this presumptive incubation, plate inocula from surface growth to a selective or selective/differential agar (e.g., thiosulfate-citrate-bile salts-sucrose [TCBS] agar) for isolation. Continue incubation of the alkaline peptone broth for 24 hours. Then once again inoculate selective/differential agar. On TCBS, colonies of *V. cholerae* (EL Tor and Classical) appear large, smooth, yellow (sucrose+), and slightly flattened with opaque centers and translucent peripheries. A sucrose-negative variant of *V. cholerae*, *V. mimicus*, produces green colonies on TCBS.

Alternatively, mCPC agar can be used; it contains modified cellobiose, polymyxin B, and colistin (see Table 21.3). The disadvantage of this medium is that polymyxin B inhibits the growth of the Classical Biotype of *V. cholerae*. Another alternative agar is Monsur's agar consisting of tauro-cholate, tellurite and gelatin (Monsur 1961). This medium is not commercially available but has the advantage that oxidase and agglutination tests can be performed on colonies directly on the plate.

Identification of Genus

First, streak at least three suspect colonies onto two Trypticase soy agar (TSA) plates containing 1% and 2% NaCl, respectively.

Second, perform an oxidase test on an 18–24-hour growth from TSA (not blood agar). Soak a piece of filter paper with oxidase reagent. Using a platinum loop, remove growth from the agar plate and apply it to the filter paper. A purple or blue color within 10 seconds, eventually turning black, is an indication of a positive reaction. Please note that all members of the genus *Vibrio* are oxidase positive except *V. metschnikovii* and *V. gazogenes*.

Third, subculture to triple sugar iron agar, Kligler iron agar, and arginine glucose slants to enable partial differentiation from *Aeromonas* and *Plesiomonas*.

Finally, inoculate two tubes of either Hugh–Leifson glucose broth or oxidation fermentation medium. These media should separate *Vibrio* and *Pseudomonas*. *Vibrio* can use glucose oxidatively and fermentatively, whereas *Pseudomonas* is limited to oxidative use only.

Confirmation of Genus

The genus *Vibrio* is confirmed by the following reactions: oxidase-positive, Gram-negative rods (straight or curved), sucrose positive (yellow) on TCBS agar, green (negative) for *V. mimicus*, acid/acid in triple sugar iron agar slants, produces acid from glucose oxidatively and fermentatively in oxidation fermentation glucose medium.

Table 21.3 Biochemical reactions of *V. cholerae* in comparison to less frequently isolated *Vibrio* spp.[a]

	V. alginolyticus	*V. anguillarum*	*V. carchariae*	*V. cholerae*[b]	*V. cincinnatiensis*	*V. damsela*	*V. fluvialis*
TCBS agar	Y	Y	Y	Y	Y	G	Y
mCPC agar	NG	NG	nd	P	nd	NG	NG
AGS medium	KA	nd	nd	Ka	nd	ng	KK
NaCl:							
0%	−[c]	−	−	+[d]	−	−	−
3%	+	+	+	+	+	+	+
6%	+	+	+	−	+	V[e]	+
8%	+	−	+	−	−	−	V
10%	+	−	−	−	−	−	−
Growth at 42°C	+	−	nd	+	−	−	V
Acid produced from:							
Sucrose	+	+	+	+	+	−	+
D-Cellobiose	−	+	+	−	+	+	+
Lactose	−	−	−	−	−	−	−
Arabinose	−	V	−	−	+	−	+
D-Mannose	+	+	+	+	+	+	+
D-Mannitol	+	+	+	+	−	−	+
Oxidase	+	+	+	+	+	+	+
ONPG	−	+	−	+	+	−	+
Voges-Proskauer	+	+	−	V	+	+	−
Arginine dihydrolase	−	+	−	−	−	+	+
Lysine decarboxylase	+	−	+	+	+	V	−
Ornithine decarboxylase	+	−	+	+	−	−	−
Sensitivity to:							
10 μg 0/129	R[f]	S[g]	R	S	R	S	R
150 μg 0/129	S	S	S	S	S	S	S
Gelatinase	+	+	+	+	−	−	+
Urease	−	−	−	−	−	−	+

(a) Adapted from Joseph et al. 1982 and Elliott et al. 1992.
(b) Includes both O1 and non-O1 serogroups.
(c) −, negative.
(d) +, positive.
(e) V, variable.
(f) R, resistant.
(g) S, susceptible.

Confirmation of Species

Pick individual colonies and inoculate onto TSA with no salt added and on one other TSA plate emended with 3% NaCl. *Vibrio cholerae* and *V. mimicus* will grow on both types of plates. Halophilic *Vibrio* spp. will grow only on the 3% NaCl–supplemented plate.

Then, inoculate the biochemical media listed in Table 21.4. For specific directions for the biochemical tests, refer to the FDA BAM manual (Elliott et al. 1992). All of the media should contain 2% NaCl in their formulations. If API strips (API 50 CH; bioMérieux, Inc.,

Table 21.4 Biochemical Characteristics for the Differentiation of the Most Frequently Isolated Vibrios.[a,b]

	Vibrio cholerae[b]	*Vibrio parahaemolyticus*	*Vibrio vulnificus*
Triple sugar iron agar	A (K rare)/A	K/A[c]	A (K rare)/A[d]
H$_2$S production	−[e]	−	−
Motility	+[f]	+	+
Growth in:			
0% NaCl	+	−	−
3% NaCl	+	+	+
6% NaCl	−	+	+
8% NaCl	−	+	−
10% NaCl	−	−	−
Acid produced from:			
Arabinose	−	+	−
Cellobiose	−	V[g]	+
Glucose	+	+	+
Lactose	−	−	+
Mannose	V	+	+
Sucrose	+	−	−
Gas from glucose	−	−	−
Decarboxylase:			
Arginine	−	−	−
Lysine	+	+	+
Ornithine	+	(+)[h]	(+)
Growth at 42°C	+	+	+
Hugh-Leifson:			
Oxidation	+	+	+
Fermentation	+	+	+
Gelatinase	+	+	+
Indole	+	+	+
ONPG	+	−	+
Oxidase	+	+	+
Susceptibility to O/129:			
10 μg	S[i]	R[j]	S
50 μg	S	S	S
Voges-Proskauer	V	−	−

(a) Adapted from Stavric and Buchanan 1995.
(b) Includes both O1 and non-O1 serogroups.
(c) K, alkaline slant.
(d) A, acid butt.
(e) −, negative.
(f) +, positive.
(g) V, variable.
(h) (+), weak response.
(i) S, susceptible.
(j) R, resistant.

Hazelwood, MD) are used, 0.85% NaCl is recommended as the diluent. See Table 21.4 for typical characteristics of *V. cholerae* as compared with *V. parahaemolyticus* and *V. vulnificus*. Table 21.3 provides the distinguishing features to differentiate *V. cholerae* frequently isolated vibrios. The minimal characteristics for identification of *Vibrio cholerae* are:

- Glucose fermentation, positive
- Cytochrome oxidase, positive
- Arginine dihydrolase, negative
- Lysine decarboxylase, positive
- Voges-Proskauer: El Tor biotype, positive
- Voges-Proskauer: Classical biotype, negative
- Growth at 42°C, positive
- Halophilism: 0%, positive
- Halophilism: 3%, positive
- Halophilism: 6%, negative
- Acid from arabinose, negative
- Acid from sucrose, positive
- 0129 susceptibility: 10 μg, susceptible
- 0129 susceptibility: 150 μg, susceptible

Differentiation of Biotypes of V. cholerae

The specific tests for differentiation of El Tor and the Classical biotypes are listed in Table 21.1. Methodologies for performing the tests are described in the FDA BAM manual (Elliott et al. 1992).

CT Assay

There are at least two very reliable tissue culture methods for assaying CT: the Y-1 mouse adrenal cell (Maneval et al. 1980) and the Chinese Hamster Ovary cell assay. If performed carefully, they are very reproducible and accurate. Typically, CT causes a rounding and enlargement of the cells after overnight incubation. Cytotoxicity causes obvious death, lysis, or detachment of the cells.

Serological Agglutination Test

Use diagnostic antisera of Group O1 and subgroup Inaba (factors AC) and Ogawa (factors AB) to serotype the O1 antigen and antisera or monoclonal antibodies to serogroup with the O139 antigen.

Vibrio parahaemolyticus

Vibrio parahaemolyticus was first isolated, identified, and described as the cause of gastroenteritis in Japan in 1951 but at the time was placed in the genus *Pasteurella*, then in a new genus *Oceanomonas*, and finally was designated as *Vibrio* by Sakazaki et al. (1963). In 1968, it was first isolated from the acquatic environment in the United States (Baross and Liston 1968, Joseph et al. 1982). *Vibrio parahaemolyticus* is a halophilic, sucrose-negative, Gram-negative rod found primarily in estuaries but that has been isolated in marine waters as well (Baross and Liston 1968, Barrow and Miller 1972, Binta et al. 1982, Urdaci et al. 1988, Levine and Griffin 1993, DePaola et al. 2003). It has been reported nearly worldwide from the environment and from human outbreaks and sporadic cases (Binta et al. 1982,

Doyle 1990). In the United States, there were approximately 15 outbreaks involving shellfish in the early 1970s (Barker 1974). Although the numbers of *V. parahaemolyticus* in fresh-caught seafood are quite low (<1 to 100 per gram), the numbers rise through delivery to restaurants, markets, and finally, to the kitchen. Thus, their ability to multiply rapidly under favorable circumstances enables them to establish infection. The infectious dose for human hosts ranges from 10^5 to 10^7 CFU (Joseph et al. 1982).

In a large outbreak involving crabs in Maryland, investigation revealed that the crabs had been recontaminated after steaming (Powell 1999). Improper cooking, under-refrigeration and cross-contamination also account for food-related *V. parahaemolyticus* infections. In Japan, infection is thought to arise primarily from consuming raw seafood, especially shirasu. Although this disease continues to rank as the second largest cause of gastroenteritis in Japan, it rarely causes outbreaks in the United States. However, isolated, sporadic cases continue to be reported. The reduction of cases in the United States probably reflects the successful consumer educational program on recontamination of seafood established by the FDA.

Most cases of disease occur in the summer months, corresponding to the rise in numbers in the aquatic environment; thus, in Japan the disease is commonly called "summer diarrhea" (Joseph et al. 1982). In tropical countries, such as Indonesia, the organism can be isolated year-round from aquatic sources, and similarly, the prevalence of *V. parahaemolyticus* in association with human diarrheal disease remains rather constant (Joseph 1973). Rather interestingly, clinical isolates of *V. parahaemolyticus* demonstrate hemolysis on a blood containing medium Wagatsuma agar (Kanagawa phenomenon), which is attributed to a thermostable direct hemolysin (TDH). Strangely, the large majority of environmental isolates, when tested for TDH, are phenotypically and genotypically negative for TDH. There have also been reports of atypical urease-positive *V. parahaemolyticus* isolated from patients (Abbott et al. 1989) and from the aquatic environment on the West Coast of the United States (Kaysner et al. 1990). In Mobile Bay, Alabama, the majority of the strains isolated were urease positive (DePaola et al. 2003). Similar urease-positive strains have also been isolated in Japan, where the relative virulence to typical *V. parahaemolyticus* has not yet been determined (Honda et al. 1992).

Isolation of *Vibrio parahaemolyticus*

The challenge of isolating *V. parahaemolyticus* in low numbers from aquatic areas and from seafoods led investigators to devise many types of isolation media, striving for greater sensitivity and selectivity. An excellent review of many of these media is provided by Twedt (1989). The agar medium preferred by most laboratory scientists is TCBS agar combined with pre-enrichment in either peptone salt, glucose-salt-Teepol or, salt polymyxin broth.

One of the preferred procedures for the isolation of *V. parahaemolyticus* from water and seafoods is very similar to those described for *V. cholerae* above, with the following modifications (Elliott et al. 1992). Homogenize 50 g of seafood in 450 mL of 3% NaCl in phosphate buffered saline, pH 7.2–7.5, in a sterile blender jar or Stomacher bag and then add the contents to a 1.0-L Erlenmeyer flask and incubate at 35°–37°C. At this point, MPN quantification can be performed on the blended suspension immediately after preparation and before incubation. Inoculate either three or five tubes containing alkaline peptone salt media (1 mL inoculum into 10 mL of broth). Then incubate 10–18 hours at 35°–37°C and observe for growth. Streak each positive tube to TCBS agar for confirmation of *V. parahaemolyticus*.

Otherwise, return to the original 1.0-L Erlenmeyer flask and inoculate TCBS agar for isolation and incubate at 35°–37°C for 24 hours. Pick three colonies and streak for purity on gelatin salt agar. Then perform biochemical and other tests as described for *V. cholerae* above.

Identification of *Vibrio parahaemolyticus*

The following characteristics provide presumptive identification of *V. parahaemolyticus*:

- Morphology, Gram-negative, asporogenous
- Triple sugar iron, alkaline slant and acid butt (no gas production)
- Hugh-Leifson, oxidation and fermentation positive
- Cytochrome oxidase, positive
- Arginine dihydrolase, positive
- Lysine decarboxylase, positive
- Voges-Proskauer, negative
- Growth at 42°C, positive
- Halophilism: 0%, positive
- Halophilism: 3%, positive
- Halophilism: 6%, positive
- Halophilism: 8%, positive
- Halophilism: 10%, ±
- Acid from arabinose, usually positive
- Acid from sucrose, negative
- 0129 susceptibility: 10 μg, resistant
- 0129 susceptibility: 150 μg, susceptible

Refer to Tables 21.3 and 21.4 for other tests to identify and confirm *V. parahaemolyticus*.

Subtyping of *V. parahaemolyticus*

Kanagawa Phenomenon

The Kanagawa reaction indicates the presence of TDH when *V. parahaemolyticus* is inoculated onto Wagatsuma agar. Although the significance of TDH in clinical disease is still questionable, a positive reaction caused by this toxin seems to correlate with pathogenicity. A zone of Beta hemolysis around a colony is considered positive.

Serotyping

Serology typing kits for *V. parahaemolyticus* usually contain 50–100 combinations of somatic (O) and capsular (K) antigen serotypes. The kits are expensive and usually are not necessary for the routine diagnostic laboratory. If serotyping is requested, the isolates should be forwarded to a reference laboratory either in the United States or Japan. For epidemiological purposes, serotyping is desirable and can contribute to investigations of outbreaks.

Other Readings

Further information on *V. parahaemolyticus* can be obtained in general reviews (Joseph et al. 1982, Morris and Black 1985, Doyle 1990) and in reviews of pathogenesis (Iida 1975, Cabassi and Mori 1976, Craun 1977, Blake et al. 1980, Lima 2001), as well as in a review of isolation and enumeration (Donovan and van Netten 1995).

Conclusions

Because of a continuing annual incidence of *V. cholerae* and *V. parahaemolyticus* in the United States and even higher frequencies in many nations of the world, the ability to isolate and identify these organisms is necessary. The migration of *V. cholerae* O1 into U.S. waters as well as the continuing presence of non O1 strains and *V. parahaemolyticus* in estuarine waters constitutes a threat to the health of target populations in nearby areas. With increasing travel to epidemic and endemic areas and the import of foods from some of those areas, food regulators and inspectors must be constantly vigilant.

Ongoing research will contribute to the efficiency of detection and identification by enabling incorporation of more molecular methods such as DNA probes, polymerase chain reaction, immunofluorescence microscopy, enzyme-linked immunosorbent assays, and ribotyping (Wong 2003).

References

Abbott, S. L., Powers, C., Kaysner, C. A., Takeda, Y., Ishibashi., M., Joseph, S. W. and Janda, J. M. 1989. Emergence of a restricted biovar of *Vibrio parahaemolyticus* as the predominant cause of *Vibrio*-associated gastroenteritis on the West Coast of the United States and Mexico. *Journal of Clinical Microbiology* **27**:2891–2893.

Anonymous. 1989. Part 9000; Microbiological examination of water, pp. 9-1–9-227. *In* Standard Methods for the Examination of Water and Wastewater, 17th ed., L. S. Clesceri, A. E. Greenberg, R. R. Trussell (Eds.). Washington, D.C.: American Public Health Association.

Baine, W. B., Massotti, N., Greco, D., Izzo, E., Zampieri, A., Angioni, G., Di Gioia, M., Gangarosa, E. J. and Pocchiari, F. 1974. Epidemiology of cholera in Italy in 1973. *Lancet* **2**:1370–1374.

Barker Jr., W. H. 1974. *Vibrio parahaemolyticus* outbreaks in the United States. *Lancet* **1**:551–554.

Baross, J. and Liston, J. 1968. Isolation of *Vibrio parahaemolyticus* from the Northwest Pacific. *Nature* **217**:1263–1264.

Barrow, G. I. and Miller, D. C. 1972. *Vibrio parahaemolyticus*: a potential pathogen from marine sources in Britain. *Lancet* **1**:485–486.

Barua, D. 1992. History of Cholera, pp. 1–36. *In* Cholera, D. Barua, W. B. Greenough III (Eds.). New York: Plenum Medical Book.

Baumann, P., Furniss, A. L. and Lee, J. V. 1984. Genus I. *Vibrio* Pacini 1854, 411[AL], pp. 518–538. *In* Bergey's Manual of Systematic Bacteriology, Vol. 1, N. R. Krieg (Ed.). Baltimore, Md.: Williams and Wilkins.

Binta, G. M., Tjaberg, T. B., Nyaga, P. N. and Valland, M. 1982. Market fish hygiene in Kenya. *Journal of Hygiene (London)* **89**:47–52.

Blake, P. A. 1994. Endemic cholera in Australia and the United States, pp. 309–319. *In Vibrio cholerae* and Cholera: Molecular to Global Perspectives, I. K. Wachsmuth, P. A. Blake, Ø. Olsvik (Eds.). Washington, D.C.: American Society for Microbiology Press.

Blake, P. A., Rosenberg, M. L., Costa, J. B., Ferreira, P. S., Guimaraes, C. L. and Gangarosa, E. J. 1977a. Cholera in Portugal. I. Modes of Transmission. *American Journal of Epidemiology* **105**:337–343.

Blake, P. A., Rosenberg, M. L., Costa, J. B., Ferreira, P. S., Guimares, C. L. and Gangarosa, E. J. 1977b. Cholera in Portugal. II. Transmission by bottled mineral water. *American Journal of Epidemiology* **105**:344–348.

Blake, P. A., Weaver, R. E. and Hollis, D. G. 1980. Diseases of humans (other than cholera) caused by vibrios. *Annual Review of Microbiology* **34**:341–367.

Bradford Jr., H. B. 1984. An epidemiological study of *V. cholerae* in Louisiana, pp. 59–72. *In* Vibrios in the Environment, R. R. Colwell (Ed.). New York: Wiley.

Bravo Farinas, L., Monte Boada, R., Valdes Ramos, F. and Dumas Valdivieso, S. 1991. Isolation and identification *Vibrio* genus microorganisms in the Quibu River. *Revista Cubana de Medicina Tropical* **43**:186–188.

Cabassi, E. and Mori, L. 1976. *Vibrio parahaemolyticus*: aetiological agent of food poisoning. *Folia Veterinaria Latina* **6**:335–354.

Cameron, J. M., Hester, K., Smith, W. L., Caviness, E., Hosty, T. and Wolf, E. S. 1977. *Vibrio cholerae*—Alabama. *Morbidity and Mortality Weekly Report* **26**:159–160.

Centers for Disease Control. 1989. Toxigenic *Vibrio cholerae* O1 infection acquired in Colorado. *Morbidity and Mortality Weekly Report* **38**:19–20.

Centers for Disease Control. 1990. Epidemiologic Notes and Reports Cholera—New York, 1991. *Morbidity and Mortality Weekly Report* **40**:516–518.

Centers for Disease Control. 1991. Cholera associated with imported coconut milk—Maryland. *Morbidity and Mortality Weekly Report* **40**:844–845.

Centers for Disease Control. 1992a. Cholera associated with an international airline flight. *Morbidity and Mortality Weekly Report* **41**:134–135.

Centers for Disease Control. 1992b. Cholera associated with international travel. *Morbidity and Mortality Weekly Report* **41**:664–667.

Centers for Disease Control. 1995. Update: *Vibrio cholerae* O1-Western hemisphere, 1991–1994, and *V. cholerae* O139-Asia, 1994. *Morbidity and Mortality Weekly Report* **44**:215–219.

Cholera Working Group. 1993. Large epidemic of cholera-like disease in Bangladesh caused by *Vibrio cholerae* O139 synonym Bengal. *Lancet* **342**:387–390.

Citarella, R. V. and Colwell, R. R. 1970. Polyphasic taxonomy of the genus *Vibrio*: polynucleotide sequence relationships among selected *Vibrio* species. *Journal of Bacteriology* **104**:434–442.

Colwell, R. R. 1970. Polyphasic taxonomy of the genus *Vibrio*: numerical taxonomy of *Vibrio cholerae*, *Vibrio parahaemolyticus* and related *Vibrio* species. *Journal of Bacteriology* **104**:410–433.

Colwell, R. R., Brayton, P., Harrington, P., Tall, B., Huq, A. and Levine, M. 1996. Viable but non-culturable *Vibrio cholerae* O1 revert to a cultivable state in the human intestine. *World Journal of Microbiology and Biotechnology* **12**:28–31.

Colwell, R. R. and Huq, A. 1994. Vibrios in the environment: viable but nonculturable *Vibrio cholerae*, pp. 117–133. *In Vibrio cholerae* and Cholera: Molecular to Global Perspectives, I. K. Wachsmuth, P. A. Blake, Ø. Olsvik (Eds.). Washington, D.C.: American Society for Microbiology Press.

Colwell, R. R., Kaper, J., and Joseph, S. W. 1977. *Vibrio cholerae*, *Vibrio parahaemolyticus*, and other vibrios: occurrence and distribution in Chesapeake Bay. *Science* **198**:394–396.

Colwell, R. R., Seidler, R. J., Kaper, J., Joseph, S. W., Garges, S., Lockman, H., Maneval, D., Bradford, H., Roberts, N., Remmers, E., Huq, I. and Huq, A. 1981. Occurrence of *Vibrio cholerae serotype* O1 in Maryland and Louisiana estuaries. *Applied and Environmental Microbiology* **41**:555–558.

Craun, G. F. 1977. Waterborne outbreaks. *Journal of the Water Pollution Control Federation* **49**:1268–1279.

Dalsgaard, A., Huss, H. H., H-Kittikun, A. and Larsen, J. L. 1995. Prevalence of *Vibrio cholerae* and *Salmonella* in a major shrimp production area in Thailand. *International Journal of Food Microbiology* **28**:101–113.

Davey, G. R., Prendergast, J. K. and Eyles, M. J. 1982. Detection of *Vibrio cholerae* in oysters, water and sediment from the Georges River. *Food Technology in Australia* **34**:334–336.

DePaola, A., Nordstrom, J. L., Bowers, J. C., Wells, J. G. and Cook, D. W. 2003. Seasonal abundance of total and pathogenic *Vibrio parahaemolyticus* in Alabama oysters. *Applied and Environmental Microbiology* **69**:1521–1526.

DePaola, A. 1981. *Vibrio cholerae* in marine foods and environmental waters: a literature review. *Journal of Food Science* **46**:66–70.

DePaola, A., Capers, G. M., Motes, M. L., Olsvik, Ø., Fields, P. I., Wells, J., Wachsmuth, I. K., Cebula, T., Koch, W. H., Khambusty, F., Kothary, M. H., Payne, W. L. and Wentz, B. A. 1992. Isolation of Latin American epidemic strain of *Vibrio cholerae* O1 from U.S. Gulf Coast. *Lancet* **339**:624.

Desmarchelier, P. M. and Reichelt, J. L. 1981. Phenotypic characterization of clinical and environmental isolates of *Vibrio cholerae* from Australia. *Current Microbiology* **5**:123–127.

Donovan, T. J. and van Netten, P. 1995. Culture media for the isolation and enumeration of pathogenic *Vibrio* species in foods and environmental samples. *International Journal of Food Microbiology* **26**:77–91.

Doyle, M. P. 1990. Pathogenic *Escherichia coli*, *Yersinia enterocolitica*, and *Vibrio parahaemolyticus*. *Lancet* **336**:1111–1115.

Dutta, A. K., Alwi, S. and Velauthan, T. 1971. A shellfish-borne cholera outbreak in Malaysia. *Transactions of the Royal Society of Tropical Medicine and Hygiene* **65**:815–816.

Elliott, E. L., Kaysner, C. A. and Tamplin, M. L. 1992. *Vibrio cholerae*, *V. parahaemolyticus*, *V. vulnificus* and other *Vibrio* spp., pp. 111–140. *In* Bacteriological Analytical Manual, Food and Drug Administration, 7th ed. Arlington, Va.: Association of Official Analytical Chemists.

Epstein, P. R. 1992. Cholera and the environment. *Lancet* **339**:1167–1168.

Eyles, M. J. and Davey, G. R. 1988. *Vibrio cholerae* and enteric bacteria in oyster-producing areas of two urban estuaries in Australia. *International Journal of Food Microbiology* **6**:207–218.

Garay, E., Arnau, A. and Amaro, C. 1985. Incidence of *Vibrio cholerae* and related vibrios in a coastal lagoon and seawater influenced by lake discharges along an annual cycle. *Applied and Environmental Microbiology* **50**:426–430.

Glass, R. I., Libel, M. and Brandling-Bennett, A. D. 1992. Epidemic cholera in the Americas. *Science* **256**:1524–1525.

Grizhebovskii, G. M., Biukhavnov, B. V., Briukhanov, A. F., Savel'ev, V. N., Artiushina, E. G., Zhilchenko, E. B., Budyka, D. A., Solodovnikov, B. V., Tikhenko, N. I., Levchenko, B. I. and Mezentsev, V. M. 2001. Comparative characterization of *Vibrio cholerae* O1, isolated from surface water reservoirs in Grozny in 1995 and 2000. *Zhurnal Mikrobiologii, Epidemiologii, i Immunobiologii* **6(Suppl.)**:48–50.

Hood, M. A. and Ness, G. E. 1982. Survival of *Vibrio cholerae* and *Escherichia coli* in estuarine water and sediment. *Applied and Environmental Microbiology* **43**:578–584.

Hood, M. A., Ness, G. E. and Rodrick, G. E. 1983. Distribution of *Vibrio cholerae* in two Florida estuaries. *Microbial Ecology* **9**:65–75.

Honda, S., Matsumoto, S., Miwatani, T. and Honda, T. 1992. A survey of urease-positive *Vibrio parahaemolyticus* strains isolated from traveller's diarrhea, sea water and imported frozen sea foods. *European Journal of Epidemiology* **8**:861–864.

Hugh, R. and Feeley, J. C. 1972. Report (1966–1970) of the subcommittee on taxonomy of vibrios to the International Committee on taxonomy of vibrios to the International Committee on Nomenclature of Bacteria. *International Journal of Systematic Bacteriology* **22**:123.

Huq, A. and Colwell, R. R. 1996. A microbiological paradox: viable but nonculturable bacteria with special reference to *Vibrio cholerae*. *Journal of Food Protection* **59**:96–101.

Huq, A., Colwell, R. R., Rahman, R., Ali, A., Chowdhury, M. A. R., Parveen, S., Sack, D. A. and Russek-Cohen, E. 1990. Detection of *V. cholerae* O1 in the aquatic environment by fluorescent monoclonal antibody and culture method. *Applied and Environmental Microbiology* **56**:2370–2373.

Huq, A., West, P. A., Small, E. B., Huq, M. I. and Colwell, R. R. 1984. Influence of water temperature, salinity, and pH on survival and growth of toxigenic *Vibrio cholerae* serovar O1 associated with live copepods in laboratory microcosms. *Applied and Environmental Microbiology* **48**:420–424.

Huq, A., West, P. A., Small, E. B., Huq, M. I., Rahman, R. and Colwell, R. R. 1983. Ecology of *Vibrio cholerae* O1 with special reference to planktonic crustacean copepods. *Applied and Environmental Microbiology* **45**:275–283.

Iida, H. 1975. Food poisoning by *Vibrio parahaemolyticus*. *Hokkaido Igaku Zasshi* **50**:533–539.

Johnston, J. M., Martin, D. L., Perdue, J., McFarland, L. M., Caraway, C. T., Lippey, E. C. and Blake, P. A. 1983. Cholera on a Gulf Coast oil rig. *New England Journal of Medicine* **309**:523–526.

Joseph, S. W. 1973. Observations on *Vibrio parahaemolyticus* in Indonesia, pp. 35–39. 1st International Symposium on *Vibrio parahaemolyticus*. Tokyo, Japan, September 17–18, 1973.

Joseph, S. W., Colwell, R. R. and Kaper, J. 1982. *Vibrio parahaemolyticus* and related halophilic Vibrios. *Critical Reviews in Microbiology* **10**:77–124.

Joseph, P. R., Tamayo, J. E., Mosley, W. H., Alvero, M. G., Dizon, J. J. and Henderson, D. A. 1965. Studies of cholera El Tor in the Philippines. 2. A retrospective investigation of an explosive outbreak in Bacolod City and Talisay, November, 1961. *Bulletin of the World Health Organization* **33**:637–640.

Kaper, J. B., Lockman, H., Colwell, R. R. and Joseph, S. W. 1979. Ecology, serology and enterotoxin production of *Vibrio cholerae* in Chesapeake Bay. *Applied and Environmental Microbiology* **37**:91–102.

Kaysner, C. A., Abeyta Jr., C., Stott, R. F., Lilja, J. L. and Wekell, M. M. 1990. Incidence of urea-hydrolyzing *Vibrio parahaemolyticus* in Willapa Bay, Washington. *Applied and Environmental Microbiology* **56**:904–907.

Kaysner, C. A., Abeyta Jr., C., Wekell, M. M., DePaola Jr., A., Stott, R. F. and Leitch, J. M. 1987. Incidence of *Vibrio cholerae* from estuaries of the United States west coast. *Applied and Environmental Microbiology* **53**:1344–1348.

Klontz, K. C., Tauxe, R. V., Cook, W. L., Riley, W. H. and Wachsmuth, I. K. 1987. Cholera after the consumption of raw oysters: a case report. *Annals of Internal Medicine* **107**:846–848.

Koch, R. 1884. An address on cholera and its bacillus. *British Medical Journal* **2**:403–407, 453–459.

Lacy, S. W. 1995. Cholera: calamatous past, ominous future. *Clinical Infectious Diseases* **20**:1409–1419.

Lee, J. V., Bashford, D. J., Donovan, T. J., Furniss, A. L. and West, P. A. 1982. The incidence of *Vibrio cholerae* in water, animals and birds in Kent, England. *Journal of Applied Bacteriology* **52**:281–291.

Levine, W. C. and Griffin, P. M. 1993. *Vibrio* infections on the Gulf Coast: results of first year of regional surveillance. Gulf Coast *Vibrio* Working Group. *Journal of Infectious Diseases* **167**:479–483.

Lima, A. A. 2001. Tropical diarrhoea: new developments in traveller's diarrhoea. *Current Opinion in Infectious Diseases* **14**:547–552.

Lin, F. Y., Morris Jr., J. G., Kaper, J. B., Gross, T., Michalski, J., Morrison, C., Libonati, J. P. and Israel, E. 1986. Persistence of cholera in the United States: isolation of *Vibrio cholerae* O1 from a patient with diarrhea in Maryland. *Journal of Clinical Microbiology* **23**:624–626.

Lipp, K. L., Huq, A. and Colwell, R. R. 2002. Effects of global climate on infectious disease: the cholera model. *Clinical Microbiology Reviews* **15**:757–770.

Madden, J. M., McCardell, B. A., and Read Jr., R. B. 1982. *Vibrio cholerae* in shellfish from U.S. coastal waters. *Food Technology* **36**:93–96.

Maneval, D. R., Joseph, S. W., Donta, S. T., Grays, R. and Colwell, R. R. 1980. A tissue culture method for the detection of bacterial enterotoxins. *Journal of Tissue Culture Methods* **6**:85–90.

McCarthy, S. A., McPhearson, R. M., Guarino, A. M. and Gaines, J. L. 1992. Toxigenic *Vibrio cholerae* O1 and cargo ships entering Gulf of Mexico. *Lancet* **339**:624–625.

McNicol, L. A. and Doetsch, R. M. 1983. A hypothesis accounting for the origin of pandemic cholera: a retrograde analysis. *Perspectives in Biological Medicine* **26**:547–552.

Merson, M. H., Martin, W. T., Craig, J. P., Morris, G. K., Blake, P. A., Craun, G. F., Feeley, J. C., Camacho, J. C. and Gangarosa, E. J. 1977. Cholera on Guam, 1974: epidemiologic findings and isolation of non-toxinogenic strains. *American Journal of Epidemiology* **105**:349–361.

Miller, C. J., Drasar, B. S. and Feacham, R. G. 1984. Response of toxigenic *Vibrio cholerae* O1 to physico-chemical stresses in aquatic environments. *Journal of Hygiene (Cambridge)* **93**:475–495.

Monsur, K. A. 1961. A highly selective gelatin-taurocholate-tellurite medium for the isolation of *Vibrio cholerae*. *Transactions. Royal Society of Medicine and Hygiene* **55**:440–442.

Moore, B. 1948. The detection of paratyphoid carriers in towns by means of sewage examination. *Monthly Bulletin of the Ministry of Health and the Public Health Laboratory Service* **7**:241–248.

Morris Jr., J. G. 1990. Non-O group 1 *Vibrio cholerae*: a look at the epidemiology of an occasional pathogen. *Epidemiologic Reviews* **12**:179–191.

Morris Jr., J. G. and the Cholera Laboratory Task Force. 1994. *Vibrio cholerae* O139 Bengal, pp. 95–102. *In Vibrio cholerae* and Cholera: Molecular to Global Perspectives, I. K. Wachsmuth, P. A. Blake, Ø. Olsvik, (Eds.). Washington, D.C.: American Society for Microbiology Press.

Morris Jr., J. G. and Black, R. E. 1985. Cholera and other vibrioses in the United States. *New England Journal of Medicine* **312**:343–350.

Pacini, F. 1854. Observazioni microscopiche e deduzioni patologiche sul cholera asiatico. *Gazette Medica Italiono* **6**:405–412.

Pavia, A. T., Campbell, J. F., Blake, P. A., Smith, J. D., McKinley, T. W. and Martin, D. I. 1987. Cholera from raw oysters shipped interstate. *Journal of the American Medical Association* **258**:2374.

Pollitzer, R. 1959. Cholera. Monograph No. 43, World Health Organization.

Powell, J. L. 1999. *Vibrio* species. *Clinics in Laboratory Medicine* **19**:537–552, vi.

Rippey, S. R. 1991. Shellfish borne disease outbreaks. North Kingstown, R.I.: Food and Drug Administration.

Roberts, N. C., Seibeling, R. J., Kaper, J. B. and Bradford Jr., H. B. 1982. Vibrios in the Louisiana Gulf Coast environment. *Microbial Ecology* **8**:299–312.

Ruiz, G. M., Rawlings, T. K., Dobbs, F. C., Drake, L. A., Mullady, T., Huq, A. and Colwell, R. R. 2000. Global spread of microorganisms by ships. *Nature* **408**:49–50.

Sakazaki, R., Iwanami, S., and Fukumi, H. 1963. Studies on the enteropathogenic, facultatively halophilic bacteria, *Vibrio parahaemolyticus*. I. Morphological, cultural and biochemical properties and its taxonomic position. *Japanese Journal of Medical Science and Biology* **16**:161–188.

Singleton, F. L., Attwell, R., Jangi, S. and Colwell, R. R. 1982. Effects of temperature and salinity on *Vibrio cholerae* growth. *Applied and Environmental Microbiology* **44**:1047–1058.

Snow, J. 1936. Snow on Cholera (being a reprint of two papers by John Snow, M.D.). New York: Commonwealth Fund. Original: 1855. On the mode of communication of Cholera, 2nd ed. London: John Churchill.

Stavric, S., and Buchanan, B. 1995. The isolation and identification of *Vibrio cholerae* O1 and non-O1 from foods. MFLP-72. Government of Canada Health Protection Branch. Polyscience Publications, Station St.-Martin, Laval, Quebec, Canada.

St. Louis, M. E., Porter, J. D., Helal, A., Drame, K., Hargrett-Bean, N., Wells, J. G. and Tauxe, R. V. 1990. Epidemic cholera in West Africa: the role of food handling and high-risk foods. *American Journal of Epidemiology* **131**:719–728.

Swaddiwudhipong, W., Akarasewi, P., Chayaniyayodhin, T., Kunasol, P. and Foy, H. M. 1990. A cholera outbreak associated with eating uncooked pork in Thailand. *Journal of Diarrhoeal Diseases Research* **8**:94–96.

Swaddiwudhipong, W., Jirakanvisun, R. and Rodklai, A. 1992. A common source foodborne outbreak of El Tor cholera following the consumption of uncooked beef. *Journal of the Medical Association of Thailand* **75**:413–417.

Tamplin, M. L. and Parodi, C. C. 1991. Environmental spread of *Vibrio cholerae* in Peru. *Lancet* **338**:1216–1217.

Twedt, R. M. 1989. *Vibrio parahaemolyticus in* Foodborne Bacterial Pathogens, pp. 543–568. *In* Foodborne Bacterial Pathogens, M. P. Doyle (Ed.). New York: Marcel Dekker.

Urdaci, M. C., Marchand, M., Marchand, M., Grimont, P. A. 1988. Species of the genus *Vibrio* associated with marine products from Arachon Bay. *Annales of the Institut Pasteur Microbiologie* **139**:351–362.

Weissman, J. B., DeWitt, W. E., Thompson, J., Muchnick, C. N., Portnoy, B. L., Feeley, J. C. and Gangarosa, E. J. 1974. A case of cholera in Texas, 1973. *American Journal of Epidemiology* **100**:487–490.

West, P. A. 1992. The ecology and survival of *Vibrio cholerae* in natural aquatic environments. *PHLS Microbiology Digest* **9**:20–23.

Wong, H.-C. 2003. Detecting and molecular typing of *Vibrio parahaemolyticus*. *Journal of Food and Drug Analysis* **11**:79–86.

Xu, H.-S., Roberts, N., Singleton, F. L., Atwell, R. W., Grimes, D. J. and Colwell, R. R. 1982. Survival and viability of non-culturable *Escherichia coli* and *Vibrio cholerae* in the estuarine and marine environment. *Microbial Ecology* **8**:313–323.

Yamai, S., Okitsu, T., Murase, T. and Katsube, Y. 1997. Studies of *Vibrio cholerae* O140 (serogroup Hakata) isolated from river water. *Kansenshogaku Zasshi* **71**:928–934.

Ziatdinov, V. B. 2002. Characterization of *Vibrio cholerae* cultures isolated in foci of cholera in the city of Kazan. *Zhurnal Mikrobiologii, Epidemiologii, i Immunobiologii* **6(Suppl.)**:78–80.

22 Molecular Methods for Microbial Detection

Suresh D. Pillai

Introduction

The ability to detect microorganisms in foods is one of the cornerstones of food microbiology, and of food safety in particular. The past century has seen significant strides in our ability to not only detect specific microorganisms but also accomplish this task in an efficient, specific, and sensitive manner. It all started with the period when microbial identification involved primarily microscopic methods. It evolved into the current technologies, which not only identify the specific microorganisms but also provide a rather comprehensive understanding of the genetic and metabolic characteristics of the organisms in question (Ye et al. 2000, Endley et al. 2001, Wei et al. 2001, Bopp et al. 2003). These revolutionary advancements have provided diagnostic tools for the research and food industry communities to obtain answers to very specific questions regarding microbial contamination scenarios, and have afforded us a better understanding of the characteristics of the contaminant in question.

This chapter provides an overview of contemporary issues that surround molecular detection of microbial pathogens. Because nucleic acid–based methods (as a result of their overall sensitivity and specificity compared to other methods) have been the target of extensive investigations over the last number of years, they will form the basis of discussion in this chapter. This chapter will also attempt to provide a "road map" for the development of detection and characterization tools for the future. The suggested improvements are presented in the context of benchmarking specifications that are required to keep pace with the evolution and re-emergence of microbial pathogens, the global distribution of foods, and the global movement of people.

Background of Microbial Detection

When one attempts to chart the evolution of microbial detection technologies, whether it is in food safety, clinical microbiology, or environmental microbiology, it is evident that major improvements in techniques closely mirror the groundbreaking developments in the basic disciplines of microbiology, molecular biology, and immunology. There are a number of reasons for the quick adoption of new detection technologies. The primary reason is the need to detect pathogens rapidly, sensitively, and specifically. The ability to rapidly detect specific organisms is critically important in clinical medicine, food, and environmental industries. A key driving force in the development of new-generation detection tools is the extraordinary amount of genomic and proteomic information that is now available. We now have a better understanding of the molecular pathogenesis of many enteric pathogens. Regulatory pressures on the food industry, such as the "zero tolerance" of *Listeria monocytogenes* and *Escherichia coli* O157:H7 in ready-to-eat meat products, for example, have spurred or "forced" the food industry to adopt some of these technologies. In 1999, the U.S.

food industry performed as many as 144.3 million microbiological tests (Strategic Consulting 2000). Out of these, 26.3% (23.5 million) were pathogen-specific tests. This was a 23% increase in microbiological tests over the 1998 estimates. The importance of regulatory pressure in catalyzing the commercialization of molecular detection tools is evident when one compares the availability of commercial kits for the food industry with those of the drinking water industry. At present, the Association of Official Analytical Chemists (AOAC)-certified DNA and polymerase chain reaction (PCR)-based kits to detect *L. monocytogenes* and *E. coli* O157:H7 are available for the food industry (Table 22.1). This is in

Table 22.1   Association of Official Analytical Chemists–Approved pathogen detection kits.

Organism	Commercial name	Detection technology	Manufacturer
Escherichia coli O157:H7	BAX	Polymerase chain reaction	Qualicon, Wilmington, Del.
E. coli O157:H7	TECRA	Visual immunoassay	TECRA International, Pty Ltd, Australia
E. coli O157:H7	Detex System MC-18 *E. coli*	Electro-immunoassay	Molecular Circuitry, Inc, Scranton, Penn.
E. coli O157:H7	EiaFoss *E. coli*	Enzyme immunoassay	Foss Electric A/S, Denmark
E. coli O157:H7	PATHATRIX *E. coli* O157 Test system	Magnetic-immunoassay	Matrix MicroScience, Ltd., United Kingdom
Campylobacter spp.	Detex System MC-18 *Campylobacter*	Electro-immunoasssay	Molecular Circuitry, Inc., Scranton, Penn.
Salmonella spp.	GENE TRAK	DNA hybridization	Neogen, Lansing, Mich.
Salmonella spp.	Reveal	Enzyme-linked immunosorbent assay	Neogen, Lansing, Mich.
Salmonella spp.	Bioline	Enzyme-linked immunosorbent assay	Bioline ApS, Denmark
Salmonella spp.	BAX	Polymerase chain reaction	Qualicon, Wilmington, Del.
Salmonella enteriditis	Reveal	Enzyme-linked immunosorbent assay	Neogen, Lansing, Mich.
Listeria spp.	VIDAS	Immunoassay	bioMérieux, Inc., Hazelwood, MD
Listeria spp.	GENE TRAK DLP	DNA hybridization assay	Neogen, Lansing, Mich.
Listeria spp.	PATHATRIX	Immunoassay	Matrix MicroScience, Ltd, United Kingdom
Listeria spp.	EiaFOSS	Enzyme-linked immunosorbent assay	Foss electric A/S, Denmark
Listeria spp.	ListerTest	Immuno-magnetic assay	Vicam, Watertown, Mass.
L. monocytogenes	BAX	Polymerase chain reaction	Qualicon, Wilmington, Del.
L. monocytogenes	Automated BAX	Polymerase chain reaction	Qualicon, Wilmington, Del.

contrast to the complete absence of any validated molecular kit for any of the bacterial, viral, or protozoan pathogens that are of concern to the drinking water industry. This is surprising given that conventional foods are normally cooked or heated, compared with drinking water, which (at least in the United States and in other developed countries) is "consumed as is." Because there are no current regulations requiring mandatory testing of drinking water for specific pathogens, there is no financial incentive for diagnostic kit manufacturers to invest resources developing kits for the drinking water industry.

Molecular Pathogen Detection

At present, the numbers of microbiological tests being performed in the U.S. food industry and around the developed world are evenly divided between traditional tests (such as isolation of selective/differential culture media) and rapid screening methods (Strategic Consulting 2000). Pathogen detection methods for foods can be broadly divided into quantitative and qualitative methods. Culture-based quantitative estimation of pathogens is still limited for the most part to Most Probable Number (MPN)-based methods, as a majority of pathogens such as *Salmonella* require an enrichment step before detection and enumeration. A majority of the molecular detection methods, however, only provide qualitative information about the presence or absence of specific pathogens. For example, the BAX-PCR (Qualicon, Wilmington, Del.) assay requires overnight enrichment before PCR amplification of the targets. Studies have shown that molecular methods involving DNA:DNA hybridization or PCR amplification are prone to interferences from background (sample) matrix components (Pillai et al. 1991, Pillai et al. 1993, Kreader 1996, Wilson 1997, Peña et al. 1999). These inhibitory substances, which in many instances have not been completely identified, inhibit the enzymatic reactions to varying degrees. The current molecular methods (e.g., BAX-PCR, designed for pathogen detection in foods) include an enrichment step that serves two purposes: first, it dilutes out the interfering substances in the food matrix, and second, it provides some assurance that the DNA sequences that were being detected were from viable cells. The inclusion of the enrichment steps, however, led to these assays being only qualitative in nature.

Nucleic acid–based detection methods can be either amplification based or probe based. The primary difference between these two categories of methods is whether or not the target sequence is enzymatically amplified. A recurring theme, however, in some of the newer methods is the integration of both the amplification and probe technologies into the same method (e.g., the TaqMan-PCR assay [Applied Biosystems, Foster City, Calif.] and Scorpion primer technology [DXS Ltd., Manchester, United Kingdom]; Lakowicz 1999, Thelwell et al. 2000).

Conventional PCR technology has undergone numerous and significant improvements, and detection kits based on PCR methods are now commercially available (Table 22.1). A salient feature of these kits is their ease of use, as all the necessary reagents are often lyophilized into a tablet form within the reaction tube. All that is now required is the addition of a known volume of an adequately enriched sample into the reaction tube and placement in a temperature cycling instrument. Current versions of the assays and instrumentation include automated gel electrophoresis modules, which reduce the potential for human errors and laboratory-based contamination issues. Self-contained portable PCR instruments are also available for field use (Belgrader et al. 2001).

In addition to the PCR-based amplification protocols, there are amplification protocols that do not involve the use of the proprietary Taq DNA polymerase technology. Kwoh et al.

(1989) developed one of the first non-PCR amplification protocols, termed the transcription-based amplification system. It involves the sequential use of reverse transcriptase (formation of cDNA), followed by RNA polymerase, which catalyzes the synthesis of multiple copies of the original RNA template. Further improvements in the transcription-based amplification system protocol resulted in the development of the 3SR (self-sustaining sequence replication) amplification technology (Guatelli et al. 1989). Because heat denaturation is typically not involved in 3SR, this amplification can be conducted under isothermal conditions, thereby alleviating the need for specialized temperature cycling instruments. However, these methods require extensive post-amplification steps, which can be cumbersome and can potentially lead to contamination errors.

Quantitative (Real-Time [RT])-PCR

The need to detect single-base substitutions within discrete regions of genomes was one of the factors that accelerated the development of quantitative PCR-based methods. Quantitative PCR methods have a variety of terminologies, including TaqMan-PCR, fluorogenic 5′ nuclease assay, and "real-time" (RT)-PCR. The primary difference between the quantitative technique and the conventional PCR method is that specifically labeled oligonucleotide probes along with conventional primers are used in the process. More important, the PCR product formation is detected during the course of the reaction, in contrast to the conventional assay. Thus, in this method the PCR product is detected in real time, as opposed to "end-point" detection in the conventional PCR assay. Because the fluorescence measurements take place during the exponential or logarithmic phase of the amplification, it is possible to quantify (in a relative sense) the formation of a specific PCR product. Appropriate standard curves need to be prepared using the sample matrix and to run alongside the experimental samples for absolute quantification. The primers and the TaqMan probes confer additional specificity as compared to the conventional PCR assay. Because the assay relies on real-time detection of product formation and it involves the detection of fluorescence, this assay is more sensitive than the conventional end-point assay. More important, as this assay does not involve any post-PCR handling, there is reduced potential for PCR contamination. Some of the instrumentation available for RT-PCR analyses permits analysis of up to 384 samples, thereby significantly enhancing the throughput. Even though the assay currently requires relatively expensive instrumentation, there is a fairly large selection of RT-PCR–related instruments to suit different budgets and needs. In addition to improvements in instrumentation, enhancements and modifications of the basic technology, using the specialized primers Amplifluor (Gaithersburg, Md.), Scorpion, and LUX (Invitrogen, Carlsbad, Calif.) fluorogenic primers, has also occurred (Thelwell et al. 2000, Nazarenko et al. 2002). These specialized primers and probes are significantly more expensive than conventional PCR reagents. However, prices have significantly dropped over the last couple of years partly because of increased demand. None of the RT-PCR methods has been AOAC approved for detecting pathogens in foods.

DNA Microarrays

There has been considerable progress in the use of DNA microarrays as detection tools (Gilles et al. 1999, Wei et al. 2001, Cho and Tiedje 2002). There are two basic types of microarrays: those that are prepared by spotting DNA probes on glass or membranes, and those that have oligonucleotides synthesized *in situ* on DNA chips. A majority of the current microarray-based studies are aimed at understanding the global gene expression patterns of microbial genes. These so-called gene expression assays rely on the immobilization of DNA probes that would be complimentary to the mRNA targets of interest. The hybridization patterns or signals of the different target genes thereby provide information on the different gene expression patterns. The primary advantage of microarrays is that thousands of specific sequences can be simultaneously screened. Though the underlying principle is based on Watson–Crick base-pairing rules, thousands of specific DNA sequences can be immobilized or synthesized on extremely small platforms. This provides a significant breakthrough in the number of samples that can be analyzed and the number of target regions that can be screened in the different samples. Hybridization patterns obtained after presenting the microarray to a sample are detected and analyzed using advanced bioinformatics tools. Issues such as use of appropriate experimental controls, normalization approaches, and independent verification of microarray data using alternate approaches are extremely critical (Bowtell and Sambrook 2003). As expected, microarray approaches for interpreting microarray data are current areas of intense research. There are a number of published studies documenting the use of microarrays for pathogen detection, pathogen characterization, and microbial pathogen gene expression patterns (Chizhikov et al. 2001, Call et al. 2003, Stintzi 2003).

Critical Issues Surrounding Molecular Methods for Food Safety Applications

Sample Volumes

At present, pathogens in foods are primarily detected by conventional culture-based methods. The reliance on culture-based methods results from a variety of factors including current mandatory federal and state regulations (which require use of culture-based methods) as well as the level of acceptance of molecular methods by the industry. The generalized protocol for the initial steps in processing food samples is that a defined amount of sample is homogenized in a suitable buffer and that an aliquot of the homogenized sample is enriched using specialized selective media. A sub-sample of the enriched culture is then plated for isolation and detection of the pathogen or pathogens of interest. Serology procedures are often used to confirm the identity of the isolated organisms. The basic concept of homogenizing or "stomaching" a large portion of the sample is advantageous because it increases the likelihood of detecting pathogens that may be at low concentrations or irregularly distributed on the sample. The use of large sample volumes for the initial homogenization also increases the likelihood that larger amounts of the potential inhibitory compounds will get co-purified with the target organisms. These inhibitory components can exert variable effects on the enzymatic reactions. Thus, using reduced volumes of the original homogenate for plating or molecular analysis can be counterproductive. This is primarily because the detection sensitivity of the overall assay is proportionally reduced. For example, let us assume that 100 gm of a meat sample contains 10 target organisms. If this sample were homogenized in 200 mL of buffer, the buffer will theoretically contain only 0.05 organisms per 1 mL. This will result in "non-detects" because the assays employed <20 mL of the original sample. In other words, only assays (molecular or conventional) that employ

20 mL or more will be capable of detecting the contaminants. To avoid significant confusion regarding the sensitivity thresholds, the detection sensitivity of the assay and the overall sensitivity of the method have to be clearly specified and understood. Very often the sensitivity threshold of an assay is erroneously calibrated by determining the lowest number of organisms that can be detected in the sample that is analyzed. The ultimate detection sensitivity of an assay should be based on the minimum number of organisms that need to be present on the sample before sample processing.

The issue of detection sensitivity is of significant importance in comparing detection assays as well as in identifying the true applicability of an assay for monitoring food safety. A closely related issue is the volume of sample that is assayed by a particular method.

As illustrated in the above example, only assays that use 20 mL of the homogenized sample would be capable of detecting the contamination. Assays that use only 1 mL or 10 mL have inadequate sensitivity and could often result in misleading "non-detects." This raises serious questions about the validity and reliability of using methods such as the 3M Petri-Film (3M Corp., St. Paul, Minn.) which uses only 1 mL of the sample to detect contamination indicators. These methods can thus only perform if there are significant levels of a contaminant in a sample or if multiple 1-mL aliquots of the same sample are analyzed. As would be expected, current protocols for use with molecular methods are significantly hampered by this sample volume limitation. This is because most of the molecular assays such as PCR and RT-PCR assays use sample volumes ranging between 2 and 10 μL/assay. As seen from the example below, microarrays and other technologies such as biosensors that are thought to be breakthrough technologies for detecting pathogens are at present significantly limited because of their "maximum sample volume" limitations. Assume a virulence gene expression (microarray-based) biosensor (which employs 50 μL sample volume for analysis) has been developed that can detect one cell of the specific pathogen. If this biosensor is to be used to detect that specific pathogen in 500 gallons of processed orange juice, it requires that either the 500 gallons be concentrated down to 50 μL or that the 500 gallons volume contain a total of 3.6×10^7 cells of the pathogen. Only under these circumstances would the biosensor detect the organism (if the 50-μL aliquot of the sample is used for the biosensor analysis). Pathogen concentrations below this threshold would be below the detection sensitivity of the biosensor and result in a false-negative result. Thus, for biosensors or microarrays or any other molecular methods to be of any practical value, the sample has to be adequately concentrated so that the contaminants in the concentrated samples are within the sensitivity thresholds of the assay. At present, rapid molecular methods are available for pathogens such as *Salmonella* and *Listeria* spp. The diagnostic kits will perform, provided a suitable sample is available. The Achilles heel of this system, however, is the inability to obtain suitable samples for analysis. The current lack of effective microbial capture and concentration methods to recover low numbers of potential targets from food and water samples into a suitable sample for analysis is a key hurdle.

Sample Concentration and Purification

Pathogen capture and concentration protocols can also present challenges for molecular assays. A number of studies have shown that when food samples are homogenized and concentrated, components that are inhibitory to molecular assays are also co-purified. These need to be removed to prevent false-negative results. Commercial products such as the Soil-DNA kit (MoBio, Solano, Calif.) and the PrepMan (Applied Biosystems, Foster City, Calif.) are available for cleaning or purifying samples. These kits are effective at reducing or removing inhibitory components (Roe 2002). It is important to note that whenever such sample purification kits are employed there is a very high likelihood for potential target nucleic acids also to be reduced. This is not of a particular concern in clinical samples that contain a large number of potential targets. However, it can be a significant problem when working with food and environmental samples (Pillai et al. 1991, Pillai and Ricke 1995). Given the variety in terms of food samples and the variability in pathogen numbers on the different types of food and food ingredients, additional research in sample processing is needed. The possible research areas include efficient sample capture and concentration methods and efficient sample purification methods. The ability to capture multiple target organisms without concentrating potential inhibitory components is a key benchmark of technologies for the future.

Regulatory Acceptance of Molecular Methods

Molecular methods, such as direct PCR assays, have often been cited as not being relevant for food safety applications because they do not provide any indication of whether the pathogen is viable or not. There is validity to both sides of the argument as to whether or not direct molecular methods are reliable indicators of the presence of viable, infectious organisms. Current AOAC-approved PCR-based methods such as BAX-PCR assays solve the issue of viability by incorporating an enrichment step before gene amplification. The AOAC Research Institute has certified more than 40 commercial pathogen and toxin test kits since 1991, when the Performance-Tested Methods Program was created (Vasavada 2001). The Performance-Tested Methods Program was initiated to provide an independent third-party verification of a detection kit's capabilities (AOAC 2003). Table 22.1 is a partial listing of the commercial kits that have been through the AOAC Performance Testing Program.

The applicability of a particular molecular method for detecting pathogens in food samples, however, also depends on the question being asked. If the question is whether or not the sample harbored specific pathogens (irrespective of their current viability) at some point in time, then it is difficult to argue against the use of molecular methods. Even though positive PCR results signify the presence of a specific microbial contamination, this does not, however, indicate any particular measure of health risks. Conversely, negative PCR results do not prove that there is a lack of contamination, nor do they provide any indication of the health risks. A negative result is very difficult to interpret without a clear delineation of the detection sensitivity of the assay (Loge et al. 2002). Detection limit is, in turn, influenced by factors such as the efficiency of target recovery, the presence of inhibitory compounds, and the volume of sample that is analyzed.

Loge et al. (2002) have recently proposed a risk-based framework for providing biological relevance to PCR assays. The framework not only provides a means of assessing the health risks associated with PCR results but also provides a tool to identify the effects of future improvements in PCR-based assays. The framework is based on a method for quantifying the detection limits of any PCR-based assay. The detection limit can be quantified using the following relationship:

$$\text{Detection limit} = \frac{RI}{V_f} \cdot \frac{S}{\%V_p}$$

Where $R = 1/\text{Re}$ (unit less), Re = recovery of organisms from the sample, I = inverse of the dilution factor necessary to remove compounds that inhibit PCR (unit less), V_f = volume of sample, S = sensitivity of the PCR-assay (colony-forming units) and $\%V_p$ = fraction of the concentrated sample analyzed by PCR.

For pathogens currently in the zero-tolerance status such as *E. coli* O157:H7 and *L. monocytogenes*, the use of direct PCR results to institute effective disinfection programs or voluntary recalls can be justified. Given the number of unknowns surrounding pathogen distribution on and within food samples, food sampling, and sample processing, it is risky to assume that PCR results are of insignificant value because positive results can originate from non-viable cells. Opponents of molecular methods often use the argument that there is a lack of significant correlation between the results of the molecular assays and the conventional assay when samples are analyzed. This lack of correlation is not surprising, as the assay results depend on the sample volumes being employed, the distribution of organisms within the original sample (thereby dictating the presence or absence of the organisms in the sample aliquots), and the detection sensitivity of the different assays. This lack of correlation is also observed when the presence or absence of indicator organisms is compared. When Endley et al. (2003) compared the presence of *E. coli* and male-specific coliphage as indicators of fecal contamination, there were differences in the occurrence level (Table 22.2). These differences could be attributed to the differences in the persistence patterns of the different fecal organisms and pathogens. These results suggest that relying only on a single indicator organism can significantly underestimate the actual presence of fecal contamination. There is a need to be able to detect specific pathogens because quantitative risk estimations are based on pathogen concentrations and not indicator organism levels. These results also highlight the point that no one method or approach is better than the other. Rather, it illustrates the complexities of how the distribution, number, and attachment of organisms within the original sample can influence different assay results. Thus, it would be erroneous to explain increased detection of target organisms by molecular methods to be a result of only naked nucleic acids or non-viable cells. A significant number of unknowns, in terms of the nonviability of infectious organisms in food samples and the factors that control the potential for these organisms to cause infection in humans, still exists.

Table 22.2 Occurrence of male-specific coliphage in relation to a bacterial indicator and a bacterial pathogen (modified from Endley et al. 2003).

Organism	Positive samples[a,b]			Total positive samples[c]
	Field	Truck	Processing shed	
MS2 coliphage	F25[d] 4% (1 of 25)	T3,T4,T5, T6 16% (4 of 25)	P5, P8, P12–P14, P17–P25 56% (14 of 25)	19 of 75
Escherichia coli	None 0%	T9, T21, T22, T24 16% (4 of 25)	P8, P18 8% (2 of 25)	6 of 75
Salmonella	F16 4% (1 of 25)	T17, T18 8% (2 of 25)	None 0%	3 of 75

(a) Detection limit = 10 organisms per carrot sample.
(b) $n = 25$.
(c) $n = 75$.
(d) Refers to sample ID.

Risk Assessment and Molecular Methods

From a regulatory standpoint, there is a growing demand that molecular methods be given serious consideration. This is partly because U.S. regulatory agencies are attempting to base all regulatory decisions on a risk-based framework. For some human viral pathogens such as Noroviruses (Norwalk-like viruses), which are responsible for over 66% of all food-borne human illnesses, culture-based methods are currently unavailable (Mead et al. 1999). Given the public health risk importance of some specific pathogens such as Noroviruses, (which are currently unculturable) in terms of foodborne illnesses, it would not be surprising that regulatory agencies accept the use of molecular methods for such pathogens in the near future. The U.S. Environmental Protection Agency has already started exploring the feasibility of adopting molecular methods for the detection of viruses in drinking water (U.S. Environmental Protection Agency 2003). Regulatory approval of molecular methods imply that strict QA/QC (quality assurance/quality control) controls have to be developed for the various techniques, as well as for the calibration of method performance and inter-laboratory validation. At present, for PCR-based methods the QA/QC procedures include the integration of internal spiked sample controls and the sequencing of PCR products.

A number of studies have shown that changing demographics, inherent changes in the pathogens themselves, and changes in food preparation, distribution, and consumption have the potential to adversely affect human health (Smith and Fratamico 1995, Alterkruse et al. 1997). Quantitative risk assessment, which ultimately will help in reducing these risks, is a possible framework for designing programs to reduce foodborne diseases (Lammerding and Paoli 1997). Hazard identification and exposure assessment are key components of quantitative microbial risk assessment. Molecular methods can be used in both hazard identification and exposure assessment.

Molecular methods will find increasing applications in risk assessment in the future. Regulatory agencies such as the U.S. Department of Agriculture and Food and Drug Administration will have to deal with the issue of emerging and re-emerging infectious organisms and agents such as Cyclospora, enteric viruses, and prions in foods. These agencies have the responsibility to identify the critical contaminants and adequately assess the potential risks involved. Regulatory agencies will have to ultimately prioritize these infectious organisms for regulatory purposes. At present, there is a tendency toward the establishment of a zero-tolerance goal for certain microbial contaminants. Indicator organisms such as *E. coli* are also used to show the possible presence of fecal contamination, thereby indirectly indicating the possible presence of pathogens. However, a number of studies have shown that using indicator organisms can be fraught with limitations because fecal coliforms and *E. coli* do not exhibit the same resistance to disinfectants and decontamination approaches as do protozoa and viruses.

Identifying the virulence function activity relationships (VFAR) of foodborne pathogens is possible with molecular methods. The National Research Council (NRC) in their study on drinking water contaminants has proposed the use of VFAR as a method to predict the ultimate virulence of microbial contaminants (NRC 2001). The concept of using a multitude of characteristics (to prioritize the contaminants in terms of virulence) rather than relying on a few characteristics has significant advantages. Because the ultimate virulence of a pathogen depends not only on the organism's virulence attributes but also on the host susceptibility, making use of multiple "descriptors" provides a science-based risk categorization process. The concept of using VFAR in the hypothetical relationship between the "descriptors" and "outcomes" is shown in Figure 22.1. The NRC has recommended that the process of identifying the critical contaminants (microbial and chemical) follow a stepwise process. The NRC recommends that the process of selecting the critical contaminant list from a preliminary critical contaminant list be based on neural network analysis of all available genomic, proteomic, and epidemiological databases. Current culture-based methods are seriously limited in their ability to provide the type of information needed for such an endeavor. Molecular methods for the detection and genetic/molecular characterization can be expected to fill this niche.

Current molecular methods, such as probe technology and gene amplifications, will get integrated into biosensor technologies in the future. A significant amount of research is currently underway in the use of gold nanoparticle probes, nanopores, signaling aptamers, and signaling DNA enzymes such as catalytic DNAs for chip-based biosensors (Authier et al. 2001, Howorka et al. 2001, Rajendran and Ellington 2002). Pathogen detection based on flow cytometry, evanescent wave technology, matrix-assisted laser desorption/ionization-time of flight mass spectrometry, immuno-magnetics, and neural networks have been published (Hutchinson 1995, Bernardo et al. 2002, Widmer et al. 2002, Chandler and Jarrell 2003, Liu et al. 2003). Biosensors and microarrays will find increasing applications in identifying the presence of specific pathogens and specific cellular components such as cell surface receptors, ribosomal RNA molecules, and DNA molecules (Bavykin et al. 2001, Call et al. 2001, Chandler and Jarrell 2003). Biosensors will be used in the future for either real-time detection of specific organisms or confirmatory purposes. A number of potential targets for biosensors have been identified. These include cell surface proteins such as porins or siderophore molecules (of which there are approximately 200,000 copies per cell), cell-surface polysaccharides such as lipopolysaccharides (approximately 2,000,000 copies per cell), ribosomal proteins and RNA in rapidly dividing cells (20,000 copies per cell), non-ribosomal RNA molecules (100–1,000 molecules per cell), and non-ribosomal proteins (3,000 molecules per cell). These rather spectacular developments have occurred at the

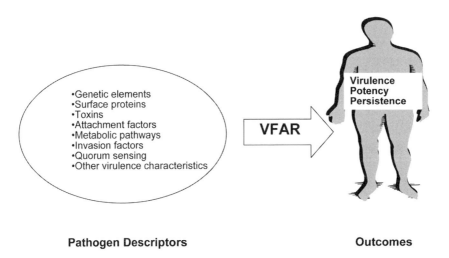

Pathogen Descriptors   **Outcomes**

Figure 22.1 The virulence function activity relationship between pathogen "descriptors" and "outcomes" adapted from National Research Council (2001).

interfacial fields of traditional scientific disciplines such as microbiology, biochemistry, and electrical engineering. These technologies have provided a plethora of choices for detecting and characterizing pathogens. It must be emphasized that many of the recent developments in detection technologies have been spurred on by the recent interest in protecting against deliberate pathogen contamination.

Personnel and Laboratory Infrastructure Needs for Molecular Methods

There is no doubt that many of the current molecular methods of pathogen detection have significant advantages over conventional microbiology methods. However, the reality is that over 90% of the current microbiology tests used in the food industry still rely on conventional agar plating techniques. Some of the concerns that the food industry has about accepting emerging molecular methods are the cost, the regulatory acceptance of the new methods, the level of technical expertise (required to perform the tests and interpret the test results), and the laboratory infrastructure required. Laboratory-based contamination very often can invalidate PCR results (Baselski 1996). Accidental human-induced cross contamination of reagents and samples is still a very serious problem facing many laboratories that employ molecular methods, especially PCR-based methods. Specific laboratory personnel training and the sample flow-through patterns in the laboratory, as well as the actual physical layout of the laboratory, can be instrumental in reducing the potential for cross contamination. As the detection methods become more specialized and sophisticated, the equipment needs and the expertise levels of a molecular/microbiology QA/QC laboratory will undoubtedly increase. Molecular methods will not totally replace conventional microbiology methods. Although molecular methods will allow for faster, sensitive, and more in-depth detection and characterization capabilities, they will carry with them an increased need for internal quality control and personnel who are adequately trained to ensure accuracy and proper interpretation of their results. In the final analysis, however, regulatory and liability pressures will ultimately dictate whether or not, and to what extent, the food industry will adopt molecular methods.

Conclusion: Strategies for the Future

With the global adoption of HACCP principles, the number of end-product tests that a typical food microbiology laboratory conducts will generally decrease. However, issues such as globalization of the food supplies, alternative processing technologies, increase in organic food consumption, increase in at-risk subpopulations, re-emerging pathogens, pathogen evolution, and an increase in consumer awareness dictate that specific pathogen tests will continue or even increase in the future. Under these scenarios, it is critical that rapid pathogen detection and characterization tools should not be allowed to languish in research laboratories. Rather, there should be a concerted effort to validate the performance characteristics of these assays in multi-laboratory round-robin trials (Pillai and Ricke 1995). There should be a close interaction between the food industry and the research communities to ensure that the molecular methods are rigorously field tested. The commercial food industry can play a significant role by providing naturally contaminated samples for assay development. This will overcome the need to rely on artificially spiked samples, which are very often used by the research community. Artificially spiked samples cannot accurately simulate naturally contaminated samples in terms of the distribution and the attachment of the contaminants within the food matrix. To cater to the industry needs for trained personnel, there is a need for appropriate college-level courses and training programs that provide hands-on experience in molecular methods. This is imperative so that the next generation of microbiologists and food scientists is adequately trained and comfortable in working with these technologies and interpreting the results of these molecular assays.

Acknowledgements

This work was supported by funds from a USDA-Initiative for Future Agriculture and Food Systems 00-52102-9637 grant, a USDA-Cooperative State Research, Education, and Extension Service grant 2001-34461-10405, and Hatch Project funds H-8708 from the Texas A&M University System.

References

Alterkruse, S. F., Cohen, M. L. and Swerdlow, D. L. 1997. Emerging foodborne diseases. *Emerging Infectious Diseases* **3**:285–293.

Association of Official Analytical Chemists. 2003. Available at: http://www.aoac.org/.

Authier, L., Grossiord, C. and Brossier, P. 2001. Gold nanoparticle-based quantitative electrochemical detection of amplified human cytomegalovirus DNA using disposable microband electrodes. *Analytical Chemistry* **73**:4450–4456.

Baselski, V. S. 1996. The role of molecular diagnostics in the clinical microbiology laboratory. *Clinical Laboratory Medical* **16**:49–60.

Bavykin, S. G., Akowski, J. P., Zakhariev, V. M., Barsky, V. E., Perov, A. N. and Mirzabekov, A. D. 2001. Portable system for microbial sample preparation and oligonucleotide microarray analysis. *Applied and Environmental Microbiology* **67**:922–928.

Belgrader, P., Young, S., Yuan, B., Primeau, M., Christel, L. A., Pourahmadi, F. and Northrup, M. A. 2001. A battery-powered notebook thermal cycler for rapid multiplex real-time PCR analysis. *Analytical Chemistry* **73**:286–289.

Bernardo, K., Pakulat, N., Macht, M., Krut, O., Seifert, H., Fleer, S., Hünger, F. and Krönke, M. 2002. Identification and discrimination of *Staphylococcus aureus* strains using matrix-assisted laser desorption/ionization-time of flight mass spectrometery. *Proteomics* **2**:747–753.

Bopp, D. J., Sauders, B. D., Waring, A. L., Ackelsberg, J., Dumas, N., Braun-Howland, E., Dziewulski, D., Wallace, B. J., Kelly, M., Halse, T., Musser, K. A., Smith, F., Morse, D. L. and Limberger, R. J. 2003. Detection, isolation and molecular subtyping of *Escherichia coli* O157:H7 and *Campylobacter jejuni* associated with a large waterborne outbreak. *Journal of Clinical Microbiology* **41**:174–180.

Bowtell, D. and Sambrook, J. 2003. DNA Microarrays. A Molecular Cloning Manual. Cold Spring Harbor, N.Y.: Cold Spring Harbor Laboratory Press.

Call, D. R., Borucki, M. K. and Besser, T. E. 2003. Mixed genome microarrays reveal multiple serotype and lineage-specific differences among strains of *Listeria monocytogenes*. *Journal of Clinical Microbiology* **41**:632–639.

Call, D. R., Brockman, F. J. and Chandler, D. P. 2001. Detecting and genotyping *Escherichia coli* O157:H7 using multiplexed PCR and nucleic acid microarrays. *International Journal of Food Microbiology* **67**:71–80.

Chandler, D. P. and Jarrell, A. E. 2003. Enhanced nucleic acid capture and flow cytometry detection with peptide nucleic acid probes and tunable-surface microparticles. *Analytical Biochemistry* **312**:182–190.

Chizhikov, V., Rasooly, A., Chumakov, K. and Levy, D. D. 2001. Microarray analysis of microbial virulence factors. *Applied and Environmental Microbiology* **67**:3258–3263.

Cho, J. C. and Tiedje, J. M. 2002. Quantitative detection of microbial genes by using DNA microarrays. *Applied and Environmental Microbiology* **68**:1425–1430.

Endley, S., Lu, L., Vega, E., Hume, M. E. and Pillai, S. D. 2003. Male-specific coliphages as an additional fecal contamination indicator for screening fresh carrots. *Journal of Food Protection* **66**:88–93.

Endley, S., Peña, J., Ricke, S. C. and Pillai, S. D. 2001. The applicability of *hns* and *fim*A primers for detecting *Salmonella* spp. in bioaerosols associated with animal and municipal wastes. *World Journal of Microbiology and Biotechnology* **17**:363–379.

Gilles, P. N., Wu, D. J., Forster, C. B., Dillon, P. J. and Chanock, S. J. 1999. Single nucleotide polymorphic discrimination by an electronic dot blot assay on semiconductor microchips. *Nature Biotechnology* **17**:365–370.

Guatelli, J. C., Gingeras, T. R. and Richman, D. D. 1989. Nucleic acid amplification in vitro: detection of sequences with low copy numbers and application to diagnosis of human deficiency virus type 1 infection. *Journal of Clinical Microbiology Review* **2**:217–226.

Howorka, S., Cheley, S. and Bayley, H. 2001. Sequence-specific detection of individual DNA strands using engineered nanopores. *Nature Biotechnology* **19**:636–639.

Hutchinson, A. M. 1995. Evanescent wave biosensors. Real-time analysis of biomolecular interactions. *Journal of Molecular Microbiology and Biotechnology* **3**:47–54.

Kreader, C. A. 1996. Relief of amplification inhibition in PCR by bovine serum albumin or T4 gene 32 protein. *Applied and Environmental Microbiology* **62**:1102–1106.

Kwoh, D. Y., Davis, G. R., Whitfield, K. M., Chappelle, H. I., DiMichele, L. J. and Gingeras, T. R. 1989. Transcription-based amplification and detection of amplified human immunodeficiency virus type 1 with a bead based sandwich hybridization format. *Proceedings of the National Academy of Sciences (USA)* **86**:1173–1177.

Lakowicz, J. R. 1999. Energy transfer, pp. 368–394. *In* Principles of Fluorescence Spectroscopy, 2nd ed., J. R. Lakowicz (Ed.) New York: Kluwer Academic.

Lammerding, A. M. and Paoli, G. M. 1997. Quantitative risk assessment: an emerging tool for emerging foodborne pathogens. *Emerging Infectious Diseases* **3**:483–487.

Liu, Y., Ye, J. and Li, Y. 2003. Rapid detection of *Escherichia coli* O157:H7 inoculated in ground beef, chicken carcass, and lettuce samples with an immunomagnetic chemiluminescence fiber-optic biosensor. *Journal of Food Protection* **66**:512–517.

Loge, F. J., Thompson, D. E. and Call, D. R. 2002. PCR detection of specific pathogens: a risk based analysis. *Environmental Science and Technology* **36**:2754–2759.

Mead, P. S., Slutsker, L., Dietz, V., McCaig, L. F., Brese, J. S., Shapiro, C., Griffin, P. M. and Tauxe, R. V. 1999. Food-related illness and death in the United States. *Emerging Infectious Diseases* **5**:607–625.

National Research Council. 2001. Classifying drinking water contaminants for regulatory consideration. Washington, D.C.: National Academy Press.

Nazarenko, I., Lowe, B., Darfler, M., Ikonomi, P., Schuster, D. and Rashtchian, A. 2002. Multiplex quantitative PCR using self-quenched primers labeled with a single fluorophore. *Nucleic Acids Research* **30**:e37.

Peña, J., Ricke, S. C., Shermer, C. L., Gibbs, T. and Pillai, S. D. 1999. A gene amplification-hybridization sensor based methodology to rapid screen aerosol samples for specific bacterial gene sequences. *Journal of Environmental Science and Health* **A34**:529–556.

Pillai, S. D. and Ricke, S. C. 1995. Strategies to accelerate the applicability of gene amplification protocols for pathogen detection in meat and meat products. *Critical Reviews in Microbiology* **21**:239–261.

Pillai, S. D., Josephson, K. L. and Pepper, I. L. 1993. Comparison of three probe labeling methods to detect PCR amplification products. *Journal Rapid Method and Autoimmunity Microbiology* **2**:299–309.

Pillai, S. D., Josephson, K. L., Bailey, R. L., Gerba, C. P. and Pepper, I. L. 1991. Rapid method for processing soil samples for polymerase chain reaction amplification of specific gene sequences. *Applied and Environmental Microbiology* **57**:2283–2286.

Rajendran, M. and Ellington, A. D. 2002. Selecting nucleic acids for biosensor applications. *Journal of Combinatorial Chemistry High Throughput Screen* **5**:263–270.

Roe, M. T. 2002. Prevalence of class 1 and class 2 integrons in poultry processing and a natural water ecosystem, Master's Thesis. College Station, Tex.: Texas A&M University.

Smith, J. L. and Fratamico, P. M. 1995. Factors involved in the emergence and persistence of food-borne diseases. *Journal of Food Protection* **58**:696–708.

Stintzi, A. 2003. Gene expression profile of *Campylobacter jejuni* in response to growth temperature variation. *The Journal of Bacteriology* **185**:2009–2016.

Strategic Consulting. 2000. Pathogen Testing in the U.S. Food Industry. Woodstock, Vt.: Strategic Consulting.

Thelwell, N., Millington, S., Solinas, A., Booth, J. and Brown, T. 2000. Mode of action and application of Scorpion primers to mutation detection. *Nucleic Acids Research* **28**:3752–3761.

U.S. Environmental Protection Agency. 2003. Workshop to develop a protocol for reliable genetic methods for the detection of viruses for use in EPA's Water Programs, January 15–17, 2003, Cincinnati, Ohio.

Vasavada, P. C. 2001. Getting really rapid test results. *Food Safety Magazine* **July**:28–38.

Wei, Y., Lee, J.-M., Richmond, C., Blattner, F. R., Rafalski, J. A. and LaRossa, R. A. 2001. High density microarray-mediated gene expression profiling of *Escherichia coli*. *The Journal of Bacteriology* **183**:545–556.

Widmer, K. W., Oshima, K. H. and Pillai, S. D. 2002. Identification of *Cryptosporidium parvum* oocysts by an artificial neural network approach. *Applied and Environmental Microbiology* **68**:1115–1121.

Wilson, I. G. 1997. Inhibition and facilitation of nucleic acid amplification. *Applied and Environmental Microbiology* **63**:3741–3751.

Ye, R. W., Tao, W., Bedzyk, L., Young, T., Chen, M. and Li, L. 2000. Global gene expression profiles of *Bacillus subtilis* grown under anaerobic conditions. *The Journal of Bacteriology* **182**:4458–4465.

23 Methods for Differentiation among Bacterial Foodborne Pathogens

Steven L. Foley and Robert D. Walker

Introduction

Knowledge of how bacterial foodborne pathogens disseminate through the food chain is important in understanding how food animals and food processing procedures contribute to foodborne illness. To track these pathogens through the food production environment, it is necessary to identify and determine their source and spread at the different stages of food production and processing. The ability to characterize the relatedness of strains and to determine the primary sources of contamination provides valuable insights into the epidemiology and natural history of foodborne pathogens and provides important tools to improve public health.

There are many ways in which specific foodborne pathogens can be differentiated within a given species of bacteria. This chapter will explore the various phenotypic and genotypic typing methods that are used to differentiate bacterial pathogens. To evaluate the use of particular typing methods, three key factors must be considered. These include discriminatory power, type-ability, and reproducibility. Discrimination is the ability of a technique to separate between non-related strains. Type-ability refers to the capacity of the technique to generate an interpretable result for each strain typed. Reproducibility is the ability of the typing technique to return the same result each time the process is repeated (Busch and Nitschko 1999). In addition, issues such as the cost of the typing method, the ease of use and interpretation, and the time required to return a result are important considerations when examining the various typing methods.

Phenotypic Methods

Historically, identification and classification of bacterial pathogens have been based on phenotypic features of the organism. Phenotypic characteristics include biochemical profiles based on the bacterium's metabolic activity, serotyping, phage typing, antimicrobial susceptibility profile, and multilocus enzyme electrophoresis. These phenotypic typing techniques are based on the expression products of particular genes that may be present in the different strains of bacteria. In contrast, genotypic techniques exploit differences in the nucleotide sequence of the genome to type bacteria and are largely independent of gene expression (Arbeit 1995). Phenotypic techniques have been in use for decades and are widely used as methods to speciate specific bacterial isolates.

Serotyping

One of the most common ways to differentiate pathogens such as *Salmonella* and *Escherichia coli* isolates is by serotyping. For example, there are over 2,400 identified serotypes of *Salmonella*. Historically, serotyping was used to divide the Salmonellae into distinct species, with one species per serotype. However, with recent advances in molecular microbiology, some microbiologists believe that the genus *Salmonella* is now divided into only two species, *S. enterica* and *S. bongor* (Brenner et al. 2000). Even with the species changes, the serotype name remains the main descriptor of the *Salmonella* isolate. For example, in reporting cases of *Salmonella* infections in the United States, the Centers of Disease Control (CDC) separates out isolates based on their serotypes; for example, *Salmonella* Typhimurium (CDC 2000). Serotyping uses differences in the somatic (O) and flagellar (H) surface antigens to separate strains into distinct serotypes (Voogt et al. 2002). For serotyping, the suspension of bacteria is mixed and incubated with a panel of antisera specific for a variety of O and H epitopes. Specific agglutination profiles are used to determine the serotype of the isolate being tested. Serotyping is also common in a number of other bacterial genera including *Campylobacter* and *E. coli* (Frost et al. 1999). For example, *E. coli* O157:H7 is defined by its agglutination profiles: O157 and H7.

Antimicrobial Susceptibility Testing

Another phenotypic technique that is used to characterize foodborne pathogens is antimicrobial susceptibility testing and the generation of antimicrobial resistance profiles. To ensure that the results are properly interpreted and are reproducible, antimicrobial susceptibility testing must be carried out using standardized testing methods such as those described by the National Committee of Clinical Laboratory Standards (NCCLS), including the appropriate use of quality control organisms (National Committee of Clinical Laboratory Standards 2002). When properly performed, the results from antimicrobial susceptibility testing can provide insight as to the relatedness of bacteria associated with an infectious disease outbreak. In addition, because there are different antimicrobial use patterns in certain production environments, geographic locations, and healthcare settings, the occurrence of resistance to certain antimicrobials may provide insight as to the animal source of particular bacterial pathogens. However, antimicrobial resistance profiles alone do not typically provide adequate discrimination for use as a single typing method. Isolates with different serotypes and from different sources may share common resistance profiles (White et al. 2001). Therefore, the use of antibiotic susceptibility profiles to characterize strains of bacteria is important for use in clinical settings for choosing the appropriate therapy and for tracking large-scale trends in antimicrobial resistance. However, an antibiotic-resistance profile is often not sufficiently discriminatory to be used as a stand-alone method to determine the relatedness of isolates.

Phage Typing

Phage typing is an additional phenotypic method that is often used to discriminate between closely related bacterial isolates. This technique, which uses the selective ability of lytic bacteriophage to infect certain strains of bacteria, has been in use for many decades (Schmieger 1999; see Chapters 8 and 27). The ability of bacteriophage to infect bacteria is the result of the molecular characteristics of the phage and phage receptor present on the surface of the bacteria (Snyder and Champness 1997). If the appropriate phage receptor is present, the phage will infect the bacterium and lyse the cell. Lysis is observed as a plaque on a lawn of bacteria growing on an agar plate. The particular phage type that is assigned to the specific strain of bacteria is dependent on whether or not specific typing phage are able to lyse the cells and form plaques in the bacterial lawns (Hickman-Brenner et al. 1991).

Phage typing schemes are available for a number of different foodborne pathogens including *Salmonella*, *E. coli*, and *Campylobacter* (Barrett et al. 1994, Frost et al. 1999, Humphrey 2001). Phage typing has been shown to be useful in the description of pandemic clones of *Salmonella*, such as *S. enterica* serovar Typhimurium definitive type 104 (DT104), which causes severe gastrointestinal illness and is typically resistant to multiple antibiotics (Humphrey 2001). In addition, phage typing has been used to differentiate *E. coli* O157:H7 isolates associated with a disease outbreak from non-outbreak strains (Barrett et al. 1994). A major drawback of phage typing is that the limited number of available phage means that many strains are untypeable. This limits the technique's capacity to distinguish among the strains (Amavisit et al. 2001). In addition, phage typing is technically demanding and requires the maintenance of multiple biologically active phage stocks. Therefore, phage typing is typically performed by reference labs with well-maintained phage stocks and specially trained personnel (Arbeit 1995).

Multilocus Enzyme Electrophoresis

Multilocus enzyme electrophoresis (MLEE) is a phenotyping technique that produces a characteristic electrophoretic profile based on variations in conserved cellular enzymes. The genes that encode many of these enzymes are housekeeping genes and are therefore, highly conserved within a genus. Differences in enzyme mobility are the result of variations in amino acid sequences leading to a change in the protein's net electrostatic charge and migration in an electric field. Therefore, MLEE provides a means to detect genetic differences in the genes encoding the enzymes. The water-soluble enzymes are separated on starch or cellulose acetate gels and visualized by the use of enzyme-specific stains, and then their gel migration patterns are compared (Selander et al. 1986). Variation in the migration patterns of multiple enzymes creates a unique profile for distinct strains that is referred to as an electrophoretic type. By comparing the electrophoretic type profile of different strains, statistical inferences can be made about their genetic relatedness.

Because the technique examines conserved gene products, MLEE is most often used for evolutionary rather than epidemiologic studies. Results obtained for bacterial typing with MLEE have demonstrated that the technique worked well to determine the relatedness of bacterial isolates (Boyd et al. 1996, Cox et al. 1996, Gordon and Lee 1999). MLEE has been used to distinguish among enteric bacteria from distinct geographic locations and, in some cases, to determine the host species in which the bacteria originated (Gordon and Lee 1999). Although MLEE has gained acceptance as a subtyping method for bacteria in general (Boyd et al. 1996, Cox et al. 1996), it is being replaced by multilocus sequence typing (MLST), a molecular typing method based on similar principles that uses DNA sequencing, making it less time consuming, more reproducible, and more portable.

Genotypic Methods

Restriction-Based Methods

Plasmid Typing

Plasmids are one of the means by which bacteria may readily share genetic material. The ease with which plasmids may be transferred from one bacterium to another makes them important elements in bacteriology and in human health (Summers 1996). Plasmids typically contain nonessential genes that impart some selective advantage to the host, such as virulence or antimicrobial resistance (Johnson et al. 2002, Tenover 2001). Thus, examining bacteria for the presence of plasmids, including the number and size of the plasmids, may be a means of detecting differences in strains of foodborne pathogens and also may provide insight into the source of the pathogens. To perform plasmid typing, plasmid DNA is isolated using a procedure that restricts the isolation of chromosomal DNA. The plasmid DNA is then separated using agarose gel electrophoresis, the gel is stained, and the profile is viewed. The number and size of the plasmids present provide a plasmid-typing profile to the strains (Nauerby et al. 2000).

Alternatively, isolated plasmid DNA can be cut with restriction enzymes to create restriction-fragment profiles that can be compared with one another to differentiate the isolates (Nauerby et al. 2000). This restriction analysis mitigates one of the major drawbacks of using plasmids for typing bacteria. The topography of a plasmid affects its electrophoretic mobility. If a plasmid is nicked or unwound, its electrophoretic migration is altered, possibly resulting in multiple bands for the same plasmid, thus confounding the analysis (Summers 1996). Other problems in using plasmid profiling as a typing method are that some strains lack plasmids and that plasmids can be mobile elements (Kumao et al. 2002). The gain or loss of plasmids can lead to problems trying to determine the relatedness of foodborne bacterial isolates. The plasmid content of a given strain is subject to changing environmental selective pressures as the pathogens spread from their original host to cause human disease.

Pulsed-Field Gel Electrophoresis (PFGE)

Electrophoresis to separate DNA fragments by size is a common technique in molecular biology and is commonly used as a means to assess strain relatedness in bacteria. Typing by standard electrophoresis using agarose or acrylamide gels is limited in that the largest fragment size that can be efficiently separated using these matrices is approximately 50 kb. In PFGE, fragments are separated in alternating electric fields. This approach can resolve DNA fragments up to 800 kb in size (Schwartz and Cantor 1984). When used in conjunction with rare-cutting restriction enzymes, PFGE profiles provide a DNA "fingerprint" that provides a more complete assessment of the entire bacterial genome. PFGE is considered by many to be the gold standard molecular typing method for bacteria (Olive and Bean 1999). It is used by public health monitoring systems such as the U.S. CDC and Prevention's PulseNet, a program to track the spread of foodborne pathogens and assist in determining sources of foodborne disease outbreaks (Swaminathan et al. 2001).

To perform PFGE, an optimal number of cells are embedded in an agarose matrix. The agarose plugs are treated with detergents or enzymes, such as sarcosine and proteinase K, to lyse the embedded cells and release the DNA. The plug is washed thoroughly to remove cellular debris and treated with a rare-cutting restriction enzyme (see Table 23.1). Following restriction enzyme treatment, the plugs are inserted into the wells of an agarose gel and the DNA separated under conditions of switching polarity. Following electrophoresis, the pattern of DNA separation is visualized by staining the gel with a fluorescent dye (Gautom 1997). The gel banding pattern from one isolate can be compared with those of other isolates, and information about the relatedness of the strains can be resolved. The determination of genetic relatedness based on PFGE banding patterns has been standardized by Tenover et al. (1995). Standardization of the technique has allowed PFGE to become a widely accepted method for comparing the genetic identity of bacteria during a disease outbreak (Swaminathan et al. 2001).

Table 23.1 Common enzymes used for pulsed-field gel electrophoresis subtyping (PulseNet).[a]

Bacterial species	Enzyme(s)
Escherichia coli (O157:H7)	*Xba*I, *Bln*I, *Spe*I
Salmonella enterica	*Xba*I, *Bln*I, *Spe*I
Campylobacter spp.	*Sma*I
Listeria monocytogenes	*Asc*I, *Apa*I, *Sma*I
Shigella spp.	*Xba*I, *Bln*I, *Spe*I
Clostridium spp.	*Sma*I

[a]Source: Centers for Disease Control 2002.

A major advantage of PFGE typing is that the method is highly reproducible for the vast majority of strains. In part, this reproducibility is the result of the lysis and of isolation of the DNA within the agarose plug. The isolated DNA is not pipetted, therefore the banding patterns generated are a result of restriction enzyme digestion, not non-specific mechanical shearing (Birren and Lai 1993). In addition, the genetic profile generated is based on the entire genome. This is an improvement over some of the PCR-based typing techniques that rely on amplification of DNA from a limited number of sequence targets to develop a genetic profile (Millemann et al. 2000, Swaminathan and Barrett 1995). If PFGE is used to compare results with other laboratories, it is important to follow standardized methods (Tenover et al. 1995), such as those proposed by groups like PulseNet (Swaminathan et al. 2001). A drawback of PFGE is that it is labor intensive, often requiring a few days to perform the procedure and interpret the results (Swaminathan and Barrett 1995). Most researchers have had good success using PFGE in molecular typing schemes for foodborne pathogens.

Restriction Fragment Length Polymorphism (RFLP) Analysis

RFLP profiles are generated by restriction-enzyme DNA digestion and visualization of banding patterns produced following gel electrophoresis. By itself, this method generates a very large number of DNA fragments that are difficult to distinguish in standard gel electrophoresis. When coupled with southern blotting using probes for repeated DNA elements, RFLP can be very informative. This is the case with ribotyping and insertion sequence (IS) RFLP typing.

IS typing has been widely used as an epidemiological tool for investigating the Mycobacteria (IS*6110*) (Cousins et al. 1998). For many enterics, genetic elements such as IS*200* sequences can be used for RFLP analysis. IS*200* are insertion sequences that are present in many *Salmonella*, as well as in some *E. coli* and *Yersinia*. These sequences are approximately 700 bp in length and are found randomly around the genome (Beuzon and Casadesus 1997). The results of IS*200* typing are varied. Some strains lack IS*200* sequences; therefore, IS*200* typing would not be effective in differentiating these isolates (Millemann et al. 2000). Among foodborne pathogens, IS typing has widely been superceded by other methods such as IS*200* repetitive element-PCR (Rep-PCR).

Ribotyping is based on the number and location of the ribosomal RNA (rRNA) gene sequences (Ling et al. 2000). Total cellular DNA is cut with frequent-cutting restriction enzymes. Common enzymes that have been used for many of the foodborne pathogens include *Pst*I, *Pvu*II, *Hind*III, and *Eco*RI (Gendel and Ulaszek 2000, Wassenaar and Newell

2000, Ling et al. 2000). For blotting, the restriction fragments from the gel are transferred to a nylon membrane and incubated with a probe homologous to the highly conserved regions of rRNA or other genes (Chisholm et al. 1999). Sequence differences in the regions flanking the rRNA lead to variability in the size of the junction fragments. This produces distinct patterns that can be used to discriminate between related strains (Gendel and Ulaszek 2000, Snipes et al. 1989).

A major advantage of ribotyping is that it can be highly reproducible. This is especially true when ribotyping is automated using the RiboPrinter Microbial Characterization System (Qualicon, Inc., Wilmington, Del.), where most of the procedure is robotically controlled. Ribotyping also generates relatively few bands, which makes analysis easier. If more information is needed, isolates can be treated with additional enzymes to increase the discriminatory value of the method (Gendel and Ulaszek 2000). A potential drawback of ribotyping arises from the limited number of rRNA genes in some bacterial groups. Also, if there is modification of the genome by methylation or other factors, it can lead to problems with DNA restriction, which in turn leads to poor band pattern resolution (Olive and Bean 1999).

Amplification-Based Methods

Amplified Fragment Length Polymorphisms (AFLP)

In addition to the techniques that rely on restriction fragment analysis, there are a number of other typing methods that use PCR to differentiate between pathogens. One of these methods is AFLP, which uses a combination of restriction digest analysis and PCR amplification to determine the relatedness of bacterial strains. The discriminatory power of AFLP is based on variability in restriction enzyme recognition sites within a portion of the genome. Following digestion with one or more enzymes, linker polynucleotides are ligated to the free DNA ends. These linkers serve as targets for PCR primers to bind and allow selective amplification of the restriction fragments. The amplified fragments are subjected to electrophoresis, and characteristic separation profiles are generated and compared with other strains (Mueller and Wolfenbarger 1999).

Often AFLP is carried out using fluorescent-labeled PCR primers, which allow detection of the fragments using an automated DNA sequencer. Because of the large number of restriction fragments that are generated, the PCR primers are designed to contain one to three additional nucleotides on their 3′-end. These interact with an unknown sequence in the restriction fragment. Thus, for each additional nucleotide that is added to the primer, the number of amplicons is reduced by about a factor of four (Savelkoul et al. 1999). Following amplification, the resulting bands are separated on a DNA sequencing gel. Following separation, an elution profile is generated using the fluorescent intensity associated with the incorporated labeled primers. The elution profiles of the isolates are compared with other profiles to determine the relatedness of each strain.

AFLP combines beneficial qualities of other typing techniques, restriction analysis, and PCR amplification. This combination amplifies the signal of the restriction fragments, and thus only a small amount of DNA is required for analysis.

Because the PCR primers are directed at the linkers, a prior knowledge of the organism's DNA sequence is not required for amplification. The fragments generated represent a wide range of different locations throughout the bacterial genome, giving a relatively good coverage of the genome to identify differences. In addition, through the use of an automated

DNA sequencer, a large number of isolates can be screened fairly rapidly, making AFLP an attractive typing technique (Desai et al. 2001). A drawback of AFLP occurs with distantly related organisms: phylogenetic inferences can be difficult to make, therefore, the technique is best suited for more closely related strains (Mueller and Wolfenbarger 1999).

Random Amplified Polymorphic DNA PCR (RAPD-PCR) and Arbitrarily Primed PCR (AP-PCR)

Other PCR-based typing methods rely on amplification profiles without DNA restriction to separate between bacterial isolates. Two of these techniques are RAPD-PCR and AP-PCR, which are similar methods of PCR subtyping. These techniques use PCR primers of random sequence. When two of the primers bind in close enough proximity to one another, they amplify the intervening portion of the genome, creating variably sized amplicons. During amplification, annealing of primers is done at low temperature because the primers have incomplete homology to the genomic binding sites. For RAPD-PCR, the primers are typically 6–10 base pairs in length, whereas for AP-PCR they are usually 20–34 bases in length (Welsh and McClelland 1990, Williams et al. 1990). In addition, for AP-PCR, a second set of primer annealing temperatures is used. The first few cycles of amplification use a lower annealing temperature (thus, lower stringency or level of sequence homology between the primer and template required for primer binding) followed by a higher temperature and stringency for the subsequent amplification cycles (Welsh and McClelland 1990). Amplicons are subjected to agarose gel electrophoresis, and characteristic banding patterns are generated that are used to subtype the strains of the bacteria.

As with other PCR-based typing methods, the results are obtained in a relatively short period of time, and only a small amount of DNA is required to perform the tests. Because of the use of random primers for electrophoresis, a prior knowledge of the DNA sequence is not required. Multiple sets of PCR primers can be used to increase discrimination of the strains (Franklin et al. 1999). Although RAPD- and AP-PCR have been used in molecular typing of foodborne pathogens, there can be difficulty in reproducing the same result over time and from one operator to the next. Because PCR reactions are done under conditions with low stringency, minor differences in reaction conditions or reagent concentrations can alter the amplification profile produced, making it difficult to interpret and compare results (Swaminathan and Barrett 1995).

Repetitive Element PCR

Typically there are a number of repeated DNA sequences throughout the genome of most bacterial species. PCR primers have been designed for many of these repetitive elements that allow for the amplification of the flanking regions of DNA. When two of these repeated elements are located in close proximity to one another, the region will be amplified (Versalovic et al. 1991). Following PCR, the amplicons are separated by gel electrophoresis and the banding patterns are then compared to one another. Differences in the sizes of the amplified fragments produce different banding patterns (Georghiou et al. 1994). A number of types of Rep-PCR techniques have been developed using various repetitive sequences; these include enterobacterial repetitive intergenic consensus PCR (ERIC-PCR) and repetitive extragenic palindromic PCR (REP-PCR). Enterobacterial repetitive intergenic consensus sequences are highly conserved repetitive sequences found in enteric bacteria, whereas repetitive extragenic palindromic elements are relatively short sequences of DNA that have highly conserved regions of palindromic DNA. Each of the elements serves as a good PCR primer target (Hulton et al. 1991, Versalovic et al. 1991). In addition, insertion sequence typing (discussed earlier) such as that done using IS*200* can also be performed as a type of Rep-PCR (Millemann et al. 2000).

One of the major advantages of this set of techniques is that the results can be obtained in a relatively short time with relatively good discrimination (Swaminathan and Barrett 1995). The techniques are fairly easy to perform and can be adapted to run on an automated DNA sequencer to facilitate analysis of the separation profiles and increase reproducibility of the results (Del Vecchio et al. 1995). Because of the utility of Rep-PCR for typing microorganisms, commercial enterprises such as Bacterial Barcodes have begun marketing semi-automated proprietary Rep-PCR systems (Bacterial Barcodes Inc., Houston, Tex.).

The results of Rep-PCR can be stored in a database library and can be available to compare with the results of other researchers (Garaizar et al. 2000). With Rep-PCR and some of the other typing techniques, the use of large databases could provide a valuable tool for characterizing foodborne pathogens. For example, if an isolate has a similar pattern to those of known geographical or host origin, it can provide potential evidence about the source of the pathogen. A drawback of Rep-PCR is that the strains being typed need to have a sufficient number of repetitive elements in close proximity to one another to generate a discriminatory profile. In addition, variations in amplification or separation conditions can cause problems with the reproducibility of results, which can cause difficulty in making relational inferences about isolates (Swaminathan and Barrett 1995).

Amplification Profiling

Amplification profiling or the amplification of a series of genes that is associated with factors such as virulence, antimicrobial resistance, and nutrient use may provide a mechanism to differentiate organisms. PCR, using sets of primer pairs, is used to assay for the presence or absence of specific genes and to measure strain relatedness. Primer pairs can be combined in multiplex PCR assays to reduce the number of reactions required. The benefit of the amplification profiling approach is that in addition to distinguishing between bacterial strains, specific genes of clinical importance can be detected (Gordon 2001). Johnson et al. (2001) used amplification profiling of virulence genes to genotype *E. coli* isolates. This technique requires sequence information on genes being examined for primers to be generated. This is becoming less of a hindrance as the number of genomes sequenced continues to rise. Examination of genes that encode products that are responsible for virulence and antimicrobial resistance may provide discrimination among strains, as these genes are often acquired to allow bacteria to survive in different environments (Conner et al. 1998). Because of environmental selective pressures, the presence of these genes may provide insight into the natural history of foodborne pathogens and may be helpful in determining the source of bacterial foodborne isolates.

Sequence-Based Methods

MLST

MLST is a relatively new molecular typing method. The foundation of this technique is dependant on different strains of bacteria having variability in the sequence of particular genes, because of mutation or recombination events, that can be used to determine the relatedness of bacteria. With MLST, multiple genes with conserved sequences are compared for nucleotide base changes. Housekeeping genes (genes required for basic cellular functions) are most often sequenced because they are present in all isolates and are not subject to strong selective pressures that can lead to relatively rapid sequence changes. However, housekeeping genes typically have sufficient variability to develop distinct alleles for different strains (Maiden et al. 1998). The alleles from each of the genes sequenced are looked at as a group, and the strain is assigned a specific sequence type. The genetic relatedness of the strains is determined by comparing differences in allele profiles that make up the sequence type. In cases in which all alleles are the same, the sequence types are identical and the two strains are defined as clonal (Enright and Spratt 1999).

MLST has a number of attractive features for use in molecular epidemiology. The technique uses specific nucleotide base changes rather than DNA fragment size to determine the genetic relatedness. Therefore, there is good reproducibility of the results, which are easily shared with others electronically. For a number of bacterial species, internet-based MLST databases have been set up to facilitate the rapid exchange of MLST results (Enright and Spratt 1999). As additional databases come on-line, they should allow public health officials to track trends in the distribution of pathogens of interest, similar to what is currently being done with the PulseNet system. This potential ability to track the distribution of isolates could be used to provide information about the sources of bacteria implicated in a foodborne infection.

An important consideration in MLST is the selection of gene targets. Genes with varying degrees of genetic difference, bracketed by stable sequences, are required. In some cases, only a few loci provide adequate discrimination among non-clonal isolates. In addition, housekeeping genes, with their low rate of genetic variability, provide a desirable sequencing target for global phylogenetic studies but may not provide enough variability to distinguish among strains involved in short-term outbreak investigations (Maiden et al. 1998). Table 23.2 lists many of the genes that have been used for typing different foodborne pathogens.

Conclusions and Future Directions

Each typing technique has strengths and weaknesses that will affect its usefulness for differentiating among non-clonal bacterial foodborne pathogens. Most of these techniques are best applied in disease outbreaks, where temporal associations and other epidemiological data are available. In surveillance systems, where information is gathered on strains from different geographical regions over extended periods of time, making inferences on ecological links between human and environmental isolates is a greater challenge. For maximal effectiveness, the typing method needs to generate highly discriminatory and reproducible results that prevent calling non-clonal strains identical.

Table 23.2 Reported gene targets for Multilocus Sequence Typing.

Bacterial species	Genes	Reference
E. coli O157:H7	*arcA, aroE, dnaE, mdh, gnd, gapA, pgm, espA, ompA*	Noller et al. 2003
Salmonella enterica	*glnA, manB, pduF,* 16sRNA gene	Kotetishvili et al. 2002
Campylobacter jejuni	*aspA, glnA, gltA, pgm, tkt, uncA*	Dingle et al. 2001
Listeria monocytogenes	*abcZ, dat, ldh, sod, cat, dapE, pgm, bglA, lhkA*	Salcedo et al. 2003

Often, multiple methods are required for effective differentiation between isolates. The order in which the methods are employed can help improve the efficiency of pathogen differentiation. The phenotypic techniques such as serotyping and antimicrobial susceptibility testing may be appropriate methods to use first. The use of genotypic techniques such as amplification profiling and sequence typing will further differentiate among the different strains. Finally, more laborious techniques such as AFLP and PFGE, which work well to distinguish between more closely related isolates, can be employed to finely differentiate between apparently clonal isolates that may be responsible for the foodborne illness. Table 23.3 provides a comparison of the important attributes for typing and how the methods compare to each other.

In addition to discriminatory ability, other factors such as the time to obtain results, the cost of the supplies, and the cost of the equipment to perform the typing method may be important for choosing methods to differentiate foodborne pathogens. Some techniques, although very discriminatory, require specialized equipment that is expensive, and thus, they are not feasible for every investigator. For example, to use MLST or AFLP as a high-throughput typing method, an investigator needs access to an automated DNA sequencer. Purchasing such equipment can be quite costly. In contrast, techniques such as RAPD-PCR and Rep-PCR require a standard thermal cycler and gel electrophoresis equipment, tools common to the molecular microbiology laboratory. In addition, some procedures require considerably more time than others to get the desired result. Techniques such as PFGE are quite discriminatory but can take a few days or longer to get results, whereas many of the

Table 23.3 Comparison of characteristics important for differentiation of bacterial foodborne pathogens.

Typing method	Discrimination	Reproducibility	Difficulty	Relative cost	Relative time
Serotyping	Low	High	High	Moderate	Moderate
Resistance profiling	Low	High	Moderate	Moderate	Moderate
Phage typing	Low	High	High	Low	Moderate
Multilocus enzyme electrophoresis	Moderate	High	High	Moderate	High
Plasmid profiling	Low	High	Low	Low	Low
Pulsed-field gel electrophoresis	High	High	High	Moderate	Moderate
Restriction fragment length polymorphism analysis /ribotyping	Moderate	High	Moderate	Moderate	Moderate
Multilocus sequence typing	Moderate/high	High	High	High	Moderate
Amplified fragment length polymorphisms	High	High	High	Moderate	Moderate
Arbitrarily primed/random amplified polymorphic DNA polymerase chain reaction	High	Low	Moderate	Low	Low
Repetitive element polymerase chain reaction	High	Low	Moderate	Low	Low

PCR-based tests can be performed in less than a day. If a rapid turnaround time is needed, time-consuming techniques, although typically more discriminatory and reproducible, may not be the best choice. With each typing technique there are trade-offs. The methods that are most time consuming and expensive tend to be the most discriminatory and provide the most accurate and reproducible measure of relatedness. Therefore, the choice of the most appropriate molecular typing method will rely on the level of differentiation needed and the resources available to perform the typing.

In the future, there may be typing methods that will allow us to differentiate among non-clonal isolates in a rapid, reproducible way with little ambiguity of results. In the interim there will likely be a trend in the development of new typing methods to improve on existing technology through further automation and standardization of techniques. A prime example of how automation has enhanced typing is through the use of the Riboprinter for ribotyping of isolates. The ribotyping steps have been automated, and the manufacturer prepares the media and chemicals required to limit the amount of user variability with the typing. The typing results are readily analyzed, and relational inferences can be drawn. A key example of how standardization can improve typing is with PFGE and the PulseNet system. Standardized methods have allowed for data generated at one location to be compared with data from others, facilitating national surveillance of foodborne diseases. Increased standardization and automation should help to reduce some of the problems that are encountered with current typing methods.

Another trend in molecular typing is the advancement of sequence-based typing methodologies. At present, MLST is in its infancy. Further advances in the selection of genes to be sequenced and the establishment of additional on-line databases to share information will help facilitate global sequence typing. As the cost of DNA sequencing is reduced, this technology will be more accessible for a wider typing audience. In addition, the sequencing of many bacterial genomes has spawned additional sequence-based typing methods. One such new method is multilocus variable number tandem repeat analysis. This analysis uses repeated sequences that can use variable copy numbers to differentiate isolates. By examining multiple repeat regions, a discriminatory pattern can often be determined. This technique shows a great deal of promise with some of the strains that appear homogeneous with PFGE, such as *E. coli* O157:H7 isolates. Multilocus variable number tandem repeat analysis is able to distinguish among many of these isolates and will prevent non-clonal strains from being called identical, which is very important in trace-back studies (Keys et al. 2002).

There will also likely be a push to reduce the amount of time it takes to go from the collection of a sample until the results of typing are obtained. Typing results will be ready in hours instead of days or weeks. Technologies such as real-time PCR and DNA microarray assays may be developed that allow for very rapid identification and differentiation of pathogens. DNA microarrays could be developed that assay for multiple virulence or other discriminatory genes. The presence/absence profiles of the various genes can be used to differentiate among isolates in a relatively short amount of time. In addition, real-time PCR could be used to screen isolates and develop presence/absence profiles to separate isolates. These rapid methods may not be as discriminatory as some of the tried and true typing methods, but they could serve as a sort of gateway to other more intensive methodologies if needed. The rapid methods could, for example, rule out potential sources in a disease outbreak and allow the investigator to focus on a more limited number of sources. Today we have a number of choices to differentiate between bacterial foodborne pathogens, with each of the methods having strengths and weaknesses. The future push in molecular subtyping will likely involve trying to overcome the current weaknesses of the available techniques to provide a more comprehensive means to differentiate foodborne pathogens.

References

Amavisit, P., Markham, P. F., Lightfoot, D., Whithear, K. G. and Browning, G. F. 2001. Molecular epidemiology of *Salmonella Heidelberg* in an equine hospital. *Veterinary Microbiology* **80**:85–98.

Arbeit, R. D. 1995. Laboratory procedures for the epidemiologic analysis of microorganisms, pp. 190–208. *In* Manual of Clinical Microbiology, P. R. Murray, E. J. Baron, M. A. Pfaller, F. C. Tenover, R. H. Yolken (Eds.). Washington, D.C.: American Society for Microbiology Press.

Barrett, T. J., Lior, H., Green, J. H., Khakhria, R., Wells, J. G., Bell, B. P., Greene, K. D., Lewis, J. and Griffin, P. M. 1994. Laboratory investigation of a multistate food-borne outbreak of *Escherichia coli* O157:H7 by using pulsed-field gel electrophoresis and phage typing. *Journal of Clinical Microbiology* **32**:3013–3017.

Beuzon, C. R. and Casadesus, J. 1997. Conserved structure of IS*200* elements in *Salmonella*. *Nucleic Acids Research* **25**:1355–1361.

Birren, B. and Lai, E. 1993. Pulsed Field Gel Electrophoresis: A Practical Guide. San Diego, Calif.: Academic Press.

Boyd, E. F., Wang, F. S., Whittam, T. S. and Selander, R. K. 1996. Molecular genetic relationships of the salmonellae. *Applied and Environmental Microbiology* **62**:804–808.

Brenner, F. W., Villar, R. G., Angulo, F. J., Tauxe, R. and Swaminathan, B. 2000. *Salmonella* nomenclature. *Journal of Clinical Microbiology* **38**:2465–2467.

Busch, U. and Nitschko, N. 1999. Methods for the differentiation of microorganisms. *Journal of Chromatography B Biomedical Sciences Applications* **722**:263–278.

Centers of Disease Control. 2000. FoodNet 1999 Final Report. Atlanta, Ga.: Centers for Disease Control. Available at: http://www.cdc.gov/foodnet/annual/1999/pdf/FoodNet__1999_Annual_Report.pdf.

Centers of Disease Control. 2002. Standardized Molecular Subtyping of Foodborne Bacterial Pathogens by Pulsed-field Gel Electrophoresis. Atlanta, Ga.: National Center for Infectious Diseases CDC Division of Bacterial and Mycotic Diseases Foodborne and Diarrheal Diseases Branch, Public Health Practice Program Office CDC Division of Laboratory Services and Association of Public Health Laboratories, Centers for Disease Control and Prevention.

Chisholm, S. A., Crichton, P. B., Knight, H. I. and Old, D. C. 1999. Molecular typing of *Salmonella* serotype Thompson strains isolated from human and animal sources. *Epidemiology and Infection* **122**:33–39.

Conner, C. P., Heithoff, D. M., Julio, S. M., Sinsheimer, R. L. and Mahan, M. J. 1998. Differential patterns of acquired virulence genes distinguish *Salmonella* strains. *Proceedings of the National Academy of the Sciences (USA)* **95**:4641–4645.

Cousins, D., Williams, S., Liebana, E., Aranaz, A., Bunschoten, A., Van Embden, J. and Ellis, T. 1998. Evaluation of four DNA typing techniques in epidemiological investigations of bovine tuberculosis. *Journal of Clinical Microbiology* **36**:168–178.

Cox, J. M., Story, L., Bowles, R. and Woolcock, J. B. 1996. Multilocus enzyme electrophoretic (MEE) analysis of Australian isolates of *Salmonella enteritidis*. *International Journal of Food Microbiology* **31**:273–282.

Del Vecchio, V. G., Petroziello, J. M., Gress, M. J., McCleskey, F. K., Melcher, G. P., Crouch, H. K. and Lupski, J. R. 1995. Molecular genotyping of methicillin-resistant *Staphylococcus aureus* via fluorophore-enhanced repetitive-sequence PCR. *Journal of Clinical Microbiology* **33**:2141–2144.

Desai, M., Threlfall, E. J. and Stanley, J. 2001. Fluorescent amplified-fragment length polymorphism subtyping of the *Salmonella enterica* serovar *enteritidis* phage type 4 clone complex. *Journal of Clinical Microbiology* **39**:201–206.

Dingle, K. E., Colles, M. F. M., Wareing, D. F. A., Ure, R., Fox, A. J., Bolton, F. E., Bootsma, H. J., Willems, R. J. L., Urwin, R. and Maiden, M. C. J. 2001. Mutilocus sequence typing system for *Campylobacter jejuni*. *Journal of Clinical Microbiology* **39**:14–23.

Enright, M. C. and Spratt, B. G. 1999. Multilocus sequence typing. *Trends in Microbiology* **7**:482–487.

Franklin, R. B., Taylor, R. D. and Mills, A. L. 1999. Characterization of microbial communities using randomly amplified polymorphic DNA (RAPD). *Journal of Microbiological Methods* **35**:225–235.

Frost, J. A., Kramer, J. M. and Gillanders, S. A. 1999. Phage typing of *Campylobacter jejuni* and *Campylobacter coli* and its use as an adjunct to serotyping. *Epidemiology and Infection* **123**:47–55.

Garaizar, J., Lopez-Molina, N., Laconcha, I., Lau, B. D., Rementeria, A., Vivanco, A., Audicana, A. and Perales, I. 2000. Suitability of PCR fingerprinting, infrequent-restriction-site PCR, and pulsed-field gel electrophoresis, combined with computerized gel analysis, in library typing of *Salmonella enterica* serovar *enteritidis*. *Applied and Environmental Microbiology* **66**:5273–5281.

Gautom, R. K. 1997. Rapid pulsed-field gel electrophoresis protocol for typing of *Escherichia coli* O157:H7 and other Gram-negative organisms in 1 day. *Journal of Clinical Microbiology* **35**:2977–2980.

Gendel, S. M. and Ulaszek, J. 2000. Ribotype analysis of strain distribution in *Listeria monocytogenes*. *Journal of Food Protection* **63**:179–185.

Georghiou, P. R., Doggett, A. M., Kielhofner, M. A., Stout, J. E., Watson, D. A., Lupski, J. R. and Hamill, R. J. 1994. Molecular fingerprinting of *Legionella* species by repetitive element PCR. *Journal of Clinical Microbiology* **32**:2989–2994.

Gordon, D. M. 2001. Geographical structure and host specificity in bacteria and the implications for tracing the source of coliform contamination. *Microbiology* **147**:1079–1085.

Gordon, D. M. and Lee, J. 1999. The genetic structure of enteric bacteria from Australian mammals. *Microbiology* **145**:2673–2682.

Hickman-Brenner, F. W., Stubbs, A. D. and Farmer III, J. J. 1991. Phage typing of *Salmonella enteritidis* in the United States. *Journal of Clinical Microbiology* **29**:2817–2823.

Hulton, C. S., Higgins, C. F. and Sharp, P. M. 1991. ERIC sequences: a novel family of repetitive elements in the genomes of *Escherichia coli*, *Salmonella typhimurium* and other enterobacteria. *Molecular Microbiology* **5**:825–834.

Humphrey, T. 2001. *Salmonella* Typhimurium definitive type 104. A multi-resistant *Salmonella*. *International Journal of Food Microbiology* **67**:173–186.

Johnson, J. R., Delavari, P., Kuskowski, M. and Stell, A. L. 2001. Phylogenetic distribution of extraintestinal virulence-associated traits in *Escherichia coli*. *Journal of Infectious Disease* **183**:78–88.

Johnson, T. J., Giddings, C. W., Horne, S. M., Gibbs, P. S., Wooley, R. E., Skyberg, J., Olah, P., Kercher, R., Sherwood, J. S., Foley, S. L. and Nolan, L. K. 2002. Location of increased serum survival gene and selected virulence traits on a conjugative R plasmid in an avian *Escherichia coli* isolate. *Avian Diseases* **46**:342–352.

Keys, C., Kemper, S. and Keim, P. 2002. MLVA: a novel typing system for *E. coli* O157:H7. Abstract No. 65221, 2002 American Society for Microbiology General Meeting, May 19–23, Salt Lake City, Utah. Available at: http://www.asmusa.org/memonly/abstracts/AbstractView.asp?AbstractID=65221. Accessed: August 28, 2003.

Kotetishvili, M., Stine, O. C., Kreger, A., Morris Jr., J. G. and Sulakvelidze, A. 2002. Multilocus sequence typing for characterization of clinical and environmental *Salmonella* strains. *Journal of Clinical Microbiology* **40**:1626–1635.

Kumao, T., Ba-Thein, W. and Hayashi, H. 2002. Molecular subtyping methods for detection of *Salmonella enterica* serovar Oranienburg outbreaks. *Journal of Clinical Microbiology* **40**:2057–2061.

Ling, J. M., Lo, N. W., Ho, Y. M., Kam, K. M., Hoa, N. T., Phi, L. T. and Cheng, A. F. 2000. Molecular methods for the epidemiological typing of *Salmonella enterica* serotype *Typhi* from Hong Kong and Vietnam. *Journal of Clinical Microbiology* **38**:292–300.

Maiden, M. C., Bygraves, J. A., Feil, E., Morelli, G., Russell, J. E., Urwin, R., Zhang, Q., Zhou, J., Zurth, K., Caugant, D. A., Feavers, I. M., Achtman, M. and Spratt, B. G. 1998. Multilocus sequence typing: a portable approach to the identification of clones within populations of pathogenic microorganisms. *Proceedings of the National Academy of the Sciences (USA)* **95**:3140–3145.

Millemann, Y., Gaubert, S., Remy, D. and Colmin, C. 2000. Evaluation of IS200-PCR and comparison with other molecular markers to trace *Salmonella enterica* subsp. *enterica* serotype Typhimurium bovine isolates from farm to meat. *Journal of Clinical Microbiology* **38**:2204–2209.

Mueller, U. G. and Wolfenbarger, L. L. 1999. AFLP genotyping and fingerprinting. *Trends in Ecology and Evolution* **14**:389–394.

National Committee of Clinical Laboratory Standards. 2002. Performance Standards for Antimicrobial Susceptibility Testing; Twelfth Informational Supplement (M100-S12)2002. Wayne, Penn.: National Committee for Clinical Laboratory Standards.

Nauerby, B., Pedersen, K., Dietz, H. H. and Madsen, M. 2000. Comparison of Danish isolates of *Salmonella enterica* serovar *enteritidis* PT9a and PT11 from hedgehogs (*Erinaceus europaeus*) and humans by plasmid profiling and pulsed-field gel electrophoresis. *Journal of Clinical Microbiology* **38**:3631–3635.

Noller, A. C., McEllistrem, M. C., Stine, O. C., Morris Jr., J. G., Boxrud, D. J., Dixon, B. and Harrison, L. H. 2003. Multilocus sequence typing reveals a lack of diversity among *Escherichia coli* O157:H7 isolates that are distinct by pulsed-field gel electrophoresis. *Journal of Clinical Microbiology* **41**:675–679.

Olive, D. M. and Bean, P. 1999. Principles and applications of methods for DNA-based typing of microbial organisms. *Journal of Clinical Microbiology* **37**:1661–1669.

Salcedo, C., Arreaza, L., Alcala, B., de la Fuente, L. and Vazquez, J. A. 2003. Development of a multilocus sequence typing method for analysis of *Listeria monocytogenes* clones. *Journal of Clinical Microbiology* **41**:757–762.

Savelkoul, P. H., Aarts, H. J., de Haas, J., Dijkshoorn, L., Duim, B., Otsen, M., Rademaker, J. L., Schouls, L. and Lenstra, J. A. 1999. Amplified-fragment length polymorphism analysis: the state of an art. *Journal of Clinical Microbiology* **37**:3083–3091.

Schmieger, H. 1999. Molecular survey of the *Salmonella* phage typing system of Anderson. *Journal of Bacteriology* **181**:1630–1635.

Schwartz, D. C. and Cantor, C. R. 1984. Separation of yeast chromosome-sized DNAs by pulsed field gradient gel electrophoresis. *Cell* **37**:67–75.

Selander, R. K., Caugant, D. A., Ochman, H., Musser, J. M., Gilmour, M. N. and Whittam, T. S. 1986. Methods of multilocus enzyme electrophoresis for bacterial population genetics and systematics. *Applied and Environmental Microbiology* **51**:873–884.

Snipes, K. P., Hirsh, D. C., Kasten, R. W., Hansen, L. M., Hird, D. W., Carpenter, T. E. and McCapes, R. H. 1989. Use of an rRNA probe and restriction endonuclease analysis to fingerprint *Pasteurella multocida* isolated from turkeys and wildlife. *Journal of Clinical Microbiology* **27**:1847–1853.

Snyder, L. and Champness, W. 1997. Molecular Genetics of Bacteria. Washington, D.C.: American Society for Microbiology Press.

Summers, D. K. 1996. The Biology of Plasmids. Oxford: Blackwell Science.

Swaminathan, B. and Barrett, T. J. 1995. Amplification methods for epidemiologic investigations of infectious disease. *Journal of Microbiological Methods* **2**:129–139.

Swaminathan, B., Barrett, T. J., Hunter, S. B. and Tauxe, R. V. 2001. PulseNet: the molecular subtyping network for foodborne bacterial disease surveillance, United States. *Emerging Infectious Diseases* **7**:382–389.

Tenover, F. C. 2001. Development and spread of bacterial resistance to antimicrobial agents: an overview. *Clinical Infectious Diseases* **33(Suppl. 3)**:S108–S115.

Tenover, F. C., Arbeit, R. D., Goering, R. V., Mickelsen, P. A., Murray, B. E., Persing, D. H. and Swaminathan, B. 1995. Interpreting chromosomal DNA restriction patterns produced by pulsed-field gel electrophoresis: criteria for bacterial strain typing. *Journal of Clinical Microbiology* **33**:2233–2239.

Versalovic, J., Koeuth, T. and Lupski, R. J. 1991. Distribution of repetitive DNA sequences in eubacteria and application to fingerprinting of bacterial genomes. *Nucleic Acids Research* **19**:6823–6831.

Voogt, N., Wannet, W. J., Nagelkerke, N. J. and Henken, A. M. 2002. Differences between national reference laboratories of the European community in their ability to serotype *Salmonella* species. *European Journal of Clinical Microbiology and Infectious Diseases* **21**:204–208.

Wassenaar, T. M. and Newell, D. G. 2000. Genotyping of *Campylobacter* spp. *Applied and Environmental Microbiology* **66**:1–9.

Welsh, J. and McClelland, M. 1990. Fingerprinting genomes using PCR with arbitrary primers. *Nucleic Acids Research* **18**:7213–7218.

White, D. G., Zhao, S., Sudler, R., Ayers, S., Friedman, S., Chen, S., McDermott, P. F., McDermott, S., Wagner, D. D. and Meng, J. 2001. The isolation of antibiotic-resistant *salmonella* from retail ground meats. *New England Journal of Medicine* **345**:1147–1154.

Williams, J. G., Kubelik, A. R., Livak, K. J., Rafalski, J. A. and Tingey, S. V. 1990. DNA polymorphisms amplified by arbitrary primers are useful as genetic markers. *Nucleic Acids Research* **18**:6531–6535.

Part V

Decontamination and Prevention Strategies

24 Chemical Methods for Decontamination of Meat and Poultry

Jimmy T. Keeton and Sarah M. Eddy

Introduction

The safety of meat and poultry products is a primary societal concern in the United States, and during recent years, reports of foodborne illnesses have intensified the public's interest for ensuring a safe food supply. In 1999, an estimated 76 million illnesses, 325,000 hospitalizations, and 5,000 deaths were attributed to foodborne diseases (Mead et al. 1999). As part of its overall food safety strategy to reduce the risk of foodborne illness associated with pathogens such as *Salmonella* spp., *Campylobacter, Escherichia coli* O157:H7, *Listeria monocytogenes, Clostridium botulinum, Clostridium perfringens, Staphylococcus aureus, Aeromonas hydrophila* and *Bacillus cereus*, the U.S. Department of Agriculture Food Safety Inspection Service (USDA-FSIS 2002a) issued a final rule on July 25, 1996 (Federal Register 1996), requiring meat and poultry establishments to adopt a science-based Hazard Analysis Critical Control Point (HACCP) system and to develop plant and product specific HACCP plans (see Chapter 20). The meat and poultry industries responded by implementing HACCP systems, with their primary objective being to make products demonstrably safe to consume.

The incidence of pathogens in the meat and poultry food supply has been influenced by several factors. Some of the more important are

- Livestock production practices that may inadvertently foster pathogen contamination;
- The emergence of "new" and antibiotic resistant pathogens in the environment;
- Increased manipulation and handling and accelerated processing of carcasses and raw materials;
- Modification of traditional processing practices and greater complexity of manufacturing procedures and equipment;
- A more complex distribution and food preparation system that increases the risk of foodborne disease;
- More discriminate and selective pathogen detection methods to improve confirmation and trace-back of contaminated product; and
- Consumer habits that represent inappropriate food handling and preparation practices.

In response to the USDA-FSIS mandate to increase the safety of meat and poultry products, numerous chemical compounds have been evaluated as ante- and postmortem decontamination agents. To be classified as an "antimicrobial," compounds must be proven effective and approved for use by the U.S. Food and Drug Administration (FDA) and the USDA-FSIS (Davidson 2002). Once proven to be safe, antimicrobials may be applied to carcass or product surfaces or, if appropriate, incorporated into the product as an ingredient. These compounds may be naturally derived or manufactured and must not conceal spoilage, but they may extend shelf life and prevent pathogen growth as a consequence of their bactericidal or bacteriostatic activity.

This chapter provides an overview of the current chemical decontamination methods used by the meat and poultry industries to prevent or reduce the incidence of microbial pathogens on animal carcasses and meat products. These antimicrobial treatments are often used in combination with other technologies to provide multiple interventions (also known as hurdle technology; Leistner 2002) for pathogens and to reduce the risk of foodborne disease. Emerging chemical decontamination methods are also presented, as these may be approved for use in the future.

Preharvest Chemical Interventions

Because of the continued outbreaks of foodborne illness resulting from the consumption of cross-contaminated, undercooked, or mishandled meat and poultry products, there is a need to develop strategies to control and eliminate these foodborne pathogens. Cattle are the primary reservoir for *E. coli* O157:H7, with feces and the hide being the most likely sources of contamination for beef (Park et al. 1999, Bacon et al. 2000, Elder et al. 2000). The incidence and level of *E. coli* O157:H7 vary with the season of the year, type of animal, anatomical site on the carcass, stage of processing, and fabrication sequence (Sofos et al. 1999). Contamination levels of *E. coli* O157:H7 have been reported to be 0.2% on steer and heifer carcasses (USDA-FSIS 1994), 0.87% in ground beef from federally inspected facilities and supermarkets (surveyed since 1995), 1.01% in ground beef purchased for the school lunch program (data obtained from the Ag Marketing Service by FSIS in 2001), and 1.8% on carcasses from four major Midwestern packing plants (Elder et al. 2000). Bacon et al. (2000) surveyed 12 beef slaughter facilities and found the incidence of *E. coli* O157:H7 to be 3.6% and 0.4% on beef hides and carcasses, respectively, before washing and 0.0% on carcasses after final decontamination. One important conclusion by Elder et al. (2000) was that the existing in-plant processing practices appeared to reduce the level of *E. coli* O157:H7 (Huffman 2002).

Salmonella spp. have been found to contaminate beef feces at levels up to 14.2%, and carcasses at levels from 0.3% to 2.6% before final decontamination (Sofos et al. 1999). *Salmonella,* however, do not pose the same degree of disease severity as *E. coli* O157:H7, but must be controlled to reduce the risk of foodborne illness. This pathogen also has been reported to reside primarily in the intestinal tract and lymph tissue of market pigs (Anderson et al. 2001), but its presence on carcasses also can vary (Currier et al. 1986). Up to 30% of market pigs may shed *Salmonella*, and this may increase during transport and holding before slaughter (Berends et al. 1996). A survey of pork carcasses in 12 slaughter plants (Zerby et al. 1998) found the incidence of *Salmonella* to range from 3.2% to 5.5% when two carcass sampling sites were combined for analysis.

Salmonella contamination of broilers entering the slaughter plant was reported to range from 3% to 5% (Lillard 1989), but the prevalence of *Salmonella*-positive broilers after processing has been observed to be as high as 36% (Green 1987). In the United States, processed poultry carcasses are often contaminated with *Salmonella* or *Campylobacter jejuni* (Park et al. 2002) by cross-contamination from carcasses following scalding, defeathering, evisceration, or post-evisceration chilling. The chiller is of particular concern as the birds are submerged in a circulating water (ca. 4°C) medium containing variable levels of chlorine (20–50 ppm; Kemp et al. 2000). Some scientists note that chlorine has limited effectiveness against *Salmonella* spp. as a result of uptake by organic matter in the chiller and of variable concentration levels throughout the daily processing cycle. Chlorine is corrosive on machinery and also may combine with organic materials to generate potential mutagens.

Rose et al. (2002) evaluated HACCP's effectiveness from 1998 to 2000 by collecting 98,204 samples and 1,502 sample sets for *Salmonella* analysis from large, small, and very small establishments. These plants produced at least one of seven raw meat and poultry products: broilers, market hogs, cows and bulls, steers and heifers, ground beef, ground chicken, and ground turkey. *Salmonella* prevalence in most product categories was lower after implementation of HACCP when compared with the pre-HACCP baseline surveys conducted by the USDA-FSIS. When survey data from all establishments were combined, they showed that >80% of the sample sets met the following *Salmonella* prevalence performance standards: 20.0% for broilers, 8.7% for market hogs, 2.7% for cows and bulls, 1.0% for steers and heifers, 7.5% for ground beef, 44.6% for ground chicken, and 49.9% for ground turkey. This decrease in *Salmonella* prevalence may be partially explained by industry improvements through HACCP implementation, improved process control, incorporation of antimicrobial interventions, and increased microbial-process control monitoring.

Consistent application of physical and chemical interventions to decontaminate carcasses in a multiple-hurdle manner has been effective for reducing or eliminating most pathogens on carcasses during the harvesting process (Bacon et al. 2000). However, because some pathogens survive decontamination procedures, additional or better intervention processes are needed to decrease the incidence of pathogens on animals before slaughter.

Feed Additives and Competitive Exclusion

Reducing pathogens in or on live animals would intuitively reduce the level of product contamination and, in turn, reduce the risk of foodborne disease (Huffman 2002, Leistner 2002). Work by Anderson et al. (2000) has demonstrated that both *E. coli* O157:H7 and *Salmonella* Typhimurium DT104 are reduced in buffered ruminal contents by approximately 1,000,000 colony-forming units (CFU)/mL when exposed to 5 mM sodium chlorate. Use of chlorate apparently is bactericidal only to microorganisms such as *E. coli* O157:H7 and *S.* Typhimurium DT 104 that possess respiratory nitrate reductase, which reduces chlorate to cytotoxic chlorite ion. Most other harmless anaerobes lack nitrate reductase and are not affected, thus preserving beneficial bacteria. Oral administration of 100 mM sodium chlorate to experimentally infected pigs also has been shown to reduce cecal concentrations of *S.* Typhimurium and may provide a means of significantly reducing pathogens before slaughter (Anderson et al. 2001).

Another method for reducing pathogen colonization of the intestines of animals, and thus reducing the level of fecal contamination, has been the use of bacterial strains that produce probiotics or grow sufficiently to exclude harmful bacteria through competitive exclusion. Probiotics are typically non-pathogenic bacteria that inhibit the growth or attachment of strains of pathogenic bacteria. Probiotic bacteria, when administered to weaned beef calves, have been shown to reduce the levels of *E. coli* O157:H7 (Zhao et al. 1998). Subsequent research with calves less than 1 week of age and pretreated orally with probiotic *E. coli* has demonstrated that fecal shedding of enterohemorrhagic *E. coli* serotypes *E. coli* O111:NM and O26:H11, but not *E. coli* O157:H7, is reduced. Colicins are specific proteins produced by strains of *E. coli* that are inhibitory against some strains of pathogenic *E. coli*. Murinda et al. (1996) identified the specific colicin, Colicin E2, and found it to inhibit 11 strains of *E. coli* O157:H7 *in vivo*. Further studies with this compound may offer a potential means for controlling pathogenic *E. coli* in the animal. Additional work with other feed additives such as seaweed extract is continuing and may ultimately result in a feed ingredient that excludes pathogens from the intestinal microflora.

A competitive exclusion spray for young chicks, Preempt (MS Bioscience, Dundee, Ill.), was approved by the FDA in 1998 and consists of a mixture of 29 unique bacteria commonly found in the gastrointestinal tract of mature chickens. When applied to young chicks shortly after hatching, the treated chick ingests the bacteria as it preens, seeding its ceca with beneficial microorganisms. The competitive bacteria then cover niches in the cecal surface so quickly and heavily that intestinal pathogens are excluded from attachment. After colonization with Preempt organisms, the prevalence of *Salmonella* declines significantly (Hume et al. 1998), thus reducing the potential for fecal contamination early in the production process.

Vaccines

An *E. coli* vaccine, originally developed by Dr. Brett Finlay at the University of British Columbia, stimulates production of antibodies to prevent *E. coli* O157:H7 from attaching to the intestinal wall of cattle. This impedes replication and reduces the number of pathogens that can be shed from the feces into the environment. *Escherichia coli* O157:H7 achieves molecular attachment to intestinal cells in cattle by embedding a protein known as Tir into the host cell membrane (Finlay 2003, Moxley et al. 2003). The pathogenic cell then joins to the intestinal cells by attaching intimin, an outer membrane protein, to the cell-embedded Tir proteins. A study conducted by Dr. Rod Moxley found that feedlot cattle vaccinated with the Canadian-developed vaccine showed a significant decline in the incidence of *E. coli* O157:H7 shedding in manure. Among the 384 cattle included in the trial, the pre-treatment prevalence of *E. coli* O157:H7 averaged 31%. Over the trial period, the average proportion of cattle shedding the bacterium after vaccination was 8.8%. The *E. coli* vaccine used in the trial is being developed by an alliance composed of the University of British Columbia, the Alberta Research Council, Saskatchewan's Vaccine and Infectious Disease Organization, and Bioniche Life Sciences Inc., which is responsible for commercialization of the vaccine.

Postharvest Decontamination

Live animals are contaminated with non-pathogenic and pathogenic microorganisms that serve as a subsequent source of carcass and meat/poultry contamination. No one method of decontamination is suitable for removing pathogens from raw materials and products (Bacon et al. 2000, Kang et al. 2001) because of surface geometries, protected sites of contamination, and inherent inefficiency of specific decontamination processes. Because of the likelihood of cross-contamination from the animal to equipment, personnel, or the building/facility surfaces through direct contact or aerosols, decontamination strategies are needed that are bactericidal and bacteriostatic. Several thermal and non-thermal systems for decontamination of carcasses are currently in use and may be classified as physical, chemical, or a combination of the two. Examples of physical systems include dehairing, spray-washing, blanching, hot water pasteurization, steam pasteurization, steam-vacuuming, atmospheric steam application, singeing, and ultrahigh pressure. Others that have been or are being considered include countercurrent scalders; electrolyzed water; pulsed electric fields; ultrasonic energy; high-energy ultraviolet light; pulsed light; sonic, infrasonic, and ultrasound; x-rays; linear accelerators; and irradiation. Some of the more important commercial technologies are covered in Chapter 25. Chemical interventions that are most often used in combination with physical systems are discussed below and include chlorine-based derivatives, organic acids, organic and inorganic compounds, bacteriocins, and emerging technologies.

Pre-evisceration

Strategies to reduce the initial pathogen load on beef animals have been to apply a bactericidal spray, such as an organic acid, to the hide just before slaughter, washing the animal, or dehairing the hide surface post-exsanguination (Sofos et al. 1999). Antimicrobial sprays and washing of the hair surface have not been studied extensively as an effective means of pathogen decontamination, but in some cases they are practiced commercially. Wetting the hair surface can result in simply spreading pathogens and is often compounded by excess feces caked into the hair. According to regulatory guidelines, animals are required to be dry or at least not dripping when slaughtered.

Chemical dehairing is a patented process (Bowling and Clayton 1992) that removes hair, dirt, and feces from the hides of beef carcasses immediately before exsanguination and evisceration. The process consists of a sequential application of 10% sodium sulfide, a water rinse at 40.5°C (3.45 bar), a second application of 10% sodium sulfide at 5.52 bar, a 3% hydrogen peroxide rinse at 3.45 bar (17 seconds) to neutralize excess sodium sulfide, a second water rinse at 40.5°C (8.28 bar), and a final 3% hydrogen peroxide rinse followed by a water rinse to reduce the pH of the hide. Evaluations of the process found that dehairing removed visible contamination from carcasses but did not reduce the overall counts of aerobic bacteria (aerobic plate count, total coliform count, *E. coli* biotype I) or the presence of *Salmonella* spp. or *L. monocytogenes* (Schnell et al. 1995). Subsequent work (Castillo et al. 1998a) did find that the chemical dehairing process applied to artificially contaminated beef hides reduced the initial numbers of *S.* Typhimurium and *E. coli* O157:H7 (5.1 and 5.3 Log_{10} CFU/cm^2, respectively) to levels below the detection limit (0.5 Log_{10} CFU/cm^2). Chemical dehairing may prove to be beneficial, but the process may only injure bacterial cells, which in turn could be a potential food safety hazard should they repair themselves (Delmore et al. 1997). Because dehairing would obviously generate considerable amounts of hydrolyzed hair and chemical residues (sodium sulfide and hydrogen peroxide), safe disposal or recycling of the waste material may become both an environmental and economic issue if this process is adopted.

Post-evisceration of Carcasses

Beef decontamination processes may include animal cleaning or disinfecting of the hide before stunning, spot-cleaning of carcasses by steam vacuuming, knife trimming of observable surface contamination, rinsing or deluging of carcasses with water (>160°F at the carcass surface), and the use of chemical sanitizing agents (e.g., organic acids, trisodium phosphate, etc.) or steam before evisceration. Subsequent post-evisceration procedures consist of applications of hot water, steam, or chemical treatments to carcass surfaces immediately before chilling. Most processes are applied at various concentrations or intensities, pressures (2–20 bar), temperatures (15°–80°C), and lengths of time (5–20 seconds; Sofos et al. 1999). Pork and poultry decontamination, however, may involve singeing, the application of hot water, organic acids, chlorine compounds or trisodium phosphate (TSP), and rapid chilling. Castillo et al. (2002) provides a comprehensive review of various cleaning and sanitizing methods for reducing microorganisms on beef carcasses, and they also note some potentially negative aspects of carcass interventions.

Organic Acids

An organic acid spray, in combination with steam vacuuming, hot water washing, or steam cabinets, is the most commonly used means of chemical decontamination of beef carcasses (Huffman 2002). Under the HACCP system, Bolton et al. (2001) recommended that the critical limits for an organic acid spray be the application of at least 500 mL of a 2.5%–10% (v/v) acid (to allow for dilution when applied to the carcass) maintained at a pH of ≤2.8 and temperature of 25°–55°C and sprayed for 35 seconds at 13.8–27.6 Pa. Acetic and lactic acids ranging in concentration from 1.5% to 2.5% are the most widely used commercially, but others approved for use by the USDA-FSIS include glucono delta lactone (D-gluconic acid δ-lactone), phosphoric acid, tartaric acid, citric acid, buffered sodium citrate (pH 5.6, citric acid), and a combination of peroxiacetic acid + octanoic acid + acetic acid + hydrogen peroxide + peroxyoctanoic acid and 1-hydroxyethylidene-1,1-diphosphonic acid (HEDP; sold by Ecolab [St. Paul, Minn.] as Inspexx; Code of Federal Regulations 2003, USDA-FSIS 2002b). Acidified calcium sulfate + an organic acid (sold by Mionix [Rocklin, Calif.] as $Safe_2O$) is approved for ready-to-eat (RTE) products and is being considered as another potential carcass decontamination agent. Under the European Union regulations, organic acids are not permitted for beef carcass decontamination because they are perceived as concealing or compensating for poor hygiene practices during slaughter (Bolton et al. 2001). Rather, they employ a non-intervention system such as the Hygiene Assessment Scheme, which is in operation in the United Kingdom and focuses on a hygienic audit of five main categories during the slaughter process. Other European non-intervention HACCP systems use on-line monitoring at four critical control points: dehiding, evisceration, removal of the spinal cord, and chilling.

The mechanism by which organic acids work is not completely known, but it is generally believed that the undissociated form of the acid, or its ester, is responsible for antimicrobial activity (Castillo et al. 2002). Weak acids that penetrate the bacterial cell membrane accumulate in the cytoplasm, and if the intracellular pH is higher than the pKa of the acid, then the protonated acid will acidify the cytoplasm, resulting in cell injury or death (Booth 1985). However, the primary mechanism or mechanisms of the organic acids' bactericidal and bacteriostatic effects have yet to be totally elucidated.

The antimicrobial activities of organic acids at equal pH were ranked by Sorrells et al. (1989) as acetic > lactic > citric > malic > hydrochloric, whereas activities based on equimolar concentration of the acids were citric > malic > lactic > acetic > hydrochloric. Thus, concentration of the acid and pH are essential components determining antibacterial activity. They also noted that heating acids to 37°C versus 10°C increased their inhibitory effect on bacteria and partially explained why organic acids are more effective on warm carcasses immediately after hide removal. Additional studies demonstrate that lactic or acetic acid sprays applied at 55°C on warm carcasses are more effective for reducing *Salmonella* and *E. coli* O157:H7 (Hardin et al. 1995, Castillo et al. 1998b). Castillo et al. (2002) notes that reductions in counts of different pathogens on beef, as reported by different authors, vary between 2 and 4.3 logs after spraying with 2% acetic acid. Hardin et al. (1995) found lactic acid to be more effective than acetic acid for reducing *E. coli* O157:H7, and equally as effective as acetic acid at inhibiting *S.* Typhimurium on beef carcasses. Spraying post-chilled *E. coli* O157:H7– and *S.* Typhimurium–contaminated beef rounds with L-lactic acid (500 mL, 30 seconds, 55°C) taken from warm carcasses that had been previously decontaminated with L-lactic acid spray (250 mL, 15 seconds, 55°C), has been shown to have an additive effect by reducing counts an additional 2.0–2.4 and 1.6–1.9 logs, respectively (Castillo et al. 2001). These counts were in addition to the 3.3–5.2 log reductions caused by the pre-chill, warm-carcass treatments. Ground beef produced from the post-chill-decontaminated rounds had significantly lower levels of the pathogens than did ground beef from the pre-chill treatment alone. Similar results were reported by Dorsa et al. (1998) when 2% acetic or 2% D,L-lactic acid treatments were applied to pre-chilled neck tissues contaminated with pathogens. Their results suggest that pre-chilling treatment of carcasses with organic acids will lower bacterial counts in ground beef produced from the trimmings. Thus, the combined application of pre-chill and post-chill organic acid treatments to beef trimmings is effective for lowering pathogen levels in ground beef.

Van Netten et al. (1995) concluded that the application of a 2% lactic acid spray at 37°C, pH 2.3 for 30 seconds to freshly slaughtered pork carcasses could effectively eliminate *Salmonella*. Additional work has shown that a 2% lactic acid spray applied for 5 seconds in combination with a hot-water wash at temperatures of 65° and 80°C will reduce mesophilic and *E. coli* bacteria on pork carcasses by approximately 2 logs (Eggenberger-Solorzano et al. 2002). As a result of the ease of an organic acid spray application, this process could be easily adapted by both small and large processing plants that wish to add additional intervention procedures to enhance the microbial safety of pork carcasses.

A comparative study of acetic, citric, lactic, malic, mandelic, propionic, and tartaric acids against *S.* Typhimurium attached to broiler skin found that, in general, acid concentrations of ≥4% were required to kill ≥2 log CFUs of the pathogen (Tamblyn and Conner 1997). Adaptation of organic acid washing of poultry carcasses, however, will require further work to overcome cost and quality obstacles for application of high acid concentrations.

A residual antibacterial effect on beef carcasses and meat cuts has been observed after spraying with lactic or acetic acid solutions. When beef carcasses and retail steaks were treated with lactic or acetic acid, *E. coli, E. coli* O157:H7, *Listeria innocua*, and *Clostridium sporogenes* aerobic plate counts increased more slowly than on untreated samples (Kotula and Thelappurate 1994, Dorsa et al. 1997), thus slowing pathogen growth and potentially reducing the risk of foodborne disease.

At low concentrations of an acid treatment (1%–2%), discoloration of the meat tissue surface is minimized; however, with applications of ≥3% of an organic acid, a noticeable decline in meat color and acceptability may become apparent. Control of the acid concentration is therefore critical with regard to both safety and product quality.

A combination of peroxyacetic acid, octanoic acid, acetic acid, hydrogen peroxide, peroxyoctanoic acid, and 1-hydroxyethylidene-1,1-diphosphonic acid is being used commercially in both poultry and beef processing plants to reduce microbial pathogens. The patented mixture (Inspexx) contains two biocides and creates oxidizing conditions in an acidic environment to effectively reduce *E. coli* O157:H7, *Salmonella*, and *L. monocytogenes* on red meat and poultry carcass surfaces (Mermelstein 2001). The mixture can be used up to a maximum concentration of 220 ppm peroxyacetic acid and 75 ppm hydrogen peroxide. Peroxyacetic and peroxyoctanoic acids also have lipophilic surface activity, enabling them to better "wet" fatty tissue and effect decontamination. The composite blend does not contain halogens and has <4 ppm phosphorous, making it more compatible with waste treatment systems because it rapidly breaks down into water, oxygen, octanoic acid, and acetic acid. Immersion and pressurized spray applications are USDA-FSIS approved for poultry processing and for decontamination of both hot and cold beef and pork carcasses. The composite solution at normal use levels has been observed to be non-corrosive to stainless steel, aluminum, and nickel, but it may corrode brass. Validation studies involving laboratory-contaminated samples and in-plant processing conditions are currently underway to establish the effectiveness of the technology for reducing pathogens.

Safe$_2$O is a patented, very acidic (pH 1.0–1.5) organic acid–calcium sulfate complex that has been shown to reduce pathogens on the surfaces of beef, poultry, and RTE meat products when used as a spray or a 30-second dip. All ingredients are affirmed as generally recognized as safe (GRAS) under the FDA code. Beef carcasses treated with acidic calcium sulfate (ACS) to determine its efficacy showed an average of 0.7 log CFU of aerobic bacteria on the surface in comparison to 3.4 logs CFU for carcasses treated with a chlorine rinse (Huffman 2002). In a study conducted by the USDA-ARS (Dickens et al. 2001), plant-run poultry carcasses were treated with water, TSP (pH 12), or ACS (pH 1.3) at 1.2 L per carcass for a simulated pre-evisceration rinse and 1.3 L per carcass for an inside/outside washer. Microbial counts ($\log_{10}$ CFU/mL) for the untreated carcasses were total aerobes, 3.99; *E. coli*, 2.68; *Salmonella*, 1.98; and *Campylobacter*, 2.47. The water spray lowered all counts to 3.11, 1.28, 0.60, and 1.38 for the respective organisms. TSP and ACS treatments reduced counts even lower for *E. coli* (0.64 and 0.52), *Salmonella* (0.24 and 0.15), and *Campylobacter* (0.15 and 0.08), respectively. Total aerobes for the TSP and Safe$_2$O were 2.31 and 2.30, respectively. The ACS poultry wash was effective as an antimicrobial treatment and reduced counts slightly more than TSP without the problems associated with phosphate disposal.

In a separate study, *L. monocytogenes* was inoculated onto the surface of frankfurters, at log 8 CFU per frank, which were then dipped for 30 seconds in ACS (Nuñez de Gonzalez et al. 2002). *Listeria monocytogenes* counts were reduced by 5.8 logs and remained at the minimum level of detection for 12 weeks of vacuum-packaged refrigerated storage. Thus, ACS was not only bactericidal but also bacteriostatic for 12 weeks. Some concerns have been raised regarding changes in the flavor of ACS-treated RTE products, but reformulation of the complex has been undertaken to lessen this effect. On the basis of the pathogenic effectiveness of ACS on raw and precooked products, this compound appears to offer potential as an antimicrobial treatment in a multiple-intervention system.

Chlorine-Based Compounds

Use of chlorinated water (20–50 ppm) to reduce the pathogen load on poultry carcasses at the pre-chill washer or in the chill tank have had mixed results (Kemp et al. 2000). Likewise, chlorine sprays or rinses rapidly applied to beef carcasses are ineffective for pathogen control because the chlorine becomes bound by organic matter and loses its antimicrobial properties (Siragusa 1995). Although approved for use by the USDA-FSIS, chlorinated water is less effective than other compounds such as hypochlorite, chlorine dioxide, acidified sodium chlorite (ASC), and cetylpyridinium chloride (CPC). In many instances, chlorinated compounds have been combined with organic acids, ozone, and other antimicrobial treatments to enhance their efficacy for destroying pathogens on meat surfaces.

Beef carcasses treated with 200 ppm hypochlorite (Chlorox, Clorox Company, Oakland, Calif.) have been shown to have lower bacterial counts, with some residual effect noted during refrigerated storage. Other chemicals (e.g., organic acids), however, are more effective against pathogens than hypochlorite, limiting its use as a carcass decontamination agent.

Chlorine dioxide is used to disinfect public water supplies and has been studied as a potential decontamination agent for carcass surfaces. This strong oxidant causes irreversible damage to fatty acids and proteins in the bacterial cell membrane, resulting in the loss of permeability and the destruction of the transmembrane ionic gradient. It has been used as a bactericide in poultry chill water at concentrations of <20 ppm and was shown to be as effective as chlorine at >20 ppm (Lillard 1979, Lillard 1980). Chlorine dioxide also has been patented as a spray treatment for pork carcasses (Cutter and Dorsa 1995), but when used at 20 ppm on beef, it was no more effective than water washing for reducing fecal contamination. Stivarius et al. (2002) applied 200 ppm chlorine dioxide to ground beef trimmings inoculated with *E. coli* and *S.* Typhimurium but noted declines of only 0.44 and 0.82 log CFU/g, respectively, in ground beef manufactured from trimmings. Application of 1% ozonated water to the trimmings was no more effective than chlorine dioxide, and both compounds caused the ground beef to become lighter in color. The oxidizing effect of chlorine dioxide and similar oxidizing agents may limit their usefulness because of quality changes in treated products.

Acidified sodium chlorite (Sanova, Alcide Corporation, Redmond, Wash.) is a broad-spectrum, acid-activated antimicrobial approved by the FDA and Environmental Protection Agency for non-food uses such as sterilization and disinfecting hospitals, dental labs, pharmaceutical clean-rooms, and teat antisepsis. It is approved by the USDA-FSIS as an antimicrobial agent for use on poultry and beef carcasses, trimmings, and organs and on processed, comminuted, or formed meat products at levels of 500 to 1,200 ppm. It must be used in combination with any GRAS acid at a level sufficient to achieve a pH of 2.3 or 2.9 depending on the meat or poultry product (USDA-FSIS 2002b). ASC has been shown to be effective against pathogens (*E. coli* O157:H7, *Listeria, Campylobacter, Salmonella*), viruses, fungi, yeast, molds, and some protozoa. Acidification of $NaClO_2$ forms $HClO_2$ (ASC), which reacts with organic matter to form a number of oxychlorous intermediates that are broad-spectrum germicides. It is these compounds that break oxidative bonds (sulfide and disulfide linkages) on the bacterial cell membrane surface and kill the cell. The reaction residue that remains is primarily chloride and chlorate salts. Because of the non-specific oxidative reactions, acquired resistance by pathogens is thought to be minimized (Kemp et al. 2000).

Kemp et al. (2000) found that a 5-second dip in 500–1,200 ppm ASC reduced total aerobes by 82.9%–90.7%, *E. coli* by 99.4%–99.6%, and total coliforms by 86.1%–98.5% on poultry carcasses before chilling. At the 1,200-ppm level, they observed ASC to cause a transient whitening of the skin surface that is lost during hydrochilling. No organoleptic changes in raw or cooked product were noted because of ASC application. In-plant testing of a water-wash procedure for poultry carcasses combined with a 1,100-ppm rinse of ASC at pH 2.5 resulted in a 99.2% reduction in *Campylobacter* from post-evisceration levels on fecal and ingestia contaminated carcasses (Kemp and Schneider 2002). Castillo et al. (1999) inoculated various regions on hot-boned beef carcasses with *E. coli* O157:H7 and

S. Typhimurium followed by a spray treatment (140 mL, 10 seconds, 69 kPa) of 1,200 ppm ASC acidified with citric acid. ASC reduced both pathogens by 4.5–4.6 logs CFU/cm², whereas water-wash reductions were only 2.3 logs CFU/cm². It was also noted that beef carcasses treated with ASC had lower pathogen levels in the areas beyond the initial site of contamination. For a direct comparison of antimicrobials, Ransom et al. (2003) applied 2% acetic acid, 0.001% acidified chlorine, 2% lactic acid, 0.02% ASC, 0.5% cetylpyridinium chloride, 1% lactoferricin B, or 0.02% peroxyacetic acid to the surface of beef carcass tissues (BCT) and beef lean pieces (LTP) inoculated with a five-strain composite of *E coli* O157:H7. Overall, they found that CPC was the most effective for reducing bacterial populations by 4.8 (BCT) and 2.1 (LTP) log CFU/cm², respectively. Of the commonly used antimicrobials, lactic acid was the most effective, reducing *E. coli* O157:H7 by 3.3 (BCT) and 1.3 (LTP) log CFU/cm². Overall, the authors of this study concluded that lactic acid and ASC were the most effective antimicrobial solutions currently approved for use (Ransom et al. 2003).

Applied as spray or dip at ambient temperature to whole beef or poultry carcasses, fresh beef trimmings, and formed and RTE products, ASC is considered a processing aid and, as its residue levels are considered insignificant, it does not require labeling. Poultry operations most often apply ASC at the end of the evisceration line before or after carcass chilling, whereas red meat operations typically applied ASC as a carcass rinse after evisceration or to trimmings immediately before grinding.

CPC (cetylpyridinium chloride) [1-hexadecylpyridinium chloride] is a water-soluble, colorless, neutral pH, quaternary ammonium compound used in oral mouthwash products, toothpastes, and throat lozenges as a bactericide. Levels of 0.05%–0.5% in mouthwashes reduce or inhibit gingivitis and biofilm and plaque formation (Cutter et al. 2000). CPC is effective because of its ability, wet, to penetrate tissue as a result of its low surface tension and hydrophillic and lipophilic properties. This enables CPC to penetrate and destroy bacterial cell walls and cell membranes. CPC kills bacteria by the interaction of basic cetylpyridinium ions reacting with the acid groups of bacteria to form weakly ionized compounds that inhibit bacterial metabolism (Kim and Slavik 1996).

At this time, CPC is neither approved for food use by the FDA nor by the USDA-FSIS for direct food contact. However, several studies have demonstrated its antimicrobial effectiveness when applied to poultry and beef carcasses. Kim and Slavik (1996) found that spraying contaminated poultry skin with 0.1% CPC reduced *Salmonella* by 0.9–1.7 logs CFU/cm², which was similar to reductions (1.0–1.6 logs CFU/cm²) when the poultry skin was immersed in CPC. Reductions of 4.87 logs CFU/cm² of *S.* Typhimurium also have been observed when CPC (0.4%) was applied to chicken skin for 3 minutes in a model system (Breen et al. 1997). When chicken carcasses were sprayed with 10% trisodium phosphate, 2% lactic acid, 0.5% CPC, or 5% sodium bisulfate in a modified inside/outside bird washer, CPC was the most effective antimicrobial spray, reducing total aerobes by 2.16 log CFU/cm² and *Salmonella* by 2 log CFU/cm² (Yang et al. 1998). Cutter et al. (2000) compared the effectiveness of a 1% CPC, 15-second spray with that of a water wash to reduce pathogens on lean- and fat-tissue beef surfaces. On lean beef, *S.* Typhimurium and *E. coli* O157:H7 were immediately reduced by a 1% CPC solution from 5–6 logs CFU/cm² to undetectable levels; whereas on fat tissues, pathogen counts were reduced to <2.5 logs CFU/cm². After 35 days of storage at 4°C, pathogen counts on lean tissue were undetectable, but counts on fat tissue remained at <1.3 log CFU/cm². Although CPC levels exceeded those for human consumption, CPC was proven to effectively reduce pathogenic bacteria. As noted previously, during a comparison of CPC and other antimicrobials CPC was the most overall effective for reducing bacterial populations (Ransom et al. 2003), whereas lactic acid and ASC were the most effective among the solutions currently approved for use.

Inorganic Compounds

TSP is a very alkaline (pH 12–13) antimicrobial ingredient approved for use as a spray or dip for raw, unchilled poultry carcasses and giblets (USDA-FSIS 2002b) and as a hog scalding agent. An 8%–12% solution may be applied to poultry carcasses within a temperature range of 65°–85°C for up to 15 seconds. Giblets do not have a temperature requirement, but they must be sprayed or dipped for a minimum of 30 seconds. TSP's antimicrobial effect is apparently the result of bacterial cell membrane disruption and of an increase in the water solubility of the bacterial DNA at high pH (Mendonca et al. 1994).

Trisodium phosphate is used commercially to decontaminate poultry carcasses and has been shown to reduce *Salmonella* by 1.6–1.8 logs when applied as a 10% solution at 50°C for 15 seconds (Slavik et al. 1994). In a companion study, *Campylobacter* on TSP-treated carcasses declined by 1.5 and 1.2 logs when stored for 1 and 6 days, respectively, at 4°C. TSP also has been shown to be effective for removing and inactivating attached *S.* Typhimurium on cooked chicken breast patties after refrigerated and frozen storage (Yoon and Oscar 2002).

A few studies have investigated the potential use of TSP solutions as a decontamination treatment for beef carcasses. Kim and Slavik (1994) rinsed fat and fascia surfaces inoculated with *E. coli* O157:H7 and *S.* Typhimurium with a 10% TSP (10°C) solution for 15 seconds. They found that the TSP treatment reduced *E. coli* O157:H7 levels by 1.35 and 0.92 logs on fat and fascia surfaces, respectively. *Salmonella* Typhimurium showed reductions of 1.39 and 0.86 logs on the same tissues. In a similar study, *S.* Typhimurium and *E. coli* O157:H7 were reduced by 1–1.5 logs on lean beef tissue previously immersed in 8%–12% solutions (25°–55°C) for up to 3 minutes (Dickson et al. 1994). *Listeria monocytogenes* was also reduced, but to a lesser extent. The concentration of TSP did not affect reductions in bacterial populations, but higher temperatures seemed more effective. Greater reductions in pathogen populations were observed on fat tissue than on lean tissue. Cutter and Rivera-Betancourt (2000) found spray treatments with 10% TSP to be the most effective for reducing *S.* Typhimurium, *S.* Typhimurium DT104, *E. coli* O157:H7, and *E. coli* O111:H8 on prerigor beef surfaces. TSP resulted in reductions of >3 logs CFU/cm^2, followed by 2% lactic and 2% acetic acid at >2 logs CFU/cm^2. Previous work by Dorsa et al. (1997) showed a 12% TSP wash to have the same degree of effectiveness for controlling *E. coli* O157:H7 and *C. sporogenes* in comparison with 3% lactic or acetic acids, but TSP was less effective for reducing aerobic plate counts, *L. innocua*, and lactic acid bacteria.

A TSP spray or dip used alone, or in concert with other pathogen interventions, is an effective means of reducing pathogens on poultry and beef carcasses. However, disposal of the highly alkaline treatment water and excessive corrosion of equipment and facilities with continuous exposure to the decontamination agent are two issues that affect the use of TSP.

High-pH treatment of beef trimmings with ammonia gas is a patented process (Roth 2002) used in the production of lean boneless beef trim (previously referred to a lean finely textured beef) to reduce pathogens in the product (Beef Products Incorporated, Dakota Dunes, S.D.). Desinewed beef trim is processed through a centrifugal separator, rapidly frozen on a drum freezer, and then chipped into $1 \times 1 \times 0.3$-cm pieces. The frozen chips are compressed (4,000 lb/in^2) into blocks and packed into boxes. A step in the process involves rapidly raising the pH of the beef to approximately 9.5 by the injection of ammonia gas into the product. Evaluation of gaseous ammonia–treated beef trim inoculated at 6 log CFU/g of *Salmonella, L. monocytogenes*, and *E. coli* O157:H7 showed that after the product has been frozen and compressed into blocks, no *Salmonella* or *E. coli* O157:H7 were detectable by enumeration or enrichment and isolation (Niebuhr and Dickson 2003). *Listeria monocytogenes* strains in the blocks were reduced by approximately 3 logs CFU/g from the initial inoculation levels. When the ammonia-treated frozen beef trim (15%) was then combined with inoculated ground beef, pathogen populations in the ground beef were reduced by approximately 0.2 logs. No detrimental changes in the ground product color, odor, or flavor were reported. Because ammonia-treated beef trim is used as a component in numerous products destined for further processing, this gasification and pH enhancement process offers an added measure of safety to those products and has the potential to retain more juices in the cooked product.

Activated Lactoferrin

Lactoferrin is a bioactive glycoprotein derived from cheese whey and skim milk that has been designated GRAS by the FDA, and is approved for use on beef carcasses and beef parts up to 2% (65.2 mg/kg beef tissue) of a water-based spray (USDA-FSIS 2002b). Naidu (2002) states that the activated form of lactoferrin (ALF), blocks foodborne pathogens such as *E. coli* O157:H7, *Salmonella* spp., *Campylobacter* spp., *Vibrio* spp., *A. hydrophila,* and *S. aureus,* as well as food spoilage microorganisms like *Bacillus* spp., *Pseudomonas* spp., and *Klebsiella* spp. In its patented activated form, ALF blocks the attachment of pathogens onto the meat surface by interacting with the pore-forming outer-membrane proteins of Gram-negative enteric pathogens, thereby reducing their growth and colonization (Naidu 2001, Naidu 2002). Because of ALF's binding ability, it can displace attached bacterial cells. It is believed to bind to the same receptor sites (e.g., collagen) as bacteria, but with greater affinity, thus displacing the bacteria. Kawasaki et al. (2000) found that lactoferrin inhibited the adhesion of *E. coli* and reduced counts by 1–2 logs along the intestinal wall of mice. This reduction was sustained up to 30 days following exposure to the organism. ALF also has the ability to strongly bind iron, thereby depriving bacteria of an essential nutrient and inhibiting growth.

Naidu (2002) reported reductions of 2.5 logs of *E. coli* O157:H7 on beef steaks treated with a 1% ALF spray, and removal of 99.9% *E. coli* O157:H7 from samples passed through a digitally simulated spray system consisting of six spray-wash steps: cold water (10 seconds), 2% lactic acid (10 seconds), hot water (180°F, 30 seconds), cold water (10 seconds), and 2% lactic acid (10 seconds), followed by a 20-second spray with 1% ALF. Venkita-narayanan et al. (1999) compared the antibacterial activity of an amino acid peptide, lacto-ferricin B (100 μg/g) (not ALF), in ground beef samples stored at 4° and 10°C and found *E. coli* O157:H7 to be reduced by approximately 0.8 log CFU/g, which was not of sufficient magnitude to merit its use in ground beef.

Sensory studies on beef steaks treated with ALF, packaged in 80% O$_2$ and 20% CO$_2$ for up to 35 days, and displayed under cool white fluorescent light at 32°–36°C showed no differences when compared with the data from nontreated steaks (Naidu 2002).

Limited information is available in the literature regarding challenge and comparative studies with activated lactoferrin and other approved antimicrobial treatments. Because activated lactoferrin is a milk-derived protein, it may induce immunoallergic reactions in some milk-protein-sensitive individuals. The number of individuals that might be affected by exposure to activated lactoferrin is unknown.

Bacteriocins

Bacteriosins are microbial metabolites that have a bacteriostatic or bactericidal effect on other microorganisms. Nisin, a small, hydrophobic peptide produced by *Lactococcus lactis* subsp. *lactis* is most effective against Gram-positive bacteria, because Gram-negative microorganisms have an additional outer cell membrane for protection. The antibacterial mechanism appears to involve inhibition of murein synthesis and disintegration of the cytoplasmic membrane, resulting in inhibition of DNA, RNA and protein synthesis. Nisin has received GRAS status by the FDA and is allowed in meat products as a component of sauces, soups, or fully cooked meat or poultry products. It may be used in sauces at no more than 600 ppm or used where meat or poultry does not exceed 50% of the product formulation (USDA-FSIS 2002b). Nisin is considered to be non-toxic and non-immunoallergenic and has been approved by the World Health Organization as a food preservative.

Nisin has been shown to inhibit growth of spoilage bacteria in several foods including milk, cheese, cottage cheese, yogurt, eggs, and canned soups and vegetables (Tu and Mustapha 2002). Meat products such as ham, bacon, frankfurters, and other cured meats have also been evaluated. Surface decontamination of beef and poultry carcasses with 5,000 Activity Units(U)/mL of nisin by Cutter and Siragusa (1994) showed reductions of 1.79–3.54 log CFU/cm^2 of *Bronchothrix thermosphacta, Carnobacterium divergens*, or *L. monocytogenes*. They concluded that nisin might be a useful sanitizer for meat surfaces. Treatment of vacuum-packaged beef with nisin or a combination of nisin and EDTA reduced *L. monocytogenes* by 2.01 and 0.99 log CFU/cm^2, respectively, when stored at 4°C and evaluated over a 30-day period (Zhang and Mustapha 1999). Reductions of *E. coli* O157:H7 on inoculated beef under the same conditions were marginal. Subsequent work by Tu and Mustapha (2002) found that surface treatment of nisin and EDTA on beef completely inhibited *B. thermosphacta* but did not reduce *S.* Typhimurium. Mustapha et al. (2002) did not find the combination of Nisaplin (nisin) (400 U/mL) and 2% lactic acid to be any more effective for reducing *E. coli* O157:H7 on vacuum-packaged beef than lactic acid alone.

Overall, the use of nisin, and perhaps other bacteriocins such as pediocin, does not appear to be as effective as other decontamination treatments against Gram-negative microorganisms, unless used in combination with other antimicrobials, for reducing these pathogens on carcass surfaces or raw meat products. In addition, limited information is available regarding the potential development of resistance by organisms exposed to nisin.

Postharvest Product Decontamination (Fresh, Frozen, RTE)

A host of chemical preservatives, antimicrobials, and ingredients with antimicrobial properties are currently approved for use in meat and poultry products (Code of Federal Regulations 2003, USDA-FSIS 2002b), but a discussion of their effectiveness is beyond the scope of this chapter.

Emerging Technologies

Decontamination of carcasses, meat trimmings, and poultry parts through the application of multiple-intervention technologies and through chemical agents (applied to carcass surfaces immediately post-slaughter) has been the most widely used approach to preventing pathogen contamination of meat and poultry foods.

Other related chemical treatments are emerging that may add additional hurdles in the HACCP system and improve food safety. Antimicrobial food packaging and edible films (including casings) are concepts being studied to simultaneously extend shelf life and improve food safety. Because the package surface is in intimate contact with a potentially contaminated product surface, treatment of package surfaces with bacteriostatic or bactericidal agents may reduce the risk of foodborne illness. Edible antimicrobial films would offer more convenience and perhaps residual antimicrobial activity for the life of the product. Current limitations for the use of antimicrobial packaging and edible films include:

- Methods used to incorporate antimicrobial agents into a packaging film or coating remain under development;
- Most studies have focused on model systems with a few tests performed on actual food systems;
- Limited information is available about the commercial ability to produce antimicrobial films;
- Even if technology were available, packaging and films may be cost prohibitive;
- Legal issues must be resolved before investment in the technology;
- Pathogens could become resistant to antimicrobial agents;
- Use of antimicrobial films may be perceived as covering up contamination at the processing plant level;
- Some compounds are not approved for food use by FDA or USDA-FSIS and would require extensive testing; and
- If an outbreak of foodborne illness occurred, who would be liable—the producer or the user of the packaging/film?

A number of basic chemistry and bioengineering laboratories (Massachusetts Institute of Technology, Northeastern University, Pace University, Long Island University, Queens College of the City University of New York, Tokyo Institute of Technology, Rensselaer Polytechnic Institute, University of Pennsylvania, University of Massachusetts–Amherst, University of Wisconsin, and Scripps Research Institute) are actively developing bonded polymer surfaces for a host of potential military, medical, or food applications. Much of the research is focused on porous materials that are carbohydrate based that could potentially lead to antibacterial surface treatments on everything from clothing to paper, glass, plastics, wood, cellulose, and possibly metals. Researchers at MIT have found that *N*-alkylated poly(4-vinyl pyridine) groups (or *N*-alkyl PVP groups) are lethal to both Gram-positive and Gram-negative bacteria when bonded to glass, polypropylene, nylon, polyethylene, and poly(ethylene terephthalate) plastics. Some compounds also have been bound to magnetic iron oxide nanoparticles. The potential offered by these new antimicrobial agents not only includes packaging but food contact surfaces such as equipment, conveyor belts, boxes, institutional uniforms, gloves, hairnets, and similar items used in the food industry. Coatings could potentially target a specific category of microorganisms or be designed to eliminate biofilm formation and build-up of microorganisms with acquired resistance to other modes of decontamination (i.e., sanitizers, and disinfectants).

Other technologies under development include silver surface bonding to stainless steel that would eliminate microbial growth on stainless steel and equipment manufactured from the coated metal. This would allow for walls, doors, equipment, and other metal-coated items to be essentially bacteria free and provide a recyclable resource.

Potential Concerns

Potential concerns with existing or developing decontamination technologies are:

- Generalized microbial reductions may reduce beneficial microorganisms such as lactic acid bacteria;
- Spraying or rinsing may cause bacterial penetration into the tissue or redistribution of the microflora depending on the application pressure;
- Pathogens may become irreversibly attached to carcass or meat surfaces because of insufficient treatment or a delay in treatment application;
- Without sufficient sanitation of equipment contact surfaces in concert with decontamination, biofilms may develop and serve as pathogen reservoirs;
- Pathogens may develop adaptive resistance with constant exposure to a specific treatment, chemical intervention, or a sublethal concentration of a decontamination agent; and
- Reductions in beneficial microflora that are antagonistic to bacterial pathogens may allow the pathogens to become more established in plants.

Conclusions

Pathogen-free meat and poultry products cannot be guaranteed, but the risk of contamination can be greatly reduced through the application of multiple-intervention decontamination systems and the development of new technologies that kill pathogens on multiple fronts. The advantages, disadvantages, and limitations of various chemical decontamination procedures have been discussed along with emerging methods for reducing or eliminating pathogens in our meat and poultry supply.

References

Anderson, R. C., Buckley, S. A., Callaway, T. R., Genovese, K. J., Kubena, L. F., Harvey, R. B. and Nisbet, D. J. 2001. Effect of sodium chlorate on *Salmonella* Typhimurium concentrations in the weaned pig gut. *Journal of Food Protection* **64**:255–258.

Anderson, R. C., Buckley, S. A., Kubena, L. F., Stanker, L. H., Harvey, R. B. and Nisbet, D. J. 2000. Bactericidal effect of sodium chlorate on *E. coli* O157:H7 and *Salmonella* Typhimurium DT 104 in rumen contents in vitro. *Journal of Food Protection* **63**:1038–1042.

Bacon, R. T., Belk, K. E., Sofos, J. N., Clayton, R. P., Regan, J. O. and Smith, G. C. 2000. Microbial populations on animal hides and beef carcasses at different stages of slaughter in plants employing multiple-sequential interventions for decontamination. *Journal of Food Protection* **63**:1080–1086.

Berends, B. R., Urlings, H. A. P., Snijders, J. M. A. and Van Knapen, F. 1996. Identification and quantification of risk factors in animal management and transport regarding *Salmonella* spp. in pigs. *International Journal of Food Microbiology* **30**:37–53.

Bolton, D. J., Doherty, A. M. and Sheridan, J. J. 2001. Beef HACCP: Intervention and non-interventions systems. *International Journal of Food Microbiology* **66**:119–129.

Booth, I. R. 1985. Regulation of cytoplasmic pH in bacteria. *Microbiological Reviews* **49**:359–378.

Bowling, R. A. and Clayton, R. P. 1992. Method for deharing animals. U.S. Patent 5,149,295, Monfort Inc., Greely, CO.

Breen, P. J., Salari, H. and Compadre, C. M. 1997. Elimination of *Salmonella* contamination from poultry tissues by cetylpyridinium chloride solutions. *Journal of Food Protection* **60**:1019–1021.

Castillo, A., Dickson, J. S., Clayton, R. P., Lucia, L. M. and Acuff, G. R. 1998a. Chemical dehairing of bovine skin to reduce pathogenic bacteria and bacteria of fecal origin. *Journal of Food Protection* **61**:623–625.

Castillo, A., Hardin, M. D., Acuff, G. R. and Dickson, J. S. 2002. Reduction of microbial contaminants on carcasses, pp. 351–382. *In* Control of Foodborne Microorganisms, V. K. Juneja, J. N. Sofos (Eds.). New York: Marcel Dekker.

Castillo, A., Lucia, L. M., Goodson, K. J., Savell, J. W. and Acuff, G. R. 1998b. Comparison of water wash, trimming, and combined hot water and lactic acid treatments for reducing bacteria of fecal origin on beef carcasses. *Journal of Food Protection* **61**:823–828.

Castillo, A., Lucia, L. M., Kemp, G. K. and Acuff, G. R. 1999. Reduction of *Escherichia coli* O157:H7 and *Salmonella* Typhimurium on beef carcass surfaces using acidified sodium chlorite. *Journal of Food Protection* **62**:580–584.

Castillo, A., Lucia, L. M., Roberson, D. B., Stevenson, T. H., Mercado, I. and Acuff, G. R. 2001. Lactic acid sprays reduce bacterial pathogens on cold beef carcass surfaces in subsequently produced ground beef. *Journal of Food Protection* **64**:58–62.

Code of Federal Regulations. 2003. Use of Food Ingredients and Sources of Radiation, 424.21, Preparation and Processing Operations, 424. *In* Code of Federal Regulations, Title 9, Food Safety and Inspection Service, Department of Agriculture, Parts 300-599. Washington, D.C.: National Archives and Records Administration.

Currier, M. M., Singleton, J. L. and Lee, D. R. 1986. *Salmonella* in swine at slaughter: incidence and serovar distribution at different seasons. *Journal of Food Protection* **49**:366–368.

Cutter, C. N. and Dorsa, W. J. 1995. Chlorine dioxide spray washes for reducing fecal contamination on beef. *Journal of Food Protection* **58**:1294–1296.

Cutter, C. N., Dorsa, W. J., Handie, A., Rodriguez-Morales, S., Zhou, X., Breen, P. J. and Compadre, C. M. 2000. Antimicrobial activity of cetylpyridinium chloride washes against pathogenic bacteria on beef surfaces. *Journal of Food Protection* **63**:593–600.

Cutter, C. N. and Rivera-Betancourt, M. 2000. Interventions for the reduction of *Salmonella* Typhimurium DT 104 and non-O157:H7 enterohemorrhagic *Escherichia coli* on beef surfaces. *Journal of Food Protection* **63**:1326–1332.

Cutter, C. N. and Siragusa, G. R. 1994. Decontamination of beef carcass tissue with nisin using a pilot scale model carcass washer. *Food Microbiology* **11**:481–489.

Davidson, P. M. 2002. Control of microorganisms with chemicals, pp. 165–190. *In* Control of Foodborne Microorganisms, V. K. Juneja, J. N. Sofos (Eds.). New York: Marcel Dekker.

Delmore Jr., R. J., Sofos, J. N., Schmidt, G. R. and Smith, G. C. 1997. Inactivation of pathogenic bacteria by the chemical dehairing process proposed for use on beef carcasses during slaughter, p. 163. *In* Proceedings of the 50th Annual Reciprocal Meats Conference, Iowa State University, Ames, IA. Savoy, Ill.: American Meat Science Association.

Dickens, J. A., Ingram, K. and Hinton Jr., A. 2001. Effect of Safe$_2$O™-poultry wash, a highly acidic calcium sulfate solution, used as a poultry wash pre and post evisceration on total aerobes, *E. Coli*, *Salmonella*, and *Campylobacter*. A Research Report from the Richard B. Russell Research Center. Athens, Ga.: Poultry Processing and Meat Quality Research.

Dickson, J. S., Nettles Cutter, C. G. and Siragusa, G. R. 1994. Antimicrobial effects of trisodium phosphate against bacteria attached to beef tissue. *Journal of Food Protection* **57**:952–955.

Dorsa, W. J., Cutter, C. N. and Siragusa, G. R. 1997. Effects of acetic acid, lactic acid and trisodium phosphate on the microflora of refrigerated beef carcass surface tissue inoculated with *Escherichia coli* O157:H7, *L. innocua* and *Clostridium sporogenes*. *Journal of Food Protection* **60**:619–624.

Dorsa, W. J., Cutter, C. N., and Siragusa, G. R. 1998. Bacterial profile of ground beef made from carcass tissue experimentally contaminated with pathogenic and spoilage bacteria before being washed with hot water, alkaline solution, or organic acid and then stored at 4 or 12°C. *Journal of Food Protection* **61**:1109–1118.

Eggenberger-Solorzano, L., Niebuhr, S. E., Acuff, G. R. and Dickson, J. S. 2002. Hot water and organic acid interventions to control microbiological contamination on hog carcasses during processing. *Journal of Food Protection* **65**:1248–1252.

Elder, R. O., Keen, J. E., Siragusa, G. R., Barkocy-Gallagher, G. A., Koohmaraie, M. and Laegreid, W. W. 2000. Correlation of entero-hemorrhagic *Escherichia coli* O157 prevalence in feces, hides and carcasses of beef cattle during processing. *Proceedings of the National Academy of Sciences (USA)* **97**:2999–3003.

Federal Register. 1996. Pathogen Reduction; Hazard Analysis and Critical Control Point (HACCP) Systems; Final Rule. United States Department of Agriculture, Food Safety and Inspection Service. Title 9 CFR Parts 304, 308, 310, 320, 327, 381, 416 and 417. *Federal Register* **61**:38805–38989.

Finlay, B. 2003. Pathogenic *Escherichia coli*: From molecules to vaccine. Abstract No. O-12. The 5th International Symposium on 'Shiga Toxin (Verocytotoxin)-Producing *Escherichia coli* Infections'. VTEC 2003, June 8–11, Edinburgh, Scotland.

Green, S. S. 1987. Results of a national survey: *Salmonella* in broilers and overflow chill tank water 1982–1984. Washington, D.C.: United States Department of Agriculture, Food Safety Inspection Service.

Hardin, M. D., Acuff, G. R., Lucia, L. M., Oman, J. S. and Savell, J. W. 1995. Comparison of methods for contamination removal from beef carcass surfaces. *Journal of Food Protection* **58**:368–374.

Huffman, R. D. 2002. Current and future technologies for the decontamination of carcasses and fresh meat. *Meat Science* **62**:285–294.

Hume, M. E., Corrier, D. E., Nisbet, D. J. and DeLoach, J. R. 1998. Early *Salmonella* challenge time and reduction in chick cecal colonization following treatment with a characterized competitive exclusion culture. *Journal of Food Protection* **61**:673–676.

Kang, D. H., Koohmaraie, M., Dorsa, W. J. and Siragusa, G. R. 2001. Development of multiple-step process for the microbial decontamination of beef trim. *Journal of Food Protection* **64**:63–71.

Kawasaki, Y., Tazume, S., Shimizu, K., Matsuzawa, H., Dosako, S., Isoda, H., Tsukiji, M., Fujimura, R., Muranaka, Y. and Isihida, H. 2000. Inhibitory effects of bovine lactoferrin on the adherence of enterotoxigenic *Escherichia coli* to host cells. *Biosciences, Biotechnology and Biochemistry* **64**:348–354.

Kemp, G. K., Aldrich, M. L. and Waldroup, A. L. 2000. Acidified sodium chlorite antimicrobial treatment of broiler carcasses. *Journal of Food Protection* **63**:1087–1092.

Kemp, G. K. and Schneider, K. R. 2002. Reduction of *Campylobacter* contamination of broiler carcasses using acidified sodium chlorite. *Dairy, Food and Environmental Sanitation* **22**:599–606.

Kim, J.-W. and Slavik, M. F. 1994. Trisodium phosphate (TSP) treatment of beef surfaces to reduce *Escherichia coli* O157:H7 and *Salmonella typhimurium*. *Journal of Food Science* **59**:20–22.

Kim, J.-W. and Slavik, M. F. 1996. Cetylpyridinium chloride (CPC) treatment on poultry skin to reduce attached *Salmonella*. *Journal of Food Protection* **59**:322–326.

Kotula, K. L. and Thelappurate, R. 1994. Microbiological and sensory attributes of retail cuts of beef treated with acetic and lactic solutions. *Journal of Food Protection* **57**:665–670.

Leistner, L. 2002. Hurdle technology, pp. 493–508. *In* Control of Foodborne Microorganisms, V. K. Juneja, J. N. Sofos (Eds.). New York: Marcel Dekker.

Lillard, S. H. 1979. Levels of chlorine and chlorine dioxide of equivalent bactericidal effect in poultry processing water. *Journal of Food Science* **44**:1594–1597.

Lillard, S. H. 1980. Effect on broiler carcasses and water of treating chiller water with chlorine or chlorine dioxide. *Poultry Science* **59**:1761–1766.

Lillard, S. H. 1989. Factors affecting the persistence of *Salmonella* during the processing of poultry. *Journal of Food Protection* **52**:829–832.

Mead, P. S., Slutsker, L., Dietz, V., McCaig, L. F., Bresee, J. S., Shapiro, C., Griffin, P. M. and Tauxe, R. V. 1999. Food-related illness and death in the United States. *Emerging Infectious Diseases* **5**:607–625.

Mendonca, A. F., Amoroso, T. L. and Knabel, S. J. 1994. Destruction of Gram-negative food-borne pathogens by high pH involves disruption of the cytoplasmic membrane. *Applied Environmental Microbiology* **60**:4009–4014.

Mermelstein, N. H. 2001. Sanitizing meat. *Food Technology* **55**:64–68.

Moxley, R., Smith, D., Klopfenstein, T., Erickson, G., Folmer, J., Macken, C., Hinkley, S., Potter, A. and Finlay, B. 2003. Vaccination and feeding a competitive exclusion product as intervention strategies to reduce the prevalence of *Escherichia coli* O157:H7 in feedlot cattle. Abstract No. O-13. The 5th International Symposium on 'Shiga Toxin (Verocytotoxin)-Producing *Escherichia coli* Infections'. VTEC 2003, June 8–11, Edinburgh, Scotland.

Murinda, S. E., Roberts, R. F. and Wilson, R. A. 1996. Evaluation of colicins for inhibitory activity against diarrheagenic *Escherichia coli* strains, including serotype O157:H7. *Applied Environmental Microbiology* **62**:3196–3202.

Mustapha, A., Ariyapitipun, T. and Clarke, A. D. 2002. Survival of *Escherichia coli* O157:H7 on vacuum-packaged raw beef treated with polylactic acid, lactic acid and nisin. *Journal of Food Science* **67**:262–267.

Naidu, A. S. 2001. Immobilized lactoferrin antimicrobial agents and their use. U.S. Patent 6,172,040 B1.

Naidu, A. S. 2002. Activated lactoferrin—a new approach to meat safety. *Food Technology* **56(3)**:40–45.

Niebuhr, S. E. and Dickson, J. S. 2003. Impact of pH enhancement of populations of *Salmonella, Listeria monocytogenes,* and *Escherichia coli* O157:H7 in boneless lean beef trimmings. *Journal of Food Protection* **66**:874–877.

Nuñez de Gonzalez, M. T., Keeton, J. T., Ringer, L. J., Lucia, L. M. and Acuff, G. R. 2002. Antimicrobial effects of surface treatments and ingredients on cured RTE meat products. A final report to the American Meat Institute Foundation, Arlington, VA. College Station: Texas A&M University.

Park, H., Hung, Y.-C. and Bracket, R. E. 2002. Antimicrobial effect of electrolyzed water for inactivating *Campylobacter jejuni* during poultry washing. *International Journal of Food Microbiology* **72**:77–83.

Park, S., Wrobo, R. W. and Durst, R. A. 1999. *Escherichia coli* O157:H7 as an emerging foodborne pathogen: a literature review. *Critical Reviews of Food Science and Nutrition* **39**:481–502.

Ransom, J. R., Belk, K. E., Sofos, J. N. Stopforth, J. D., Scanga, J. A. and Smith, G. C. 2003. Comparison of intervention technologies for reducing *Escherichia coli* O157:H7 on beef cuts and trimmings. *Food Protection Trends* **23**:24–34.

Rose, B. E., Hill, W. E., Umholtz, R., Ransom, G. M. and James, W. O. 2002. Testing for *Salmonella* in raw meat and poultry products collected at federally inspected establishments in the United States, 1998 through 2000. *Journal of Food Protection* **65**:937–947.

Roth, E. 2002. Apparatus for reducing microbe content in food-stuffs by pH and physical manipulation. U.S. patent 6,389,838.

Schnell, T. D., Sofos, J. N., Littlefield, V. G., Morgan, J. B., Gorman, B. M., Clayton, R. P. and Smith, G. C. 1995. Effects of postexanguination dehairing on the microbial load and visual cleanliness of beef carcasses. *Journal of Food Protection* **58**:1297–1302.

Siragusa, G. R. 1995. The effectiveness of carcass decontamination systems for controlling the presence of pathogens on the surfaces of meat animal carcasses. *Journal of Food Safety* **15**:229–238.

Slavik, M. F., Kim, J.-W. Pharr, M. D., Raben, D. P., Tsai, S. and Lobsinger, C. M. 1994. Effect of trisodium phosphate on *Campylobacter* attached to post-chill chicken carcasses. *Journal of Food Protection* **57**:324–326.

Sofos, J. N., Belk, K. E. and Smith, G. C. 1999. Processes to reduce contamination with pathogenic microorganisms in meat. Proceedings of the 45th International Congress of Meat Science and Technology, Vol. 45, No. 2, Yokohama, Japan, pp. 596–605. White Paper, pp. 1–20. Fort Collins: Colorado State University.

Sorrells, K. M., Enigl, D. C. and Hatfeld, J. H. 1989. Effect of pH, acidulant, time and temperature on the growth and survival of *Listeria monocytogenes*. *Journal of Food Protection* **52**:571–573.

Stivarius, M. R., Pohlman, F. W., McElyea, K. S. and Apple, J. K. 2002. Microbial, instrumental color and sensory color and odor characteristics of ground beef produced from beef trimmings treated with ozone or chlorine dioxide. *Meat Science* **60**:299–305.

Tamblyn, K. C. and Conner, D. E. 1997. Bactericidal activity of organic acids against *Salmonella typhimurium* attached to broiler chicken skin. *Journal of Food Protection* **60**:629–633.

Tu, L. and Mustapha, A. 2002. Reduction of *Brochothrix thermosphacta* and *Salmonella* serotype Typhimurium on vacuum-packaged fresh beef treated with nisin and nisin combined with EDTA. *Journal of Food Science* **67**:302–306.

U.S. Department of Agriculture, Food Safety Inspection Service. 1994. Microbiological Baseline Data Collection Program: Steers and Heifers, October 1992–September 1993. Washington, D.C.: United States Department of Agriculture. Available at: http://www.aphis.usda.gov/vs/ceah/cahm.

U.S. Department of Agriculture, Food Safety Inspection Service. 2002a. Pathogen Reduction; Hazard Analysis and Critical Control Point (HACCP) Systems. Title 9 Code of Federal Regulations, Parts 304, 308, 310, 320, 327, 381, 416 and 417. Washington, D.C.: Government Printing Office.

U.S. Department of Agriculture, Food Safety Inspection Service. 2002b. Safe and Suitable Ingredients used in the production of Meat and Poultry Products. FSIS Directive 7120.1. Washington, D.C.: United States Department of Agriculture, Food Safety Inspection Service.

Van Netten, P., Mossel, D. A. A. and Huis In't Veld, J. 1995. Lactic acid decontamination of fresh pork carcasses: a pilot plant study. *International Journal of Food Microbiology* **25**:1–9.

Venkitanarayanan, K. S., Zhao, T. and Doyle, M. P. 1999. Antibacterial effect of lactoferricin B on *Escherichia coli* O157:H7 in ground beef. *Journal of Food Protection* **62**:747–750.

Yang, Z., Li, Y. and Slavik, M. 1998. Use of antimicrobial spray applied with and inside-outside birdwasher to reduce bacterial contamination on prechilled chicken carcasses. *Journal of Food Protection* **61**:829–832.

Yoon, K. S. and Oscar, T. P. 2002. Survival of *Salmonella typhimurium* on sterile ground chicken breast patties after washing with salt and phosphates and during refrigerated and frozen storage. *Journal of Food Science* **67**:772–775.

Zerby, H. N., Belk, K. E., Sofos, J. N., Schmidt, G. R. and Smith, G. C. 1998. Microbiological sampling of hog carcasses. A Report to the National Pork Producers Council, Des Moines, Iowa. Fort Collins: Colorado State University.

Zhang, S. and Mustapha, A. 1999. Reduction of *Listeria monocytogenes* and *Escherichia coli* O157:H7 numbers on vacuum-packaged fresh beef treated with nisin or nisin combined with EDTA. *Journal of Food Protection* **62**:1123–1127.

Zhao, T., Doyle, M. P., Harmon, B. G., Brown, C. A., Mueller, P. O. E. and Parks, A. H. 1998. Reduction of carriage of enterohemorrhagic *Escherichia coli* O157:H7 in cattle by inoculation with probiotic bacteria. *Journal of Clinical Microbiology* **36**:641–647.

25 Decontamination Systems

M. Elena Castell-Perez and Rosana G. Moreira

Introduction

The increased occurrence of pathogens that cause major foodborne disease outbreaks (e.g., *Listeria monocytogenes, Escherichia Coli* O157:H7, *Salmonella*) and the presence of chemical residues in foods are the main concern of food producers and processors worldwide. These challenges have led to increased global demand for actions to improve the safety of raw and manufactured foods. One consequence of this effort is the active participation of food processors in the application of intervention strategies to decontaminate their products.

An efficient decontamination system should reduce bacterial numbers or eliminate harmful chemicals without any detrimental changes to the appearance or nutritional value of the product. There are a number of technologies currently available for use as food decontamination systems. Some of these technologies are already widely implemented in the industry. Others need additional research before commercial industrialization occurs. Some of these technologies are more promising than others regarding their effectiveness for reducing pathogens, enhancing food quality, and reducing costs. Some are concerned with preventing contamination, whereas others are designed with the goal of reducing the numbers of microorganisms in the final product (inactivation regimes) by destroying pathogens (thermal and non-thermal methods) and removing pathogens (e.g., by washing or spraying). The nature of the inactivation regime depends on the target organism and the physical properties of the food system (Stewart et al. 2002).

Decontamination systems are classified as conventional heat-based (thermal) methods and non-thermal methods. Each type of technology has its advantages and disadvantages. However, there is no single technology that will allow the high-quality production of every food product while ensuring safety. Research continues to evaluate and refine those processes that are most effective for each type of product. Scientific evidence suggests that the greatest potential may fall on decontamination systems consisting of a combination of two or more technologies, sometimes with limited thermal heating.

This chapter describes the current and future technologies for use in pathogen decontamination of the food supply. Emphasis is given to antimicrobial intervention systems, although some of the technologies may also ensure chemical safety. The primary focus is on the equipment and theoretical mechanisms behind these decontamination methods and processing criteria and their overall effectiveness.

Thermal Technologies

The use of heat is by far the most common technology used to ensure microbial safety of foodstuffs. Several technologies have been developed and applied to reduce bacteria on the surface of products such as meat and poultry carcasses and fresh produce. These include water washing, chemical washing, blanching, and steam or hot-water pasteurization.

Hot-Wash Spraying Systems

Washing is a fundamental unit operation in the processing of meat and produce. In practice, carcasses are decontaminated with hot water via spray-washing cabinets through which carcasses are passed automatically (Sofos and Smith 1998). Spray washing is mainly used to remove soil residue and chemical or biological contaminants from the surface of the product. The advantages of spray washing over washing by dipping, soaking, or gravity rinse are the increased energy efficiency, the reduced volume of water use and wastewater generation, and the reduced water uptake by produce (Belk 2000).

Spray washing is usually considered to be a control point—even a critical control point—by many meat and produce packers. There are reports of small but significant reductions in bacterial numbers after washing the products. However, a reason for concern is that the use of inappropriate equipment or excessive spray pressures may lead to bacterial penetration into the food or to spreading and redistribution on the product. These issues can be addressed by properly selecting and controlling the key processing factors, including water temperature, water pressure, method and duration of application, and the interval between the contamination occurring and washing (Belk 2000). Studies on meat indicate that optimal water pressures should range from 1,379 to 2,070 kPa to be effective in reducing microbial counts. A water flow rate of 7.5 liters per minute is usually recommended. It is standard procedure to use water at temperatures greater than 74°C to reduce bacterial counts (2–3 log CFU/cm^2) on carcass tissue. Effectiveness of the treatment increases as temperatures approach 80° to 85°C (Belk 2000). It must be noted that these operating parameters may be too high for decontamination of fresh produce, and modifications should be made (Pordesimo et al. 2002).

An alternative method to destroy bacteria in the interior of produce is to use mild heat treatments (such as blanching) to penetrate the tissue of the produce. A typical procedure is to place the produce inside a wire basket, submerge the basket in water (55°–95°C, 15 seconds), and quickly cool the produce (Breidt et al. 2000). Similar blanching procedures may be developed to reduce bacteria cell counts on a wide variety of fruits and vegetables without greatly affecting quality.

Hot-Water Pasteurization Systems

Several technologies have been developed for surface decontamination of meat products using hot water. Gill et al. (1995) developed an apparatus for pasteurizing polished, uneviscerated pig carcasses using water at 85°C. A fully commercial version of the apparatus was tested in the dressing line of a pork packing plant. However, adaptation to treat fully dressed carcasses may not be a commercially viable option because of the risk of damage to the appearance of the carcasses by the heat treatment (Gill et al. 1997).

At present, there is available an effective and economical method for microbial decontamination of animal carcasses before chill storage in a commercial slaughter operation (Anderson and Gangel 1999). The technology consists of placing the carcass in a hot-water pasteurization chamber and spraying it with recirculation water at a minimum temperature of 74°C over its entire surface for a minimum time sufficient to kill bacteria (at least 5 seconds). Recirculation water is collected and then delivered with a pump to a filtering step to remove any particulates. The filtered water is combined in a storage vessel with fresh water and then heated with a heat exchanger. Incoming water should be at least between 82° and

91°C. Typical water flow rates range within 560 to 1,324 L/min. The entire carcass surface is then subjected to acid application with a pressure pump and spray system (3.8–7.6 L/min of 1.5%–2.5% v/v of lactic or acetic acid at 43° to 54°C). The microbiological effectiveness of the technology has been confirmed (Anderson and Gagel 1999).

Advantages of this technology are the use of recirculated water and treatment of the complete carcass, as opposed to other methods that provide only "spot" treatment for specific areas of the carcass, such as the steam-vacuum carcass spot system developed by Kentmaster (Kentmaster Manufacturing Company, Inc., Monrovia, CA; Avens et al. 2002).

Steam Technology

A heat-based decontaminatiom system is successful when it can raise the food surface temperature rapidly to a value greater than 70°C, at which temperature pathogens are killed (James et al. 2000a). The heat must then be quickly removed to prevent it from penetrating into the food and cooking it. One such method of rapidly heating surfaces is the application of steam. The main advantage is the large amount of heat transferred to the food when steam condenses, increasing the surface temperature rapidly. Steam pasteurization is suitable for decontamination of food products because it does not require the application of chemicals or disposal of waste products. The development and commercial applications of surface decontamination methods for meat products using steam are well documented (Sofos and Smith 1998, Avens et al. 2002).

Steam Pasteurization Systems

This technology consists of the application of a controlled treatment of saturated or superheated steam to the surface of the product, instantaneously raising the surface temperature to 88°C for 10 seconds. Immediately following the steam treatment, the products are chilled using cold water. A specific application of this decontamination method is the exposure of animal carcasses to pressurized steam. A patented process, the Frigoscandia (Frigoscandia Equipment, Bellevue, WA) steam pasteurization system, is currently used by several packers (Wilson et al. 2000).

Commercially, steam pasteurization reduces bacterial counts by applying pressurized steam to the surface of carcasses for about 6 seconds (longer exposure periods may cause discoloration). The method destroys pathogens on meat carcasses in three steps: removal of surface water from the meat, passage through a steam-heating chamber to quickly heat the exterior of the meat, and rapid chilling the surface of the meat. A meat conveyor carries the meat throughout the three steps. Steam is continually pumped into the steam chamber through a pipe such that a positive pressure is created within the chamber (pressure differential is about 489 Pa). Heat is drawn into the surface of the meat through steam contact. The condensation of steam onto the surface of the meat produces a transfer of heat energy (latent heat), which causes rapid heating of the meat surface, effectively destroying any pathogens. It is recommended that the meat be kept within the steam chamber for 2.5 to 30 seconds, with 10 seconds being optimal (Wilson et al. 2000). The surface of the meat is heated 1–5 μm deep at about 71°–92°C during this time. These conditions are sufficient to reduce coliform bacteria, *Salmonella,* and other pathogens. The time within the steam chamber is set by the speed of the conveyor combined with the length of the steam chamber (Wilson et al. 2000).

Advantages of this system include reduced water and energy requirements. Some drawbacks are the high capital outlay required for installation and the fact that only a few units have been installed in large U.S. plants (Wilson et al. 2000).

Vacuum-Steam-Vacuum System

Application of steam to the surface of meat, poultry, and fresh produce can be optimized if the process is very rapid. The principle behind this idea is the removal of the thin layer of air and water present on the surface of the product. A new process, called the vacuum-steam-vacuum surface intervention process, consists of an apparatus that surrounds the product with a vacuum to withdraw all interfering layers of air and water, then flushes it with a quick burst of 143°C steam, and then cools it with a vacuum (Goldberg et al. 2001). The reapplication of the vacuum cools the food surface, preventing any thermal damage. The apparatus is capable of evacuating the surface, applying steam, and evacuating the surface again in less than 1 second. This technology has been successful for eliminating *Listeria inocua* from hot dogs and raw catfish and has been adapted to inactivate bacteria on the surface of fruits and vegetables (Morgan et al. 1996, Kozempel et al. 2001).

The simplest design is one chamber in one rotor. A cylindrical chamber, used for a broiler carcass, should be about 200 mm in diameter and 240 mm deep. The steam generator is charged with deionized water and the water boiled for 30 min to remove air. The vacuum receiver is adjusted to 7 kPa and its condenser coil cooled to 4°C. Process time is between 1 and 1.1 seconds (Kozempel et al. 2001). Advantages include the ability to follow high-speed process lines and low processing cost. This technology kills *Salmonella* and other harmful pathogens very quickly (in 25 milliseconds, the machine kills 99.99% of bacteria). However, the unit has not been suitable for on-line treatment.

Atmospheric Steam Systems

Modern poultry processing lines run at very high speeds—up to 12,000 birds in an hour. Such high speeds make the implementation of pressurized steam decontamination systems difficult because they do not easily integrate in continuous lines. An alternative is the use of atmospheric steam systems. The basic principle is to use a vessel, with an open base, permanently filled with steam. The steam remains in the vessel because it is less dense than the surrounding air (James et al. 2000b). Potential benefits over pressurized steam are simplicity, lower capital cost and ease of maintenance, reduced safety risk, and shorter processing time. However, operation is at one condensation temperature (10°C), and there is no integrated cooling following heat treatment.

Although numerous studies have been conducted to establish the efficacy of steam treatment as an antimicrobial intervention system (James et al. 2000a, James et al. 2000b, James et al. 2002), there is still no evidence that these systems are superior to the pressurized steam systems. More research is needed to determine the efficiency of this technology as a food decontamination system.

Non-Thermal Technologies

Most new technologies are considered non-thermal because their antimicrobial action does not require the use of a heat treatment. Non-thermal treatments are currently regarded as promising alternative decontamination systems by the food industry. These emerging technologies have been researched and applied to a broadening array of food processes and products.

Some non-thermal techniques are gaining in popularity and have emerged as full-blown commercial processes, such as high-pressure processing (HPP), pulse-electric field pasteurization (PEF), ozone-treatment, and irradiation (gamma and electron beam). Other technologies are at various developmental stages, including sonication, ultraviolet light, active packaging, and surface coatings (Majchrowicz 1999, Tiller et al. 2001).

HPP Systems

Over the last 20 years, use of HPP as a food preservation method has been pursued in an effort to produce safe foods with a reasonable shelf life. Typically, the food is submerged in a liquid (usually water) contained in a vessel that generates pressure by pumping more liquid into the pressure vessel or by reducing the volume of the pressure chamber. This exposure of foods to high pressures lasts for a short period of time, usually ranging from a few seconds to several minutes (Institute of Food Technologists 2000).

Decontamination is the major application for which this technology is known and accepted in the food industry. The use of HPP as an antimicrobial method has been shown to yield good-quality foods from an economically viable process (Ludwing et al. 1992, Erkmen and Karataş 1997).

In an excellent overview of the technology, Tewari et al. (1999) described the basic principles of HPP and critically evaluated the major research done on various applications of the technology around the world. The mechanism of action is simple: microorganisms are killed by damage of the membrane, as high pressure (100–1,000 MPa) disrupts secondary and tertiary structures of macromolecules (e.g., proteins and polysaccharides) and alters their structural and functional integrity. For most foods, 10-minute exposures to pressures in the range of 250–300 MPa result in "cold pasteurization." This "non-thermal" process reduces the losses in sensory quality (texture and flavor) and nutrient content while ensuring a significant reduction in microbial populations (Institute of Food Technologists 2000). Very high pressure (709–911 MPa) is necessary to damage bacterial spores. However, pressure higher than 500 MPa can inactivate food enzymes and alter texture, color, and physiochemical properties of foods (Kalchayanand et al. 1998). In contrast, this technology can be a powerful tool in the study of protein folding and prion inactivation similar to that responsible for bovine spongiform encephalopathy (see Chapter 14) from foods. Pressure alters the equilibrium of the interactions that stabilize the folded three-dimensional shape of proteins. The extent of denaturation by pressure depends on the structure of the protein, the pressure range, and other external parameters such as temperature, pH, and solvent composition (Casolari 1998).

In HPP systems, the food can be treated by direct and indirect methods, using water as a pressure-transmitting medium. For prepackaged foods, the technology is a batch process. For pumpable liquids, semi-continuous or continuous systems are used. In both cases, the pressure treatment is achieved in cycles. An initial time is needed to reach the desired working pressure (come up time), after which the pressure is maintained for the required processing time (holding time). Finally, the pressure is released, taking only a few seconds (release time) to complete (Barbosa-Cánovas and Gould 2000). The major suppliers of HPP equipment are ABC Pressure Systems (Sweden) and Flow International Corp. (Kent, WA).

The units supplied by both companies have differences in capacity, achievable pressures, and capital costs.

Batch production ensures product integrity by placing it in a final flexible consumer package before the application of ultrahigh pressure (UHP; Avomex; Sabinas, Mexico). The packages are loaded into a handling basket that is then placed inside the pressure vessel. The pressure vessel applies water pressure of 689 MPa and temperatures as high as 70°C. Batch processing can accommodate both liquid or solid products. However, its low throughput capacity may be a disadvantage.

A patented concept of HPP compatible with in-line production methods for pumpable foods such as sauces, juices, and purées is available. This continuous method uses a device called an isolator, inside which a separator partitions the food from the UHP water source. Pumping of the food into each isolator is performed under computer control. The food is then pressurized with UHP water, held for a short period of time, and then pumped out of the isolator into a clean or aseptic filling station. Increased system production rate is achieved through the use of additional isolators. The isolator system also provides continuous automated monitoring of critical points to assure product sanitation. The choice of packaging material for the in-line process is virtually unlimited because filling is performed after pressurization (Tewari et al. 1999). The energy requirement for production is comparable to conventional process technologies. Because the food product is not discharged under pressure, the product departs the system at the same temperature as it entered.

For some products, a combined moderate initial temperature and UHP are used to achieve the desirable effect. This combination treatment has been shown to be important in the inactivation of spores for shelf-stable low-acid foods. This is an area of emphasis and new research for the food industry.

Some advantages of HPP over traditional thermal processes include reduced process times; minimal heat damage problems; retention of freshness, flavor, texture, and color; and minimal undesirable functionality alterations (Tewari et al. 1999). However, spore inactivation is a major challenge for HPP technology. Methods used to achieve full inactivation of spores using HPP are yet to be developed. In addition, there is a need to develop and standardize HPP process parameters with respect to microbial inactivation, because none exist (Tewari et al. 1999).

Electric Field Systems

Three preservation technologies in food processing use electricity: ohmic heating, oscillatory magnetic fields, and high electric field pulses/pulsed electric fields. Ohmic heating is an "old" new technology. Unilever has patented a method for ohmic heating of muscle foods. It does inactivate microorganisms, but it also reduces food quality because of the presence of chemical products from electrolysis of the food. Oscillatory magnetic fields are applied to foods sealed in plastic bags. Frequencies between 5 and 100 pulses for a total duration of 25–100 milliseconds are typically used (Jones et al. 1998).

PEF technology is one of the most promising non-thermal methods to prevent microbial contamination of liquid foods. The process consists of the application of high electric field strengths (10–70 kV/cm) applied in pulses (10–50) of short duration (1–30 ms) to liquid and pumpable foods placed between two electrodes. The result is a membrane potential across the bacterial cell wall of more than 1.0 V, which is sufficient to lyse or damage the cell irreparably. The inactivation of various microbes, including *E. coli*, *Lactobacillus brevis*,

P. fluorescens, and *Bacillus cereus* spores, has been found to be dependent on field strength and treatment times that are unique to each species. Because this method has little effect on proteins, enzymes, or vitamins, it is perfectly suited for food processing applications (Barbosa-Cánovas et al. 1998).

The application of PEF to a fluid food may be a static (batch) or continuous operation, of which the latter is the most attractive for the food processing industry. A PEF continuous system may be operated by passing the food through the treatment chamber once, multiple times (stepwise), or in recirculation. The typical components of a PEF processing equipment are a high-voltage generator for power supply (ca. 30 kV); an energy storage capacitor; a pulse generator; a switching system; a treatment chamber through which food is pumped between two electrodes; a unit for control and monitoring of voltage, current, and field strength; a cooling system; and product delivery and packaging systems (Barbosa-Cánovas et al. 1998). The major function of the treatment chamber is to provide uniform field strength throughout the cell so that there is no cold spot as well as no electric breakdown of food caused by low or high electric field enhancement. Some systems use multiple treatment chambers to eliminate these problems.

A batch system was developed for microbial inactivation in solid and semi-solid foods (Zhang et al. 1996). The apparatus comprises a detachable processing vessel held within a supporting electrifier, which applies pulsed electric fields to the processed food. The processing vessel encloses the treatment chamber, and two electrodes face both sides of the chamber, separated at a specific distance. Modifications of the described features are also possible. The number of pulses applied will vary depending on the system and the type of food product being treated. Treatment sessions having 2–100 positive and negative pulses will typically provide adequate inhibiting effects on microbes. Temperature of the processing vessel should be controlled within the range of 4°–40°C.

Mittal et al. (2000) patented a method for electrically treating foodstuffs flowing through a treatment chamber with low-energy (0.21–25 Joules/pulse), high-voltage instant-charge-reversal electrical pulses (up to 120 kV/cm, 1–5 ms). This type of pulse was more effective in terms of microbial inactivation than other commonly used pulse waveforms.

Process parameters of importance when selecting a pulse field system include critical electric field strength, treatment times, and desired inactivation of particular microorganisms (Zhang et al. 1995). Advances in PEF technology are one step behind HPP decontamination, but they are catching up as a result of the process being continuous. The capital and running costs are also lower than HPP. One of the limitations of this technology is the paucity of commercial systems. Some manufacturers are PurePulse Technologies, Inc., and Thomson-CSF, both in the United States. In addition, application is restricted to foods that can withstand high electric fields (they should be homogeneous liquids with low electrical conductivity for the continuous systems). The presence of bubbles in the liquid product may lead to a nonuniform treatment as well as operational and safety problems. This method may be combined with other methods to enhance its efficacy.

Ozone Systems

The U.S. Food and Drug Administration recently approved the use of ozone in the processing of food. The potential utility of ozone to ensure safety in the food industry lies in the fact that ozone is 52% stronger than chlorine and has been shown to be effective over a much wider spectrum of microorganisms than chlorine and other disinfectants. In addition, unlike other disinfectants, ozone leaves no chemical residue and degrades to molecular oxygen on reaction (Kim et al. 1999b).

Ozonation has been proven to be effective for the inactivation of bacteria (*E. Coli* O157:H7, *Salmonella* Typhimurium) on several muscle foods and fish (Kim and Yousef 2000) and fresh produce (Kim et al. 1999a). At present, meat, poultry, and seafood processing plants are considering ozonation as a food safety measure, and fresh produce and minimally processed food producers are looking at ozonated washing systems for decontamination. However, inactivation of bacteria by ozone does depend on the type of microorganism and the tested food.

Ozone effectively kills microorganisms through oxidation of their cell membranes. Bacteria are more susceptible to ozone than viruses, followed by vegetative fungi and bacterial spores. Not only are there varying levels of susceptibility to ozone among organisms if exposed to ozone under the same conditions, but these conditions by themselves or in concert together exert a wide range of effects on ozone activity, which subsequently influences ozone treatment effectiveness. Some of the more common factors include pH, metallic ions in water, chemical oxidation demand (COD), and total organic carbon (TOC). Therefore, before ozone can be applied successfully as a decontamination system in the food processing industry, patterns of microbial inactivation by ozone should be elucidated (Kim and Yousef 2000).

There are several available commercial ozone generator units (Ozonia, Duebendorf, Switzerland; Lynntech, Inc., Bryan, TX). Ozone can be generated by irradiating an oxygen-containing gas with ultraviolet light or corona discharge. Ultraviolet-light ozone generators have limited application because of their minimal ozone generation capacity. Corona discharge units have significant space requirements. An alternative, the electrochemical ozone generator, was developed by Lynntech, Inc. (Denvir et al. 2001). The only requirements of this technology are sources of electrical power and water, and it eliminates the gas drying process. In addition, it does not form gaseous toxic byproducts (i.e., NOx). The ozone is generated on site and on demand, thus eliminating ozone waste. Direct current is applied between the anode and cathode, which are positioned on either side of a Nafion (E. I. Dupont de Nemours and Co., Inc., Old Hickory Tennessee, TN) 117-proton-exchange membrane. When deionized water is fed to the anode side, where the electrode is coated with lead dioxide, two water oxidation reactions take place: oxygen evolution and ozone formation. The electrochemical ozone generator consists of several cells stacked in series, with the cell components positioned between two titanium endplates that incorporate ports for liquids and gasses to enter and leave the cell. Typically the geometric area of each electrode is 100 cm^2 (Denvir et al. 2001). Audy et al. (2002) recently developed a method to decontaminate raw meat using ozone.

Although ozonation shows great potential for use as a food decontamination technology, many of the developed technologies require additional research before they are ready for commercial application.

Irradiation

The effectiveness of ionizing and electron beam irradiation on food microbial inactivation is well documented (Murano 1995, Molins et al. 2001). This technology will not be discussed here as it is the main subject of Chapter 28.

Other Decontamination Systems

Totally new technologies are beginning to be tested. For example, there is tremendous opportunity for use of intense pulsed-light systems as decontamination systems for the food industries. This technique is currently used for tumors and other medical applications; however, it may have the potential to sterilize surfaces of food items just before they are packaged. Advantages include high processing speed, cost-effectiveness, and ability to be incorporated into future production lines (Wekhof 2002).

The feasibility of using electric fields (alternating current) to inactivate *E. coli* and bacterial spores in media such as saline water and orange juice has been demonstrated (Uemura 2000, Uemura and Isobe 2002, Uemura and Isobe 2003). This technology can be applied to any other liquid food because the system is able to run continuously and is easy to scale up for commercial use, compared with high electric field pulses.

The synergistic effect of combined technologies has been and continues to be evaluated. Potentially successful combinations include combining heat treatment with high pressure; combining electric fields with high temperature; combining ultrasound/pressure/heat treatment; combining ozone with ultraviolet light; supercritical fluid extraction; and combining vacuum-steam-vacuum and ionizing radiation.

Microbial pathogens could be weakened or even inactivated through sonication treatment. Ultrasound technology consists of the transmission of mechanical waves through a food material at frequencies above 18 MHz. A transducer transforms electrical energy into mechanical energy, which in turn is amplified and transmitted through the material to generate traveling sonic energy waves.

Microbiological inactivation has been tested in continuous flow using a 1-kW pilot-scale ultrasonic system operating at 20 kHz and a maximum amplitude of 35–45 m. Studies have shown that inactivation of vegetative organisms (e.g., *E. coli* spp. or *Lactobacillus* spp.) is increased when ultrasound is used in combination with heat (Zenk 2002). For the bactericidal effect, the applied ultrasonic intensity was 50 W/cm^2, which corresponds to a specific energy input from 70 to 80 kJ/kg (19–22 kWh/t).

The toxicity of prion peptides responsible for bovine spongiform encephalopathy is closely related to their aggregation state. Once the aggregation state is disrupted, their infectivity is affected. Although ultrasound lacks the power and versatility to inactivate microorganisms, the technique can be used in combination with ozone and ultraviolet light primarily to enhance prion inactivation in foods (Casolari 1998, Dealler and Lacey 1990).

Recent developments in materials science may be of great promise for new food decontamination methods. An example is the impregnation of the surface of food contact surfaces such as plastics and glass with antimicrobial agents (Tiller et al. 2001). Research on modifications of stainless steel surfaces may also prove to be useful for food industry applications.

Conclusions and Future Needs

The future of alternative food decontamination technologies is promising. However, technical and regulatory barriers need to be overcome to achieve successful industrial implementation.

Whether the new decontamination technologies contribute significantly to the improvement of safety of food products will be gradually determined as surveillance data on microbial foodborne illness are collected. However, to be considered successful, the technologies have to achieve at least the same food safety standards as those of conventional thermal processes. Understanding the mechanism of microbial inactivation involved on each technology would allow using them in the most suitable conditions. Each of these technologies could be used alone or in combination with other technologies to optimize product quality, processing time, and microbial inactivation.

The challenge for the future is not only to develop more effective decontamination methods. There is still quite a need to develop and standardize process parameters with respect to kinetics of microbial inactivation or elimination. This assessment must be carried out for all foodborne pathogens in a wide variety of food products. Results will allow for improved suitability of the technologies to specific food types (e.g., solid vs. liquid).

References

Anderson, G. W. and Gangel, M. G. 1999. Method and apparatus for antimicrobial treatment of animal carcasses. U.S. Patent No. 5,980,375.

Audy, S., Laberge, F., Steiner, E. F. and Yuan, J. T. Y. 2002. Food disinfection using ozone. U.S. Patent No. 6,485,769 B2.

Avens, J. S., Albright, S. N., Morton, A. N., Prewitt, B. E., Kendall, P. A. and Sofos, J. N. 2002. Destruction of microorganisms on chicken carcasses by steam and boiling water immersion. *Food Control* **13**:445–450.

Barbosa-Cánovas, G. V. and Gould, G. W. 2000. Innovations in Food Processing. New York: Marcel Dekker.

Barbosa-Cánovas, G. V., Pothakamury, U. R., Palau, E. and Swanson, B. G. 1998. Non-Thermal Preservation of Foods. New York: Marcel Dekker.

Belk, K. E. 2000. Beef Decontamination Technologies. Beef Facts 1-5. Washington, D.C.: National Cattlemen's Beef Association.

Breidt, F., Hayes, J. S. and Fleming, H. P. 2000. Reduction of microflora of whole pickling cucumbers by blanching. *Journal of Food Science* **65**:1354–1358.

Casolari, A. 1998. Heat resistance of prions and food processing. *Food Microbiology* **15**:59–63.

Dealler, S. F. and Lacey, R. W. 1990. Transmissible spongiform encephalopathies: the threat of BSE to man. *Food Microbiology* **7**:253–279.

Denvir, A. J., McKenzie, K. S., Rogers, T. D., Miller, D. R., Hitchens, G. D. and Andrews, C. C. 2001. Method of food decontamination by treatment with ozone. U.S. Patent No. 6,171,625 B1.

Erkmen, O. and Karataş, Ş. 1997. Effect of high hydrostatic pressure on *Staphylococccus aureus* in milk. *Journal of Food Engineering* **33**:257–262.

Gill, C. O., Bedard, D. and Jones, T. 1997. The decontamination performance of a commercial apparatus for pasteurizing polished pig carcasses. *Food Microbiology* **14**:71–79.

Gill, C. O., McGinnis, D. S., Bryant, J. and Chabot, B. 1995. Decontamination of commercial, polished pig carcasses with hot water. *Food Microbiology* **12**:143–149.

Goldberg, N. M., Radewonuk, R., Kozempel, M. F. and Morgan, A. I. 2001. Method and Apparatus for Surface Treatment of Materials. U.S. Patent No. 6,245,294 BI.

Institute of Food Technologists. 2000. Emerging Microbiological Food Safety Issues. Implications for Control in the 21st Century. Institute of Food Technologists Expert Report. Chicago: Institute of Food Technologists.

James, C., Goksoy, E. O., Corry, J. E. L. and James, S. J. 2000a. Surface pasteurization of poultry meat using steam at atmospheric pressure. *Journal of Food Engineering* **45**:111–117.

James, C., Lechevalier, V. and Ketteringham, L. 2002. Surface pasteurization of shell eggs. *Journal of Food Engineering* **53**:193–197.

James, C., Thorton, J. A., Ketteringham, L. and James, S. I. 2000b. Effect of steam condensation, hot water or chlorinated hot water immersion on bacterial numbers and quality of lamb carcasses. *Journal of Food Engineering* **43**:219–225.

Jones, R. D., Stephenson, P. R. and Wilding, P. 1998. Method of heat processing of solid food. U.S. Patent No. 5,789,006.

Kalchayanand, N, Sikes, A., Dunne, C. P. and Ray, B. 1998. Factors influencing death and injury of foodborne pathogens by hydrostatic pressure-pasteurization. *Food Microbiology* **15**:207–214.

Kim, J.-G. and Yousef, A. E. 2000. Inactivation kinetics of foodborne spoilage and pathogenic bacteria by ozone. *Journal of Food Science* **63**:521–528.

Kim, J.-G., Yousef, A. E. and Chism, G. W. 1999a. Use of ozone to inactivate microorganisms on lettuce. *Journal of Food Safety* **19**:17–34.

Kim, J.-G., Yousef, A. E. and Dave, D. 1999b. Application of ozone for enhancing the microbiological safety and quality of foods: a review. *Journal of Food Protection* **62**:1071–1087.

Kozempel, M. F., Marshall, D. L., Radewonuk, E. R., Scullen, O. J., Goldberg, N. and Bal'A, M. F. A. 2001. A rapid surface intervention process to kill *Listeria* innocua on catfish using cycles of vacuum and steam. *Journal of Food Science* **66**:1012–1016.

Ludwing, H., Bieler, C., Hallbauer, K. and Scigalla, W. 1992. Inactivation of microorganisms by hydrostatic pressure, Vol. 224, pp. 25–32. *In* High Pressure and Biotechnology, C. Balny, R. Hayashi, K. Heremans, P. Masson (Eds.). London: Colloques INSERM/John Libbey Eurotext.

Majchrowicz, A. 1999. Innovative technologies could improve food safety. *FoodReview* **22**:16–20.

Mittal, G. S., Ho, S. Y. W., Cross, J. D. and Griffiths, M. W. 2000. Method and apparatus for electrically treating foodstuffs for preservations. U.S. Patent No. 6,093,432.

Molins, R. A., Motarjemi, Y. and Kaferstein, F. K. 2001. Irradiation: a critical control point in ensuring the microbiological safety of raw foods. *Food Control* **12**:347–356.

Morgan, A. I., Radewonuk, E. R. and Scullen, O. J. 1996. Ultra high temperature, ultra short time surface pasteurization of meat. *Journal of Food Science* **61**:1216–1218.

Murano, E. A. 1995. Microbiology of irradiated foods. *In* Food Irradiation—A Sourcebook, E. A. Murano (Ed.). Ames: Iowa State University Press.

Pordesimo, L. O., Wilkerson, E. G., Womac, A. R. and Cutter, C. N. 2002. Review: process engineering variables in the spray washing of meat and produce. *Journal of Food Protection* **65**:222–237.

Sofos, J. N. and Smith, G. C. 1998. Nonacid meat decontamination technologies: model studies and commercial applications. *Journal of Food Microbiology* **44**:171–188.

Stewart, C. M., Tompkin, R. B. and Cole, M. M. 2002. Food safety: new concepts for the new millennium. *Innovative Food Science and Emerging Technologies* **3**:105–112.

Tewari, G., Jayas, S. D. and Holley, R. A. 1999. High pressure processing of foods: an overview. *Science des Aliments* **19**:619–660.

Tiller, J. C., Liao, C. J., Lewsi, K. and Klibanov, A. M. 2001. Designing surfaces that kill bacteria on contact. *Proceedings of the National Academy of Sciences (USA)* **98**:5981–5985.

Uemura, K. 2000. Continuous liquid pasteurizing apparatus and a method therefor. U.S. Patent No. 6,056,884.

Uemura, K. and Isobe, S. 2002. Developing a new apparatus for inactivating *Escherichia coli* in saline water with high electric field AC. *Journal of Food Engineering* **53**:203–207.

Uemura, K. and Isobe, S. 2003. Developing a new apparatus for inactivating *Bacillus subtilis* spores in orange juice with a high electric field AC under pressurized conditions. *Journal of Food Engineering* **56**:325–329.

Wekhof, A. 2002. Pulsed UV to sterilize food packaging and to preserve food stuffs, pp. 4–88. Ayr, Ontario: Newsletter of International UV Association.

Wilson, R. C., Leising, J. D., Strong, J., Hocker, J. and O'Connor, J. 2000. Apparatus for steam pasteurization of food. U.S. Patent No. 6,019,033.

Zenk, M. 2002. AIF-Project: AiF-FV 11395 N. Combined application of ultrasound and temperature for energy-saving and mild preservation of liquid food. Berlin: Department of Food Biotechnology and Food Process Engineering, University of Technology, pp. 189–192.

Zhang, Q., Barbosa-Cánovas, G. V. and Swanson, B. G. 1995. Engineering aspects of pulsed electric field pasteurization. *Journal of Food Engineering* **25**:261–281.

Zhang, Q., Qin, B.-L., Barbosa-Cánovas, G. V., Swanson, B. G. and Pedrow, P. D. 1996. Batch mode food treatment using pulsed electric fields. U.S. Patent No. 5,549,041.

26 Control of *Listeria monocytogenes* in Ready-to-Eat Foods

Douglas L. Marshall

Introduction

The Gram-positive bacterium *Listeria monocytogenes* is ubiquitous in nature, being found frequently in both raw and ready-to-eat (RTE) foods. It is also comparatively resistant to harsh environmental conditions such as heat, cold, and chemicals. Therefore, consumption of the bacterium is a frequent occurrence. Despite this exposure, human disease caused by the bacterium is very rare (Notermans et al. 1998). Listeriosis outbreaks are almost always associated with consumption of either raw or cooked RTE foods by high-risk individuals (immunocompromised, elderly, pregnant women and their fetuses, and newborns).

There are eight factors that contribute to listeriosis outbreaks: inadequate thermal processing, refrigeration temperature too high, inadequate cleaning and sanitation, inadequate product flow through the processing plant, inadequate personal hygiene, shelf life too long, inadequate environmental monitoring and control, and inadequate end-product testing. Foods associated with listeriosis outbreaks have one major contributing factor that makes them high risk as a vehicle; namely, the ability to support growth of the bacterium during extended refrigerated shelf life. Foods with short chill shelf life (<5 days) are generally not considered at risk because the bacterium will grow slowly under proper refrigeration. That said, improper refrigeration, where temperatures above 5°C can allow proliferative growth of the bacterium, could lead to infectious doses within a short period of time. An example of the importance of proper refrigeration is shown in Figure 26.1, where small increases in

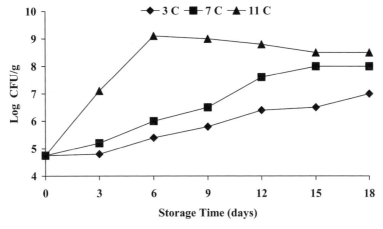

Figure 26.1 Effect of storage temperature on growth of *Listeria monocytogenes* on ready-to-eat chicken nuggets (from Marshall et al. 1991).

chill temperature had dramatic influence on proliferative ability of the bacterium (Marshall et al. 1991). Retail display units are notoriously poor at keeping product adequately chilled.

Inadequate thermal processing is caused by using too low of a heat treatment or using too short of a process time. Poor cleaning and sanitation is usually traced to a constant battle between production managers and sanitation managers. Sanitation takes time and expense, both of which rob productivity. Often overlooked are point sources of contamination. Inadequate product flow can compromise food safety through a failure to segregate raw from cooked products and a failure to limit flow of personnel within the processing environment. Inadequate personal hygiene is linked to improper husbandry of garments and outerwear, including gloves. Maintenance personnel also can be a source of contamination. Thus, equipment should be resanitized after contact. Proper hand washing followed by alcohol-based wipes can be a valuable control measure. Likewise, use of iodophor-containing hand-dip stations can reduce contamination.

As mentioned previously, product shelf life can be frequently too long, resulting in proliferative growth of *L. monocytogenes*. It is suggested that product shelf life be based on food safety concerns rather than food quality concerns. Processors are encouraged to conduct inoculated pack studies to assess the potential for outgrowth of *L. monocytogenes* during normal and abusive temperature storage of each product.

Inadequate environmental monitoring is a contributing factor in many listeriosis outbreaks. Routine failure to detect and control hot spots is risky behavior. Floors, drains, and food-contact surfaces (conveyors, slicers, and peelers) are recommended locations for monitoring. In addition, end-product testing can be useful to detect gross contamination but is of limited value in detecting sporadic contamination.

To control the potential for the bacterium to either grow slowly in long–shelf life foods or to grow rapidly during temperature abuse, three main points of control are discussed in this chapter. These points are hereafter referred to as the 3E approach (exposure, exclusion, and elimination), which can provide food processors with the necessary arsenal to deal with *L. monocytogenes* contamination of foods. The first control method is to monitor the processing environment and foods in exposed areas likely to contribute to *L. monocytogenes* contamination. The benefit of rapid detection methods becomes evident for processors, as they are able to either quickly recall contaminated product or to reevaluate processing conditions when positive samples are indicated. The second control method, exclusion, is used to prevent contamination of the food product. Exclusion involves both environmental control to prevent cross contamination of *Listeria* free foods and processing control to eliminate contamination in the food. Food processing plants can produce raw products with low *L. monocytogenes* incidence levels (ca. 2%), but the same product sampled from retail outlets might show substantially higher incidence of contamination (ca. 70%). The last control method is to build elimination barriers into products that have a history of causing listeriosis outbreaks. These barriers can be either biological, chemical, or physical hurdles or a combination of these.

Future efforts in product and process development should greatly affect the safety of our food supply in regard to *L. monocytogenes*. Because substantial archival material is available at present on the characteristics of the bacterium (Ryser and Marth 1999) and post-processing controls (Tompkin et al. 1999), the focus of this chapter is on presenting recent advances that may provide a snapshot of future 3E approaches to control this deadly pathogen in RTE foods.

Listeria monocytogenes Incidence and Survival in Foods

Listeria monocytogenes incidence in foods is well described (Ryser and Marth 1999). More recently, 5% of Belgian retail cooked meats were contaminated with *L. monocytogenes* (Uyttendaele et al. 1999). They found higher incidence levels (6.6%) on sliced ham and bacon than on their unsliced counterparts (1.6%), which implicates meat slicers as a possible cross contamination source. Likewise, minced meats had higher incidence levels (6.1%) than did whole cooked products (4%). Even higher incidence levels were found on dry fermented sausages (12%). RTE foods available at retail shops and restaurants had a 9% *L. monocytogenes* contamination rate (de Simon and Ferrer 1998), with 1% of the samples containing high levels of contamination (>100 colony-forming units [CFU]/g). Cheese products, particularly soft-style cheeses, are frequently contaminated with *L. monocytogenes*. For example, fresh soft cheese produced in Brazil had a *L. monocytogenes* contamination rate of 41% (da Silva et al. 1998). Smoked seafoods also can carry a high *L. monocytogenes* incidence rate (14%), with incidence levels usually higher in cold smoked fish rather than hot smoked products (Heinitz and Johnson 1998). Methods to control *L. monocytogenes* in cold smoked salmon have been reviewed (Huss et al. 1995, Duffes 1999). Uyttendaele et al. (1999) found a high incidence of *L. monocytogenes* in mayonnaise-based salads (21%) and in prepared vegetarian meals (12%).

In most studies, all *L. monocytogenes* isolated from foods contain at least one, but more frequently more than one, pathogenicity property. These pathogenic markers include production of hemolysin, phospholipase, and the ability to be cytotoxic. Most studies demonstrate that >99% of isolates are hemolysin and phospholipase positive (Amoril and Bhunia 1999). Most hemolysin positive cultures are also cytotoxic. According to the U.S. Centers for Disease Control and Prevention, the number of reported *L. monocytogenes* cases for 2000 and 2001 were 755 and 613, respectively (Centers for Disease Control 2002, Centers for Disease Control 2003a).

The ability of *L. monocytogenes* to survive and grow in foods is key to its ability to cause human infections. One important factor that enhances its ability to tolerate acid stress in foods is the ability to mount an acid-tolerance response (Gahan et al. 1999). This response enables the bacterium to both survive high acid challenges in certain foods (fermented milks, meats, and vegetables) and to survive the high acid shock experienced during stomach transit of potential human victims. The importance of low-temperature storage of RTE peeled oranges was demonstrated by Pao et al. (1998). *Listeria monocytogenes* was able to grow on the surface of peeled oranges (surface pH > 6.0) stored at 24°C but not at 4°C. Similarly, the bacterium was unable to grow on a variety of packaged fresh cut vegetables when stored at 4°C, but it was able to multiply at higher temperatures (Farber et al. 1998).

Rapid Detection Methods

Methods to rapidly and specifically expose contaminated foods and food-contact surfaces are important for reactive responses by food processors and regulators. Elimination therefore depends on identifying areas needing remediation. Environmental monitoring within food processing environments can be valuable to detect contamination sites that harbor *L. monocytogenes*. In fact, floor-drain incidence correlates well with product contamination and has been suggested as a useful indicator site for overall plant hygiene (Rorvik et al. 1997).

Improved culture techniques have focused on a better ability to inhibit close relatives of *L. monocytogenes* and yet increase recovery levels and speed. Addition of nalidixic acid, acriflavine, and fosfomycin to selective media increases time to recovery by 24 hours but improves selectivity (Jacobsen 1999). *Listeria monocytogenes* blood agar improves differentiation of β-hemolytic *L. monocytogenes* from other *Listeria* spp. (Johansson 1998). A rapid colony lift immunoassay has been described that allows detection and quantification within 24 hours (Carroll et al. 2000). This method was more accurate than a most probable number method for quantification.

Rapid detection of *L. monocytogenes* from enrichment cultures can be accomplished using polymerase chain reaction (PCR) methods (Cox et al. 1998, Manzano et al. 1998, Stewart and Gendel 1998, Norton and Batt 1999) and antibody antigen reactions such as enzyme-linked immunosorbent assays (ELISA) or latex agglutination assays (Matar et al. 1997). An automated ELISA coupled with a nonradioactive DNA probe specific for *L. monocytogenes* was able to detect 96% of high-inoculum (10–100 CFU/25 g sample) and 89% of low-inoculum (1–10 CFU/25 g sample) spiked raw and partially processed foods (Kerdahi and Istafanos 2000). Others have reported similar results using PCR-ELISA (Scheu et al. 1999) and immunomagnetic separation followed by PCR techniques (Hudson et al. 2001).

A multiplex DNA PCR reaction based on the *iap* gene was described that allows for simultaneous detection and differentiation of all *Listeria* species (Bubert et al. 1999). The method may be useful to track contamination trends within food processing environments, where non-monocytogenes listeriae may be a contamination problem. A novel quantitative PCR method to rapidly detect and enumerate *L. monocytogenes* was described by Wang and Hong (1999). Duffy et al. (1999) developed a surface-adhesion PCR method targeting the listerolysin O gene for rapid confirmation of the presence of *L. monocytogenes* in enriched meat samples.

Typing *L. monocytogenes* isolates allows public health investigators the ability to link outbreak strains with food and environmental strains and allows processors the ability to track contamination events within the food processing environment (see Chapter 23). Several typing methodologies are available and include amplified fragment length polymorphism (AFLP; Aarts et al. 1999) and serotyping coupled with pulsed field gel electrophoresis (PFGE; Johansson et al. 1999, Miettinen et al. 1999). PFGE was used to track the source of *L. monocytogenes* contamination in a fish-processing plant (Autio et al. 1999). Effective control procedures were implemented once the contamination sources were identified. Serotyping and PFGE were used to link smoked mussels as a cause of a New Zealand listeriosis outbreak (Brett et al. 1998). Phage typing and PFGE were used to link human listeriosis strains to salmon and rainbow trout consumption (Loncarevic et al. 1998; see also Chapter 23). A comparison of five typing methods (serotyping, zymotyping, ribotyping, random amplified polymorphic DNA, and PFGE) revealed that ribotyping and PFGE were most discriminating and thus probably more useful for epidemiological investigations (Kerouanton et al. 1998).

Biopreservative Control Methods

Several naturally occurring lactic acid bacteria were found that inhibited *L. monocytogenes* growth in vacuum-packed cooked, sliced meats (Bredholt et al. 1999) and modified atmosphere stored mungbean spouts (Bennik et al. 1999). Similarly, a *Lactococcus lactis* subsp.

diacetylactis strain was able to prevent growth of *L. monocytogenes* during storage of a Spanish-style soft cheese (Mendoza-Yepes et al. 1999).

A novel approach to colonize food-contact surfaces with "good" bacteria as a method to control *L. monocytogenes* colonization was studied by Leriche et al. (1999). Their approach used a nisin-producing *L. lactis* strain to preinoculate stainless steel before exposure to *L. monocytogenes*. Results demonstrated that low *L. monocytogenes* populations could be adequately controlled on surface biofilms but that high inoculum levels resulted in survival. It remains to be seen how such a surface preconditioning could be implemented in food processing plants. Presumably surfaces would be cleaned and sanitized using usual techniques followed by a surface spray of a desirable preconditioning agent such as *L. lactis* before processing startup.

Non-monocytogenes listeriae are capable of producing bacteriophage and bacteriocins that are inhibitory to *L. monocytogenes* (Kalmokoff et al. 1999). Use of these strains as biocontrol agents either as direct microbial competitors or as a source of active agents remains underexploited (see Chapter 27). Interestingly, Bal'a and Marshall (1998) previously found that a mixed strain culture of *L. monocytogenes* failed to grow on inoculated catfish. Their explanation for this observation was that perhaps there was strain-to-strain antagonism occurring that prevented proliferation of the entire culture. This presumption was confirmed by Kalmokoff et al. (1999), who revealed the presence of bacteriocin-like substances produced by some *L. monocytogenes* strains that were inhibitory to other *L. monocytogenes* strains.

Nisin has been shown to reduce heat resistance of *L. monocytogenes* in liquid whole egg (Knight et al. 1999) and in cold packed lobster (Budo-Amoako et al. 1999). Some strains of *L. monocytogenes* are resistant to nisin; however, when nisin is combined with heat, inactivation of these strains happens faster than when heat is used alone (Modi et al. 2000). Further combination of nisin with other naturally occurring antimicrobials may defeat the ability of nisin-resistant strains to grow. For example, nisin combined with ALTA 2341 was shown to inhibit *L. monocytogenes* growth in vacuum- or modified atmosphere–packaged smoked salmon (Szabo and Cahill 1999). Likewise, lactate combined with ALTA 2341 extended the lag phase of *L. monocytogenes* on cooked poultry but did not prevent growth (Barakat and Harris 1999). Schillinger et al. (1998) showed that nisin-resistant *L. monocytogenes* was sensitive to other bacteriocins (sakacin and enterocin) and their bacteriocin-producing cultures (*Lactobacillus sake* and *Enterococcus faecium*). Synergistic inhibition of *L. monocytogenes* was seen in skim milk treated with the nisin and lactoperoxidase system (Zapico et al. 1998). Nisin combined with high CO_2 headspace atmosphere delayed but did not prevent growth of *L. monocytogenes* on cold-smoked salmon in a manner greater than when nisin was used alone (Nilsson et al. 1997). Thus, the combination of more than one biopreservative appears warranted to prevent outgrowth of resistant variants.

Enterocin was claimed to be bactericidal to *L. monocytogenes* in yogurt (Laukova et al. 1999) and soy milk (Laukova and Czikkova 1999). Another *Enterococcus* bacteriocin from *E. mundtii* appears to be bactericidal to *L. monocytogenes* (Bennik et al. 1998). The bacteriocin Lacticin has activity against *L. monocytogenes* in infant milk formulations, resulting in at least a 99% population reduction (Morgan et al. 1999). Lacticin is also active in cottage cheese (McAuliffe et al. 1999). Both the lantibiotic bacteriocins nisin and lacticin have decreased activity against acid-adapted *L. monocytogenes* (van Schaik et al. 1999). Therefore, the question is raised whether or not the use of such hurdles is appropriate when there is high potential of acid-adapted *L. monocytogenes* being a food contaminate.

In a comparative study, Meghrous et al. (1999) demonstrated that pediocin was more effective than two nisin preparations. Pediocin was found to prevent growth of *L. monocytogenes* during Cheddar cheese manufacturing and aging (Buyong et al. 1998). Others have found that pediocin and enterocin were more active against *L. monocytogenes* than were nisin and lacticin (Cintas et al. 1998).

Reuterin, a bacteriocin produced by *Lactobacillus reuteri* was able to reduce *L. monocytogenes* populations on cooked pork (El-Ziney et al. 1999) and in milk and cottage cheese (El-Ziney and Debevere 1998). The inhibitory property of *Carnobacterium piscicola* and its bacteriocin may be useful to prevent growth of *L. monocytogenes* on vacuum-packaged meat (Schobitz et al. 1999) and smoked fish (Duffes et al. 1999, Nilsson et al. 1999).

Chemical Control Methods

One of the critical factors controlling the ability of *L. monocytogenes* to cross-contaminate RTE foods is the ability of the bacterium to establish residence on food-contact surfaces, leading to the formation of biofilms (Oh and Marshall 1995a, Oh and Marshall 1996). Therefore, chemical disinfection of these surfaces becomes important as an in-plant control method (Fatemi and Frank 1999). Most current sanitation practices are able to control the pathogen if properly employed. Sanitation crew laziness, ignorance, and negligence can be significant factors contributing to lack of *L. monocytogenes* control in processing environments.

The virulence properties of *L. monocytogenes* are affected by presence of iron in the growth medium (Fisher and Martin 1999). This suggests that chelation of iron in foods or on environmental harborages may be able to decrease the virulence of contaminating *L. monocytogenes*.

The bacteriostatic properties of sodium lactates and acetates in preventing growth of *L. monocytogenes* on extended-shelf-life RTE meats are well known (Bedie et al. 2001). Slices of luncheon meat surface treated with sodium benzoate, sodium propionate, potassium sorbate, or sodium diacetate were unable to support growth of *L. monocytogenes* during refrigerated storage (Islam et al. 2002). An edible film was impregnated with lauric acid or nisin and was found to prevent growth of *L. monocytogenes* on turkey bologna slices (Dawson et al. 2002).

The outer surfaces of RTE fresh produce can be decontaminated using high levels (2,000 ppm) of chlorine (Beuchat et al. 1998). It is known that *L. monocytogenes* can grow in refrigerated precooked RTE vegetables. As such, additional barriers are needed to prevent proliferative growth during storage and during abused temperature conditions. Addition of citric acid may be one such barrier (Thomas et al. 1999). Likewise, the outer surfaces of meat animals can harbor *L. monocytogenes*. Examples of chemical treatments using trisodium phosphate dips (Figure 26.2) and lactic acid dips (Figure 26.3) demonstrate the utility of carcass decontamination to reduce the population of attached *L. monocytogenes*. The importance of combined chemical treatments was demonstrated in smoked salmon, where salt and smoke combine to inhibit the bacterium (Niedziela et al. 1998). Smoke products of high phenolic content are more effective at controlling *L. monocytogenes* than are low-phenolic smokes.

Physical Control Methods

Freezing is generally ineffective at inactivating *L. monocytogenes* (Chou et al. 1999). An obvious control method such as heating that is improperly applied can lead to contamina-

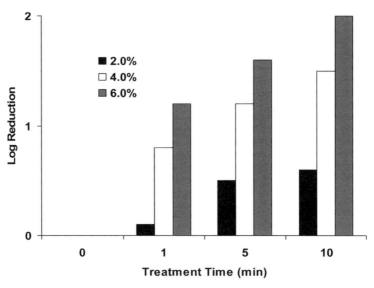

Figure 26.2 Effect of trisodium phosphate concentration on inactivation of *Listeria monocytogenes* on catfish skin after 10 minutes of exposure (from Kim and Marshall 2002).

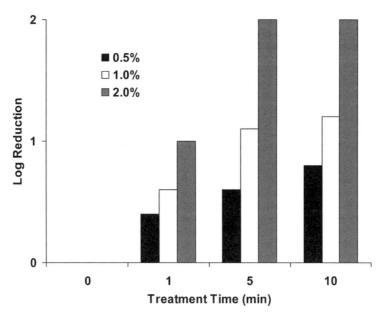

Figure 26.3 Effect of lactic concentration on inactivation of *Listeria monocytogenes* on catfish skin after 10 minutes of exposure (from Kim and Marshall 2001).

tion of finished product, where the source of contamination was raw materials. Inadequate cooking (core temperature below 70°C) was confirmed as the primary reason for contaminated processed meats (Samelis and Metaxopoulos 1999). Heating packaged RTE deli meats to 90°C for 2 minutes can provide a 2 log *L. monocytogenes* reduction (Muriana et al. 2002). Likewise, low heating temperatures (55°C) used in mozzarella cheese stretching may allow for survival of contaminating *L. monocytogenes* (Kim et al. 1998). However, when the stretching temperature was higher (>66°C) cell counts decreased during mozzarella cheese manufacturing. Thus, simple changes in thermal processing parameters, using either higher temperatures or longer times, may be sufficient to inactivate the pathogen during manufacturing.

An example of the importance of proper refrigeration as a control parameter was provided by Juneja et al. (1998), who showed that *L. monocytogenes* was able to grow prolifically on vacuum-packed fresh, peeled potatoes stored at 15° and 28°C but was unable to multiply during storage at 4°C. In an apparent contradiction, fermented dry sausages that were stored at room temperature had more rapid die-off of *L. monocytogenes* than sausages stored at 4°C (Nissen and Holck 1998). Therefore, to ensure product safety, inoculated pack studies are recommended to define growth/death kinetics for individual foods.

High-energy pulsed light has been considered for control of *L. monocytogenes* (MacGregor et al. 1998). Questions remain of whether such treatments have sufficient penetrating power to inactivate cells in cracks and cavities on food or food-contact surfaces. High pressure (345 MPa) combined with moderate heat (50°C) can achieve greater than 8 log reductions of *L. monocytogenes* within 5 minutes of exposure (Alpas et al. 1999). Likewise, high pressure combined with bacteriocins can have from 1 to 5 logs greater bacterial count reductions than pressure alone (Kalchayanand et al. 1998). High-voltage pulsed electric fields combined with moderate heat (50°C) were able to achieve a 5 log reduction of *L. monocytogenes* in milk with different fat contents (Reina et al. 1998). Manothermosonication is the combination of high pressure (200 kPa), ultrasonication (117 micrometer wave length), and heat (40°C). This treatment combination was able to inactivate *L. monocytogenes*, with D values of 1.5 minutes (Pagan et al. 1999).

RTE foods packaged in modified atmospheres rich in carbon dioxide not only have extended refrigerated shelf life but also have an antiproliferative effect on the growth of *L. monocytogenes* (Marshall et al. 1991, Marshall et al. 1992, Oh and Marshall 1995b, Pothuri et al. 1996). Although the bacterium is usually able to grow in high-CO_2 environments, its growth rate is significantly slower than in vacuum or oxygen-containing atmospheres. Conversely, the presence of oxygen in the package headspace increases effectiveness of thermal (Dorsa and Marshall 1995, George et al. 1998) and irradiation (Thayer and Boyd 1999) inactivation treatments. Irradiation doses of 3 kGy are sufficient to inactivate moderate populations of *L. monocytogenes* on RTE meats (Thayer et al. 1998). Like all inactivation methods, the effectiveness of irradiation is population-size dependant (Andrews et al. 1995).

Use of predictive models to evaluate growth and survival trends of *L. monocytogenes* is becoming increasingly valuable for food processors. Recent examples include an inactivation model for combined temperature, pH, salt, and phosphate barriers (Juneja and Eblen 1999). A unique approach for studying various combinations of factors was described by Bal'a and Marshall (1996). Their approach used the double-gradient plate technique to analyze the combined interaction of pH, salt, monolaurin, and temperature on growth of *L. monocytogenes*. Results of their work are shown in Figures 26.4–26.7, where monolaurin activity is clearly seen to increase with high salt and low pH. A simple graphic display of

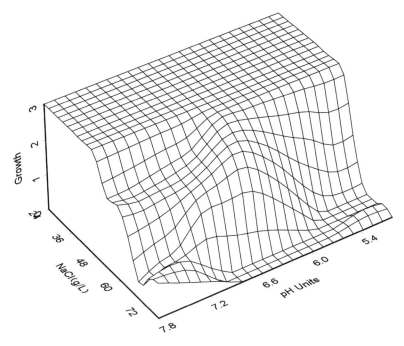

Figure 26.4 Growth profile of *Listeria monocytogenes* on two-dimensional salt and pH gradient plates incubated at 35°C for 48 hours in the presence of 0 μg of monolaurin per milliliter (from Bal'a and Marshall 1996).

growth domains of the bacterium was produced showing the combined effect of the four variables (Figure 26.8). A central composite model was used to evaluate growth parameters as affected by salt, sodium lactate, sodium diacetate, and product moisture of RTE meats (Seman et al. 2002).

Conclusions

To maintain an adequate *L. monocytogenes* control program, processors should continually reevaluate their hazard analysis critical control point programs (see Chapter 20). This evaluation should include an assessment of *L. monocytogenes* contamination of source materials, efficacy of kill steps, postkill controls, the ability of cross-contaminating *L. monocytogenes* to proliferate on products, and examination of testing records. A number of on-line resources are available to provide information to consumers, from the Centers for Disease Control (2003b), the Food and Drug Administration (2003), and the U.S. Department of Agriculture Food Safety Inspection Service (2003b). An on-line guideline for processors is also available from the U.S. Department of Agriculture Food Safety Inspection Service (1999). Two computer-based modeling programs are available: Food Micromodel (Leatherhead, Surrey) and Pathogen Modeling Program (Eastern Regional Research Center, Agricultural Research Service, U.S. Department of Agriculture, Wyndmoor, Penn.). Care should be taken when considering outputs of computer models because real-world naturally contaminated pack studies rarely give the same growth/death kinetics as the computer models (Dalgaard and Jorgensen 1998). Computer modeling can be a useful starting

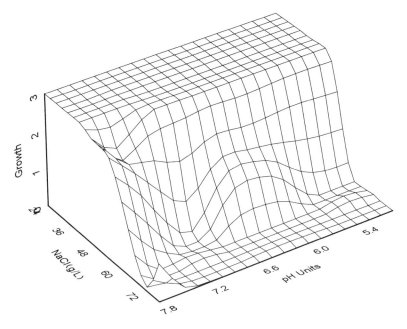

Figure 26.5 Growth profile of *Listeria monocytogenes* on two-dimensional salt and pH gradient plates incubated at 35°C for 48 hours in the presence of 2 μg of monolaurin per milliliter (from Bal'a and Marshall 1996).

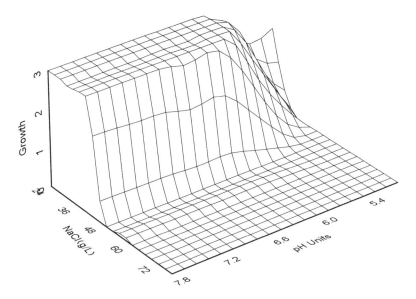

Figure 26.6 Growth profile of *Listeria monocytogenes* on two-dimensional salt and pH gradient plates incubated at 35°C for 48 hours in the presence of 4 μg of monolaurin per milliliter (from Bal'a and Marshall 1996).

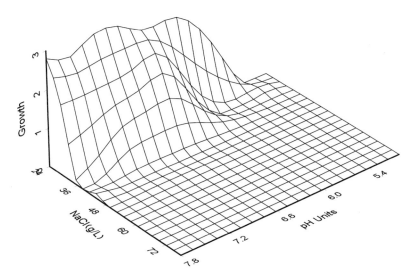

Figure 26.7 Growth profile of *Listeria monocytogenes* on two-dimensional salt and pH gradient plates incubated at 35°C for 48 hours in the presence of 8 μg of monolaurin per milliliter (from Bal'a and Marshall 1996).

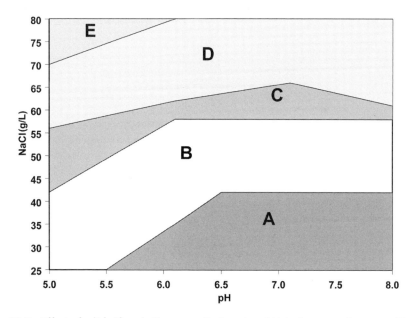

Figure 26.8 Effect of salt (g/L) and pH on growth domains of *Listeria monocytogenes* with monolaurin concentrations of (A) 8 μg/mL, (A+B) 4 μg/mL, (A+B+C) 2 μg/mL, (A+B+C+D) 0 μg/mL, and (E) area of no growth at 0 μg/mL (from Bal'a and Marshall 1996).

point to assess risk and to gauge safety of current and new product formulations; however, it is recommended that inoculated pack studies be performed to confirm safety.

Abundant *L. monocytogenes* control methods are available to food processors. Failure to either use these methods or to use them incorrectly sets the stage for listeriosis outbreaks. The wide variety of methods available suggests to this commentator that no high-risk food should be processed and stored in a manner that allows for growth of the pathogen during the expected shelf life of the product. That the rate of listeriosis has declined over the last two decades implies that most prudent processors have heeded this advice.

References

Aarts, J. J. M., Hakemulder, L. E. and van Hoef, A. M. A. 1999. Genomic typing of *Listeria monocytogenes* strains by automated laser fluorescens analysis of amplified fragment length polymorphism fingerprint patterns. *International Journal of Food Microbiology* **49**:95–102.

Alpas, H., Kalchayanand, N., Bozoglu, F., Sikes, A., Dunne, C. P. and Ray, B. 1999. Variation in resistance to hydrostatic pressure among strains of food-borne pathogens. *Applied and Environmental Microbiology* **65**:4248–4251.

Amoril, J. G. and Bhunia, A. K. 1999. Immunological and cytopathogenic properties of *Listeria monocytogenes* isolated from naturally contaminated meats. *Journal of Food Safety* **19**:195–207.

Andrews, L. S., Marshall, D. L. and Grodner, R. M. 1995. Radiosensitivity of *Listeria monocytogenes* at various temperatures and cell concentrations. *Journal of Food Protection* **58**:748–751.

Autio, T., Hielm, S., Miettinen, M., Sjöberg, A.-M., Aarnisalo, K., Björkroth, J., Mattila-Sandholm, T. and Korkeala, H. 1999. Sources of *Listeria monocytogenes* contamination in a cold-smoked rainbow trout processing plant detected by pulsed-field gel electrophoresis typing. *Applied and Environmental Microbiology* **65**:150–155.

Bal'a, M. F. A. and Marshall, D. L. 1996. Use of double gradient plates to study combined effects of salt, pH, monolaurin, and temperature on *Listeria monocytogenes*. *Journal of Food Protection* **59**:601–607.

Bal'a, M. F. A. and Marshall, D. L. 1998. Organic acid dipping of catfish fillets: effect on color, microbial load, and *Listeria monocytogenes*. *Journal of Food Protection* **61**:1470–1474.

Barakat, R. K. and Harris, L. J. 1999. Growth of *Listeria monocytogenes* and *Yersinia enterocolitica* on cooked modified-atmosphere-packaged poultry in the presence and absence of a naturally occurring microbiota. *Applied and Environmental Microbiology* **65**:342–345.

Bedie, G. K., Samelis, J., Sofos, J. N., Belk, K. E., Scanga, J. A. and Smith, G. C. 2001. Antimicrobials in the formulation to control *Listeria monocytogenes* postprocessing contamination on frankfurters stored at 4°C in vacuum packages. *Journal of Food Protection* **64**:1949–1955.

Bennik, M. H. J., Vanloo, B., Brasseur, R., Gorris, L. G. M. and Smid, E. J. 1998. A novel bactericin with a YGNZGV motif from vegetable-associated *Enterococcus mundtii*: full characterization and interaction with target organisms. *Biochimica and Biophysica Acta* **1373**:47–58.

Bennik, M. H. J., van Overbeek, W., Smid, E. J. and Gorris, L. G. M. 1999. Biopreservation in modified atmosphere stored mungbean sprouts: the use of vegetable-associated bacteriocinogenic lactic acid bacteria to control the growth of *Listeria monocytogenes*. *Letters in Applied Microbiology* **28**:226–232.

Beuchat, L. R., Nail, B. V., Adler, B. B. and Clavero, M. R. S. 1998. Efficacy of spray application of chlorinated water in killing pathogenic bacteria on raw apples, tomatoes, and lettuce. *Journal of Food Protection* **61**:1305–1311.

Bredholt, S., Nesbakken, T. and Holck, A. 1999. Protective cultures inhibit growth of *Listeria monocytogenes* and *Escherichia coli* O157:H7 in cooked, sliced, vacuum- and gas-packaged meat. *International Journal of Food Microbiology* **53**:43–52.

Brett, M. S. Y., Short, P. and McLauchlin, J. 1998. A small outbreak of listeriosis associated with smoked mussels. *International Journal of Food Microbiology* **43**:223–229.

Bubert, A., Hein, I., Rauch, M., Lehner, A., Yoon, B., Goebel, W. and Wagner, M. 1999. Detection and differentiation of *Listeria* spp. by a single reaction based on multiplex PCR. *Applied and Environmental Microbiology* **65**:4688–4692.

Budo-Amoako, E., Ablett, R. F., Harris, J. and Delves-Broughton, J. 1999. Combined effect of nisin and moderate heat on destruction of *Listeria monocytogenes* in cold pack lobster meat. *Journal of Food Protection* **62**:46–50.

Buyong, N., Kok, J. and Luchansky, J. B. 1998. Use of genetically enhanced, pediocin-producing starter culture, *Lactococcus lactis* subsp. *lactis* MM217, to control *Listeria monocytogenes* in Cheddar cheese. *Applied and Environmental Microbiology* **64**:4842–4845.

Carroll, S. A., Carr, L. E., Mallinson, E. T., Lamichanne, C., Rice, B. E., Rollins, D. M. and Joseph, S. W. 2000. Development and evaluation of a 24-hour method for the detection and quantification of *Listeria monocytogenes* in meat products. *Journal of Food Protection* **63**:347–353.

Centers for Disease Control. 2002. Reported cases of notifiable diseases, by month—United States, 2000. *Morbidity and Mortality Weekly Report* **49**:1–102.

Centers for Disease Control. 2003a. Reported cases of notifiable diseases, by month—United States, 2001. *Morbidity and Mortality Weekly Report* **50**:1–108.

Centers for Disease Control. 2003b. Listeriosis. Available at: http://www.cdc.gov/ncidod/dbmd/diseaseinfo/listeriosis_g.htm.

Chou, C. C., Cheng, S. J., Wang, Y. C. and Chung, K. T. 1999. Behavior of *Escherichia coli* O157:H7 and *Listeria monocytogenes* in tryptic soy broth subjected to various low temperature treatments. *Food Research International* **32**:1–6.

Cintas, L. M., Casaus, P., Fernandez, M. F. and Hernandez, P. E. 1998. Comparative antimicrobial activity of enterocin L50, pediocin PA-1, nisin A and lactocin S against spoilage and foodborne pathogenic bacteria. *Food Microbiology* **15**:289–298.

Cox, T., Frazier, C., Tuttle, J., Flood, S., Yagi, L., Yamashiro, C. T., Behari, R., Paszko, C. and Cano, R. J. 1998. Rapid detection of *Listeria monocytogenes* in dairy samples using a PCR-based fluorogenic 5′ nuclease assay. *Journal of Industrial Microbiology and Biotechnology* **21**:167–174.

Dalgaard, P. and Jorgensen, L. V. 1998. Predicted and observed growth of *Listeria monocytogenes* in seafood challenge tests and in naturally contaminated cold-smoked salmon. *International Journal of Food Microbiology* **40**:105–115.

da Silva, M. C. D., Hofer, E. and Tibana, A. 1998. Incidence of *Listeria monocytogenes* in cheese produced in Rio de Janeiro, Brazil. *Journal of Food Protection* **61**:354–356.

Dawson, P. L., Carl, G. D., Acton, J. C. and Han, I. Y. 2002. Effect of lauric acid and nisin-impregnated soy-based films on the growth of *Listeria monocytogenes* on turkey bologna. *Poultry Science* **81**:721–726.

de Simon, M. and Ferrer, M. D. 1998. Initial number, serovars and phagevars of *Listeria monocytogenes* isolated in prepared foods in the city of Barcelona (Spain). *International Journal of Food Microbiology* **44**:141–144.

Dorsa, W. J. and Marshall, D. L. 1995. Influence of lactic acid and modified atmosphere on thermal destruction of *Listeria monocytogenes* in crawfish tail meat homogenate. *Journal of Food Safety* **15**:1–9.

Duffes, F. 1999. Improving the control of *Listeria monocytogenes* in cold smoked salmon. *Trends in Food Science and Technology* **10**:211–216.

Duffes, F., Leroi, F., Boyaval, P. and Dousset, X. 1999. Inhibition of *Listeria monocytogenes* by *Carnobacterium* spp. strains in a simulated cold smoked fish system at 4°C. *International Journal of Food Microbiology* **47**:33–42.

Duffy, G., Cloak, O. M., Sheridan, J. J., Blair, I. S. and McDowell, D. A. 1999. The development of a combined surface adhesion and polymerase chain reaction technique in the rapid detection of *Listeria monocytogenes* in meat and poultry. *International Journal of Food Microbiology* **49**:151–159.

El-Ziney, M. G. and Debevere, J. M. 1998. The effect of reuterin on *Listeria monocytogenes* and *Escherichia coli* O157:H7 in milk and cottage cheese. *Journal of Food Protection* **61**:1275–1280.

El-Ziney, M. G., van den Tempel, T., Debevere, J. and Jakobsen, M. 1999. Application of reuterin produced by *Lactobacillus reuteri* 12002 for meat decontamination and preservation. *Journal of Food Protection* **62**:257–261.

Farber, J. M., Wang, S. L., Cai, Y. and Zhang, S. 1998. Changes in populations of *Listeria monocytogenes* inoculated on packaged fresh-cut vegetables. *Journal of Food Protection* **61**:192–195.

Fatemi, P. and Frank, J. F. 1999. Inactivation of *Listeria monocytogenes/Pseudomonas* biofilms by peracid sanitizers. *Journal of Food Protection* **62**:761–765.

Fisher, C. W. and Martin, S. E. 1999. Effects of iron and selenium on the production of catalase, superoxide dismutase, and listeriolysin O in *Listeria monocytogenes*. *Journal of Food Protection* **62**:1206–1209.

Food and Drug Administration. 2003. *Listeria monocytogenes*. Available at: http://www.cfsan.fda.gov/~mow/chap6.html.

Gahan, C. G. M., Gahan, G. M. and Hill, C. 1999. The relationship between acid stress responses and virulence in *Salmonella typhimurium* and *Listeria monocytogenes*. *International Journal of Food Microbiology* **50**:93–100.

George, S. M., Richardson, L. C. C., Pol, I. E. and Peck, M. W. 1998. Effect of oxygen concentration and redox potential on recovery of sublethally heat-damaged cells of *Escherichia coli* O157:H7, *Salmonella enteritidis* and *Listeria monocytogenes*. *Journal of Applied Microbiology* **84**:614–621.

Heinitz, M. L. and Johnson, J. M. 1998. The incidence of *Listeria* spp., *Salmonella* spp., and *Clostridium botulinum* in smoked fish and shellfish. *Journal of Food Protection* **61**:318–323.

Hudson, J. A., Lake, R. J., Savill, M. G., Scholes, P. and McCormick, R. E. 2001. Rapid detection of *Listeria monocytogenes* in ham samples using immunomagnetic separation followed by polymerase chain reaction. *Journal of Applied Microbiology* **90**:614–621.

Huss, H. H., Ben-Embarek, P. K. and Jeppesen, V. F. 1995. Control of biological hazards in cold smoked salmon production. *Food Control* **6**:335–342.

Islam, M., Chen, J., Doyle, M. P. and Chinnan, M. 2002. Effect of selected generally recognized as safe preservative sprays on growth of *Listeria monocytogenes* on chicken luncheon meat. *Journal of Food Protection* **65**:794–798.

Jacobsen, C. N. 1999. The influence of commonly used selective agents on the growth of *Listeria monocytogenes*. *International Journal of Food Microbiology* **50**:221–226.

Johansson, T. 1998. Enhanced detection and enumeration of *Listeria monocytogenes* from foodstuffs and food-processing environments. *International Journal of Food Microbiology* **40**:77–85.

Johansson, T., Rantala, L., Palmu, L. and Honkanen-Buzalski, T. 1999. Occurrence and typing of *Listeria monocytogenes* strains in retail vacuum-packaged fish products and in a production plant. *International Journal of Food Microbiology* **47**:111–119.

Juneja, V. K. and Eblen, B. S. 1999. Predictive thermal inactivation model for *Listeria monocytogenes* with temperature, pH, NaCl, and sodium pyrophosphate as controlling factors. *Journal of Food Protection* **62**:986–993.

Juneja, V. K., Martin, S. T. and Sapers, G. M. 1998. Control of *Listeria monocytogenes* in vacuum-packaged pre-peeled potatoes. Journal of Food Science **63**:911–914.

Kalchayanand, N., Sikes, A., Dunne, C. P. and Ray, B. 1998. Factors influencing death and injury of foodborne pathogens by hydrostatic pressure-pasteurization. *Food Microbiology* **15**:207–214.

Kalmokoff, M. L., Daley, E., Austin, J. W. and Farber, J. M. 1999. Bacteriocin-like inhibitory activities among various species of *Listeria*. *International Journal of Food Microbiology* **50**:191–201.

Kerdahi, K. F. and Istafanos, P. F. 2000. Rapid determination of *Listeria monocytogenes* by automated enzyme-linked immunoassay and nonradioactive DNA probe. *Journal of the Association of Official Analytical Chemists International* **83**:86–88.

Kerouanton, A., Brisabois, A., Denoyer, E., Dilasser, F., Grout, J., Salvat, G. and Picard, B. 1998. Comparison of five typing methods for the epidemiological study of *Listeria monocytogenes*. *International Journal of Food Microbiology* **43**:61–71.

Kim, J. and Marshall, D. L. 2001. Effect of lactic acid on *Listeria monocytogenes* and *Edwardsiella tarda* attached to catfish skin. *Food Microbiology* **18**:589–596.

Kim, J. and Marshall, D. L. 2002. Influence of catfish skin mucus on trisodium phosphate inactivation of attached *Salmonella* Typhimurium, *Edwardsiella tarda*, and *Listeria monocytogenes*. *Journal of Food Protection* **65**:1146–1151.

Kim, J., Schmidt, K. A., Phebus, R. K. and Jeon, I. J. 1998. Time and temperature of stretching as critical control points for *Listeria monocytogenes* during production of mozzarella cheese. *Journal of Food Protection* **61**:116–118.

Knight, K. P., Bartlett, F. M., McKellar, R. C. and Harris, L. J. 1999. Nisin reduces the thermal resistance of *Listeria monocytogenes* Scott A in liquid whole egg. *Journal of Food Protection* **62**:999–1003.

Laukova, A. and Czikkova, S. 1999. The use of enterocin CCM 4231 in soy milk to control the growth of *Listeria monocytogenes* and *Staphylococcus aureus*. *Journal of Applied Microbiology* **87**:182–186.

Laukova, A., Czikkova, S., Dobransky, T. and Burdova, O. 1999. Inhibition of *Listeria monocytogenes* and *Staphylococcus aureus* by enterocin CCM 4231 in milk products. *Food Microbiology* **16**:93–99.

Leriche, V., Chassaing, D. and Carpentier, B. 1999. Behaviour of *L. monocytogenes* in an artificially made biofilm of a nisin-producing strain of *Lactococcus lactis*. *International Journal of Food Microbiology* **51**:169–182.

Loncarevic, S., Danielsson-Tham, M. L., Gerner-Smidt, P., Sahlstrom, L. and Tham, W. 1998. Potential sources of human listeriosis in Sweden. *Food Microbiology* **15**:65–69.

MacGregor, S. J., Rowan, N. J., McIlvaney, L., Anderson, J. G., Fouracre, R. A. and Farish, O. 1998. Light inactivation of food-related pathogenic bacteria using a pulsed power source. *Letters in Applied Microbiology* **29**:416–420.

Manzano, M., Cocolin, L., Cantoni, C. and Comi, G. 1998. A rapid method for the identification and partial serotyping of *Listeria monocytogenes* in food by PCR and restriction enzyme analysis. *International Journal of Food Microbiology* **42**:207–212.

Marshall, D. L., Andrews, L. S., Wells, J. H. and Farr, A. J. 1992. Influence of modified atmosphere packaging on the competitive growth of *Listeria monocytogenes* and *Pseudomonas fluorescens* on precooked chicken. *Food Microbiology* **9**:303–309.

Marshall, D. L., Wiese-Lehigh, P. L., Wells, J. H. and Farr, A. J. 1991. Comparative growth of *Listeria monocytogenes* and *Pseudomonas fluorescens* on precooked chicken nuggets stored under modified atmospheres. *Journal of Food Protection* **54**:841–843,851.

Matar, G. M., Hayes, P. S., Bibb, W. F. and Swaminathan, B. 1997. Listeriolysin O-based latex agglutination test for the rapid detection of *Listeria monocytogenes* in foods. *Journal of Food Protection* **60**:1038–1040.

McAuliffe, O., Hill, C. and Ross, R. P. 1999. Inhibition of *Listeria monocytogenes* in cottage cheese manufactured with a lacticin 3147-producing starter culture. *Journal of Applied Microbiology* **86**:251–256.

Meghrous, J., Lacroix, C. and Simard, R. E. 1999. The effects on vegetative cells and spores of three bacteriocins from lactic acid bacteria. *Food Microbiology* **16**:105–114.

Mendoza-Yepes, M. J., Abellan-Lopez, O., Carrion-Ortega, J. and Marin-Iniesta, F. 1999. Inhibition of *Listeria monocytogenes* and other bacteria in Spanish soft cheese made with *Lactococcus lactis* subsp. *diacetylactis*. *Journal of Food Safety* **19**:161–170.

Miettinen, M. K., Björkroth, K. J. and Korkeala, H. J. 1999. Characterization of *Listeria monocytogenes* from an ice cream plant by serotyping and pulsed-field gel electrophoresis. *International Journal of Food Microbiology* **46**:187–192.

Modi, K. D., Chikindas, M. L. and Montville, T. J. 2000. Sensitivity of nisin-resistant *Listeria monocytogenes* to heat and the synergistic action of heat and nisin. *Letters in Applied Microbiology* **30**:249–253.

Morgan, S. M., Galvin, M., Kelly, J., Ross, R. P. and Hill, C. 1999. Development of a lacticin 3147-enriched whey powder with inhibitory activity against foodborne pathogens. *Journal of Food Protection* **62**:1011–1016.

Muriana, P. M., Quimby, W., Davidson, C. A. and Grooms, J. 2002. Postpackage pasteurization of ready-to-eat deli meats by submersion heating for reduction of *Listeria monocytogenes*. *Journal of Food Protection* **65**:963–969.

Niedziela, J. C., MacRae, M., Ogden, I. D. and Nesvadba, P. 1998. Control of *Listeria monocytogenes* in salmon; antimicrobial effect of salting, smoking and specific smoke compounds. *Food Science and Technology* **31**:155–161.

Nilsson, L., Gram, L. and Huss, H. H. 1999. Growth control of *Listeria monocytogenes* on cold-smoked salmon using a competitive lactic acid bacteria flora. *Journal of Food Protection* **62**:336–342.

Nilsson, L., Huss, H. H. and Gram, L. 1997. Inhibition of *Listeria monocytogenes* on cold-smoked salmon by nisin and carbon dioxide atmosphere. *International Journal of Food Microbiology* **38**:217–227.

Nissen, H. and Holck, A. 1998. Survival of *Escherichia coli* O157:H7, *Listeria monocytogenes* and *Salmonella Kentucky* in Norwegian fermented, dry sausage. *Food Microbiology* **15**:273–279.

Norton, D. M. and Batt, C. A. 1999. Detection of viable *Listeria monocytogenes* with a 5′ nuclease PCR assay. *Applied and Environmental Microbiology* **65**:2122–2127.

Notermans, S., Dufrenne, J., Teunis, P. and Chackraborty, T. 1998. Studies on the risk assessment of *Listeria monocytogenes*. *Journal of Food Protection* **61**:244–248.

Oh, D. H. and Marshall, D. L. 1995a. Destruction of *Listeria monocytogenes* biofilms on stainless steel using monolaurin and heat. *Journal of Food Protection* **58**:251–255.

Oh, D. H. and Marshall, D. L. 1995b. Influence of packaging method, lactic acid and monolaurin on *Listeria monocytogenes* in crawfish tailmeat homogenate. *Food Microbiology* **12**:159–163.

Oh, D. H. and Marshall, D. L. 1996. Monolaurin and acetic acid inactivation of *Listeria monocytogenes* attached to stainless steel. *Journal of Food Protection* **59**:249–252.

Pagan, R., Manas, P., Raso, J. and Condon, S. 1999. Bacterial resistance to ultrasonic waves under pressure at nonlethal (manosonication) and lethal (manothermosonication) temperature. *Applied and Environmental Microbiology* **65**:297–308.

Pao, S., Brown, G. E. and Schneider, K. R. 1998. Challenge studies with selected pathogenic bacteria on freshly peeled Hamlin orange. *Journal of Food Science* **63**:359–362.

Pothuri, P., Marshall, D. L. and McMillin, K. W. 1996. Combined effects of packaging atmosphere and lactic acid on growth and survival of *Listeria monocytogenes* in crayfish tail meat at 4°C. *Journal of Food Protection* **59**:253–256.

Reina, L. D., Jin, Z. T., Zhang, Q. H. and Yousef, A. E. 1998. Inactivation of *Listeria monocytogenes* in milk by pulsed electric fields. *Journal of Food Protection* **61**:1203–1206.

Rorvik, L. M., Skjerve, E., Knudsen, B. R. and Yndestad, M. 1997. Risk factors for contamination of smoked salmon with *Listeria monocytogenes* during processing. *International Journal of Food Microbiology* **37**:215–219.

Ryser, E. T. and Marth, E. H. (Eds.) 1999. *Listeria*, Listeriosis, and Food Safety, 2nd ed. New York: Marcel Dekker.

Samelis, J. and Metaxopoulos, J. 1999. Incidence and principal sources of *Listeria* spp. and *Listeria monocytogenes* contamination in processed meats and a meat processing plant. *Food Microbiology* **16**:465–477.

Schillinger, U., Chung, H. S., Keppler, K. and Holzapfel, W. H. 1998. Use of bacteriocinogenic lactic acid bacteria to inhibit spontaneous nisin-resistant mutants of *Listeria monocytogenes* Scott A. *Journal of Applied Microbiology* **85**:657–663.

Scheu, P., Gasch, A. and Berghof, K. 1999. Rapid detection of *Listeria monocytogenes* by PCR-ELISA. *Letters in Applied Microbiology* **29**:416–420.

Schobitz, R., Zaror, T., Leon, O. and Costa, M. 1999. A bacteriocin from *Carnobacterium piscicola* for the control of *Listeria monocytogenes* in vacuum-packaged meat. *Food Microbiology* **16**:249–255.

Seman, D. L., Borger, A. C., Meyer, J. D., Hall, P. A. and Milkowski, A. L. 2002. Modeling the growth of *Listeria monocytogenes* in cured ready-to-eat processed meat products by manipulation of sodium chloride, sodium diacetate, potassium lactate, and product moisture content. *Journal of Food Protection* **65**:651–658.

Stewart, D. and Gendel, S. M. 1998. Specificity of the BAX polymerase chain reaction system for detection of the foodborne pathogen *Listeria monocytogenes*. *Journal of the Association of Official Analytical Chemists International* **81**:817–822.

Szabo, E. A. and Cahill, M. E. 1999. Nisin and ALTA 2341 inhibit the growth of *Listeria monocytogenes* on smoked salmon packaged under vacuum or 100% CO_2. *Letters in Applied Microbiology* **28**:373–377.

Thayer, D. W. and Boyd, G. 1999. Irradiation and modified atmosphere packaging for the control of *Listeria monocytogenes* on turkey meat. *Journal of Food Protection* **62**:1136–1142.

Thayer, D. W., Boyd, G., Kim, A., Fox, J. B. and Farrell Jr., H. M. 1998. Fate of gamma-irradiated *Listeria monocytogenes* during refrigerated storage on raw or cooked turkey breast meat. *Journal of Food Protection* **61**:979–987.

Thomas, C., Prior, O. and O'Beirne, D. 1999. Survival and growth of *Listeria* species in a model ready-to-use vegetable product containing raw and cooked ingredients as affected by storage temperature and acidification. *International Journal of Food Science and Technology* **34**:317–324.

Tompkin, R. B., Scott, V. N., Bernard, D. T., Sveum, W. H. and Gombas, K. S. 1999. Guidelines to prevent postprocessing contamination from *Listeria monocytogenes*. *Dairy, Food and Environmental Sanitation* **19**:551–562.

U.S. Department of Agriculture Food Safety Inspection Service. 1999. *Listeria* guidelines for industry. Available at: http://www.fsis.usda.gov/OA/topics/lmguide.htm.

U.S. Department of Agriculture Food Safety Inspection Service. 2003. *Listeria monocytogenes*. Available at: http://www.fsis.usda.gov/OA/topics/lm.htm.

Uyttendaele, M., de Troy, P. and Debrevere, J. 1999. Incidence of *Listeria monocytogenes* in different types of meat products on the Belgian retail market. *International Journal of Food Microbiology* **53**:75–80.

van Schaik, W., Gahan, C. G. M. and Hill, C. 1999. Acid-adapted *Listeria monocytogenes* displays enhanced tolerance against the lantibiotics nisin and lacticin 3147. *Journal of Food Protection* **62**:536–539.

Wang, C. and Hong, C. 1999. Quantitative PCR for *Listeria monocytogenes* with colorimetric detection. *Journal of Food Protection* **62**:35–39.

Zapico, P., Medina, M., Gaya, P. and Nunez, M. 1998. Synergistic effect of nisin and the lactoperoxidase system on *Listeria monocytogenes* in skim milk. *International Journal of Food Microbiology* **40**:35–42.

27 Bacteriophage: Potential Role in Food Safety

William E. Huff, Gerry R. Huff, Narayan C. Rath, Janice M. Balog, and Annie M. Donoghue

Introduction

Bacteriophage are ubiquitous viruses that infect and kill bacteria, giving them tremendous potential to prevent and treat bacterial diseases and to eliminate foodborne pathogens on agriculture products. Bacteriophage were discovered in the early 1900s independently by both Twort (1915) and d'Herelle (1917). Early after the discovery of bacteriophage, their potential to prevent and treat bacterial diseases was investigated with mixed results. The inconsistency of bacteriophage in treating bacterial infections, coupled with the discovery of antibiotics, led to diminished interest in bacteriophage therapy. At present, there is renewed interest in this therapy as a result of concern over the consequences of antibiotic resistance in the treatment of human and animal bacterial infections. In addition, there is growing concern that the use of antibiotics in animal production poses a threat to human health through the emergence of antibiotic-resistant bacteria in animal production systems that make it difficult to treat human infections. The relative importance of antibiotic use in animal production versus human medicine in respect to the emergence of antibiotic-resistant bacteria that present a threat to human health is a long and continuing debate. However, there is serious consideration being given to restricting the use of antibiotics in animal production, which could have serious consequences to animal health and could increase foodborne pathogen levels on animals coming to the processing plants. Therefore, there is a real need to find alternatives to antibiotics that can be used to effectively control and treat animal diseases and to reduce the levels of foodborne pathogens on animal products. For the last several years we have been investigating the concept that bacteriophage can provide an effective alternative to antibiotics in animal production.

Bacteriophage Biology

Bacteriophage are typical viruses that have a protein coat that encloses a nucleic acid, which can either be DNA or RNA. There are two general types of bacteriophage, virulent and temperate, that differ in their life cycle. Virulent bacteriophage kill bacteria through a multistep process. First they adsorb to bacteria through recognition of specific attachment sites (receptors) on the surface of the bacteria (Figure 27.1). Their nucleic acid (either DNA or RNA) is then injected into the bacterium and viral replication is then achieved in

---

Mention of a trade name, proprietary product, or specific equipment does not constitute a guarantee or warranty by the U.S. Department of Agriculture and does not imply its approval to the exclusion of other products that may be suitable.

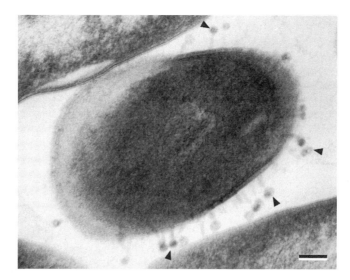

Figure 27.1 Electron micrograph of a bacterium from a mixed culture of intestinal organisms with attached bacteriophage. Multiple bacteriophage (arrowheads) are attached to the outside of the cell. The electron micrograph was provided courtesy of Robert E. Drolesky, U.S. Departmet of Agriculture, Agricultural Research Service, Southern Plains Agricultural Research Center, College Station, Tex. Bar = 100 nm.

the bacterium. The bacteria are destroyed through lysis, resulting in an average release of 50–200 daughter particles. Temperate bacteriophage do not immediately replicate in the bacterium they infect but coexist within the bacterium as a prophage (viral nucleic acid inserted into the bacterial genome) that is replicated along with the bacterium, thereby converting the bacterium to a lysogenic strain. When lysogenic bacteria are stressed, the prophage can become activated, resulting in replication of the virus and in killing of the bacteria through lysis.

Bacteriophage are ubiquitous in nature, presumably co-evolving with bacteria, and it is thought that all bacteria can be infected with bacteriophage. Bacteriophage only infect bacteria and not plant or animal cells. Bacteriophage are generally very specific, often infecting only a single strain within a species of bacteria, although some bacteriophage do have a larger host (target) bacterial range that can cross generic boundaries; these are characterized as polyvalent bacteriophage. An excellent general text on bacteriophage was published by Adams (1966).

Bacteriophage Therapy and Food Safety Research

Almost immediately after the discovery of bacteriophage there was interest in using them to prevent and treat bacterial infections. There were successful applications of bacteriophage therapeutics, but therapeutic efficacy of bacteriophage was inconsistent, and research and product development was for the most part discontinued with the development of antibiotics. There were exceptions to this trend, mainly in Eastern Europe. The Eliava Institute in Tblisi, Georgia, of the former Soviet Union, has continued bacteriophage research to the present. The Eliava Institute was established in 1923 by Giorgi Eliava, a former student of

Felix d'Herelle. The Russian research on bacteriophage in human medicine was reviewed by Alisky et al. (1998). In addition to the continued research on bacteriophage conducted at the Eliava Institute, there are a number of papers on work done in Poland using bacteriophage therapy in human medicine (Ślopek et al. 1981, Ślopek et al. 1984, Ślopek et al. 1985, Ślopek et al. 1987, Weber-Dąbrowska et al. 1987).

There is a growing interest in the use of bacteriophage to control bacterial infections. The ability of bacteriophage to control *Escherichia coli*–induced diarrhea in calves, piglets, and lambs was demonstrated by research conducted by H. W. Smith (Smith and Huggins 1983, Smith et al. 1987), who also demonstrated the ability of phage to treat *E. coli* infections in mice (Smith and Huggins 1982). Barrow et al. (1998) demonstrated the ability of bacteriophage to protect chickens from an intramuscular challenge by *E. coli* when bacteriophage were simultaneously injected at different sites. Soothill (1992) found that bacteriophage would protect mice from infection with *Acinetobacter baumanii* and *Pseudomonas aeruginosa*. Vancomycin-resistant *Enterococcus faecium* is a serious and growing nosocomial infection. Biswas et al. (2002) demonstrated that bacteriophage could rescue mice from a lethal challenge with vancomycin-resistant *Enterococcus faecium*. Bacteriophage provided effective control of a disease in the ayu, a fish raised in Japan, caused by *Pseudomonas plecoglossicida* (Park et al. 2000). Matsuzaki et al. (2003) was able to protect mice from a lethal injection of *Staphylococcus aureus*.

Vibrio vulnificus is an opportunistic pathogen of humans that can result in septicemia and that is often associated with the consumption of oysters. Cerveny et al. (2002) were able to demonstrate that bacteriophage had therapeutic value in the treatment of both localized and systemic infections with *Vibrio vulnificus* in a mice model.

Colibacillosis is a serious problem in poultry production, characterized by airsacculitis, pericarditis, and perihepatitis. This severe respiratory infection in poultry is one of the major reasons for the use of antibiotics by the poultry industry. We have been able to use bacteriophage to both prevent and treat this severe *E. coli* respiratory infection in poultry (Huff et al. 2002a, Huff et al. 2002b, Huff et al. 2003).

The above discussion clearly demonstrates that bacteriophage have tremendous potential and proven efficacy as therapeutic agents, proven in both human clinical settings and animal disease research models. We believe the research strongly supports the concept of bacteriophage therapy being developed as an alternative to antibiotics. There are a number of excellent reviews on this subject (Carlton 1999, Duckworth and Gulig 2002, Summers 2001, Barrow and Soothill 1997).

There is significant research on the use of bacteriophage to control foodborne pathogens such as *Salmonella, Listeria monocytogenes, E. coli* O157:H7, and *Campylobacter* on agricultural products. Research demonstrating the efficacy of bacteriophage to reduce *L. monocytogenes* on fresh-cut produce was demonstrated by Leverentz et al. (2003). Lytic bacteriophage have been shown to decrease *Salmonella* and *Campylobacter* contamination on chicken skin (Goode et al. 2003). There is a significant research effort being conducted on controlling *Salmonella* on poultry products (Higgins 2002, Higgins et al. 2002). Research on *E. coli* O157:H7 is being conducted in cattle and swine at a number of locations including Auburn University (S. Price, personal communication) and the U.S. Department of Agriculture, Agricultural Research Service, Laboratory in College Station, Texas (T. Callaway, personal communication). *Campylobacter* is a significant contaminant of poultry products, and research on the use of bacteriophage to reduce *Campylobacter* contamination of poultry products is being pursued at the U.S. Department of Agriculture,

Agricultural Research Service, Russell Research Center in Athens, Georgia (G. Siragusa, personal communication).

Bacteriophage Methods

The first step in establishing a bacteriophage research program is to isolate bacteriophage to the bacteria of interest. Bacteriophage are ubiquitous and plentiful in nature and are relatively easy to isolate. However, there are a few basic concepts to keep in mind in a bacteriophage discovery program. The choice of where to look for bacteriophage to any given bacteria is directed by where the bacteria exist in nature. The traditional environmental sources to explore for bacteriophage have been the primary settling tanks of municipal wastewater treatment plants, soil, hospital wound dressings, animal beddings, and surface waters.

The adsorption of bacteriophage to bacteria is a critical step in the bacteriophage replication cycle. Bacteriophage adsorption is random, driven by probability, sensitive to the ionic environment, and fragile. Therefore, when working with bacteriophage, gentle agitation is necessary to facilitate chance bacteriophage/bacteria meetings, but vigorous agitation may disrupt adsorption. It appears that the divalent cation concentration in this process is important, and in all the work we do, we incorporate a 5 mM concentration of $MgSO_4$ into our media.

Another important consideration is that bacteriophage replicate best in actively growing bacterial cultures; all of our work is conducted with fresh, actively growing cultures.

Our research interests have been to isolate bacteriophage to enteric bacteria associated with poultry diseases and contamination of poultry products (Huff et al. 2002a, Huff et al. 2002b, Huff et al. 2003). Therefore, the environmental sources we have used in our bacteriophage discovery research have been water samples taken from the primary settling tanks of municipal wastewater treatment plants and poultry-processing plants. The water samples are filter sterilized by initial filtration through coarse filter paper followed by filtration through 0.2-μm membrane filters. Because bacteriophage can bind to membrane filters, the membrane filters are protein coated with 10% fetal calf serum before filtration. These water samples can be stored at refrigeration temperatures, but bacteriophage activity in these samples does diminish with time. The targeted bacteria are grown in a double-strength nutrient-rich medium to the mid- to late log phase of its growth cycle. The water sample suspected of containing bacteriophage is then added to the actively growing culture and incubated with gentle agitation for 24 hours to expand the bacteriophage titers. These bacteriophage enrichment cultures are then centrifuged to sediment the bacteria and then filter sterilized. The bacteriophage enrichment cultures are serially diluted and plated using the soft agar overlay plating technique (Huff et al. 2002b). We have tried different plating techniques in our laboratory for bacteriophage, but the soft agar overlay technique works best for us. We tried simply dropping suspected bacteriophage preparations onto a lawn of bacteria. In our hands, this technique does not work well because of the disruption of the bacterial lawn with the application of the sample.

We perform the soft agar overlay technique by first making Petri plates that contain our base medium. Before making the soft agar plates, the base medium plates are heated at 45°C. Soft agar overlay tubes are prepared with a 0.2% agar medium and 5 mM $MgSO_4$ at a volume of 1.5 mL, adjusting the concentrations and allowing for the addition of 1 mL of the sample. These tubes are maintained at 50°C, inoculated with 0.1 mL of the target bacteria and 1 mL of the suspected bacteriophage solution, mixed, and evenly poured onto the base medium plates. These plates are then incubated overnight and examined for plaques, which are clear zones created by the replication of the bacteriophage and consequent killing of the bacteria.

This method of bacteriophage isolation has worked very well for us. There are a few things we have learned using this procedure that might be helpful. It is important that the enrichment cultures are diluted over a wide range. We routinely make 10-fold dilutions of our enrichment tubes out to 10^{-8}, and plate the four highest dilutions. A control plate should be made that contains only the bacteria. After the incubation period, if plaques (clear zones) are not observed, but the plates look grainy or different than the control plate, the plates may contain bacteriophage at high concentrations such that distinct plaque formation is not observed. We characterize these plates as grainy or diffuse plates. If this is observed, we go back to the original environmental enrichment culture and further dilute the sample, plating these more dilute samples. If the original plates look identical to the control plates at the original dilutions plated, the enrichment culture may contain bacteriophage at titers below those that were originally plated. In this case, we go back to the original enrichment culture and plate the lower dilutions.

Once clear distinct plaques can be seen, we isolate the bacteriophage by stabbing the plaque with an inoculation needle, rinse the needle in phosphate buffered saline, make dilutions of this rinse, and plate the dilutions using the soft agar overlay technique. Bacteriophage are purified this way by repeating this procedure for at least three passages.

After bacteriophage are isolated, they can be stored for short periods of time refrigerated or for longer periods of time frozen. However, some bacteriophage that we have isolated lose activity when stored in phosphate-buffered saline at refrigeration temperatures.

Once bacteriophage have been isolated it is usually necessary to amplify the bacteriophage to obtain high titers and larger volumes of the bacteriophage to be used in a research program. We have used two methods of bacteriophage amplification, and the one that works best for us is amplifying bacteriophage in liquid cultures. We initially amplified our bacteriophage by taking a high-titer bacteriophage solution, which gave us diffuse plates when plated using the soft agar overlay technique. These diffuse plates were flooded with 10–12 mL phosphate buffered saline. The plates were allowed to stand for 2–4 hours with periodic gentle agitation. The liquid was then removed, centrifuged, and filter sterilized. The resulting liquid obtained from these flooded plates is a good storage liquid to maintain bacteriophage at refrigeration temperatures, and our bacteriophage titers can be maintained in this liquid for a couple of months. However, this procedure is cumbersome, and we have abandoned it, replacing it with amplification performed in liquid cultures. Basically, we inoculate a broth culture that is incubated to mid-log phase growth, which is then inoculated with the bacteriophage that we want to amplify and incubated overnight. As in any fermentation-type procedure, inoculation levels, incubation conditions, and incubation duration must be optimized to maximize bacteriophage amplification. After the incubation period is complete, the broth is centrifuged and filter sterilized. Bacteriophage activity in this broth is stable at refrigerated temperatures for a couple of months, at least for the bacteriophage that we have isolated to date.

Bacteriophage Potential and Limitations

Bacteriophage kill bacteria; therefore, their potential to be used as therapeutic agents and to improve food safety appears to be obvious. Why this potential of bacteriophage has not been realized is not at all clear to these authors.

Although polyvalent bacteriophage can be found that infect bacteria across genera, in general, bacteriophage are very specific, with activity to strains within species of bacteria. This is often cited for the inconsistent results obtained by early researchers as they attempted to develop bacteriophage therapeutics. In comparison, most antibiotics have broad specificity. This allows them to be used without isolating the causative agent. Although in theory, the causative agent of a disease should be isolated and antibiotic sensitivities conducted before antibiotic therapy, in practice, this is only done when antibiotic therapy has not been effective.

Early researchers developing bacteriophage therapeutics may not have had a full appreciation for the specificity of bacteriophage and did not realize that the disease for which they were trying to develop a bacteriophage treatment could be caused by many serotypes or, more appropriately, phage types of the species of bacteria targeted. Today we understand that not all members of a species of bacteria are the same, and therefore, treatment of a disease with bacteriophage requires a cocktail of bacteriophage effective against most if not all the phage types of bacteria that are known to cause the disease. Alternatively, the disease agent may be isolated, and the bacteriophage treatment can then be customized to the specific bacteria causing the disease on a case-by-case basis. If a disease is caused by a few phage types of the bacteria, it would seem to be practical to develop a bacteriophage therapeutic product that contained a number of bacteriophage that were lytic to all of the known strains of the bacteria that cause the disease, a cocktail of bacteriophage. This would decrease the need to isolate the bacteria from the infection and perform bacteriophage-sensitivity assays before treatment. If the number of phage types that cause a disease is large, a therapeutic bacteriophage cocktail may not be feasible. However, a library of bacteriophage to all of the known pathogenic strains of bacteria could be established and treatment could be customized to the phage type of the bacteria involved in the infection. This would require isolation of the bacteria and phage-sensitivity assays performed before treatment.

Regulatory or product approval is also a major consideration in the development of bacteriophage therapeutic products. If a single bacteriophage product can not be developed that would effectively treat all of the pathogenic strains of the bacteria of interest, then it might be necessary to have each bacteriophage in the library approved for use, which would be costly and cumbersome. Because not many bacteriophage therapeutic products have gone through the regulatory process, it is uncertain how bacteriophage therapeutic products would be regulated. Certainly this issue would be greatly simplified if polyvalent bacteriophage could be discovered with efficacy to treat most if not all of the high-profile pathogenic strains of bacteria.

The issue of bacteriophage specificity does present challenges to the development of effective bacteriophage products. However, specificity of bacteriophage is also an advantage of bacteriophage over antibiotics. Because antibiotics have broad specificity, prolonged use of antibiotics can disrupt the natural bacterial flora, with serious health consequences. Bacteriophage specifically target the pathogenic bacteria of interest without harm to the beneficial bacteria.

Bacteria develop resistance to bacteriophage just as they do to antibiotics. However, resistance to bacteriophage does not occur at a faster rate than resistance to antibiotics (Carlton 1999). The development of effective bacteriophage products does have an advantage over antibiotics with respect to the development of bacterial resistance. It is possible, and in fact relatively easy, to develop bacteriophage therapeutics that may prevent or delay bacterial resistance. This can be done by isolating the resistant bacteria to a single bacteriophage and finding bacteriophage that infect the resistant bacteria. This can be done repeatedly, with the resulting bacteriophage therapeutic product a cocktail of bacteriophage developed to anticipate and be effective against any resistant strains that emerge.

Bacteriophage can act as a vehicle for bacterial genes in a process known as transduction. General transduction can occur at random when bacterial genes are packaged inside virulent (lytic) bacteriophage. After a bacterium is infected with a bacteriophage, the bacteria's DNA is broken into pieces by the bacteriophage. These pieces of the bacteria's DNA can then be mistakenly packaged within a bacteriophage. If the pieces of the bacteria's DNA are functional genes and the lytic bacteriophage is able to infect, but not kill, a subsequent bacterial cell that it infects, then the second bacterium can be transformed by the genetic material delivered to it by the bacteriophage. In this way bacteriophage may deliver virulence or antibiotic resistance determinants to bacteria it infects, transforming the infected bacterium from a non-virulent to a virulent strain, or from a cell sensitive to antibiotics to a cell that is resistant to antibiotics. An additional type of transduction is called specialized transduction. Specialized transduction occurs when temperate (lysogenic) bacteriophage are excised from the bacteria's genome and adjacent bacterial genes on either side of the bacteriophage may be attached to the bacteriophage genome, replicated, and packaged into the intact bacteriophage. This bacteriophage will deliver the genes to the bacterium it infects, which then can transform the infected bacterium.

This ability of bacteriophage to transport bacterial genes from one bacterium to another and consequently transform bacteria is a major issue in the development of any bacteriophage product. Because virulent (lytic) as opposed to temperate (lysogenic) bacteriophage would appear to be the likely bacteriophage to be used in bacteriophage products, generalized transduction would be the major concern with these products. It would be necessary to ensure that bacteriophage therapeutics would either not be capable of or unlikely to be able to transform bacteria with considerable concern over virulent and antibiotic-resistance traits. This could be accomplished through the genetic characterization of bacteriophage products to ensure that they do not contain bacterial genes. Another, less analytical, approach would be to amplify bacteriophage to be used in bacteriophage products in strains of bacteria that were non-virulent and antibiotic sensitive. This would increase the chances that virulence traits and antibiotic-resistance genes would not be transferred by bacteriophage because they were not traits of the strain used to develop the bacteriophage product. This would require the ability to isolate bacteriophage that could be amplified in non-pathogenic, antibiotic-sensitive strains of bacteria and still be effective against the virulent bacteria to be targeted by the bacteriophage product. Bielke et al. (2003) has demonstrated that this is a viable approach in bacteriophage used to control *Salmonella*.

Bacteriophage are diverse with respect to their environmental fragility, but in general, they may be inactivated by the digestive process. This presents a problem with the development of bacteriophage products to be used to target enteric bacteria of food safety concern in preharvest animals. There are at least three approaches that can be used to solve this problem. It may be possible to isolate bacteriophage that are resistant to the digestive processes. This can be accomplished by giving bacteriophage orally and by reisolating the bacteriophage from excreta. If this method of bacteriophage isolation is repeated, it may be possible to isolate bacteriophage that can survive the digestive processes to reach its targeted enteric bacteria. Another approach would be to coat the bacteriophage, allowing them to reach their enteric targets. The third approach to deliver bacteriophage to targeted enteric bacteria would be to precondition the animal through the use of antacids before the oral administration of bacteriophage.

A single bacteriophage infects a targeted bacterium, amplifies within the bacterium, and thus, when the bacterium is lysed, it may release 50–200 active bacteriophage, which in turn can infect 50–200 more bacteria. Because of the exponential growth of bacteriophage, it would be anticipated that bacteriophage used as therapeutics would continue to amplify until all the targeted bacteria were killed, which would cause the phage to stop producing, thus creating a self–regulating system. This undoubtedly has contributed to the success of bacteriophage therapeutics demonstrated over the years. However, the success of bacterio-phage therapeutics is titer dependent, and normally 10^9 to 10^{12} bacteriophage per mL are administered to obtain maximum therapeutic value. It is not understood why relatively high *in vivo* levels of bacteriophage are required to demonstrate disease treatment efficacy in light of the exponential growth of bacteriophage. There are many possible explanations that include sequestering of the bacteria within tissues, thus limiting the chance meeting with the targeted bacteria; nonspecific protein binding of bacteria or bacteriophage that might inhibit bacteriophage adsorption to the bacteria; inactivation of bacteriophage by the ani-mal's immune system; and the physical dynamics of the physiology of an animal may reduce the opportunity for the chance meeting of bacteria with bacteriophage.

Another issue with bacteriophage is the efficacy of repeated use of bacteriophage to treat a chronic disease. It would be anticipated that an immunological response would occur directed at the systemic administration of bacteriophage. If an immunological response to bacteriophage inhibited the virus's activity, it could diminish the efficacy of bacteriophage therapy with repeated treatment with the same bacteriophage. However, there is little pub-lished research on this potential limitation of bacteriophage therapy.

Bacteriophage represent a natural way to control bacteria and can be considered as inte-grated pest management on a microscale. It would be anticipated that this natural process would be more acceptable to the general public than use of chemicals to control bacteria. However, the fact that bacteriophage are natural viruses does present issues with protection of capital investment in the development of bacteriophage therapeutics through the patent process. One potential of bacteriophage lies in the relatively simple genome. Genetic engi-neering of bacteriophage to improve their therapeutic value would appear to be a real oppor-tunity (Bernhardt et al. 2001). One possibility is to use bacteriophage as a vector to target specific pathogenic bacteria delivering the genetic information that would kill these bacte-ria without bacteriophage replication. This would solve several problems with bacterio-phage. It would greatly reduce the risk of bacteriophage transduction and the consequent transformation of bacteria and would allow patent protection of the genetically modified bacteriophage. However, it would be anticipated that genetically modifying bacteriophage would reduce the general public acceptance of this technology.

Conclusions

It has been known since the early 1900s that bacteriophage kill bacteria. The potential of bacteriophage to treat diseases and control foodborne pathogenic bacteria appears obvious. With the continued emergence of pathogenic bacteria resistant to antibiotics, and restric-tions on the use of antibiotics in animal production systems, there is renewed interest in the use of bacteriophage as an alternative to antibiotics. There are a number of issues that must be solved before bacteriophage products can be practical alternatives to antibiotics. Some of the issues addressed in this chapter include the specificity of bacteriophage, emergence of resistance to bacteriophage, transduction of bacterial genes and consequent transformation of infected bacteria, uncertainty over the efficacy of repeated bacteriophage therapy, unknown regulatory requirements of bacteriophage products, and how to protect proprietary rights to bacteriophage products. The advantages of bacteriophage compared with antibiotics include their specificity, exponential growth, cost of development, lack of effects on plant or animal cells, ease in solving the emergence of resistance, potential for genetic modification to improve bacteriophage efficacy, and ubiquitous and plentiful nature. There is a real need for concerted research to determine how we can take nature's own way to control bacteria and make it a tool to use in our continuing efforts to control pathogenic bacteria. The concept that bacteriophage can be used to control bacteria has been examined for over 90 years. The real challenge is in developing practical bacteriophage products to control pathogenic bacteria of interest in agriculture products both pre- and postharvest. Bacteriophage will not replace antibiotics, but they have potential to provide a viable alternative to antibiotics.

Acknowledgements

We acknowledge the U.S. Department of Agriculture, Agricultural Research Service, for its support of this research program. W.E.H. would also like to thank all the coauthors of this chapter, with special thanks to his wife and coauthor, Gerry Huff, who has always provided both support of my research program and uninhibited criticism, which is invaluable.

References

Adams, M. H. (Ed.). 1966. Bacteriophages, 2nd ed. New York: Interscience Publishers.

Alisky, J., Iczkowski, K., Rapoport, A. and Troitsky, N. 1998. Bacteriophages show promise as antimicrobial agents. *Journal of Infection* **36**:5–15.

Barrow, P., Lovell, M. and Berchieri Jr., A. 1998. Use of lytic bacteriophage for control of experimental *Escherichia coli* septicemia and meningitis in chickens and calves. *Clinical and Diagnostic Laboratory Immunology* **5**:294–298.

Barrow, P. A. and Soothill, J. S. 1997. Bacteriophage therapy and prophylaxis: rediscovery and renewed assessment of potential. *Trends in Microbiology* **5**:268–271.

Bernhardt, T. G., Wang, I.-N., Struck, D. K. and Young, R. 2001. A protein antibiotic in the phage Qβ virion: diversity in lysis targets. *Science* **292**:2326–2329.

Bielke, L. R., Higgins, S. E., Guenther, K. L., Nava, G. M., Tellez, G. I., Donoghue, D. J., Donoghue, A. M. and Hargis, B. M. 2003. Determination of *Salmonella* host range of selected bacteriophages which exhibit increased host specificity. *Poultry Science* **82(Suppl. 1)**:42.

Biswas, B., Adhya, S., Washart, P., Paul, B., Trostel, A. N., Powell, B., Carlton, R. and Merril, C. R. 2002. Bacteriophage therapy rescues mice bacteremic from a clinical isolate of vancomycin-resistant *Enterococcus faecium*. *Infection and Immunity* **70**:204–210.

Carlton, R. M. 1999. Phage therapy: past history and future prospects. *Archivum Immunologiae et Therapiae Experimentalis* **47**:267–274.

Cerveny, K. E., DePaola, A., Duckworth, D. H. and Gulig, P. A. 2002. Phage therapy of local and systemic disease caused by *Vibrio vulnificus* in iron-dextran-treated mice. *Infection and Immunity* **70**:6251–6262.

d'Herelle, F. 1917. Sur un microbe invisible antagonists des *bacilles dysenteriques*. *Comptes Rendus de L'Academie des Sciences de Paris* **165**:373–375.

Duckworth, D. H. and Gulig, P. A. 2002. Bacteriophages potential treatment for bacterial infections. *Biodrugs: Clinical Immunotherapies, Biopharmaceuticals and Gene Therapy* **16**:57–62.

Goode, D., Allen, V. M. and Barrow, P. A. 2003. Reduction of experimental *Salmonella* and *Campylobacter* contamination of chicken skin by application of lytic bacteriophage. *Applied and Environmental Microbiology* **69**:5032–5036.

Higgins, J. P. 2002. Use of specific bacteriophage treatment to reduce *Salmonella* in poultry products. Master's thesis, University of Arkansas, Fayetteville.

Higgins, J. P., Higgins, S. E., Guenther, K. L., Huff, W. E. and Hargis, B. M. 2002. Evaluation of bacteriophage treatment as a method to reduce culturable *Salmonella* in poultry carcass rinse water. *Poultry Science* **81 (Suppl. 1)**:130.

Huff, W. E., Huff, G. R., Rath, N. C., Balog, J. M. and Donoghue, A. M. 2002a. Prevention of *Escherichia coli* infection in broiler chickens with a bacteriophage aerosol spray. *Poultry Science* **81**:1486–1491.

Huff, W. E., Huff, G. R., Rath, N. C., Balog, J. M. and Donoghue, A. M. 2003. Evaluation of aerosol spray and intramuscular injections of bacteriophage to treat an *Escherichia coli* respiratory infection. *Poultry Science* **82**:1108–1112.

Huff, W. E., Huff, G. R., Rath, N. C., Balog, J. M., Xie, H., Moore Jr., P. A. and Donoghue, A. M. 2002b. Prevention of *Escherichia coli* respiratory infection in broiler chickens with bacteriophage (SPR02). *Poultry Science* **81**:437–441.

Leverentz, B., Conway, W. S., Camp, M. J., Janisiewicz, W. J., Abuladze, T., Yang, M., Saftner, R. and Sulakvelidze, A. 2003. Biocontrol of *Listeria moncytogenes* on fresh-cut produce by treatment with lytic bacteriophage and a bacteriocin. *Applied and Environmental Microbiology* **69**:4519–4526.

Matsuzaki, S., Yasuda, M., Nishikawa, H., Kuroda, M., Ujihara, T., Shuin, T., Shen, Y., Jin, Z., Fujimoto, S., Nasimuzzaman, M. D., Wakiguchi, H., Sugihara, S., Sugiura, T., Koda, S., Muraoka, A. and Imai, S. 2003. Experimental protection of mice against lethal *Staphylococcus aureus* infection by novel bacteriophage φMR11. *The Journal of Infectious Diseases* **187**:613–624.

Park, S. C., Shimamura, I., Fukunaga, M., Mori, K. and Nakai, T. 2000. Isolation of bacteriophages specific to a fish pathogen, *Pseudomonas plecoglossicida*, as a candidate for disease control. *Applied and Environmental Microbiology* **66**:1416–1422.

Ślopek, S., Durlakowa, I., Weber-Dąbrowska, B., Dąbrowski, M. and Kucharewicz-Krukowska, A. 1984. Results of bacteriophage treatment of suppurative bacterial infections: III. Detailed evaluation of the results obtained in further 150 cases. *Archivum Immunologiae et Therapiae Experimentalis* **32**:317–335.

Ślopek, S., Durlakowa, I., Weber-Dąbrowska, B., Dąbrowski, M. and Kucharewicz-Krukowska, A. 1987. Results of bacteriophage treatment of suppurative bacterial infections in the years 1981–1986. *Archivum Immunologiae et Therapiae Experimentalis* **35**:569–583.

Ślopek, S., Durlakowa, I., Weber-Dąbrowska, B., Kucharewicz-Krukowska, A., Dąbrowski, M. and Bisikiewicz, R. 1981. Results of bacteriophage treatment of suppurative bacterial infections: II. Detailed evaluation of the results. *Archivum Immunologiae et Therapiae Experimentalis* **31**:293–327.

Ślopek, S., Kucharewicz-Krukowska, A., Weber-Dąbrowska, B. and Dąbrowski, M. 1985. Results of bactiophage treatment of suppurative bacterial infections. VI. Analysis of treatment of suppurative staphylococcal infections. *Archivum Immunologiae et Therapiae Experimentalis* **33**:261–273.

Smith, H. W. and Huggins, M. B. 1982. Successful treatment of experimental *Escherichia coli* infections in mice using phage: Its general superiority over antibiotics. *Journal of General Microbiology* **128**:307–318.

Smith, H. W. and Huggins, M. B. 1983. Effectiveness of phages in treating experimental *Escherichia coli* diarrhoea in calves, piglets and lambs. *Journal of General Microbiology* **129**:2659–2675.

Smith, H. W., Huggins, M. B. and Shaw, K. M. 1987. The control of experimental *Escherichia coli* diarrhoea in calves by means of bacteriophages. *Journal of General Microbiology* **133**:1111–1126.

Soothill, J. S. 1992. Treatment of experimental infections of mice with bacteriophages. *Journal of Medical Microbiology* **37**:258–261.

Summers, W. C. 2001. Bacteriophage therapy. *Annual Review of Microbiology* **55**:437–451.

Twort, F. W. 1915. An investigation on the nature of ultramicroscopic viruses. *Lancet* **2**:1241–1243.

Weber-Dąbrowska, B., Dąbrowski, M. and Ślopek, S. 1987. Studies on bacteriophage penetration in patients subjected to phage therapy. *Archivum Immunologiae et Therapiae Experimentalis* **35**:563–568.

28 Food Irradiation

Suresh D. Pillai

Introduction

Foodborne infections cause over 76 million cases of human illness in the United States alone. This translates to an estimated $7 billion annual impact on the U.S. economy. It is also estimated that over 5,000 people die each year in the United States from foodborne illnesses. As a point of reference, the number of foodborne associated deaths in the United States per year exceeds the number of those who died as a result of the terrorist activities on September 11, 2001. Even though major advancements have taken place to improve food safety across the proverbial "farm-to-fork" continuum, the current level of human illness and the resulting economic effect from pathogen-contaminated foods is unacceptable for a country such as the United States. Obviously, the effect of foodborne illnesses in developing and underdeveloped regions of the world, if tallied, would be at near catastrophic levels.

Foodborne illness is preventable. It can be prevented by improved food production methods, improved food processing technologies, and improved food preparation and consumption habits within the home. A number of food processing technologies have been developed and employed in recent years. However, none of the technologies has had the same level of promise and, unfortunately, the level of criticism as food irradiation. The aim of this chapter is to highlight the salient features of the technology, identify some of the current technical challenges in using this technology, and attempt to provide a "science-based roadmap" into the future to help harness this technology. The discussion does focus to some extent on electron beam (E-beam) irradiation primarily because of its potential to gain widespread acceptance. This chapter does not attempt to discuss the politics or the economics of food irradiation.

Food Irradiation Technology

Food irradiation involves the use of ionizing radiation to destroy pests and pathogens on food and agricultural products. There is probably no other food processing technology that has been so extensively researched as food irradiation. The finding that ionizing radiation could destroy bacteria occurred in 1904, and the technology was evaluated as early as 1921 for destroying trichinae in pork (Josephson 1983). Today, in 2003, we have federally approved irradiation protocols for a variety of food products including uncooked meat and poultry products. Federal approvals permitting the use of irradiation for ready-to-eat (RTE) products are anticipated soon.

Ionizing radiation, as the name implies, can be defined as radiation that has enough energy to remove electrons from atoms, thereby leading to the formation of ions. There are different types of ionizing radiation such as X-rays, gamma-rays, and beta-rays. The irradiation sources that have been internationally approved for food processing are gamma-rays produced from the radioisotopes cobalt-60 (1.17 and 1.33 MeV) and cesium-137

(0.662 MeV), machine-generated electron-beams (maximum energy 10 MeV), and X-rays (maximum energy 5 MeV; Codex Alimentarius Commission 1984). The facilities that provide ionizing radiation share common features such as the irradiation source or sources, shielding, product handling systems, and safety/control systems (Olson 1995). However, there are unique aspects to the construction and the engineering specifications of the facilities themselves, depending on the type of ionizing radiation that is produced. For example, gamma facilities such as those employing cobalt-60 have specific characteristics to protect workers and the surrounding environment from the radioactive isotopes and for storing the isotope material under water when not in use. Electron (E-beam; beta-ray) and X-ray facilities do not have the same shielding requirements as cobalt-60 facilities. However, they still contain a significant amount of electrical circuitry, cooling systems, and ozone attenuation capabilities. It must be kept in mind that irradiation facilities basically "sell" irradiation doses. They do not sell or guarantee pathogen kill. It is incumbent on the customer to independently develop methods and verify and validate the necessary irradiation doses that can achieve the desired levels of pathogen inactivation for their respective product.

Irrespective of the source and the facility providing the irradiation, all types of ionizing radiation destroy biological entities by essentially the same process. Ionizing radiation causes "breaks" in RNA or the DNA double helix. It is believed that ionizing radiation disrupts normal cellular activity by damaging the nucleic acids by "direct" and "indirect" effects. Specifically, single- or double-stranded breaks occur on the DNA or RNA. Ionizing radiation does not discriminate between pathogens and non-pathogens, and so both the indigenous normal flora and pathogens on a food product can be inactivated. The evidence that RNA is also a target for ionizing radiation is the lethal effects observed when RNA-containing viruses are subjected to these types of irradiations.

The accelerated electrons during E-beam irradiation damage the nucleic acids by direct "hits." In addition, damage to the nucleic acids can also occur when the radiation ionizes an adjacent molecule, which in turn reacts with the genetic material. Water is very often the adjacent molecule that ends up producing a lethal product. Ionizing radiation causes water molecules to lose an electron, producing H_2O^+ and e^-. These products react with other water molecules to produce a number of compounds including hydrogen and hydroxyl radicals, molecular hydrogen, oxygen, and hydrogen peroxide (Arena 1971). These products in turn react with other water molecules, with nucleic acids, and with other biologically sensitive molecules. The most reactive by-products arising from the hydrolysis of water are the hydroxyl radicals (OH•) and hydrogen peroxide (H_2O_2). These molecules are known to react with nucleic acids and the chemical bonds that bind one nucleic acid to another in a single strand as well as with the bonds that link the adjacent base pair to an opposite strand. Because the precise locations of where the ionization of water occurs and where the direct DNA hits occur are random, the location of DNA damage from direct and indirect effects is also random. The indirect effects can also cause single- and double-stranded breaks between the nucleic acid molecules. Studies have shown that direct-ionization strand breakages are not related to the nucleic acid sequence (Razskazovskiy et al. 2003). Even though biological systems do have a capacity to repair both single- and double-stranded breaks of the DNA backbone (Bartek and Lukas 2003), the damage occurring from ionizing radiation is probably so extensive that bacterial repair of radiation damage at doses used in food irradiation is nearly impossible.

Microbial cells, depending on their physiological state, can exhibit varying resistance to ionizing radiation. Bacterial cells in logarithmic growth phase have multiple copies of their genomes in one cell. This is an important point to consider, because the presence or absence of multiple genome copies and the physiological status of the cell can dictate whether the cell can ultimately reproduce; that is, the probability of "survivors" in a pathogen population exposed to ionizing irradiation. Similarly, the formation of bacterial endospores can enhance the radiation resistance of these organisms.

Irradiation Sensitivity of Specific Pathogens on Selected Foods

Red Meat

Time and temperature abuse in conjunction with under-cooking of products are the primary causes of human infections. The pathogens that are of primary concern in meat and meat products are the toxigenic *Escherichia coli*, *Salmonella* spp., and *Yersinia* spp., as well as parasites such as *Toxoplasma gondii*, *Taenia* spp., and *Trichinella spiralis*. Red meat can be irradiated at low doses to eliminate bacterial pathogens and parasites. The D_{10} values for selected bacterial pathogens in various meats are shown in Table 28.1. The D_{10} value is the irradiation dose required to destroy 90% of the organisms. Mathematically, it is the reciprocal of the slope of the inactivation curve and is a measure of the inactivation rate. The Food and Drug Administration (FDA) has already approved the use of irradiation for microbial control on fresh (chilled) and frozen meat and poultry products (Table 28.2).

Poultry, Turkey, and Egg Products

Studies show that a very high percentage of fresh and frozen poultry products sold at retail harbor *Campylobacter jejuni*. Both *Salmonella* and *Campylobacter* are the primary pathogens of concern in fresh and frozen poultry either because of under-cooking or cross-contamination events. Irradiation is probably the only physical process for removing

Table 28.1 D_{10} values for selected bacterial pathogens in meat products under specified temperature conditions (adapted from Molins 2001).

Product	Temperature (°C)	Organism	D_{10} value	References
Ground beef patties	5	*Escherichia coli* O157:H7	0.27–0.38	Lopez-Gonzalez et al. 1999
Ground beef patties	−15	*E. coli* O157:H7	0.32–0.63	Lopez-Gonzalez et al. 1999
Ground beef	3	*Salmonella enteritidis*	0.55–0.78	Tarkowski et al. 1984
Beef	5	*E. coli* O157:H7	0.30	Thayer et al. 1995
Beef	3	*Yersinia enterocolitica*	0.10–0.21	Tarkowski et al. 1984
Beef	5	*Staphylococcus aureus*	0.46	Thayer et al. 1995
Ground beef	5	*Campylobacter jejuni*	0.16	Lambert and Maxcy 1984
Deboned meat	5	*Bacillus cereus* spores	2.56	Thayer and Boyd 1994
Beef	5	*Listeria monocytogenes*	0.45	Thayer et al. 1995

Table 28.2 Food and Drug Administration–approved irradiation doses for meat and poultry.

Product	Maximum dose (kGy)
Uncooked chilled meat	4.5
Uncooked frozen meat	7.0
Uncooked chilled poultry meat	3.0
Uncooked frozen poultry meat	3.0

pathogens from poultry meat other than heat treatment. Table 28.3 is a listing of the D_{10} values of selected pathogens on poultry meat under fresh and frozen conditions. The FDA has already approved the use of irradiation to destroy pathogens in chilled and frozen raw poultry products.

Seafood

Irradiation doses that have been approved for poultry can achieve at least a 3-log reduction of pathogens that are relevant to seafood such as *Vibrio parahaemolyticus* (see Chapter 21) and *Aeromonas hydrophila*. There is significant interest in developing E-beam irradiation protocols for shellfish such as oysters to eliminate pathogens such as *Vibrio vulnificus*. However, significant challenges still remain in developing appropriate dosimetry of oysters because of the "half-shell" and "full-shell" configuration at the retail level. Because irradiation can interact with fats in foods, sensory characteristics of the irradiated seafood is also another area that needs to be thoroughly investigated. Table 28.4 is a listing of the D_{10} values of selected pathogens in prawns.

Table 28.3 D_{10} values for specific pathogens on poultry, turkey, and egg products (adapted from Molins 2001).

Product	Temperature (°C)	Organism	D_{10} value (kGy)	References
Turkey breast meat	5	*Staphylococcus aureus*	0.45	Thayer et al. 1995
Ground turkey	5	*Campylobacter jejuni*	0.19	Lambert and Maxcy 1984
Ground turkey	30	*Campylobacter jejuni*	0.16	Lambert and Maxcy 1984
Ground turkey	−30	*Campylobacter jejuni*	0.29	Lambert and Maxcy 1984
Turkey breast meat	5	*Salmonella* spp.	0.71	Thayer et al. 1995
Poultry (air packed)	0	*Salmonella Heidelberg*	0.24	Licciardello et al. 1970
Poultry (vacuum packed)	0	*Salmonella Heidelberg*	0.39	Licciardello et al. 1970
Egg powder	5	*Salmonella enteritidis*	0.6	Matic et al. 1990

Table 28.4 D_{10} values of selected enteric bacteria in prawns (adapted from Murano 1995).

Pathogen	D_{10} value (kGy)
Vibrio spp.	0.11
Staphylococcus sp.	0.29
Salmonella spp.	0.48

RTE Foods

The presence of enteric pathogens in RTE products have resulted in product recalls totaling millions of dollars. A number of disease outbreaks have been attributed to the direct ingestion of pathogen-contaminated RTE foods. The key pathogens that are of concern in RTE foods are *Listeria monocytogenes* and *Salmonella* spp. A cumulative 10-year (1990–1999) prevalence of *L. monocytogenes* and *Salmonella* spp. in RTE foods was conducted by the USDA-FSIS (Table 28.5). None of the samples tested during the survey were positive for *E. coli* O157:H7 or staphylococcal enterotoxins. A final ruling by the FDA on the application to use ionizing radiation for RTE foods is anticipated.

Fruits, Vegetables, Herbs, and Sprouts

The early work on the use of irradiation on fruits and vegetables was to extend shelf life and as a quarantine treatment for export/import purposes. Even though quarantine-based irradiation is widely used (e.g., to export fruits from Hawaii onto the United States mainland), there is considerable interest at this time in using ionizing radiation to eliminate human enteric pathogens on these products. Fruits and vegetables become contaminated with enteric pathogens at various stages of production, beginning at the farm-level (Endley et al. 2003). Recent studies have also documented the internalization of enteric pathogens into fruits and vegetables. There have been foodborne outbreaks associated with the consumption of contaminated apples, mangoes, and tomatoes. Under this scenario, it is evident that conventional intervention strategies will not be able to address internalized pathogens. Ionizing radiation is probably one of the few technologies that can be used to eliminate pathogens in fruits and vegetables once they are internalized. However, significant research

Table 28.5 Cumulative prevalence of *Listeria monocytogenes* and *Salmonella* spp. in selected ready-to-eat foods (1990–1999; Levine et al. 2001).

Product	*L. monocytogenes*	*Salmonella* spp.
Jerky	0.52%	0.31%
Cooked, uncured poultry products	2.12%	0.10%
Large-diameter cooked sausages	1.31%	0.07%
Small-diameter cooked sausages	3.56%	0.20%
Cooked beef/roast beef/cooked corned beef	3.09%	0.22%
Salads/spreads/pates	3.03%	0.05%
Sliced ham/luncheon meat	5.16%	0.22%

related to preserving the sensory attributes of fruits and vegetables under such treatment conditions still remains to be done. Herbs and sprouts such as cilantro, parsley, and bean sprouts, which are very popular, are extremely vulnerable to enteric organisms such as *E. coli* O157:H7. Bari et al. (2003) have reported that an irradiation dose of 2.0 kGy in combination with dry heat eliminated *E. coli* O157:H7 completely from alfalfa and mung bean seeds, whereas a 2.5-kGy dose of irradiation was required to eliminate the pathogen completely from radish seeds. While studying broccoli seeds and sprouts, Rajkowski et al. (2003) reported that the germination percentage decreased at a dose level of 4 kGy, whereas the yield ratio, sprout length, and thickness decreased at the 2-kGy dose level. The radiation doses required to inactivate *Salmonella* spp. and strains of *E. coli* O157:H7 were different than previously reported values and depended on whether the strains were originally isolated from vegetables. The D_{10} values for the vegetable and non-vegetable *Salmonella* spp. isolates were 1.10 kGy and 0.74 kGy, respectively, whereas the D_{10} values for the vegetable and non-vegetable strains of *E. coli* O157:H7 were 1.11 kGy and 1.43 kGy, respectively. Thayer et al. (2003) reported that 2 kGy is capable of eliminating *Salmonella* spp. and *E. coli* O157:H7 while still retaining satisfactory yields of alfalfa sprouts.

Detection of Irradiated Foods

The detection of irradiated foods is an important issue both from the regulatory and consumer points of view. For many years, the detection of irradiated foods was not a high priority because it was assumed that all irradiation would be performed under regulated and licensed facilities and that appropriate documentation would follow each and every irradiated product. However, current regulations, especially in Europe, have driven the need for specific "tools" to detect irradiated products. Some key criteria for irradiation detectors include the ability to "discriminate" irradiated from non-irradiated foods; they should be specific to irradiation treatments, dose-dependant, and accurate. Some of the practical criteria include cost, simplicity, speed, and the ability to detect nondestructively (Stewart 2001). The currently available methods are based on physical, chemical, biological, and microbiological changes. The physical methods include electron spin resonance (ESR) spectroscopy (Raffi and Stocker 1996), luminescence measurement (Heide et al. 1989), viscosity measurement (Hayashi et al. 1996a), and electrical impedance measurement (Hayashi et al. 1996b). The ESR spectroscopy detects species with unpaired electrons or free radicals. Studies have been conducted demonstrating that ESR can be used to detect a variety of irradiated foods, especially those containing cellulose and crystalline sugars. The technique is specific, rapid, simple, and nondestructive. Instrumentation costs are still a major hurdle. Even though chemiluminescence and thermoluminescence have both been studied as tools for detecting irradiated foods, thermoluminescence has become a preferred method. This method is based on the principle that electrons in the excited state return to the ground state when thermally stimulated. There is a requirement that food samples are carefully prepared for thermoluminescence analysis. There have been major improvements in this technology, especially in the use of light rather than heat to release radiation-induced trapped energy. The development of photostimulated luminescence has matured and is considered a validated method in Europe. Viscosity measurement-based irradiation detection is based on the principle that molecular structures of polymers are changed under irradiation conditions and that, as a result, there is a change in viscosity. In addition to the physical methods, there are some reliable chemical methods for detecting irradiated products such as measurements of hydrocarbons (Schulzki et al. 1995) and 2-alkylcyclobutanones (Lembke et al. 1995). DNA-based detection methods include the DNA "Comet Assay" (Cerda 1998), agarose electrophoresis of mitochondrial DNA (Marchioni et al. 1992), and immunologic detection of modified DNA bases (Deeble et al. 1994).

Current Technical Challenges

Low-Dose Irradiation and Dosimetry

Under E-beam irradiation conditions, the amount of dose delivered to a product is controlled by adjusting the amount of time the product is in the "beam." This is controlled by adjusting the speed of the conveyor belt that transports the product through the beam. There are practical limits to the conveyor belt speeds for achieving low doses of irradiation (L. Braby, personal communication). Achieving low doses of irradiation will become very important when RTE products and multicomponent foods are irradiated to eliminate any trace of microbial contamination and retain sensory qualities.

In conjunction with low-dose irradiation capabilities, dosimeters that can measure such low doses are urgently required. Reliable dosimeters that can measure from 10 to 1,000 Gy are currently a serious technical limitation (Braby 2003). Dosimetry at these levels is critically important for quarantine treatments and for RTE foods. A scenario can be envisioned in which a particular food product may only need to be irradiated at low doses to guarantee elimination of potential pathogens and to retain sensory attributes. Without an accurate means of measuring and validating low doses, definitive conclusions about the received dose and the thresholds causing sensory changes cannot be made.

Detection of Irradiated Foods

The ability to detect irradiated foods will be important not only for regulatory purposes but also as an assurance of the irradiation process itself. Though there are promising methods such as ESR spectroscopy, photostimulated luminescence, and 2-alkylcyclobutanone-based methods, the ability to detect irradiation especially at low levels is critically important for the reasons mentioned earlier.

Future Research Directions

Inactivation Kinetics of Pathogens in Naturally Contaminated Foods

A number of studies have been published documenting the D_{10} values for specific pathogens in a variety of foods. Many of these studies have used pathogenic strains that have been previously implicated in disease outbreaks. This is important because studies have shown that there are intrinsic differences in the resistance of strains depending on the environment from which they were isolated. In addition, irradiation studies to determine the D_{10} values should be conducted using food products that have been naturally contaminated rather than relying on artificial contamination in the laboratory. This is critically important because the association of pathogens to the food matrix, and the potential "shielding" of the organisms by the food components, cannot be accurately simulated in the laboratory using normal inoculation methods. The food manufacturing facilities should be the source of such samples for validation purposes. It is incumbent on researchers that they employ such types of samples for irradiation experiments. Only then will the results be truly applicable and relevant. Failure to use naturally contaminated samples for developing molecular pathogen detection tools is partly responsible for the present situation: Even a decade or more after the introduction of molecular tools, we are still faced with issues such as "inhibitory compounds," "inability to detect naturally contaminated samples," and so forth. There are differences in the dose rate of E-beam radiation as compared with gamma radiation. Hence, the differences in the inactivation kinetics of the various enteric pathogens under E-beam irradiation as compared with gamma-ray irradiation should also be elucidated. This will permit the customization of irradiation protocols depending on the source of ionizing radiation that is available to a particular food processor.

Inactivation Kinetics of Foodborne Viruses

Chapter 8 provides an overview of the significance of viral contamination of foods. It is obvious that a variety of enteric viruses including rotavirus, enteric adenoviruses, noroviruses (human caliciviruses known previously as Norwalk and Norwalk-like viruses), hepatitis virus type A, hepatitis virus type E, and astrovirus can be transmitted through foods (see Chapter 8). Recent studies suggest that even viruses are sensitive to E-beam radiation at levels significantly lower than those observed under cobalt-60 irradiation (Pillai and Espinosa 2003). More studies are needed to identify the conditions that can eliminate viral pathogens in RTE foods and in foods that are minimally processed and yet highly vulnerable to human contamination sources.

Direct and Indirect Effects of Ionizing Irradiation

E-beam irradiation is assumed to inactivate microorganisms by either direct damage or indirect damage to their nucleic acids. However, there are still lingering questions regarding the actual contribution of direct versus indirect damage to nucleic acids. The importance of indirect damage to nucleic acids is evident when one analyzes the inactivation of viruses. Studies in our laboratory and those of others have shown that viruses, even though they are very small, are quite sensitive to E-beam irradiation. Studies in our laboratory using Poliovirus type 1 in different matrices have shown that the D_{10} value ranges between 1.83 and 2.82, depending on the matrix. Peptone was found to shield the viruses from rapid inactivation. These studies were conducted using suspensions containing approximately 10,000 virus particles per milliliter, which in reality is a relatively small number of targets for direct attacks. The "shielding effects" or "scavenger activity" exhibited by peptone suggests that in some situations, the indirect effects of irradiation may be the primary mode of action. This, however, needs to be further studied and verified. Additional research to delineate the precise mechanisms of irradiation-induced inactivation is needed because this can allow the incorporation of specific "quenching" molecules placed directly in the food, the matrix, or the packaging materials to attain or prevent a certain desired level of nucleic acid damage. This can be particularly important when attempting to develop low-dose irradiation protocols on multicomponent foods that may contain these scavenger molecules, which may inadvertently reduce the desired dose.

Microbial Stress Conditions and Irradiation Sensitivity

Irradiation sensitivity of organisms in a particular food depends on the irradiation dose, the numbers and type of organisms in the food, the composition of the food, the food preservation methods that have been used, and the temperature at which the food is stored. The D_{10} values of enteric bacteria such as *Salmonella*, *L. monocytogenes*, and *E. coli* O157:H7 are different depending on whether the food is fresh or frozen and on the suspending matrix. This difference is caused by the presence of frozen water molecules in frozen foods, which are not as effective at exchanging their energies with adjoining molecules compared with water molecules in the unfrozen state. In addition to these parameters, recent studies have shown that understanding the physiological state of the cell is critically important when evaluating its radiation resistance. Buchanan et al. (1999) have reported that acid-resistant *E. coli* O157:H7 can withstand higher levels of ionizing irradiation compared with non-acid-resistant *E. coli* strains. They also report that different strains of the same pathogen can exhibit significant differences in radiation sensitivity, which is presumably a reflection of the physiological status of the different strains.

Microbial cells in starvation mode can also exhibit increased resistance to irradiation. Because starved or moribund cells have a significantly reduced number of DNA replication forks, the potential targets for DNA damage are subsequently reduced. Stress-induced proteins and other cellular components such as lipid- and protein-rich foods may either protect the cells or enhance DNA repair under optimal conditions. Studies have also shown that carbon monoxide in modified atmosphere packaging (MAP) and hydrogen peroxide treatments can also protect microorganisms against ionizing irradiation to varying degrees. The precise mechanism of protection or repair needs to be elucidated so that appropriate strategies could be adopted when irradiating such foods. A number of other stress factors such as osmotic stress, heat stress, and alkali stress can also enhance radiation resistance. It is thus essential that, when D_{10} values are developed on foods, the possibility of the factors influencing the behavior of pathogens and indigenous organisms be taken into consideration.

Potential Pathogen Regrowth

The potential for pathogen regrowth in a food matrix in which the background (indigenous) microbial population is reduced because of irradiation can be an issue depending on the food matrix, the pathogen in question, and the storage conditions. Thayer and Boyd (2000) have reported that *L. monocytogenes* did not multiply any faster during storage at 7°C on irradiated raw ground turkey compared with the growth exhibited on the non-irradiated product when packaged under MAP conditions. Further research using more-sensitive quantitative methods is needed to accurately identify whether survivors within an irradiated population have increased multiplication rates in the absence of normal indigenous microflora or whether they undergo any sort of competitive advantage.

Low-Dose Irradiation and Dosimetry

Methods for setting accelerator current to lower values to achieve lower dose rates in the conventional belt-speed range need to be explored. The minimum current that can be achieved by typical linear accelerators should also be determined. Because capital cost of the facility is a major consideration in terms of the cost of radiation processing, a study of the net cost of processing using beam-current reduction and X-ray irradiation should be conducted (L. Braby, personal communication). If X-ray irradiation is significantly more efficient when low doses are required, the relative biological effectiveness of X-rays relative to electrons for bacterial inactivation and other relevant endpoints must be determined.

Improved dosimeters, more sensitive and reproducible radiochromic film, an increase in the sensitivity of ESR measurements, or a variation of thermoluminescent dosimetry that could measure radiation doses in the range of 10–1,000 Gy will be needed as the range of products and the objectives of irradiation expand. A dosimeter certification service, similar to that currently existing for radiation protection dosimeters, will be needed to ensure consistent results at different processing and research facilities. This will require development of standardized techniques for placing dosimeters and also development of standard methods for reading dosimeters (Braby 2003)

Organoleptic Attributes

There is an urgent need for standardization in terms of sensory changes or organoleptic attributes of irradiated product as it relates to irradiation sources, irradiation conditions, dosimetry, and product profiles. Without such standardization, it would be difficult to compare and analyze irradiation results. There is also a need to objectively characterize and quantify adverse or positive changes in these attributes analytically. There are significant costs associated with employing trained sensory panels. The use of multidimensional gas-chromatography (Microanalytics, Round Rock, Tex.) for analytical determination of the formation of specific odoriferous compounds in irradiated foods needs to be explored.

Multicomponent Foods

Once federal approvals for RTE foods are obtained, there will be a significant set of opportunities to use food irradiation for multicomponent foods such as RTE meals and dishes. The issues of dosimetry, pathogen reduction, and sensory issues will be extremely significant in these types of foods because of the anticipated differences in the food matrix, the potential varying pathogen loads, the types of pathogens that could be encountered, and the critical need to retain the sensory attributes of the packaged meals.

Product Packaging

Research is needed on the next generation of packaging materials that can retard possible negative changes in sensory attributes or enhance desired sensory attributes during irradiation. Song et al. (2003) reported that MAP in combination with irradiation can improve the chemical (nitrosamine concentrations) and microbiological safety of cooked sausage. Ahn et al. (2003) reported that irradiation in an optimal combination with MAP can enhance desirable color characteristics in sausage. Recent studies have also shown that MAP in combination with a hot-water treatment can reduce undesirable changes in fresh-cut iceberg lettuce (Fan and Sokorai 2002, Fan et al. 2003). Research in antimicrobial coatings and antioxidant additions can provide avenues that could potentially extend the product lines for which irradiation becomes a viable option. The development of packaging material that can visually denote an irradiated product or dose range, or that can detect adverse changes in a product, can also find commercial application.

Conclusions

Food irradiation will become a mainstay of food processing similar to pasteurization. Consumers are going to demand pathogen-free poultry meat and poultry products and RTE foods. Given that many of the foodborne pathogens originate within farm animals, and as these organisms are almost a part of the normal flora of these animals, it is quite unlikely that we can totally eradicate these pathogens at the preharvest stage. Postharvest pathogen reduction is probably the only reliable solution. However, irradiation should only be used as the final step of a validated HACCP protocol within the food industry. Using irradiation to circumvent currently employed disinfection and intervention strategies within the food processing plants would be disastrous. We should anticipate and plan for the use of food irradiation similar to our use of milk pasteurization, where only Grade A milk is pasteurized for human consumption. This will guarantee a microbiologically safe product in which the society at large believes and that it supports.

Acknowledgements

This work was supported by funds from a USDA-Initiative for Future Agriculture and Food Systems grant 00-52102-9637, a USDA-Cooperative State Research, Education, and Extension Service grant 2001-34461-10405, and Hatch Project H-8708 funds from the Texas A&M University System.

References

Ahn, H. J., Jo, C., Lee, J. W., Kim, J. H., Kim, K. H. and Byun, M. W. 2003. Irradiation and modified atmosphere packaging effects on residual nitrite, ascorbic acid, nitrosomyoglobin, and color in sausage. *Journal of Agricultural Food Chemistry* **51**:1249–1253.

Arena, V. 1971. Ionizing Radiation and Life; An Introduction to Radiation Biology and Biological Radiotracer Methods. St. Louis, Mo.: Mosby.

Bari, M. L., Nazuka, E., Sabina, Y., Todoriki, S. and Isshiki, K. 2003. Chemical and irradiation treatments for killing *Escherichia coli* O157:H7 on alfalfa, radish, and mung bean seeds. *Journal of Food Protection* **66**:767–774.

Bartek, J. and Lukas, J. 2003. Chk1 and Chk2 kinases in checkpoint control and cancer. *Cancer Cell* **3**:421–429.

Braby, L. A. 2003. The underlying principles of E-beam technology. Annual Meeting of the Institute of Food Technologists, June 13–16, Chicago, IL.

Buchanan, R. L., Edelson, S. G. and Boyd, G. 1999. Effects of pH and acid resistance on the radiation resistance of enterohemorrhagic *Escherichia coli*. *Journal of Food Protection* **62**:219–228.

Cerda, H. 1998. Detection of irradiated frozen meat with the comet assay. Interlaboratory test. *Journal of the Science of Food and Agriculture* **76**:435–442.

Codex Alimentarius Commission. 1984. Codex General Standards for Irradiated Foods and Recommended International Code of Practice for the Operation of Radiation Facilities Used for the Treatment of Foods, CAC, Vol XV E-1. Rome: Food and Agricultural Organization.

Deeble, D. J., Christiansen, J. F., Jones, M., Tyreman, A. L., Smith, C. J., Beaumont, P. C. and Williams, J. H. 1994. Detection of irradiated food based on DNA base changes. *Food Science Technology Today* **8**:96–98.

Endley, S., Lu, L., Vega, E., Hume, M. E. and Pillai, S. D. 2003. Male-specific coliphages as an additional fecal contamination indicator for screening fresh carrots. *Journal of Food Protection* **66**:88–93.

Fan, X. and Sokorai, K. J. 2002. Changes in volatile compounds of gamma-irradiated fresh cilantro leaves during cold storage. *Journal of Agricultural and Food Chemistry* **50**:7622–7626.

Fan, X., Toivonen, P. M., Rajkowski, K. T. and Sokorai, K. J. 2003. Warm water treatment in combination with modified atmosphere packaging reduces undesirable effects of irradiation on the quality of fresh-cut iceberg lettuce. *Journal of Agricultural and Food Chemistry* **51**:1231–1236.

Hayashi, T., Todoriki, S. and Koyhama, K. 1996a. Applicability of viscosity measurements to the detection of irradiated peppers, pp. 215–228. *In* Detection Methods for Irradiated Foods—Current Status, C. H. McMurray, E. M. Stewart, R. Gray, J. Pearce (Eds.). Special Publication No. 171. Cambridge: Royal Society of Chemistry (Great Britain).

Hayashi, T., Todoriki, S., Otobe, K. and Sugiyama, J. 1996b. Detection of irradiated potatoes by impedance measurements, pp. 202–214. *In* Detection Methods for Irradiated Foods—Current Status, C. H. McMurray, E. M. Stewart, R. Gray, J. Pearce (Eds.). Special Publication No. 171. Cambridge: Royal Society of Chemistry (Great Britain).

Heide, L., Guggenberger, R. and Bogl, K. W. 1989. Identification of irradiated spices with luminescence measurement: a European intercomparison. *Journal of Agricultural and Food Chemistry* **38**:2160–2163.

Josephson, E. S. 1983. An historical review of food irradiation. *Journal of Food Safety* **5**:161–190.

Lambert, J. D. and Maxcy, R. B. 1984. Effect of gamma radiation on *Campylobacter jejuni*. *Journal of Food Science* **49**:665–667.

Lembke, P., Bornert, J. and Engelhardt, H. 1995. Characterization of irradiated food by SFE and GC-MSD. *Journal of Agricultural and Food Chemistry* **43**:38–45.

Levine, P., Rose, B., Green, S., Ransom, G. and Hill, W. 2001. Pathogen testing of ready-to-eat meat and poultry products collected at federally inspected establishments in the United States, 1990–1999. *Journal of Food Protection* **64**:1188–1193.

Licciardello, J. J., Nickerson, J. T. R. and Goldblith, S. A. 1970. Inactivation of *Salmonella* in poultry with gamma radiation. *Poultry Science* **49**:663–675.

Lopez-Gonzalez, V., Murano, P. S., Brennan, R. E. and Murano, E. A. 1999. Influence of various commercial packaging conditions on survival of *Escherichia coli* O157:H7 to irradiation by electron beam versus gamma rays. *Journal of Food Protection* **62**:10–15.

Marchioni, E., Tousch, M., Zumsteeg, V., Kuntz, F. and Hasselmann, C. 1992. Alterations of mitochondrial DNA: a method for the detection of irradiated beef liver. *Radiation Physical Chemistry* **40**:485–488.

Matic, S., Minokovic, V., Katusin-Razem, B. and Razem, D. 1990. The eradication of *Salmonella* in egg powder by gamma irradiation. *Journal of Food Protection* **53**:111–114.

Molins, R. A. 2001. Irradiation of meats and poultry, pp. 131–192. *In* Food Irradiation: Principles and Applications, R. A. Molins (Ed.). New York: Wiley.

Murano, E. A. 1995. Microbiology of irradiated foods, pp. 29–61. *In* Food Irradiation: A Source Book, E. A. Murano (Ed.). Ames: Iowa State University Press.

Olson, D. G. 1995. Irradiation processing, pp. 3–27. *In* Food Irradiation: A Source Book, E. A. Murano (Ed.). Ames: Iowa State University Press.

Pillai, S. D. and Espinosa, I. Y. 2003. E-beam inactivation of RNA and DNA containing viruses. Abstract No. Q-273. Annual Meeting of the American Society for Microbiology, May 18–22, Washington, D.C.

Raffi, J. and Stocker, P. 1996. Electron paramagnetic resonance detection of irradiated foodstuffs. *Applied Magnetic Resonance* **10**:357–373.

Rajkowski, K. T., Boyd, G. and Thayer, D. W. 2003. Irradiation D-values for *Escherichia coli* O157:H7 and *Salmonella* spp. on inoculated broccoli seeds and effects of irradiation on broccoli sprout keeping quality and seed viability. *Journal of Food Protection* **66**:760–766.

Razskazovskiy, Y., Debije, M. G., Howerton, S. B., Williams, L. D. and Bernhard, W. A. 2003. Strand breaks in x-irradiated crystalline DNA: alternating CG oligomers. *Radiation Research* **160**:334–339.

Schulzki, G., Spiegelberg, A., Bogl, K. W. and Schreiber, G. A. 1995. Detection of radiation-induced hydrocarbons in baked sponge cake prepared with irradiated liquid egg. *Radiation Physical Chemistry* **46**:765–769.

Song, I. H., Kim, W. J., Jo, C., Ahn, H. J., Kim, J. H. and Byun, M. W. 2003. Effect of modified atmosphere packaging and irradiation in combination on content of nitrosamines in cooked pork sausage. *Journal of Food Protection* **66**:1090–1094.

Stewart, E. M. 2001. Detection methods for irradiation foods, pp. 347–386. *In* Food Irradiation: Principles and Applications, R. A. Molins (Ed.). New York: Wiley.

Tarkowski, J. A., Stoffer, S. C. C., Beumer, R. R. and Kampelmacher, E. H. 1984. Low dose gamma irradiation of raw meat. 1. Bacteriological and sensory quality effects in artificially contaminated samples. *International Journal of Food Microbiology* **1**:13–23.

Thayer, D. W. and Boyd, G. 1994. Control of enterotoxic *Bacillus cereus* on poultry or red meats and in beef gravy by gamma radiation. *Journal of Food Protection* **57**:758–764.

Thayer, D. W. and Boyd, G. 2000. Reduction of normal flora by irradiation and its effect on the ability of *Listeria monocytogenes* to multiply on ground turkey stored at 7 degrees C when packaged under a modified atmosphere. *Journal of Food Protection* **63**:1702–1706.

Thayer, D. W., Boyd, G., Fox Jr., J. B., Lakritz, L. and Hampson, J. W. 1995. Variations in radiation sensitivity of foodborne pathogens associated with the suspending meat. *Journal of Food Science* **60**:63–67.

Thayer, D. W., Rajkowski, K. T., Boyd, G., Cooke, P. H. and Soroka, D. S. 2003. Inactivation of *Escherichia coli* O157:H7 and *Salmonella* by gamma irradiation of alfalfa seed intended for production of food sprouts. *Journal of Food Protection* **66**:175–181.

29 Clay-Based Interventions for the Control of Chemical and Microbial Hazards in Food and Water

Henry J. Huebner, Paul Herrera, and
Timothy D. Phillips

Introduction: The Problem in Perspective

Food and water supplies destined for human consumption can contain complex and highly variable mixtures of chemicals and microbes that have been linked to disease in humans and that have been targeted as significant food safety and environmental concerns (Moore et al. 1979, McConnell 1980, Smith et al. 1982, Stalling et al. 1983). These foodborne and water-borne contaminants encompass a highly diverse family consisting of bacteria, viruses and other microbes, molds and mycotoxins, exogenous aromatic and polynuclear aromatic hydrocarbons (PAHs), halogenated aromatic hydrocarbons, polychlorinated dibenzo-*p*-dioxins, chlorinated phenols, heavy metals, pesticides, herbicides, fungicides, insecticides, nematocides, phthalates, and so forth (Bievenue and Beckman 1967, McConnell and Moore 1979, Exon 1984, Donnelly et al. 1987, Harvey et al. 1991a, Harvey et al. 1991b, Mueller et al. 1989, Mueller et al. 1991, United States Environmental Protection Agency 1997, Gevao and Jones 1998). The problem in water is further complicated by the necessary (and federally mandated) treatment with chlorine to destroy bacteria and to maintain adequate disinfection throughout the distribution system. During the conventional processing of drinking water (especially from surface water supplies), a variety of toxic byproducts known as trihalomethanes are formed when dissolved organic matter reacts with chlorine (Bull et al. 1995). These endogenous contaminants in drinking water have been shown to be toxic and potentially carcinogenic (Bull et al. 1995, Doyle et al. 1997, Lilly et al. 1997, Swan et al. 1998). It's well-established in toxicology that "the dose makes the poison" (Deichmann et al. 1986, p. 210), therefore, interventions that decrease exposure levels of humans and animals to toxins and microbes in the environment are credible strategies for diminishing risk and preventing disease.

The frequent and widespread occurrence of chemical and microbial contaminants in food and water underscores the critical need for interventions to control endogenous and exogenous chemicals and microbes (before and after processing of food and consumable water). One type of intervention that is practical, sustainable, and cost-effective is the use of clay-based materials for the remediation of food and water.

Since earliest recorded history, the unique properties and beneficial uses of clay minerals have been well-documented. In many instances, animals and humans have been reported to eat clay (geophagy), purportedly for various reasons, including detoxification of foodstuffs and the prevention of illness (Phillips et al. 1995). Many of these clays vary considerably in their molecular structure and in their ability to sequester toxic chemicals and microbes. Recent clay-based interventions for the control of chemical and microbial hazards in food and water are discussed in this chapter.

Clay-Based Interventions for the Control of Aflatoxins

Dietary Inclusion of Aflatoxin Enterosorbents

The aflatoxins are a group of carcinogenic mycotoxins, produced primarily by *Aspergillus flavus* and *Aspergillus parasiticus* fungi, that are often detected in foods and agricultural commodities. These compounds are heat stable and can survive a variety of food processing procedures; thus, aflatoxins can occur as "unavoidable" contaminants of most foods and feeds (particularly those derived from maize and peanuts). Of the four naturally occurring aflatoxins (B_1, B_2, G_1, and G_2), aflatoxin B_1 (AfB_1) is the most toxic and has been shown to disrupt genes involved in carcinogenesis (McMahon et al. 1986, McMahon et al. 1987) and tumor suppression (Aguilar et al. 1993). In addition, several studies suggest that low-level exposure to aflatoxins may cause suppression of the immune system and increased susceptibility to disease (Richard et al. 1978, Pestka and Bondy 1994).

Many strategies are available for managing aflatoxins in agricultural commodities, the simplest of which requires isolation and destruction of the contaminated source. This approach, however, is often not practical because alternative food supplies may not be available or replacement supplies may not be affordable. As a result, the use of detoxification and decontamination procedures has become a viable option for remediating aflatoxin-contaminated foodstuffs. Optimally, remediation procedures should effectively destroy, inactivate, or remove the mycotoxin from the commodity; should not result in the deposition of toxic substances, metabolites, or by-products in the commodity; should retain nutrient value and feed acceptability of the product or commodity; should not cause significant alterations in the physical properties of a commodity; and should destroy fungal spores, if possible (Park et al. 1988). Importantly, detoxification or decontamination procedures should be readily available, easily used, and inexpensive (Council for Agricultural Science and Technology 2003).

One of the most practical approaches for the prevention of aflatoxicoses in livestock involves the incorporation of clays or various "binding agents" into diets contaminated with these toxins. These additives reduce the bioavailability of the toxin in the gastrointestinal tract; that is, they serve as sequestering agents (enterosorbents) of the toxins, thus reducing uptake and distribution to the blood and target organs. Early studies using phyllosilicate minerals as sorbents have shown that certain clays could bind AfB_1 in solution (Masimango et al. 1979). Bleaching clays that had been used to process canola oil were also shown to lessen the effects of T-2 toxin, a trichothecene mycotoxin (Carson and Smith 1983, Smith 1984).

In the first enterosorbent studies with aflatoxins, NovaSil (a processed calcium montmorillonite clay from Engelhard Corporation, Cleveland, Ohio) was shown to bind AfB_1 with high affinity and high capacity *in vitro* and *in vivo*. When added to the diet, NovaSil (NS) significantly protected broiler and Leghorn chicks from the toxic effects of dietary aflatoxin exposure (Phillips et al. 1987, Phillips et al. 1988).

Following these initial findings, the efficacy of NS and certain hydrated sodium calcium aluminosilicates clays for aflatoxins has been confirmed in a variety of young animals including rodents, chicks, turkey poults, ducklings, lambs, pigs, and mink (Colvin et al. 1989, Harvey et al. 1989, Kubena et al. 1990a, Kubena et al. 1990b, Phillips et al. 1990, Bonna et al. 1991, Harvey et al. 1991a, Harvey et al. 1991b, Kubena et al. 1991, Phillips et al. 1991, Jayaprakash et al. 1992, Kubena et al. 1993, Lindemann et al. 1993, Voss et al. 1993, Harvey et al. 1994, Phillips et al. 1994, Smith et al. 1994, Marquez and de Hernandez

1995, Phillips et al. 1995, Abdel-Wahhab et al. 1998, Ledoux et al. 1999). NS has also been shown to decrease the bioavailability of radiolabeled aflatoxins and reduce aflatoxin residues in poultry (Davidson et al. 1987), rats (Sarr et al. 1995) and pigs (Beaver et al. 1990). Importantly, aflatoxin M_1 (an oxidative metabolite of AfB_1) in the milk of lactating dairy cattle and goats was reduced when NS was incorporated into the contaminated diets (Ellis et al. 1990, Smith et al. 1994, Harvey et al. 1991a).

Numerous *in vitro* studies have assessed the sorption of aflatoxins onto the surface of NS and other hydrated sodium calcium aluminosilicate clays (Ellis et al. 1990, Ellis et al. 1991, Sarr 1992, Phillips et al. 1995, Ramos and Hernández 1996, Grant et al. 1998, Grant and Phillips 1998, Phillips 1999, Phillips et al. 2002). NS, in aqueous solution, has been shown to tightly and preferentially sorb AfB_1 and similar analogs of AfB_1 that contain an intact β-dicarbonyl system in their molecular structure (Phillips et al. 1988, Sarr 1992). Equilibrium isotherms for the sorption of AfB_1 and similar ligands onto NS indicated high affinity, high capacity, regio- and stereo-specificity, and favorable thermodynamics (Grant 1998, Phillips et al. 2002). For mechanistic details, the reader is referred to a recent review by Phillips et al. (2002).

Similar to the *in vitro* results, animal studies also indicated a preference of NS for aflatoxins. When included in the diet of animals, NS did not significantly prevent the toxicity of other common (and structurally diverse) mycotoxins (e.g., zearalenone, deoxynivalenol, T-2 toxin, ochratoxin A, cyclopiazonic acid, ergotamine, and fumonisins). For example, the use of clay in mink fed zearalenone helped to alleviate some of the fetotoxicity but did not reduce the hyperestrogenic effects (Bursian et al. 1992). In addition, supplementing swine diets with clay at 0.5% and 1.0% w/w did not influence the average daily weight gain of pigs exposed to deoxynivalenol. In poultry studies, the inclusion of NS clay in the diet as an enterosorbent did not significantly prevent the adverse effects of T-2 toxin (Kubena et al. 1990a), ochratoxin A (Huff et al. 1992), diacetoxyscirpenol (Kubena et al. 1993), cyclopiazonic acid (Dwyer et al. 1997), and fumonisins (Lemke 2000). Other studies have shown that clay (2.0% in the diet) did not protect rats or sheep from fescue toxicosis (Chestnut et al. 1992), although *in vitro* experiments demonstrated favorable sorption characteristics of a predominant ergot alkaloid, ergotamine, for NS (Huebner et al. 1999).

A variety of physical, chemical, and biological interventions for the remediation of aflatoxin-contaminated food and feed is available. Clay-based enterosorbents that are selective for aflatoxins (i.e., NS) offer a practical and economically feasible solution to the problem: NS is unique in that it diminishes the bioavailability of aflatoxins from the gastrointestinal tract, thus minimizing exposure and health risks to the consumer. Importantly, all enterosorbents should be routinely evaluated to confirm their efficacy and ensure their safety. In addition, clay-based sorbents (and other sorbent materials) added to the diet should be rigorously tested in sensitive animal models to determine their potential for nutrient interactions and toxicity.

Affinity Probes for Aflatoxin Testing

The health risks associated with chronic exposure to aflatoxins, particularly AfB_1, have fostered considerable interest in the occurrence of aflatoxins in foods and feeds and the extent to which they pose elevated hazards. In response to this problem, governmental agencies worldwide have instituted detection and monitoring programs and set contamination limits for aflatoxins in grains and food products to ensure the safety and health of humans and animals.

Effective monitoring and surveillance of aflatoxins in foods and agricultural products is dependent on the use of specified analytical procedures for identifying and measuring these mycotoxins. Aflatoxin analysis typically involves three steps: extraction of aflatoxins from a test sample, cleanup of the sample extract to remove co-extracted compounds that interfere with the determination of aflatoxins, and detection and quantification of aflatoxins in the purified sample extract (Jaimez et al. 2000).

Extract cleanup procedures are important not only for reducing interfering compounds but also for concentrating aflatoxins from a large sample volume. Although extract cleanup methods serve to facilitate aflatoxin analysis in many applications, they become mandatory when modern sensitive analytical methods are used for detecting very low levels of aflatoxins (Scott and Trucksess 1997, Jaimez et al. 2000).

One of the most common extract cleanup methods available (i.e., immunoaffinity column cleanup) offers specificity, reproducibility, speed, and simplicity (Scott and Trucksess 1997). At present, immunoaffinity cleanup procedures offer many advantages over other methods of purification; however, higher costs (because of the preparation and requirements of monoclonal antibodies) may prohibit their use in repeated analyses and surveillance of food and feed for aflatoxins, especially in developing countries.

The unique chemical characteristics and high affinity of NS clay for aflatoxins have been recently applied in the development of an NS-based inorganic affinity probe (IAP-NS) for aflatoxins (Huebner and Phillips 2003). In these studies, nanostructuring techniques were employed to construct inorganic (clay-based) affinity media for the cleanup and purification of AfB_1 from aqueous solutions and grain samples. In this research, thin films of NS clay differing in arrangement were electrostatically anchored onto the surface of quartz particles using a cationic polyelectrolyte to produce multiple formulations.

Initial studies using prototypical affinity columns suggested that IAP-NS with larger NS aggregate sizes and increased NS layer coatings enhanced AfB_1 sorption capacities. In contrast, average recoveries of AfB_1 bound onto IAP-NS decreased in formulations with larger NS aggregate sizes and increased NS layer coatings. Similar trends were observed in breakthrough studies in which IAP-NS formulated with larger NS aggregate sizes and multiple layer coatings also exhibited increased breakthrough capacities for AfB_1. Experimental results showed that 50% breakthrough capacities for AfB_1 using a single-layer coating ranged from 13 to 78 µg, whereas breakthrough capacities for AfB_1 using a triple-layer coating ranged from 16 to 140 µg. These studies demonstrated that IAP-NS columns can be formulated to offer narrowly defined, reproducible capacity ranges that are ideal for use in analytical applications. Additional experiments showed that recoveries of AfB_1 from spiked rice and corn followed linear trends that were highly correlated with the range of spike levels. AfB_1 recoveries from rice samples spiked at 10, 50, 100, 200, and 400 ng/g averaged 43%, 43%, 56%, 61%, and 63%, respectively. Similarly, AfB_1 recoveries from corn samples spiked at 10, 50, 100, 200, and 400 ng/g were determined to be 43%, 34%, 41%, 45%, and 49%, respectively. Importantly, the recoveries of AfB_1 from both series of spiked grains exhibited a good correlation with the range of spike levels ($r^2 = 0.9991$ and 0.9974, respectively).

In other studies, commercially available immunoaffinity columns and IAP-NS columns were compared for total aflatoxin recoveries using an aflatoxin positive corn sample. On the basis of LC analysis of column eluate residues, AfB_1 was the only congener detected in any of the eluates from either cleanup procedure. IAP-NS columns provided approximately 1/2 the average recovery of AfB_1 (8.4 ± 0.5 ng vs. 15.7 ± 0.3 ng) from naturally contaminated corn under the experimental conditions of the tests. Although IAP-NS offered lower recoveries for AfB$_1$ as compared with immunoaffinity columns, the recoveries were consistent and purified extracts were void of interfering compounds as determined by liquid chromatography with fluorescence detection.

Biological-based cleanup methods are typically characterized by a high affinity (selective) and low capacity (saturable) for target analytes. In contrast, IAP-NS methods are selective and offer both high affinities and high capacities for AfB$_1$. These materials may provide an alternative and cost-effective approach for the selective cleanup, purification and concentration of aflatoxins in contaminated grains. Additional research may optimize extract cleanup and elution procedures for use with IAP-NS columns and further assess aflatoxin recoveries from a variety of contaminated media.

Clay-Based Interventions for Water Purification

Chemical Contaminants

The occurrence of complex mixtures of toxic chemicals in drinking water constitutes a substantial environmental and public health hazard. In the United States, PAHs and chlorinated phenols originating from the uncontrolled release of wood-preserving wastes extensively contribute to groundwater and aquifer contamination. Because of their persistence in the environment, the removal of trace, yet biologically significant, levels of PAHs and chlorinated phenols from water continues to be an active area of research.

An important strategy offering potential for groundwater remediation is the use of chemically modified clay minerals to sorb contaminants from aqueous media. This area of research has focused largely on the use of quaternary ammonium compounds to modify clay, resulting in hydrophobic clay minerals (organoclays) with increased affinities and capacities for a variety of organic compounds (Boyd et al. 1988, Srinivasan and Fogler 1990a, Srinivasan and Fogler 1990b, Brixie and Boyd 1994, Smith and Jaffe 1994, Dentel et al. 1995, Li et al. 1996, Sheng et al. 1996, Xu et al. 1997).

To date, the widespread application of clays as filtration media has been severely limited because of the poor hydraulic conductivity associated with clay minerals (i.e., most clays are not amenable to large-scale processing of groundwater via in-line columns or fixed-bed filtration systems). To address this problem, research in our laboratory has focused on methods whereby natural and chemically modified clays can be immobilized onto solid substrates. The porous nature of these clay-based composites enhances their overall water permeability and flow characteristics in filtration applications and facilitates the movement of organic contaminants through the matrix (Phillips 1999, Huebner et al. 2000, Ake et al. 2001).

The efficacy of porous organoclay composites for the remediation of pentachlorophenol and PAHs in groundwater has recently been investigated at an active Superfund site (Ake et al. 2003). Effluent from both an oil–water separator and a bioreactor (BIO) were passed through fixed-bed filtration columns containing either an organoclay composite or granular activated carbon (GAC). Samples of treated water were collected over time and extracted, and target contaminants were quantified via gas chromatography/mass spectrometry. Results indicated that organoclay composites (i.e., cetylpyridinium-exchanged montmorillonite immobilized onto sand) were able to reduce the contaminant load from the oil–water separator effluent stream by 97%. Concentrations of pentachlorophenol and benzo[*a*]pyrene were significantly decreased (i.e., >99% decrease). Additional studies involved the filtration of BIO effluent through columns containing composite or an equiva-lent amount of GAC. Results showed that the organoclay composite reduced the majority of BIO contaminants, although it was less effective in remediating lower–molecular weight PAHs. In contrast, GAC was more effective in removing the lower–molecular weight PAHs from BIO effluent, with the exception of naphthalene and pentachlorophenol. Overall, these findings have delineated differences between organoclay- and GAC-based sorbents for groundwater remediation and suggest that combinations of immobilized organoclay and GAC may be more effective for the treatment of contaminated water than either sorbent used alone.

These organoclay-based composites, used in column filtration systems, will facilitate the "field-practical" purification of water contaminated with a wide variety of hazardous organic chemicals. These materials have several advantages: fixed-bed filtration columns can effectively use high-affinity organoclays for the sorption of PAHs and chlorophenols without the need for sophisticated batch remediation systems; depending on the type of organic contaminant or mixture of contaminants, multisorbent columns can be tailor-made for the remediation of a particular source of groundwater; and water purification columns can be easily installed into existing water treatment facilities. Preliminary research clearly demonstrates the effectiveness and potential of sand-immobilized organoclay for the cleanup of creosote-contaminated water.

Microbial Contaminants

Clean water is essential in many aspects of food production, including washing, heating, cooling, and cooking of food products. In the poultry processing industry, water is used in scalding baths to facilitate the removal of feathers, to wash carcasses, and to chill carcasses in preparation for refrigeration (Hagstad and Hubbert 1986). Because potable water is scarce, the cost of obtaining usable water and disposing of wastewater represents a substan-tial economic expenditure. As a result, considerable interest has been focused on the poten-tial of using recycled or reconditioned water in carcass production processes (Hamza et al. 1978, Chang and Sheldon 1989, Chang et al. 1989, Palumbo et al. 1997). Of primary con-cern to poultry processors who use reconditioned water is the removal of bacteria, especial-ly the *Salmonella* species, from water used in scalding and pre-chilling stages where poten-tial cross-contamination of uninfected carcasses can occur.

The physical and chemical properties of montmorillonite clays, including high surface areas and high cationic exchange capacities, make them ideal sorbents for multiple applica-tions; however, these types of clays are less than ideal for removing bacteria from water. Under physiologic conditions, the surfaces of many silicate minerals as well as bacteria are negatively charged, representing a large energy barrier to the adsorption of bacteria onto clay surfaces (Stotzky 1980). This problem can be resolved by altering the surface chemistry of montmorillonite clays by exchanging predominant interlayer cations with positively charged organic compounds (Koh and Dixon 2001). For example, the exchange of cations within the gallery (interlayer) of montmorillonite clays with quaternary amine surfactants having long alkyl chains results in the production of hydrophobic organoclays. As the con-centration of organic cations exchanged onto the clay increases, hydrophobic interactions between the alkyl chains attract more organic cations to the clay surface. Further addition of organic cations, in excess of the exchange capacity of the clay, results in an organoclay with a positive (instead of negative) surface charge (Xu et al. 1997). Organoclays have proven useful in the sorption of aromatic and halogenated hydrocarbons as well as other organic contaminants that the parent (hydrophilic) clays would not normally sequester.

Previous research has shown that exchanging montmorillonite clays with quaternary amine compounds, such as cetylpyridinium chloride (CP), increases their ability to sorb and inactivate *Salmonella* in aqueous solution (Herrera et al. 2000). In studies with *Salmonella enteritidis* concentrations as high as 4.0×10^8 colony forming units (CFU)/mL, various CP-exchanged montmorillonite clays were shown to significantly reduce bacterial loads in test solutions as measured by colony plate counting methods (Figure 29.1A). To determine whether the antibacterial activity of the modified clays was the result of CP dissociating from the surface and affecting bacteria in suspension, CP desorption studies were performed. Washed organoclay pellets retained their antibacterial properties by 90%–95%, whereas supernatants from the washed pellets showed very little antibacterial activity. These findings suggest that the antibacterial activity of CP-exchanged clays is mainly localized on the surface of these clays and is not the result of the release of CP into solution. For confirmation, CP-exchanged clays treated with *Salmonella* were visualized by electron microscopy to characterize the interaction of bacteria at the surface of the clay particles. Figures 29.1B and 29.1C show electron micrograph images of parent and CP-exchanged clays after treatment with *Salmonella* suspensions. The micrographs show that bacterial sorption is strongly favored by CP-exchanged clay surfaces as opposed to surfaces of parent clays. In further studies, the effect of CP exchange rates on the antibacterial activity of modified clays was determined. Bacterial binding studies were performed using clays exchanged with amounts of CP that were from 0.5 to 3.0 times the estimated cation exchange capacity of the parent clay (i.e., 90 cmol$_{charge}$/kg). Results showed that as the exchange rate of CP on clay increased, the reduction of bacterial counts also notably increased.

Overall, experimental results demonstrated that *Salmonella* sorption to CP-exchanged clays reached a plateau within 12 hours and remained stable and irreversible for up to 24 hours. Organoclays were also evaluated at varying pHs, temperatures, and concentrations of organic material to delineate optimal antibacterial activity. It was determined that pH did not significantly influence the antibacterial activity of the modified clays. Conversely, as temperature decreased the rate of bacterial reduction also decreased.

In column filtration studies, various organoclays were immobilized onto the surface of sand (as a solid support) to produce a porous composite media for the sorption and removal of bacteria from aqueous solution. Immobilization techniques produced composite sorbents that displayed a heterogeneous distribution of organoclay. Depending on the type of clay and exchangeable cation, the immobilized organoclays were shown to significantly reduce *Salmonella* colony plate counts by levels ranging from 17% to 84%. Importantly, in studies conducted with *Salmonella* concentrations ranging from 3.0×10^4 to 3.0×10^8 CFU/mL, it was determined that sorption or filtration efficiency was dependent on bacterial concentration of the suspension being treated.

The affinity of bacteria for organoclays may involve both electrostatic attraction and hydrophobic interactions. Under physiologic conditions, bacterial cell walls are negatively charged because of functional groups such as carboxylates present in lipoproteins at the surface (Breen et al. 1995). At physiological pHs, montmorillonites exhibit a net negative charge (Stotzky 1980). Under these conditions, bacteria would not normally be significantly attracted to these clays. However, by exchanging clays with CP at rates greater than the cation exchange capacity, a material that is both hydrophobic and positively charged is produced. The positive charge of these exchanged clays may enhance their ability to attract

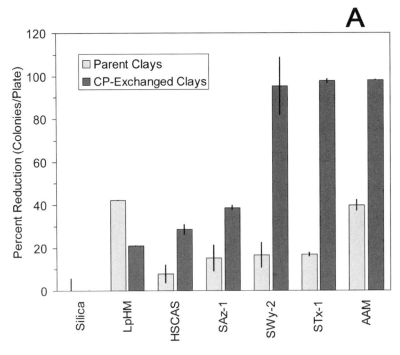

Figure 29.1 (A) Illustration of the percentage reduction in *Salmonella enteritidis* plate counts by parent and cetylpyridinium chloride (CP)–exchanged materials. Sorbents included silica, low-pH montmorillonite (LpHM), hydrated sodium calcium aluminosilicate (HSCAS), the calcium montmorillonites STx-1 and SAz-1, the sodium montmorillonite SWy-2, and acid activated montmorillonite (AAM). Electron micrographs of parent clay STx-1 and CP*STx-1 after treatment with *Salmonella* are shown in (B) and (C). The image of the parent clay STx-1 (B) shows no sorption of bacteria on clay surfaces, whereas numerous rod-shaped bacteria are evident on the surface of the CP–exchanged STx-1 clay (C). Modified from Herrera et al. (2000).

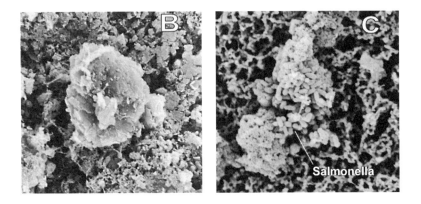

bacteria through electrostatic interactions. In addition, there may be hydrophobic attractions between the hydrophobic alkyl chains of CP and lipophilic components of the bacterial cell walls, such as lipoproteins, liposaccharides, and phospholipids.

The preliminary findings of these studies suggest that both hydrophobic and electrostatic interactions may be important factors in the attraction of bacteria to modified clay surfaces. These studies suggest that CP-exchanged clays could facilitate the removal of bacteria from various water sources and that the antibacterial activity of the organoclay sorbent may be enhanced by increasing the amount of cationic surfactant exchanged into the gallery of the parent clay. The antibacterial efficacy of immobilized organoclays is dependent on the concentration of bacteria suspended in the filtrate; thus, filtration columns may be best suited for water-polishing applications with low to moderate bacterial loads. Although this research focused on remediating *S. enteritidis* in wastewaters produced during meat processing, future work should address other infectious agents and diverse chemical applications.

Conclusions

Clays have been used for a variety of agricultural, industrial, and medicinal purposes for centuries. Since earliest recorded history, the Chinese, Egyptians, Greeks, and Romans were well aware of the numerous beneficial properties of clay minerals. As time progresses, scientific advances in our understanding of clay-based sorption processes will help to facilitate the development of novel molecular applications whereby clay minerals can be used advantageously.

For example, clay-based enterosorbent technologies may ultimately be applicable and sustainable for the protection of human foods in developing countries by decreasing aflatoxin exposure and by improving human health. From an analytical standpoint, clays may be instrumental in the development of rapid diagnostic tests for the detection and monitoring of exposure biomarkers present in urine, blood, or body tissues. In addition, the sorptive and selective properties of clays could be harnessed for the manufacture of organic/inorganic microarrays or biochips useful for the detection of genetic polymorphisms, gene identification, bacteria and virus identification, or cell immobilization.

In environmental settings, multifunctional composite sorbents based on hydrophilic clays or hydrophobic organoclays may eventually have widespread acceptance for the filtration and clarification of hazardous chemicals, heavy metals, and bacteria/viruses from contaminated drinking water. Similarly, the application of organoclays as dietary enterosorbents/sanitizing agents in poultry and other livestock before slaughter may effectively reduce the number of gastrointestinal microorganisms including *S. enteritidis* and *E. coli*. As a result, bacterial contamination during processing may be significantly minimized.

The research presented in this chapter clearly demonstrates that clay-based technologies offer innovative, field-practical, and economical solutions to contemporary issues and future challenges attributed to chemical and microbial contamination in our environment and food supplies.

References

Abdel-Wahhab, M. A., Nada, S. A., Farag, I. M., Abbas, N. F. and Amra, H. A. 1998. Potential protective effect of HSCAS and bentonite against dietary aflatoxicosis in rat: with special reference to chromosomal aberration. *Natural Toxins* **6**:211–218.

Aguilar, F., Hussain, S. P. and Cerutti, P. 1993. Aflatoxin B$_1$ induces the transversion of G → T in codon 249 of the p53 tumor suppressor gene in human hepatocytes. *Proceedings of the National Academy of Sciences (USA)* **90**:8586–8590.

Ake, C. L., Mayura, K., Huebner, H., Bratton, G. R. and Phillips, T. D. 2001. Development of porous clay-based composites for the sorption of lead from water. *Journal of Toxicology and Environmental Health (Part A)* **63**:459–475.

Ake, C. L., Wiles, M. C., Huebner, H. J., Donnelly, K. C., McDonald, T. J., Richardson, M. B., Cosgriff, D. and Phillips, T. D. 2003. Porous organoclay composite for the sorption of polycyclic aromatic hydrocarbons and pentachlorophenol from groundwater. *Chemosphere* **51**:835–844.

Beaver, R. W., Wilson, D. M., James, M. A. and Haydon, K. D. 1990. Distribution of aflatoxins in tissues of growing pigs fed an aflatoxin-contaminated diet amended with a high affinity aluminosilicate sorbent. *Veterinary and Human Toxicology* **32**:16–18.

Bievenue, A. and Beckman, H. 1967. Pentachlorophenol: a discussion of its properties and its occurrence as a residue in human and animal tissues. *Residue Reviews* **19**:83–134.

Bonna, R. J., Aulerich, R. J., Bursian, S. J., Poppenga, R. H., Braselton, W. E. and Watson, G. L. 1991. Efficacy of hydrated sodium calcium aluminosilicate and activated charcoal in reducing the toxicity of dietary aflatoxin to mink. *Archives of Environmental Contamination and Toxicology* **20**:441–447.

Boyd, S. A., Shaobai, S., Lee, J. F. and Mortland, M. M. 1988. Pentachlorophenol sorption by organoclays. *Clays and Clay Minerals* **36**:125–130.

Breen, P. J., Compadre, C. M., Fifer, E. K., Salari, H., Serbus, D. D. and Lattin, D. L. 1995. Quaternary ammonium compounds inhibit and reduce the attachment of viable *Salmonella typhimurium* to poultry tissues. *Journal of Food Science* **60**:1191–1196.

Brixie, J. M. and Boyd, S. A. 1994. Organic chemicals in the environment: treatment of contaminated soils with organoclays to reduce leachable pentachlorophenol. *Journal of Environmental Quality* **23**:1283–1290.

Bull, R. J., Birnbaum, L. S., Cantor, K. P., Rose, J. B., Butterworth, B. E., Pegram, R. and Tuomisto, J. 1995. Water chlorination: essential process or cancer hazard? *Fundamental and Applied Toxicology* **28**:155–166.

Bursian, S. J., Aulerich, R. J., Cameron, J. K., Ames, N. K. and Steficek, B. A. 1992. Efficacy of hydrated sodium calcium aluminosilicate in reducing the toxicity of dietary zearalenone to mink. *Journal of Applied Toxicology* **12**:85–90.

Carson, M. S. and Smith, T. K. 1983. Role of bentonite in prevention of T-2 toxicosis in rats. *Journal of Animal Science* **57**:1498–1506.

Chang, S. Y., Toledo, R. T. and Lillard, H. S. 1989. Clarification and decontamination of poultry chiller water for recycling. *Poultry Science* **68**:1100–1108.

Chang, Y. H. and Sheldon, B. W. 1989. Effect of chilling broiler carcasses with reconditioned poultry prechiller water. *Poultry Science* **68**:656–662.

Chestnut, A. B., Anderson, P. D., Cochran, M. A., Fribourg, H. A. and Gwinn, K. D. 1992. Effects of hydrated sodium calcium aluminosilicate on fescue toxicosis and mineral absorption. *Journal of Animal Science* **70**:2838–2846.

Colvin, B. M., Sangster, L. T., Hayden, K. D., Bequer, R. W. and Wilson, D. M. 1989. Effect of high affinity aluminosilicate sorbent on prevention of aflatoxicosis in growing pigs. *Veterinary and Human Toxicology* **31**:46–48.

Council for Agricultural Science and Technology. 2003. Mycotoxins: risks in plant, animal, and human systems. Report No. 139. Ames, Iowa: Council for Agricultural Science and Technology Task Force.

Davidson, J. N., Babish, J. G., Delaney, K. A., Taylor, D. R. and Phillips, T. D. 1987. Hydrated sodium calcium aluminosilicate decreases bioavailability of aflatoxin in the chicken. *Poultry Science* **66**:89–94.

Deichmann, W. B., Henschler, D., Holmstedt, B. and Keil, G. 1986. What is there that is not poison? A study of the *Third Defense* by Paracelsus. *Archives of Toxicology (Berlin)* **58**:207–213.

Dentel, S. K., Bottero, J. Y., Khatib, K., Demougeot, H., Duguet, J. P. and Anselme, C. 1995. Sorption of tannic acid, phenol, and 2,4,5-trichlorophenol on organoclays. *Water Research* **29**:1273–1280.

Donnelly, K. C., Brown, K. W. and Kampbell, K. 1987. Chemical and biological characterization of hazardous industrial waste. *Mutation Research* **180**:31–42.

Doyle, T. J., Zheng, W., Cerhan, J. R., Hong, C.-P., Sellers, T. A., Kushi, L. H. and Folsom, A. R. 1997. The association of drinking water source and chlorination by-products with cancer incidence among postmenopausal women in Iowa: a prospective cohort study. *American Journal of Public Health* **87**:1168–1176.

Dwyer, M. R., Kubena, L. F., Harvey, R. B., Mayura, K., Sarr, A. B., Buckley, S., Bailey, R. H. and Phillips, T. D. 1997. Effects of inorganic adsorbents and cyclopiazonic acid in broiler chickens. *Poultry Science* **76**:1141–1149.

Ellis, J. A., Bailey, R. H., Clement, B. A. and Phillips, T. D. 1991. Chemisorption of aflatoxin M$_1$ from milk by hydrated sodium calcium aluminosilicate. *Toxicologist* **11**:96. (Abstract).

Ellis, J. A., Harvey, R. B., Kubena, L. F., Bailey, R. H., Clement, B. A. and Phillips, T. D. 1990. Reduction of aflatoxin M_1 residues in milk utilizing hydrated sodium calcium aluminosilicate. *Toxicologist* **10:**163. (Abstract).

Exon, J. H. 1984. A review of chlorinated phenols. *Veterinary and Human Toxicology* **26:**508–520.

Gevao, B. and Jones, K. C. 1998. Kinetics and potential significance of polycyclic aromatic hydrocarbon desorption from creosote-treated wood. *Environmental Science and Technology* **32:**640–646.

Grant, P. G. 1998. Investigation of the mechanism of aflatoxin B_1 adsorption to clays and sorbents through the use of isothermal analysis. Ph.D. dissertation, Texas A&M University, College Station, Tex.

Grant, P. G., Lemke, S. L., Dwyer, M. R. and Phillips, T. D. 1998. Modified Langmuir equation for S shaped and multi-site isotherm plots. *Langmuir* **14:**4292–4299.

Grant, P. G. and Phillips, T. D. 1998. Isothermal adsorption of aflatoxin B_1 on HSCAS clay. *Journal of Agricultural and Food Chemistry* **46:**599–605.

Hagstad, H. V. and Hubbert, W. T. 1986. Food production technology: the food chain, pp. 3–66. *In* Food Quality Control: Food of Animal Origin, H. V. Hagstad (Ed.). Ames: Iowa State University Press.

Hamza, A., Saad, S. and Witherow, J. 1978. Potential for water reuse in an Egyptian poultry processing plant. *Journal of Food Science* **43:**1153–1161.

Harvey, R. B., Kubena, L. F., Elissalde, M. H., Corrier, D. E. and Phillips, T. D. 1994. Comparison of two hydrated sodium calcium aluminosilicate compounds to experimentally protect growing barrows from aflatoxicosis. *Journal of Veterinary Diagnostic Investigation* **6:**88–92.

Harvey, R. B., Kubena, L. F., Phillips, T. D., Corrier, D. E., Elissalde, M. H. and Huff, W. E. 1991a. Diminution of aflatoxin toxicity to growing lambs by dietary supplementation with hydrated sodium calcium aluminosilicate. *American Journal of Veterinary Research* **52:**152–156.

Harvey, R. B., Kubena, L. F., Phillips, T. D., Huff, W. E. and Corrier, D. E. 1989. Prevention of aflatoxicosis by addition of hydrated sodium calcium aluminosilicate to the diets of growing barrows. *American Journal of Veterinary Research* **50:**416–420.

Harvey, R. B., Phillips, T. D., Ellis, J. A., Kubena, L. F., Huff, W. E. and Petersen, H. D. 1991b. Effects of aflatoxin M_1 residues in milk by addition of hydrated sodium calcium aluminosilicate to aflatoxin contaminated diets of dairy cows. *American Journal of Veterinary Research* **52:**1556–1559.

Herrera, P., Burghardt, R. C. and Phillips, T. D. 2000. Adsorption of *Salmonella enteritidis* by cetylpyridinium-exchanged montmorillonite clays. *Veterinary Microbiology* **74:**259–272.

Huebner, H. J., Lemke, S. L., Ottinger, S. E., Mayura, K. and Phillips, T. D. 1999. Molecular characterization of high affinity, high capacity clays for the equilibrium sorption of ergotamine. *Food Additives and Contaminants* **16:**159–171.

Huebner, H. J., Mayura, K., Pallaroni, L., Ake, C. L., Lemke, S. L., Herrera, P. and Phillips, T. D. 2000. Development and characterization of a carbon-based composite for decreasing patulin levels in apple juice. *Journal of Food Protection* **63:**106–110.

Huebner, H. J. and Phillips, T. D. 2003. Clay-based affinity probes for the selective cleanup and determination of aflatoxin B_1 utilizing nanostructured montmorillonite on quartz. *Journal of the Association of Official Analytical Chemists International* **86:**534–539.

Huff, W. E., Kubena, L. F., Harvey, R. B. and Phillips, T. D. 1992. Efficacy of a hydrated sodium calcium aluminosilicate to reduce the individual and combined toxicity of aflatoxin and ochratoxin A. *Poultry Science* **71:**64–69.

Jaimez, J., Fente, C. A., Vazquez, B. I., Franco, C. M., Cepeda, A., Mahuzier, G. and Prognon, P. 2000. Application of the assay of aflatoxins by liquid chromatography with fluorescence detection in food analysis. *Journal of Chromatography A* **882:**1–10.

Jayaprakash, M., Gowda, R. N. S., Vijayasarathi, S. K. and Seshadri, S. J. 1992. Adsorbent efficacy of hydrated sodium calcium aluminosilicate in induced aflatoxicosis in broilers. *Indian Journal of Veterinary Pathology* **16:**102–108.

Koh, S.-M. and Dixon, J. B. 2001. Preparation and application of organo-minerals as sorbents of phenol, benzene and toluene. *Applied Clay Science* **18:**111–122.

Kubena, L. F., Harvey, R. B., Huff, W. E., Corrier, D. E., Phillips, T. D. and Rottinghaus, G. E. 1990a. Ameliorating properties of a hydrated sodium calcium aluminosilicate on the toxicity of aflatoxin and T-2 toxin. *Poultry Science* **69:**1078–1086.

Kubena, L. F., Harvey, R. B., Huff, W. E., Yersin, A. G., Elissalde, M. H. and Witzel, D. A. 1993. Efficacy of a hydrated sodium calcium aluminosilicate to reduce the toxicity of aflatoxin and diacetoxyscirpenol. *Poultry Science* **72:**51–59.

Kubena, L. F., Harvey, R. B., Phillips, T. D., Corrier, D. E. and Huff, W. E. 1990b. Diminution of aflatoxicosis in growing chickens by dietary addition of a hydrated sodium calcium aluminosilicate. *Poultry Science* **69:**727–735.

Kubena, L. F., Huff, W. E., Harvey, R. B., Yersin, A. G., Elissalde, M. H., Witzel, D. A., Giroir, L. E., Phillips, T. D. and Petersen, H. D. 1991. Effects of a hydrated sodium calcium aluminosilicate on growing turkey poults during aflatoxicosis. *Poultry Science* **70**:1823–1830.

Ledoux, D. R., Rottinhaus, G. E., Bermudez, A. J. and Alonso-Debolt, M. 1999. Efficacy of a hydrated sodium calcium aluminosilicate to ameliorate the toxic effects of aflatoxin in broiler chicks. *Poultry Science* **78**:204–210.

Lemke, S. L. 2000. Investigation of clay-based strategies for the protection of animals from the toxic effects of selected mycotoxins. Ph.D. dissertation, Texas A&M University, College Station, Tex.

Li, J., Smith, J. A. and Winquist, A. S. 1996. Permeability of earthen liners containing organobentonite to water and two organic liquids. *Environmental Science and Technology* **30**:3089–3093.

Lilly, P. D., Ross, T. M. and Pegram, R. A. 1997. Trihalomethane comparative toxicity: acute renal and hepatic toxicity of chloroform and bromodichloromethane following aqueous gavage. *Fundamental and Applied Toxicology* **40**:101–110.

Lindemann, M. D., Blodgett, D. J., Kornegay, E. T. and Schurig, G. G. 1993. Potential ameliorators of aflatoxicosis in weanling/growing swine. *Journal of Animal Science* **71**:171–178.

Marquez, R. N. and de Hernandez, I. T. 1995. Aflatoxin adsorbent capacity of two Mexican aluminosilicates in experimentally contaminated chick diets. *Food Additives and Contaminants* **12**:431–436.

Masimango, N., Remacle, J. and Ramaut, J. 1979. Elimination, par des argiles gonflantes de L'aflatoxine B$_1$ des milieus contamines. *Annales de la Nutrition et de l'Alimentation* **33**:137–147.

McConnell, E. E. and Moore, J. A. 1979. Toxicopathology characteristics of halogenated aromatic hydrocarbons. *Annals of the New York Academy of Sciences* **320**:138–150.

McConnell, E. E. 1980. Acute and chronic toxicity, carcinogenesis, reproduction, teratogenesis and mutagenesis in animals, pp. 109–150. *In* Halogenated Biphenyls, Triphenyls, Naphthalenes, Dibenzodioxins and Related Products, R. D. Kimbrough (Ed.). Amsterdam: Elsevier/North Holland.

McMahon, G., Davis, E. and Wogan, G. N. 1987. Characterization of c-Ki-ras oncogene alleles by directing sequencing of enzymatically amplified DNA from carcinogen-induced tumors. *Proceedings of the National Academy of Sciences (USA)* **84**:4974–4978.

McMahon, G., Hanson, L., Lee, J. J. and Wogan, G. N. 1986. Identification of an activated c-Ki-ras oncogene in rat liver tumors induced by aflatoxin B$_1$. *Proceedings of the National Academy of Sciences (USA)* **83**:9418–9422.

Moore, J. A., McConnell, E. E., Dalgard, D. W. and Harris, M. W. 1979. Comparative toxicity of three halogenated dibenzofurans in guinea pigs, mice, and rhesus monkeys. *Annals of the New York Academy of Science* **320**:151–163.

Mueller, J. G., Chapman, P. J. and Pritchard, P. H. 1989. Creosote contaminated sites: their potential for bioremediation. *Applied Environmental Microbiology* **57**:1277–1285.

Mueller, J. G., Lantz, S. E., Blattman, B. O. and Chapman, P. J. 1991. Biodegradation of creosote and pentachlorophenol in contaminated groundwater: chemical and biological assessment. *Environmental Science and Technology* **25**:1045–1055.

Palumbo, S. A., Rajkowski, K. T. and Miller, A. J. 1997. Current approaches for reconditioning process water and its use in food manufacturing operations. *Trends in Food Science Technology* **8**:69–74.

Park, D. L., Lee, L. S., Price, R. L. and Pohland, A. E. 1988. Review of the decontamination of aflatoxins by ammoniation: current status and regulation. *Journal of the Association of Official Analytical Chemists International* **71**:685–703.

Pestka, J. J. and Bondy, G. S. 1994. Mycotoxin-Induced Immune Modulation, pp. 163–182. *In* Immunotoxicology and Immunopharmacology, J. H. Dean, M. I. Luster, A. E. Munson, I. Kimber (Eds.). New York: Raven.

Phillips, T. D., Clement, B. A. and Park, D. L. 1994. Approaches to reduction of aflatoxins in foods and feeds, pp. 383–406. *In* The Toxicology of Aflatoxin: Human Health, Veterinary and Agricultural Significance, L. D. Eaton, J. D. Groopman (Eds.). New York: Academic Press.

Phillips, T. D., Clement, B. A., Kubena, L. F. and Harvey, R. B. 1990. Detection and detoxfication of aflatoxins: prevention of aflatoxicosis and aflatoxin residues with hydrated sodium calcium aluminosilicates. *Veterinary and Human Toxicology* **32**:15–19.

Phillips, T. D., Kubena, L. F., Harvey, R. B., Taylor, D. R. and Heidelbaugh, N. D. 1987. Mycotoxin hazards in agriculture: new approach to control. *Journal of the American Veterinary Medical Association* **190**:1617–1622.

Phillips, T. D., Kubena, L. F., Harvey, R. B., Taylor, D. R. and Heidelbaugh, N. D. 1988. Hydrated sodium calcium aluminosilicate: a high affinity sorbent for aflatoxin. *Poultry Science* **67**:243–247.

Phillips, T. D., Lemke, S. L. and Grant, P. G. 2002. Characterization of clay-based enterosorbents for the prevention of aflatoxicosis, pp. 157–171. *In* Advances in Experimental Medicine and Biology: Mycotoxins and Food Safety, Vol. 504, J. W. DeVries, M. W. Trucksess, L. S. Jackson (Eds.). New York: Kluwer Academic/Plenum Publishers.

Phillips, T. D., Sarr, A. B. and Grant, P. G. 1995. Selective chemisorption and detoxification of aflatoxins by phyllosilcate clay. *Natural Toxins* **3**:204–213.

Phillips, T. D., Sarr, A. B., Clement, B. A., Kubena, L. F. and Harvey, R. B. 1991. Prevention of aflatoxicosis in farm animals via selective chemisorption of aflatoxin, pp. 223–237. *In* Mycotoxins, Cancer and Health, Vol. 1, G. A. Bray, D. H. Ryan (Eds.). Baton Rouge: Louisiana State University Press.

Phillips, T. D. 1999. Dietary clay in the chemoprevention of aflatoxin-induced disease. *Toxicological Sciences* **52**:118–126.

Ramos, A. J. and Hernández, E. 1996. In vitro aflatoxin adsorption by means of a montmorillonite silicate. A study of adsorption isotherms. *Animal Feed Science and Technology* **62**:263–269.

Richard, J. L., Thurston, J. R. and Pier, A. C. 1978. Effects of mycotoxins on immunity, pp. 801–817. *In* Toxins: Animal, Plant, and Microbial, P. Rosenberg (Ed.). New York: Pergamon.

Sarr, A. B. 1992. Evaluation of innovative methods for the detection and detoxification of aflatoxin, Ph.D. dissertation. Texas A&M University, College Station, Tex.

Sarr, A. B., Mayura, K., Kubena, L. F., Harvey, R. B. and Phillips, T. D. 1995. Effects of phyllosilicate clay on the metabolic profile of aflatoxin B_1 in Fischer-344 rats. *Toxicology Letters* **75**:145–151.

Scott, P. M. and Trucksess, M. W. 1997. Application of immunoaffinity columns to mycotoxin analysis. *Journal of the Association of Official Analytical Chemists International* **80**:941–949.

Sheng, G., Xu, S. and Boyd, S. A. 1996. Cosorption of organic contaminants from water by hexadecyl-trimethylammonium exchanged clays. *Water Research* **30**:1483–1489.

Smith, E. E., Phillips, T. D., Ellis, J. A., Harvey, R. B., Kubena, L. F., Thompson, J. and Newton, G. 1994. Dietary hydrated sodium calcium aluminosilicate reduction of aflatoxin M_1 residue in dairy goat milk and effects on milk production and components. *Journal of Animal Science* **72**:677–682.

Smith, J. A. and Jaffe, P. R. 1994. Benzene transport through landfill liners containing organophilic bentonite. *Journal of Environmental Engineering* **120**:1559–1577.

Smith, R. M., O'Keefe, P. W., Hilker, D. R., Jelus-Tyror, B. L. and Aldous, K. M. 1982. Analysis for 2,3,7,8-tetrachlorodibenzofuran and 2,3,7,8-tetrachlorodibenzo-*p*-dioxin in a soot sample from a transformer explosion in Binghamton, New York. *Chemosphere* **8**:715–720.

Smith, T. K. 1984. Spent canola oil bleaching clays: potential for treatment of T-2 toxicosis in rats and short-term inclusion in diets of immature swine. *Canadian Journal of Animal Science* **64**:725–732.

Srinivasan, K. R. and Fogler, H. S. 1990a. Use of inorgano-organo clays in the removal of priority pollutants from industrial wastewater: structural aspects. *Clays and Clay Minerals* **38**:277–286.

Srinivasan, K. R. and Fogler, H. S. 1990b. Use of inorgano-organoclays in the removal of priority pollutants from industrial wastewaters: adsorption of benzo(*a*)pyrene and chlorophenols from aqueous solutions. *Clays and Clay Minerals* **38**:287–293.

Stalling, D. L., Smith, L. M., Petty, J. D., Hogan, H. W., Johnson, J. L., Rappe, C. and Buser, H. R. 1983. Residues of polychlorinated dibenzo-*p*-dioxins and dibenzofurans in Laurentian Great Lakes fish, pp. 221–240. *In* Human and Environmental Risks of Chlorinated Dioxins and Related Compounds, R. E. Tucker, A. L. Young, A. P. Gray (Eds.). New York: Plenum.

Stotzky, G. 1980. Surface interactions between clay minerals and microbes, viruses, and soluble organics, and the probable importance of these interactions to the ecology of microbes in soil, pp. 231–247. *In* Microbial Adhesion to Surfaces, R. C. W. Berkeley (Ed.). Chichester: Halsted.

Swan, S. H., Waller, K., Hopkins, B. and DeLorenze, G. 1998. Trihalomethanes in drinking water and spontaneous abortion. *Journal of Epidemiology* **9**:134–140.

United States Environmental Protection Agency. 1997. Test methods for evaluating solid waste. Office of Solid Waste and U.S. Department of Commerce, National Technical Information Service (PB97-501928).

Voss, K. A., Dorner, J. W. and Cole, R. J. 1993. Amelioration of aflatoxicosis in rats by volclay NF-BC, microfine bentonite. *Journal of Food Protection* **56**:595–598.

Xu, S., Sheng, G. and Boyd, S. A. 1997. Use of organoclays in pollution abatement. *Advances in Agronomy* **59**:25–62.

Part VI

Risk Analysis

30 Food Safety Risk Communication and Consumer Food-Handling Behavior

Wm. Alex McIntosh

Introduction

Although consumers are increasingly aware that foodborne illnesses are a problem, they continue to blame others for the problem, unaware of the major role they themselves play in spreading foodborne pathogens. Surveys continue to show that <20% of consumers believe that unsafe foods result from unsafe food-handling practices in the home (Food Marketing Institute 2002, Williamson et al. 1992). This perception results from the consumer's lack of knowledge, particularly regarding safe food-handling procedures at home, and from the consumer's exposure to mass media stories that tend to focus on industry-associated food safety miscues. Some stories focus on a particular event such as a recall of contaminated meat; others are more in-depth exposés of food mishandling in restaurant kitchens. Stories about mishandling food in the home are much rarer. In general, consumers have a high level of awareness and some knowledge, but they lack a more basic understanding of food safety principles that might be used to protect them. Furthermore, it is well known that increased knowledge does not automatically translate into behavior change (Bennett and Murphy 1997, Rogers 1983). Most food-handling practices are habitual; changing those practices is difficult and requires breaking old habits. To change habits, people must be highly motivated. The purpose of this chapter is to discuss a number of issues associated with the consumer's food-handling practices, in particular the knowledge/awareness distinction, the lack of connection between knowledge change and behavior change, the problem of habit, and the need for motivation for behavior change to take place. Mechanisms for improving the consumer's food-handling practices are also discussed.

Awareness Knowledge and Its Problems

Rogers (1983) suggested that knowledge comes in several forms, including awareness, knowledge of principles, and how-to knowledge. These forms of knowledge have a particularly useful application in the study of consumer food behavior (Guthrie et al. 1999). Awareness of a food problem (e.g., the existence of *Escherichia coli*) can be thought of as a precursor to knowledge of principles (e.g., *E. coli* is associated with undercooked hamburger) and knowledge of how to deal with the problem (e.g., deal with *E. coli* by properly handling and cooking hamburger meat). Given the mass media's close attention to nutrition and food safety (Woodburn and Rabb 1997), we should expect a high percentage of consumers to express awareness of foodborne microbes. In fact, some studies of awareness bear this out. For example, The Food Marketing Institute (2002) found that 83% of people polled believe that bacteria/germs constituted a health risk.

Less than 15% of Californians were found to be completely confident in the food safety practices of supermarkets, but only 50% of these respondents considered bacterial contamination a "major concern"(Bruhn and Schutz 1999). Fein et al. (1995) found that less than half of their pooled sample believed that foodborne illnesses were common, and only 29% of those who thought that they had experienced a foodborne illness in the past recognized the names of five to eight pathogens; 20% of those who believed that they had never had a foodborne illness also were aware of the names of five to eight pathogens. Ninety-nine percent of Oregonians had heard of *Salmonella* Typhimurium (Woodburn and Raab 1997), but only 7% had heard of *Campylobacter*. A national telephone survey conducted in 1992–1993 found that 80% of respondents had heard of *Salmonella* Typhimurium, 75% heard of *Clostridium botulism*, and 4.7% heard of *Campylobacter jejuni* or *C. coli* (Altekruse et al. 1996). Fewer respondents, however, were able to connect these pathogens to specific foods. Only 23% knew of *Clostridium botulism*'s food vehicle.

Forty-six percent of Texans indicated awareness of the dangers associated with the degree of cooking hamburger patties (McIntosh et al. 1994). However, only about half of the respondents identified undercooking as the reason for this danger; nearly one-third associated danger with overcooking, believing that overcooking increased their chances of cancer. The Texas study did find that the more aware respondents were of the various types of foodborne illnesses, the more aware they were that there are dangers associated with cooking hamburger meat. In addition, in many of these studies, awareness of foodborne illness or problematic cooking practices were not associated with either principle or how-to knowledge.

Furthermore, awareness does not always lead to proper behavior. McIntosh et al. (1994) found that awareness of the dangers associated with cooking hamburger was only moderately associated with how the respondents currently cooked hamburger. Awareness of the dangers had no association with the willingness to cook hamburger to a greater degree of doneness. Those most willing to change already preferred well-done hamburger meat. Altekruse et al. (1996) found that although 80% of their respondents were aware of the need to wash their hands or to change/wash their cutting boards after cutting raw meat, only two-thirds of those respondents actually did so. Despite mistrust in the food safety record of supermarkets, the respondents in the Bruhn and Schutz (1999) study were more likely to minimize the health risk by reducing the amount of fat in their diets (73%) rather than potential bacterial contamination (58%). Nineteen percent of these respondents dealt with concerns about the safety of poultry by buying less of this food item.

The lack of relationship between awareness on one hand, and principles, how-to knowledge, and behavior on the other, suggests that consumers may retain only superficial levels of knowledge. However, there may be methodological reasons for this lack of observed relationships as well.

Methodological Issues in Studying Knowledge

Studies vary in terms of the degree to which respondents are found to be knowledgeable about foodborne pathogens and related issues. Part of the lack of similarity may be the result of differences in sampling design; part has to do with the timing of the survey (surveys conducted during those times in which the mass media has devoted a greater than average frequency of reporting on food safety tend to demonstrate a greater knowledge of the respondents); and part may be related to differential wording of the surveys themselves. One such difference is the use of questions regarding awareness of foodborne illnesses as proxies for questions regarding principles or how-to knowledge. Having heard of *E. coli* is not the same thing as knowing the origins of *E. coli* in food or how to deal with it. In addition, awareness questions may be more prone to social-desirability biases than knowledge questions (Fowler 1995). Research on "psuedo attitudes" has demonstrated that as many as one-third of the respondents in a survey may express an attitude toward something about which they have never heard (Bishop et al. 1989). Less prone to social desirability are knowledge questions that deal with principles or practices, for these questions are usually framed in terms of a set of answer choices, only one of which is the correct answer. Awareness questions tend to have "yes-no" response formats and there is no right or wrong answer involved. Thus respondents may claim to be familiar with a number of foodborne illnesses while actually possessing no real knowledge of them.

Principle and Practice Knowledge

In a study of low-income Arizonans, Meer and Misner (2000) found that although 67% correctly identified cross-contamination as a potential source of problems (principle knowledge), a substantial percentage believed that they could determine the safety of a food by either smelling or tasting it (practice knowledge). Oregonians appeared to have greater principle knowledge than residents of other states in terms of the kinds of foods that posed a risk of foodborne illness: 88% identified meat, fish, poultry, milk, eggs, and salads. Fifty-six percent knew that "thorough cooking" was the way to avoid *Salmonella,* and 59% knew that such cooking was also the means for dealing with *E. coli* (Woodburn and Raab 1997). However, a few respondents believed that either thorough washing or freezing would effectively deal with *Salmonella* or *E. coli*, and nearly one-third believed that "food could not be made safe." Those who were concerned about food safety possessed more principle knowledge. Williamson et al. (1992) studied people's knowledge of food typically associated with various foodborne illnesses. Fifty-eight percent of the individuals interviewed knew that infectious cuts could be caused by *Staphylococcus*; 75% correctly associated *Salmonella* with raw poultry and eggs. Knowledge of the correct food-handling procedures was less widespread; only one-third could identify the proper way of storing leftover stew, and only 54% knew that knives and cutting boards used to cut raw meat should be washed with soap and water before using them to cut vegetables.

In a recent study by Anderson et al. (2000) only 55% correctly identified that eating a foil-covered potato left out overnight was risky and 54% acknowledged that thawing chicken 24 hours in advance was risky. The same study found that respondents believed that the amount of time needed to properly wash one's hands was 56 seconds (as opposed to the recommended 20 seconds). Gettings and Kiernan (2001) found that many of their elderly respondents believed that sight or touch were sufficient indicators of doneness in foods.

Finally, we can not assume that consumers' knowledge of all aspects of food safety is consistently high simply because those consumers have adequate knowledge of some aspects of food safety issues.

The Link Between Knowledge and Behavior

Of course our interest in knowledge is not for knowledge's sake, but because we believe that it will lead to improved behavior. Numerous studies have found that existing knowledge as well as efforts to increase knowledge often have little effect on behavior. This is true in the areas of nutrition knowledge and nutrition behavior. One problem associated with this area of research has to do with the reliance on self-reported behavior. In studies that include questions about food safety knowledge, respondents may alter their answers to food safety behavior questions to show greater consistency of their answers to the knowledge questions. Research by Anderson et al. (2000) not only asked consumers questions regarding such issues as the length of time food preparers should wash their hands before handling food, they also asked these consumers how often they washed their own hands on average. The researchers then videotaped those consumers' meal preparation behavior in the consumers' kitchens. The results of comparing the answers to the questions and the behavior observation demonstrate that respondents claims about the length of time they usually spent washing their hands greatly exaggerated their actual behavior (respondents indicated that they should wash their hands for 56 seconds, and 89% indicated they exceeded that amount; however, on average they actually spent less than 5 seconds washing their hands).

Years of effort to change knowledge and attitudes have often ended without having much effect on behavior (Rogers 1995, Gochman 1997). In fact, consumers' awareness of, and knowledge about, foodborne illnesses have increased considerably over the last decade, although concomitant changes in their food-handling behavior have not occurred (Anonymous 2001). What appears to be most effective is the change in particular beliefs that may hinder or facilitate behavior change such as perceptions of costs and benefits of change or the seriousness of the consequences of non-change. In addition to knowing, consumers must also be motivated. Motivation will be needed to overcome the inertia of habit. The Health Belief Model, which has a long history of use in studies of health behavior change, accounts for such beliefs and motivations.

Uses of the Health Belief Model

The Health Belief Model was developed by researchers in the Public Health Service as a means of investigating the source of "widespread failure of people to participate in programs to prevent or detect disease" (Strecher et al. 1997, p. 73). The model was developed around several assumptions, including first, that people value health, and second, that they believe that their actions may affect their well-being. The latter, however, was said to be a function of the seriousness of a given threat to their health, the perception of personal susceptibility to the threat, existing "courses of action" enabling them to reduce their risk, or the benefits outweighing the costs of action (Strecher et al. 1997). As Gochman (1997) adds, "the individual must also be aware of the risk" (p. 5). This awareness as well as efforts to reduce the risk may occur because of a cue or trigger factor. This could include exposure of an individual to a risk that has led to an unwanted outcome (e.g., an illness) and to general information about the risk and its consequences (e.g., through a newspaper article).

Seriousness of the Threat

A number of studies have included measures of seriousness, usually in the form of questions about how concerned the respondents are about food safety issues. The results of such surveys show that the consumer indeed perceives food safety as a serious issue and that many perceive "food spoilage" the reason for that concern. However, many studies indicate that equal proportions of respondents perceive "chemical residues" and "fat content" as equally serious problems. More sophisticated studies have been done in which the seriousness of foodborne illnesses has been compared with the seriousness of other threats. For example, it would appear that other food safety concerns involving food production and food processing, such as worries about hormones in meat, food additives, pesticide residues, and irradiation, are perceived to be more threatening than is microbial contamination (Brewer et al. 1994, Sparks and Shepard 1994). Schafer et al. (1993a) used a measure of the degree of seriousness that consumers lend to food safety; however, their scale of questions included consumer fears about food processing. They found that respondents who perceived serious threats to health from microbial contamination or from chemical residues engaged in more food safety behaviors.

Susceptibility

In addition to seriousness of the threat, consumers also must believe that they themselves are at risk. Many people view particular conditions or behaviors as particularly risky yet discount the risk to themselves. This is often the case because the individual perceives that they infrequently find themselves in risk-taking conditions or behaviors. They also may indicate that the risk is created by others rather than themselves. This is the case with risky food behaviors; despite efforts to educate the consumer regarding the importance of food safety practices in the home, many continue to claim that either the food industry or restaurants are the primary cause of foodborne illnesses (Food Marketing Institute 2002).

Schafer et al. (1993a) investigated the effects of perceived risk (situational stress—threat, loss, and challenge) of the health threat of unsafe food. People generally cope with threats with irrational approaches such as "faith, wishful thinking, fatalism, avoidance, or rational action such as learning more about food safety" (Schafer et al. 1993a, p. 387). Susceptibility to foodborne illnesses was measured by the degree to which respondents felt personally threatened by unsafe food. Schafer et al. (1993a) found that those who felt personally threatened by foodborne illnesses were more likely to report that they would use rational means for coping with the threat of foodborne illness.

Personal Experience

The Health Belief Model does not take into account the possibility that the individual may have already experienced the very health problem that behavior change was meant to avoid. This is most likely the result of the nature of the health problems to which the Health Belief Model has been applied. Many of the health problems are chronic in nature, such as cardiovascular disease. However, a person can contract and recover from a foodborne illness multiple times. It is not unreasonable to suggest that those with prior experiences with a given illness are likely to perceive that disease's seriousness and their susceptibility to that disease differently from those who have not experienced the ailment. Those who believe that they had recently contracted a foodborne illness are more aware and somewhat more knowledgeable. For example, Fein et al. (1995) found that 42% of those who perceived that they had experienced a foodborne illness perceived that food poisoning was very common; only 28% of those who had not had a foodborne illness felt this way.

We must be cautious about people's claims of having had a foodborne illness. Fein et al. (1995) suggested that apparently many who believe that they had such an illness probably had not. Such reports should not be a surprise given that lay perceptions often differ from those of medical professionals (Lau and Hartman 1983).

Finally, there is some evidence that suggests that those with a perceived previous experience with a foodborne illness actually engage in fewer food safety practices than those reporting no such experience (Fein et al. 1995). This suggests that although the perceived risk of contracting a foodborne illness may increase for having experienced what was believed to be a foodborne illness in the past, those who reported mild or inaccurate symptoms may have concluded that the consequences of such an illness are not very severe.

Benefits/Barriers to Change

Few studies have actually investigated perceptions of benefits consumers would obtain from changing their food-handling practices. However, several studies have included barriers to change in the form of perceived costs of such change. Schafer et al. (1993b) measured barriers to change with an item that inquired about the respondents' willingness to sacrifice food safety in return for lower food costs. The study found that individuals who were willing to sacrifice food safety for lower food prices were more likely to cope irrationally with the threat of unsafe food (they perceived that food safety was out of their control, tried not to think about the safety of the food eaten, and felt that concerns about food safety are overreactions). Other studies show that some consumers would be willing to pay more in exchange for safer food (Fox et al. 1995), whereas others are not willing (Lin and Milon 1995, Frenzen et al. 2000).

Safe food preparation does not necessarily significantly increase the amount of time needed to prepare food. However, people's perceptions of time expenditures are often grossly out of touch with actual time expenditures. It may seem to consumers that taking the additional time to wash their hands, clean cutting boards and cutting instruments, and so forth, may be too inconvenient. A study of elderly food preparers found that time expenditure was perceived as a barrier to practicing safer food-handling procedures (Gettings and Kiernan 2001). Mothersbaugh et al. (1993) found, however, that persons with more nutrition knowledge were less likely to let perceptions of time constraints interfere with the practice of recommended dietary practices.

Finally, a cost often overlooked in behavior change models is the effort involved in making the change. Such effort involves changing established habits and is often considerable.

Habits

"A habit is a behavior that has been performed repeatedly so that it is performed automatically without necessity for a conscious decision that involves the consideration of at least one other alternative course of action" (Maddux and DuCharme 1997, p. 145). Most behaviors began as intentional and later become habits. Tooth brushing and wearing seatbelts tend to become part of automatic cognitive processes that are set in motion by situational cues. The problem habits suggest that once they have become automatic, they are far less amenable to changes in beliefs, attitudes, or intentions. McIntosh et al. (1994) found that consumer habits include preferred modes of cooking meats and that knowledge of foodborne risks may be insufficient to overcome those habits. The recent Research Triangle Institute report on food safety points out that forgetfulness lies behind people's failure to properly wash their hands before preparing food (Anonymous 2001). This forgetfulness may be simply habit.

Cues to Behavior: Food Scares

Health belief researchers suggest that behavior change is often assisted by a cue or trigger. This can come in the form of a warning from a physician about the seriousness of a health problem and the patient's susceptibility to that problem. For many others, cues come from the mass media. When news outlets report a foodborne illness outbreak or the endemic presence of *Salmonella* in chicken is published, "food scares" are said to have occurred. Under these circumstances, consumer perception of risk may outstrip actual risk, with the mass media acting as a catalyst in the process (Beardsworth 1990). As the consumer becomes sensitized to a high degree, a "moral panic" ensues; in other words, there becomes a "spiral of rising public concern and intensifying media attention which accompanies these events" (Beardsworth 1990, p. 11). Once sensitized, an audience reacts, leading to further media attention, and thus further reaction. However, as other social problems appear on the horizon, the media switches attention, as does the audience. Thus, perceived risk and panic regarding problems such as foodborne illnesses may lessen as the public turns its attention to a more recent crisis. The problems of food scares are made worse by the fact that the science of nutrition, at least in the mind of the consumer, sometimes produces conflicting advice regarding topics such as dietary recommendations (Guthrie et al. 1999). In addition, the wide variety of food items available to consumers may inadvertently contribute to fear because there are more things to be worried about. Woodburn and Raab (1997) suggest that knowledge of foodborne illnesses is greater immediately after an outbreak of foodborne illness. From May to July 1995, nutrition and food safety was the eighth most frequent topic in national/local news outlets (Woodburn and Raab 1997). Furthermore, consumers frequently report that their chief source of information about food safety is the mass media (Woodburn and Raab 1997, McIntosh et al. 1994).

Issues Related to Amplification

Some have argued that journalists exaggerate risks associated with new food technologies (Maney and Plutzer 1996). These researchers found that newspaper editors were generally no more concerned with "levels of pesticide use in American agriculture, genetically engineered fruits and vegetables, or meat products whose shelf life has been extended due to exposure to radiation" than were Congressional staffers or scientists/engineers (Maney and Plutzer 1996, p. 46). However, 79% of the scientists rated irradiation as safe, whereas only slightly over half of the journalists did so.

Despite the tendency to exaggerate risks associated with food technologies on the part of the mass media, research supports the hypothesis that messages about the seriousness of a condition such as a foodborne illness that induce fear in the receiver will more likely lead to behavior change (Rogers and Prentice-Dunn 1997).

Improving Knowledge and Changing Behavior

Given the discussion of the Health Belief Model presented above, any intervention aimed at changing food safety behavior at home must first change beliefs about the seriousness of foodborne illnesses, susceptibility to foodborne illnesses, and benefits and costs of behavior change to avoid a foodborne illness.

Some have suggested that food labels might serve as a vehicle for messages about how to handle the food item safely. Hunter (1994) claimed that because many people do not read current food labels at point of purchase, such labels may be ineffective. She also notes that the consumer most likely needs proper cooking or storage information after the food has been purchased. There is little research that supports product labeling as an effective means of behavior change. This is supported by the fact that warning labels on cigarette packages have made little headway toward reducing smoking (Bennett and Murphy 1997). Effective food labeling would have to emphasize susceptibility to and the seriousness of foodborne illness and the benefits and costs of changing food-handling practices at home. The elements of such a message would undoubtedly need a great deal of space, making the size of such food labels unacceptably large from the food manufacturers' point of view.

Another approach that has demonstrated some success is group discussions. Trained group leaders organize consumers into groups to discuss food safety. Some researchers have found that such group discussions lead to increased knowledge and expressions of a willingness to change food-handling practices at home (Reicks et al. 1994). Others have found that such group discussions increase consumers' favorability toward the use of food irradiation (Sapp and Herrod 1992). However, organizing small groups from large numbers of consumers to discuss changing personal food safety behavior is not feasible in terms of logistical and economic costs.

The most practical and cost-effective means of reaching the consumer would appear to be public information campaigns. These campaigns might consist of news releases and extended articles in widely read magazines. They should include information aimed at increasing perceptions of the seriousness of foodborne illnesses, people's susceptibility to such illnesses, and behaviors that lessen the risk of such illnesses. The development of such an information campaign could involve a cooperative effort on the part of the food industry and the federal government, as consumers have expressed trust in both industry and government as sources of food safety information (Peters et al. 1997).

Another less expensive intervention with perhaps long-term consequences is to use the public school system as a vehicle for change. Many states require students to take a health education class into which information regarding food safety could be inserted. A further advantage of this strategy is that children and adolescents are in the process of forming food habits that will carry over into adulthood.

Increasing trust in information sources may be another way to influence consumers. "Perceptions of trust and credibility of environmental risk communications depended on: (1) perceptions of knowledge and expertise; (2) perceptions of openness and honesty; and (3) perceptions of concern and care" (Peters et al. 1997, p. 44). Recent focus group research indicates that the consumers' level of trust in the food industry has increased (Anonymous 2001). This suggests that the food industry could potentially play a role in disseminating information about safe food-handling practices. Consumers trust government in terms of its commitment to the health and safety of the public and because of the knowledge of its food safety experts (Peters et al. 1997).

Conclusions

Consumers are aware of food safety problems and are concerned about them, yet are not making improvements in their food-handling behavior. The following recommendations are made to ameliorate this situation:

(1) Studies that continue to explore knowledge need to distinguish between awareness knowledge and principles and how-to knowledge. Researchers should consider measuring only principles and how-to knowledge, given the biased nature of awareness questions.

(2) Studies of self-reported safe food-handling behavior may also be biased, as respondents overestimate the extent of their efforts. Observations of actual food-handling behavior are recommended.

(3) These studies should employ theoretical models that provide the basis for predicting food safety behavior. The Health Belief Model, with its emphasis on perceptions of susceptibility, severity, and costs versus benefits of behavior change, has shown promise in predicting food safety behavior.

(4) Efforts to increase safe food-handling by the consumer needs not only to increase knowledge but also to change unsafe behaviors. This means a greater focus on susceptibility and severity of foodborne illnesses, using the public school system as well as public health messages that emphasize the threat that foodborne illnesses pose to the average person.

(5) Efforts should be made by the food industry and the federal government to work cooperatively in the development and dissemination of these public health messages. As the mass media is a frequent source of information for many adults, these messages could be presented in the form of public service announcements.

References

Altekruse, S. F., Street, D. A., Fein, S. B. and Levy, A. S. 1996. Consumer knowledge of foodborne microbial hazards and food-handling practices. *Journal of Food Protection* **59**:287–294.

Anderson, J. B., Schuster, T., Gee, E., Hansen, K., Mendenhall, V. T. and Volk, A. R. 2000. A camera's view of consumer food safety practices. Final report to the U.S. Food and Drug Administration. Logan, Utah: Department of Nutrition and Food Science, Utah State University.

Anonymous. 2001. Changes in Consumer Knowledge, Behavior, and Confidence since the 1996 PR/HACCP Final Rule. PR/HACCP Rule Evaluation Report. Research Triangle Park, N.C.: Research Triangle Institute.

Beardsworth, A. D. 1990. Trans-science and moral panics: understanding food scares. *British Food Journal* **92**:11–16.

Bennett, P. and Murphy, S. 1997. Psychology and Health Promotion. Philadelphia, Penn.: Open University Press.

Bishop, G. F., Oldendick, R. W., Tuchfarber, A. J. and Bennett, S. E. 1989. Pseudo-Opinions on Public Affairs, pp. 425–436. *In* Survey Research Methods: A Reader, E. Singer, S. Presser (Eds.). Chicago: University of Chicago Press.

Brewer, M. S., Sporuls, G. K. and Russon, C. 1994. Consumer attitudes toward food safety issues. *Journal of Food Safety* **14**:63–76.

Bruhn, C. and Schutz, H. G. 1999. Consumer food safety knowledge and practices. *Journal of Food Safety* **19**:73–87.

Fein, S. B., Lin, C.-T. J. and Levy, A. B. 1995. Foodborne illness: Perceptions, experience, and prevention behaviors in the United States. *Journal of Food Protection* **58**:1405–1411.

Food Marketing Institute. 2002. Trends in the United States: Consumer Attitudes and the Supermarket, 2002. Washington, D.C.: Research Department, Food Marketing Institute.

Fowler Jr., F. J. 1995. Improving Survey Questions: Design and Evaluation. Thousand Oaks, Calif.: Sage.

Fox, J., Shogren, J., Hayes, D. and Kliebenstein, J. 1995. Experimental auctions to measure willingness to pay for food safety, pp. 115–128. *In* Valuing Food Safety and Nutrition, J. A. Caswell (Ed.). Boulder, Colo.: Westview.

Frenzen, P. D., Majchronwicz, A., Buzby, J. C. and Imhoff, B. 2000. Consumer acceptance of irradiated meat and poultry products. *In* Issues in Food Safety Economics. Agriculture Information Bulletin No. 757. Washington, D.C.: Economic Research Service, United States Department of Agriculture.

Gettings, M. A. and Kiernan, N. E. 2001. Practices and perceptions among seniors who prepare meals at home. *Journal of Nutrition Education* **33**:148–154.

Gochman, D. S. 1997. Health behavior research: definitions and diversity, pp. 3–20. *In* Handbook of Health Behavior Research, Vol. I: Personal and Social Determinants, D. S. Gochman (Ed.). New York: Plenum Press.

Guthrie, J., Derby, B. M. and Levy, A. S. 1999. What people know and don't know about nutrition, pp. 243–280. *In* America's Eating Habits: Changes and Consequences. Agriculture Information Bulletin No. 750, E. Frazao (Ed.). Washington, D.C.: Economic Research Service, United States Department of Agriculture.

Hunter, B. T. 1994. Labels for food safety education? *Consumer's Research Magazine* **77**:8–9.

Lau, R. R. and Hartman, K. A. 1983. Common sense representations of common illnesses. *Health Psychology* **2**:167–185.

Lin, C.-T. J. and Milon, W. 1995. Contingent valuation of health risk reductions for shellfish products, pp. 83–114. *In* Valuing Food Safety and Nutrition, J. A. Caswell (Ed.). Boulder, Colo.: Westview.

Maddux, J. E. and DuCharme, K. A. 1997. Behavioral intensions in theories of health behavior, pp. 133–152. *In* Handbook of Health Behavior Research, Vol. I: Personal and Social Determinants, D. S. Gochman (Ed.). New York: Plenum Press.

Maney, A. and Plutzer, E. 1996. Scientific information, elite attitudes, and the public safety debate over food safety. *Policy Studies Journal* **24**:42–56.

McIntosh, W. A., Christensen, L. B. and Acuff, G. R. 1994. Perceptions of risks of eating undercooked meat and willingness to change cooking practices. *Appetite* **22**:83–96.

Meer, R. R. and Misner, S. L. 2000. Food safety knowledge and behavior of expanded food and nutrition education program participants in Arizona. *Journal of Food Protection* **63**:1725–1731.

Mothersbaugh, D. L., Herrman, R. O. and Warland, R. H. 1993. Perceived time pressure and recommended dietary practices. *Journal of Consumer Affairs* **27**:106–126.

Peters, R. G., Covello, V. T. and McCallum, D. B. 1997. The determinants of trust and credibility in environmental risk communication: an empirical study. *Risk Analysis* **17**:43–54.

Reicks, M., Bosch, A., Herman, M. and Krinke, U. B. 1994. Effectiveness of a food safety teaching strategy promoting critical thinking. *Journal of Nutrition Education* **26**:97–100.

Rogers, E. M. 1983. Diffusion of Innovations, 3rd ed. New York: Free Press.

Rogers, E. M. 1995. Diffusion of Innovations, 4th ed. New York: Free Press.

Rogers, R. W. and Prentice-Dunn, S. 1997. Protection-motivation theory, pp. 113–132. *In* Handbook of Health Behavior Research, Volume I: Personal and Social Determinants, D. Gochman (Ed.). New York: Plenum Press.

Sapp, S. G. and Harrod, W. J. 1992. Consumer acceptance of irradiated food: a study of symbolic adoption. *Journal of Consumer Studies and Home Economics* **14**:133–145.

Schafer, R. B., Schafer, E., Bultena, G. and Hoiberg, E. 1993a. Coping with a health threat: a study of food safety. *Journal of Applied Social Psychology* **23**:386–394.

Schafer, R. B., Schafer, E., Bultena, G. L. and Hoiberg, E. 1993b. Food safety: an application of the health belief model. *Journal of Nutrition Education* **25**:17–24.

Sparks, P. and Shepard, R. 1994. Public perceptions of the potential hazards associated with food-production and food consumption: an empirical study. *Risk Analysis* **14**:799–806.

Strecher, V. J., Champion, V. L. and Rosenstock, I. M. 1997. The Health Belief Model and health behavior, pp. 71–92. *In* Handbook of Health Behavior Research, Vol. I: Personal and Social Determinants, D. S. Gochman (Ed.). New York: Plenum Press.

Williamson, D. M, Gravani, R. B. and Lawless, H. T. 1992. Correlating food safety knowledge with home food-preparation practices. *Food Technology* **46**:94–100.

Woodburn, M. J. and Rabb, C. A. 1997. Household food preparers' food-safety knowledge and practices following widely publicized outbreaks of foodborne illness. *Journal of Food Protection* **60**:1105–1109.

31 Addressing Microbial Food Safety Issues Quantitatively: A Risk Assessment Approach

Kristina D. Mena, Joan B. Rose, and Charles P. Gerba

Introduction

The occurrence of foodborne illnesses in the United States has increased in recent years, although the continued underreporting of such events makes it difficult to fully realize the associated public health impact. It is estimated, however, that 76 million illnesses occur in the United States each year, resulting in as many as 325,000 hospitalizations and 5,000 deaths (Mead et al. 1999). Greater than 200 types of foodborne illnesses are caused by either microorganisms such as bacteria, viruses, and protozoan parasites or by metals, toxins, and prions. This diversity of types and causative agents of foodborne disease contributes to the difficulty in identifying preventative strategies, especially as each foodborne disease agent can be transmitted through different routes. In addition, the lack of reporting of foodborne illness cases as well as the impact of foodborne pathogens yet to be identified makes surveillance difficult at best.

A systematic approach to addressing food safety issues has become necessary to unravel the complicated and intertwined factors that contribute to foodborne disease. Risk assessment is an approach that was developed by the National Research Council (NRC) in the 1970s as a method to evaluate hazards in the environment and to form a partnership between science and government that would educate the government on current scientific issues (NRC 1983). Risk assessment is the first component of the risk analysis process, followed by risk management and risk communication (NRC 1994; see also Figure 31.1). The goal of developing a risk analysis framework was to have an objective process available that used scientific information to evaluate the human health risks associated with exposure to environmental hazards; this evaluation would then lead to risk management decisions that would minimize the risks (such as the development of hazard analysis critical control point progams).

This methodology has proven to be a useful tool in interpreting and critically analyzing data obtained from environmental epidemiological studies conducted in the United States and throughout the world that address the occurrence of microorganisms and chemicals in the environment, including the food and water environment (Regli et al. 1991, Rose et al. 1991, Rose et al. 1995, Jaykus 1996, Lammerding and Paoli 1997). Many governmental agencies and organizations, such as the Codex Alimentarius Commission (1999), have adopted this process to address microbial food safety risks. In addition, enthusiasm for using risk assessment in the food safety arena has increased as a result of the Sanitary and Phyto-Sanitary Agreement (SPS) of the World Trade Organization (WTO) and the General Agreement on Tariffs and Trade (GATT; Marks et al. 1998). Microbial risk assessment

Figure 31.1   The risk analysis framework.

provides the opportunity for microbial standards that are developed and endorsed to be science-based rather than (perhaps) arbitrary and subjective.

The risk assessment paradigm involves four basic steps: hazard identification, dose-response assessment, exposure assessment, and risk characterization (NRC 1983). Research-based information is incorporated in the risk assessment process and, where data are lacking, assumptions are made and applied. Each of the first three steps is an integral part of the fourth step, risk characterization. The process is dynamic, making this approach applicable to specific hazards or to populations. The following sections discuss how food safety issues are addressed within each of the four steps of the paradigm.

Quantitative Microbial Risk Assessment: Its Components and Application for Foodborne Pathogens

Identifying the Hazard

This first step in the risk assessment framework provides qualitative information regarding the microorganism and the host–microorganism interaction to be used in the other steps. Available laboratory and field data as well as information from epidemiological studies are reviewed and evaluated to determine the role of the microorganism as a hazard, or in other words, its capability of causing adverse harm. In most cases, evidence of a microorganism as a hazard is already apparent, which is what initiated the risk assessment. (Traditionally with chemical agents, the hazard identification is an exploratory step at which the risk assessment may stop if it is determined that the chemical poses no human health threat.) The hazard identification step encompasses a broad spectrum of characteristics regarding the microorganism, including a complete description of the range of human illnesses (both acute and chronic) associated with exposure, severity ranges of those health outcomes, a description of the populations affected. Morbidity, mortality, and hospitalization ratios are also determined here as well as information on endemic and epidemic disease.

Completing the hazard identification step can be challenging if there are limited data available regarding the microorganism and its interaction with hosts, particularly if epidemiological studies are lacking. It is also imperative that the risk assessor critically analyze such studies, which include surveillance studies and outbreak investigations, for appropriateness for inclusion in a risk assessment. Establishing causality (regarding a microorganism's responsibility in initiating infection and disease) when evaluating epidemiological investigations can be difficult, particularly during foodborne outbreak investigations that take place after the event has occurred. Often, food samples are not available for testing, and illness cases are not always identified or specifically diagnosed.

Even with the challenges in identifying etiological agents of foodborne disease, the documentation of foodborne illness cases in the United States is increasing; in particular, illnesses associated with the consumption of fresh produce. This may be the result of several factors: consumer awareness of healthy eating, changes in food production and processing practices, increased awareness of outbreak events, and the emergence of new foodborne pathogens.

Table 31.1 provides a list of some microorganisms that have been implicated as causative agents of foodborne disease. In most cases the associated illnesses involve the gastrointestinal tract (diarrhea, dysentery, vomiting, and nausea), but other health outcomes may result, such as botulism, listeriosis, hepatitis, and even death in the immunocompromised.

Identifying emerging pathogens as foodborne agents of disease before a significant public health effect is challenging. This is particularly important when considering foodborne pathogens within the international trade market. As mentioned earlier, risk assessments of

Table 31.1 Examples of foodborne pathogens.

Bacteria
Bacillus cereus *Campylobacter jejuni* *Clostridium botulinum* *Escherichia coli* O157:H7 *Listeria monocytogenes* *Salmonella typhi* (and non-typhoid) *Shigella* spp.
Viruses
Astrovirus Hepatitis A Norwalk and Norwalk-like Rotavirus
Protozoa
Cryptosporidium parvum *Cyclospora cayetanensis* *Giardia lamblia* *Toxoplasma gondii* *Trichinella spiralis*

microorganisms are usually initiated after the microbe has been observed as a hazard. One approach to overcome this challenge has been termed "structure-activity relationship analyses," where the aim is to conduct a comparison between a particular disease agent whose associated health outcomes are yet to be determined and those agents that may be similar in other ways (Jaykus 1996). The author emphasizes the need and importance of the hazard identification step to be "proactive" rather than "reactive" to better protect public health, particularly when considering emerging pathogens (Jaykus 1996). Some foodborne pathogens posing the greatest public health risks today (e.g., *Escherichia coli* O157:H7 and *Cyclospora cayetanensis*) were only relatively recently identified as causative agents of foodborne disease.

Assessing Exposure

In its simplest definition, exposure assessment determines the intensity and frequency of human exposure to a hazard. The goal is to generate a probability of exposure, and when considering microbial foodborne hazards, there are many factors to consider when conducting an exposure assessment. These factors include source of exposure, population exposed, event or events leading to exposure, magnitude of exposure, and duration of exposure. Information regarding source (what type or types of food) and population characterization as well as exposure frequency may be obtained from the hazard identification step. The phenomenon of secondary transmission that is characteristic of microbial hazards needs to be considered here as well.

To address and describe the event or series of events that would need to take place for a particular outcome to occur (e.g., exposure to a foodborne pathogen or a resultant adverse human health outcome), scenarios are built with regard to food production, processing, and preparation pathways. These "scenario trees" can be built regarding specific microbial hazards and have been employed to address the likelihood of certain endpoints of interest occurring (Jaykus 1996, Marks et al. 1998).

Factors related to the food vehicle, the population exposed, and the survivability of the microorganism determine the magnitude (How many organisms are ingested?) and duration (How often?) of exposure. Food processing, distribution, and preparation influence the potential for and outcome of a microorganism–host interaction. Specifically, the shelf life of the food product, the distribution practices, and the consumption patterns within and among populations all contribute to the magnitude and duration of exposure to a foodborne pathogen. Population demographics and sociocultural factors (food preparation practices and consumption patterns) also play a role in the likelihood of exposure to a microbial hazard and the likelihood of a resulting adverse health outcome.

Quantitative data on the occurrence of specific pathogens in food are needed to fully address the question of magnitude of exposure. These data may be obtained from surveillance studies or perhaps outbreak investigations when possible; however, quantitative data for specific pathogens are lacking because of the complexities of investigating outbreak situations as described previously in this chapter. Furthermore, data would need to be obtained using sensitive and specific laboratory techniques to detect a potentially low number of a specific pathogen present in a food sample.

Predictive microbiology is an approach to estimate—or predict—the growth or death of microorganisms because of environmental factors (Walls and Scott 1997). This is a very active research area within the food microbiology field, particularly with bacteria. The numbers of microorganisms present can change throughout the food production and processing pathways. In predictive microbiology, mathematical models are developed that incorporate variables that (may) influence microorganism survivability to predict microbial growth or die-off. Mathematical models are developed from laboratory-based research, where microorganisms are grown and observed under different conditions (variables) and the data are then fit to a mathematical equation. The complex array of variables includes the composition of the food product itself, the microbial flora of the food, and effects of food processing and preparation, which can introduce different pH and temperature environments, to list a few.

As discussed above, quantitative data are needed on the occurrence of specific microorganisms in food. Sensitive laboratory methods to detect low numbers of microorganisms in food need to be developed or modified to be quick, cost-effective, and capable of detecting specific infectious microorganisms of interest.

The applicability of microbial survivability models developed using predictive microbiology also needs to be assessed and further investigations of variables that influence microbial growth, survival, and inactivation to be performed. This type of research will encompass an assessment of production, processing, and preparation practices that influence microbial food quality. Furthermore, research needs to not only address the characteristics of specific microorganisms of interest but also evaluate the unique conditions present resulting from particular types of foods and how those foods are processed (e.g., produce vs. meat, etc.).

Emerging Issues Associated with Dose–Response

One of the key steps in quantitative microbial risk assessment is defining the outcome associated with a specific exposure. The dose–response assessment is the relationship between the magnitude of the exposure (dose) to the biological agent and the frequency of associated effects. This assessment is accomplished via the development of a dose–response curve based on experimental data. A mathematical characterization of the relationship between the dose administered and probability of infection or disease in the exposed population can be determined. Therefore, in general terms one may answer the question of how likely an infection is, based on the numbers of microorganisms ingested.

For foodborne microorganisms, most of the data sets have been developed in humans, although animal data have been used on occasion. The microorganisms were measured in doses that are routinely used to count the specific microbes in the laboratory, such as colony counts on agar media for bacteria, plaque counts in cell culture for viruses (often underestimating the dose), and direct microscopic counts of cysts/oocysts for the protozoa (potentially overestimating the dose). Numerous subjects were given a range of doses, and the outcomes were determined in each group at each dose. The outcomes were "colonization," which defines infection, disease, or symptoms, or "antibody response" in the individual exposed. In many studies both disease and infection were measured as endpoints. In many cases, however, less-virulent strains of the microorganisms and healthy human adults were used. Despite these limitations in estimation of the dose, the methods used were similar to those used to detect these same microorganisms in food samples, and natural routes of exposure were used; that is, direct ingestion, inhalation, or contact.

Numerous dose–response data sets now exist for foodborne pathogens (Haas et al. 1999a). Most have focused on the bacteria, and several other issues have remained unresolved, including (Haas 2002) types of models and the suggestion of a threshold, low-dose extrapolation, and validation of the models.

Issues Surrounding Model Selection

One of the more controversial areas surrounding microbial modeling is the choice of the model that is used to fit the data. There are a number of alternative models that can be used to model the dose–response data including exponential, beta-Poisson, probit, logistic, and Gompertz (Coleman and Marks 1998). The beta-Poisson model or the exponential model (Table 31.2) has been used predominantly. The exponential or the beta-Poisson models provide a statistically significant improvement in fit over the log-normal model when predicting a no-threshold effect (Haas 1983, Haas et al. 1999a, Teunis and Havelaar 2000).

The mechanistic features of a dose–response model should capture several features associated with the infection process: first, there is heterogeneity in the dose in a population at low doses; second, the distribution of microorganisms between doses is random; third, each organism has an independent and identical survival probability; and fourth, the minimum dose possible is 1. In the simplest form this was shown to be the exponential dose–response relationship:

$$P_i = 1 - \exp(-rN),$$

Table 31.2   Some dose–response models for foodborne pathogens.

Microorganism	Model	Animal	Reference
Bovine spongiform encephalopathy	Exponential $r = 129$	Mice	Gale 1998
Campylobacter	beta-Poisson $\alpha = 0.145$ $N_{50} = 896$	Humans	Haas et al. 1999a
Cryptosporidium	Exponential $r = 0.0042$	Humans	Haas et al. 1996
Escherichia coli	beta-Poisson $\alpha = 0.1952$ $N_{50} = 3.01 \times 10^7$	Humans (included enteroinvasive, enterotoxigenic and enteropathogenic strains)	Haas et al. 1999a
E. coli O157:H7	beta-Poisson $\alpha = 0.49$ $N_{50} = 5.96 \times 10^5$	Infant rabbits	Haas et al. 2000
Listeria	beta-Poisson $\alpha = 0.25$ and 0.14 $N_{50} = 270$ and 410	Mice (two assessments showing two fits)	Haas et al. 1999b
Salmonella	beta-Poisson $\alpha = 0.3126$ $N_{50} = 2.36 \times 10^4$	Humans (nontyphoid, multiple strains)	Haas et al. 1999a
Shigella	beta-Poisson $\alpha = 0.21$ $N_{50} = 1.12 \times 10^3$	Human *flexnerri* and *dysenteriae* (pooled)	Haas et al. 1999a
Rotavirus	beta-Poisson $\alpha = 0.2531$ $N_{50} = 6.17$	Human	Haas et al. 1993, Rose and Sobsey 1993

where P_i equals the probability of infection from a single-dose exposure, N is the exposure—as measured in number of organisms—and r equals a constant associated with the organism's probability of causing infection.

The beta-Poisson distribution assumes that there is some variation in terms of the pathogen–host survival probability (r factor in the exponential formulation). The model also assumes that there is some diversity in the microbial population and the ability to survive and transcend the host defenses as well as variation in the host–response to the microbial insult. The beta-Poisson model is defined as:

$$P_i = 1 - (1 + N/\beta)^{-\alpha}$$

α and β are two parameters that define this variability in the dose–response curve, and they are organism-specific. P_i equals the probability of infection from a single-dose exposure, N is the number of organisms ingested, and β can be further defined as:

$$\beta = N_{50}/(2^{1/\alpha} - 1),$$

which results in the following equation for the beta-Poisson model:

$$P_i = 1 - [1 + N/N_{50}(2^{1/\alpha} - 1)]^{-\alpha}.$$

These have often been described as single-hit models, which predict the potential for a single organism to initiate an infection (known as non-threshold). The choice of these models has been shown to be critical, particularly as these are used to extrapolate risks to low levels of exposure. Teunis and Havelaar (2000) have suggested that the beta-Poisson is not a single-hit model and that discrepancies are greatest at the low-dose level and in evaluating the upper confidence level of the dose–response. This becomes important in uncertainty analysis and in decision-making.

The beta-Poisson is empirical in the sense that it is based on data, yet it is mechanistic in that it has underpinnings based on mechanism. Other empirical models include the log-logistic, log-probit, and Weibull. These have the same mathematical form as a cumulative probability distribution function and have been explored as dose–response models but do not have an underlying mechanistic basis. Best-fitting techniques such as maximum-likelihood methods (Haas et al. 1999a) have shown that these are inferior to the beta-Poisson model.

Low-Dose Extrapolation

Most of the data sets for microorganisms, particularly for bacteria, did not test low doses. The lowest dose evaluated for non-typhoid *Salmonella*, for example, was between 10^3 and 10^4 colony-forming units (Haas et al. 1999a). When comparing the data from the Weibull, beta-Poisson, probit, and logistic models at low doses (1–10 colony-forming units) the differences were about three orders of magnitude, with the beta-Poisson providing the lowest risk and with Weibull providing the highest risk. Data for viruses and protozoa do have low doses in the data sets, and thus the extrapolation to the single exposure and likely outcome is not as problematic.

Buchanan et al. (2000) have described microbiological factors, host factors, and food matrix factors that need to be considered in the establishment of a mechanistic model for dose–response and infection (Table 31.3). They evaluated the gastric acidity barrier and attachment/infectivity process associated with *E. coli*. They were able to predict an increasing survival of bacteria with increasing pH of the stomach and, therefore, the increase in risk of attachment and infection.

Validation of the Models

The plausibility of the various dose–response models requires testing in some type of framework. Fazil (1996) found that outbreak data, if well investigated, could be used to provide a scenario for evaluation of the models. This type of validation was accomplished for *Listeria* and *E. coli* O157:H7 (Haas et al. 1999b, Haas et al. 2000). In foodborne outbreaks the exposure often takes place during a single meal; thus, adequate information can be obtained on attack rates, exposures, and consumption rates. If properly sampled, the concentration of the microorganism can be determined and the probable inoculum size can be used to determine the doses. Uncertainties will remain; often asymptomatic cases are not investigated or reported, the food sample is consumed and there is no record of the contamination level, or the levels were determined in a raw product and the outbreak took place after improper cooking.

Development of New Data

Dose–response data sets are not exhaustive in terms of answering questions regarding the host–microbe interaction. More research using human and animal studies will be needed to further address the mechanisms of dose–response, including virulence, strain variation, immunity, and multiple exposures. Animal dose–response data have been reported for *E. coli* O157:H7 and *Listeria*. New risk estimates have been made for bovine spongiform encephalopathy prions associated with mad cow disease (Gale 1998; see Chapter 14). Foodborne viral disease has received little attention (see Chapter 8), although risks associated with consumption of raw shellfish have been estimated as high as 1 in 10 (Rose and Sobsey 1993). *Cryptosporidium* dose–response models exist (Haas et al. 1996), and more recently, the potential application of cell culture has been used to examine variability in dose–response (Slifko et al. 2002), but this protozoan is not often seen in foodborne outbreaks (Slifko et al. 2000). The more important foodborne protozoa are *Cyclospora* and *Toxoplasma*, and data on dose–response are needed for these parasites. Thus, more attention will need to be paid to the data, estimates, types of models, and validation of those mod-

Table 31.3 Factors for consideration in the development of dose–response models (Buchanan et al. 2000).

Factor	Influences an increase or decrease in probability of infection
Host	Genetic backgrounds, general health, nutritional status, age, immune status, stress levels, and prior exposures
Microorganism	Class of the organism (infectious, toxico-infectious and toxigenic), attachment factors, and virulence factors
Food matrix factors	Buffering capacity, liquids rapid transit from the stomach, emulsified foods with fat droplets, e.g., ice cream (all increasing the potential probability of infection)

els in the future. Given the interest globally in using the risk assessment framework and the large differences between models, it will be important to address these issues surrounding dose–response modeling.

Characterizing Risk

The risk characterization step combines all of the information learned from the first three steps and, in a quantitative risk assessment, computes a numerical estimate of risk. This estimate of risk reflects both the likelihood that a microorganism will cause adverse health outcomes within a population and the severity of those outcomes. Different human health endpoints of interest can be addressed. Not only can probability of infection resulting from exposure to a pathogen be determined but also risks of illness and death can be calculated by incorporating morbidity and mortality ratios in the model. The information presented in this step of the risk assessment framework is actually both qualitative and quantitative in nature, as it reflects the information gathered in the previous steps including (but not limited to) the factors that affect the growth and survivability of the pathogen and the variability in the human response to exposure.

Computed risk estimates may provide a distribution of risk that represents a range of exposures (contamination of a food) or may be that presented as a point-estimate of risk that was calculated based on worst-case exposure scenarios to provide a risk estimate that is protective of populations, particularly vulnerable populations.

An ultimate goal of a risk assessment is to provide the foundation for science-based decision making in the regulatory arena. Risk managers can evaluate the information obtained from a risk assessment to determine the appropriateness for use. A risk manager may deem it appropriate to compare calculated point estimates of risk, or distributions of risk, to other such risk estimates for other microbes when considering regulatory options; however, this should be conducted with caution. Because the risk assessment process is dynamic and can be applied to many types of foods, microorganisms, and populations, making such comparisons may not be meaningful or appropriate.

Uncertainty, Variability, and Assumptions

A degree of uncertainty and variability exists in all risk assessments. Uncertainty can result because of lack of data or unknown parameters. Variability refers to parameters that are changeable or that have much heterogeneity. In either situation, this leads to parameters in a risk assessment that are either not precisely defined because of the lack of data or that are difficult to define because of variation. A risk characterization may include a sensitivity analysis that determines how a computed risk estimate is influenced by altering a value used in the risk assessment process. This allows a risk manager to identify which risk values (be it a dose–response parameter or an exposure parameter) contribute the greatest influence on the risk estimate.

Monte Carlo analysis is a tool that enables a risk assessor to address uncertainty and variability in a risk estimate (Burmaster and Anderson 1994). The risk assessor is able to develop a distribution of risk from point estimates. This is accomplished through many simulations that input random values for each exposure variable; for example, based on a defined probability distribution (i.e., realistic values) for each exposure within the risk assessment model. This results in a final probability distribution of the risk. This final probability distribution computed from many iterations (deterministic calculations), which sampled many distributions, reflects the possible input combinations. Monte Carlo analysis has improved the confidence and value placed on computer risk estimates within a risk characterization.

It is imperative that any assumptions made throughout the risk assessment process (because of lack of or limited data) be acknowledged and considered during the risk characterization. A risk manager needs to understand the assumptions that were incorporated throughout a risk assessment as well as how those assumptions influenced the risk characterization to appropriately interpret the risk estimate or estimates.

Escherichia coli *O157:H7 in Hamburger: A Case Study*

Many quantitative microbial risk assessments have been conducted for specific pathogens in specific food products. Marks et al. (1998) conducted a risk assessment of *E. coli* O157:H7 involving different exposure scenarios. The following four sections are a brief summary of their endeavor to provide an example of such an assessment. The aim of this risk assessment of *E. coli* O157:H7 was to determine the probability of adult illness per meal.

Hazard Identification

Escherichia coli O157:H7 was already identified as a hazard before this risk assessment. This type of *E. coli* is classified as an enterohemorrhagic *E. coli* (EHEC) from which infections can result in diarrhea (bloody and non-bloody), hemolytic uremic syndrome (HUS), and thrombotic thrombocytopenic purpura (TTP). This pathogen has been implicated in outbreaks in the United States involving ground beef, resulting in severe illnesses and deaths among children. *Escherichia coli* O157:H7 has also been associated with illness cases caused by exposure to contaminated apple juice and cider, lettuce, and recreational waters.

Exposure Assessment

For the exposure assessment step, the investigators created three scenarios to evaluate: a "baseline" scenario (assumes adequate refrigeration after grinding of hamburger), an "abuse" scenario (involves two periods of time at which adequate temperature was not maintained—one prior to cooking, one after cooking), and an "intervention" scenario (assumes imposed processing procedures reduced the occurrence and density of the pathogen in raw hamburger patties). Three different internal cooking temperatures were employed within each scenario to represent rare, medium, and well-done hamburgers. The investigators used data by Juneja et al. (1997) to model the change of microbial populations during cooking.

Estimates were made regarding the percentage occurrence of *E. coli* O157:H7 in 25-g portions of raw hamburger, the density of *E. coli* O157:H7 per gram in positive samples, and the number of *E. coli* O157:H7 microorganisms in a serving. The investigators relied on existing data for these estimations.

Dose–Response Assessment

Quantitative data regarding dose–response assessment for *E. coli* O157:H7 was not available. The investigators selected *Shigella* data (based on certain criteria related to genetics, pathogenesis, and transmission) to be used as a surrogate. The beta-Poisson (non-threshold) model was applied to determine the illness probability as well as a threshold model.

Risk Characterization

The investigators calculated mean, median, and 95% probability intervals for probability of illness for the three scenarios. The 95% probability intervals for the baseline scenario covered five orders of magnitude for the models evaluated. It was also observed that assuming even a small threshold significantly changed the calculated risk estimates. The authors acknowledged the assessment as an example of what could be conducted. The investigators concluded that some of the input parameters may not be representative and that more data were needed in some areas, including data regarding *E. coli* O157:H7 occurrence in hamburger meat and on the farm, the survivability of *E. coli* O157:H7 during cooking, and the ability to model disease severity.

Conclusions

Quantitative microbial risk assessment has many useful applications for the food industry. As described above, it can provide quantitative information regarding human health risks associated with pathogens, which will provide a science-based approach to formulating regulations. This interdisciplinary process initially helps define a (potential) problem whether it is directed at a particular microorganism or food product or food-processing practice. The framework provides a standardized process for addressing food production and processing procedures, which is particularly important when considering international trade. Perhaps more important, quantitative microbial risk assessment provides an opportunity for identifying information gaps that can then direct future research.

Traditionally, good manufacturing practices (GMPs) and hazard analysis critical control point systems (see Chapter 20) have effectively addressed microbial food-quality issues. However, they do not provide the proactive approach needed to combat evolving issues such as the adoption of new food production and processing practices, the emergence of foodborne pathogens, and the ever-changing global food market. Quantitative microbial risk assessment provides the approach necessary to predict the public health significance of such trends before an effect is observed.

References

Buchanan, R. L., Smith, J. L. and Long, W. 2000. Microbial risk assessment: dose-response relations and risk characterization. *International Journal of Food Microbiology* **58**:159–172.

Burmaster, D. E. and Anderson, P. D. 1994. Principles of good practice for the use of Monte Carlo techniques in human health and ecological risk assessment. *Risk Analysis* **14**:477–481.

Codex Alimentarius Commission. 1999. Principles and Guidelines for the Conduct of a Microbiological Risk Assessment. CAC/GL-30. Rome: Food and Agriculture Organization.

Coleman, M. and Marks, H. 1998. Topics in dose-response modeling. *Journal of Food Protection* **61**:1550–1559.

Fazil, A. M. 1996. Quantitative risk assessment model for *Salmonella*. Master's thesis, Environmental Studies Institute, Drexel University, Philadelphia, Penn.

Gale, P. 1998. Quantitative BSE risk assessment: relating exposures to risk. *Letters in Applied Microbiology* **27**:239–242.

Haas, C. N. 1983. Estimation of risk due to the doses of microorganisms: a comparison of alternative methodologies. *American Journal of Epidemiology* **188**:573–582.

Haas, C. N. 2002. Progress and data gaps in quantitative microbial risk assessment. *Water Science and Technology* **46**:277–284.

Haas, C. N., Crockett, C. S., Rose, J. B., Gerba, C. P. and Fazil, A. M. 1996. Assessing the risk posed by oocysts in drinking water. *Journal of the American Water Works Association* **88**:131–136.

Haas, C. N., Rose, J. B. and Gerba, C. P. (Eds.). 1999a. Quantitative Microbial Risk Assessment. New York: Wiley.

Haas, C. N., Rose, J. B., Gerba, C. P. and Regli, S. 1993. Risk assessment of virus in drinking water. *Risk Analysis* **13**:545–552.

Haas, C. N., Thayyar-Madabusi, A., Rose, J. B. and Gerba, C. P. 1999b. Development and validation of dose-response relationship for *Listeria monocytogenes*. *Quantitative Microbiology* **1**:89–102.

Haas, C. N., Thayyar-Madabusi, A., Rose, J. B. and Gerba, C. P. 2000. Development of a dose-response relationship for *Escherichia coli* O157:H7. *International Journal of Food Microbiology* **1**:1–7.

Jaykus, L.-A. 1996. The application of quantitative risk assessment to microbial food safety risks. *Critical Reviews in Microbiology* **22**:279–293.

Juneja, V. K., Snyder, O. P., Williams, A. C. and Marmer, B. S. 1997. Thermal destruction of *Escherichia coli* O157:H7 in hamburger. *Journal of Food Protection* **10**:1163–1166.

Lammerding, A. M. and Paoli, G. M. 1997. Quantitative risk assessment: an emerging tool for emerging foodborne pathogens. *Emerging Infectious Diseases* **3**:483–487.

Marks, H. M., Coleman, M. E., Lin, C.-T. J. and Roberts, T. 1998. Topics in microbial risk assessment: dynamic flow tree process. *Risk Analysis* **18**:309–328.

Mead, P. S., Slutsker, L., Dietz, V., McCaig, L. F., Bresee, J. S., Shapiro, C., Griffin, P. M. and Tauxe, R. V. 1999. Food-related illness and death in the United States. *Emerging Infectious Diseases* **5**:607–625.

National Research Council. 1983. Risk Assessment in the Federal Government: Managing the Process. Washington, D.C.: National Academy Press.

National Research Council. 1994. Science and Judgment in Risk Assessment. Washington, D.C.: National Academy Press.

Regli, S., Rose, J. B., Haas, C. N. and Gerba, C. P. 1991. Modeling the risk from *Giardia* and viruses in drinking water. *Journal of the American Water Works Association* **83**:76–84.

Rose, J. B., Haas, C. N. and Regli, S. 1991. Risk assessment and control of waterborne giardiasis. *American Journal of Public Health* **81**:709–713.

Rose, J. B., Haas, C. N. and Gerba, C. P. 1995. Linking microbiological criteria for foods with quantitative risk assessment. *Journal of Food Safety* **15**:121–132.

Rose, J. B. and Sobsey, M. D. 1993. Quantitative risk assessment for viral contamination of shellfish and coastal waters. *Journal of Food Protection* **56**:1042–1050.

Slifko, T. R., Huffman, D. E., Bertrand, D., Owens, J. H., Jakubowski, W., Haas, C. N. and Rose, J. B. 2002. Comparison of animal infectivity and cell culture systems for evaluation of *Cryptosporidium parvum* oocysts. *Experimental Parasitology* **101**:97–106.

Slifko, T. A., Smith, H. V., Rose, J. B. 2000. Emerging parasite zoonoses associated with water and food. *International Journal of Parasitology* **30**:1379–1393.

Teunis, P. F. and Havelaar, A. H. 2000. The Beta Poisson dose-response model is not a single-hit model. *Risk Analysis* **20**:513–520.

Walls, I. and Scott, V. N. 1997. Use of predictive microbiology in microbial food safety risk assessment. *International Journal of Food Microbiology* **36**:97–102.

32 How to Manage Risk—The Way Forward

Ewen C. D. Todd

Introduction

Managing risk has always been done on an intuitive basis, with recommendations being made by individuals or *ad hoc* teams. With an increasing demand for science-based decisions, there is a need for a more structured approach to management. In addition, within this more complex environment, it is essential to establish clear risk management goals along with agreement on the definitions used. This will avoid misunderstandings on what is being requested and who is involved at the different stages in the management process. Different agencies within a country and between countries may use similar but not identical definitions. The Codex Alimentarius Commission (CAC 2002) recommends definitions for international use. The following chapter material is derived from a variety of sources as well as from the author's thoughts (Food and Agriculture Organization [FAO] 1997, the Presidential/Congressional Commission on Risk Assessment and Risk Management 1997, CAC 1999, World Health Organization [WHO] 1999, International Commission on Microbiological Specifications for Foods 2002, Krewski et al. 2002; Health Canada, unpublished data).

Risk Analysis Framework

There should be an overall framework that helps managers make decisions. This is important for regulatory authorities to better balance risks and benefits across a society in an acceptable way and to make the decision-making process more explicit by being more open and consistent (Krewski et al. 2002).

The prime objective of any food safety management strategy is to maintain or improve the safety of the public from exposure to contaminants in food. Several components are key to achieving this goal, including focusing on the specific problem or issue; involving all interested parties (stakeholders) at all stages in the process; using a collaborative and integrated approach for managers, assessors, and stakeholders to work together for increased efficiency, consistency of decisions, avoidance of duplication of effort, and identification of gaps in the data and policies; being creative and flexible in responding to different issues; having clearly defined roles and responsibilities to ensure that all involved in the process know what is expected and what commitments have been made; making the best use of sound science for an evidence-based approach to decision making; making the process thorough but transparent; considering a precautionary approach to take timely and appropriately preventative measures, even if there is not enough information to clearly identify and control the risks; and having a communications strategy prepared before final decisions are made.

The risk assessment and risk management teams should play a role in issue identification and in the need to exchange information throughout the entire decision-making process, but their roles are distinct. The risk assessment team assesses risk based on the science and identifies potential risk management options or strategies that are related to the level of risk considering its probability and severity. The risk management team considers the results of risk assessments, together with economic, social, political, and ethical considerations, and uses this information to make risk management decisions. Thus, their roles are different enough that the managers should not affect the risk assessment process although it is necessary to maintain some flexibility throughout the decision-making process.

Structure of the Framework

A risk management decision-making framework requires more than conducting a risk assessment and communicating the results. The Framework for Environmental Health Risk Management (Presidential/Congressional Commission on Risk Assessment and Risk Management 1997) consists of a series of six interconnected and interrelated steps. These include the following:

(1)   Defining the problem and putting it into proper context: characterization of a food-borne disease problem or a potential problem caused by microorganisms or their toxins (this also would apply to chemical hazards); and identification of the appropriate risk managers with the authority and the involvement of stakeholders.

(2)   Analyzing the risks associated with the problem in context: this is the risk assessment stage with hazard identification, dose response assessment, exposure assessment, and risk characterization steps; risk perceptions and local or regional information need to be considered if the problem is specific to one area; and this process should generate enough information for managers to make decisions.

(3)   Examining options for addressing the risks: these are considering regulatory or other options for managing the risk; the feasibility to carry out these options; analyzes for cost-effectiveness and cost distributions among the affected parties; and determination of any possible new risks that might arise from the proposed management actions.

(4)   Make decisions about which options to implement: decisions are based on the best scientific, economic, and other technical information; actions take into account all the sources of the hazard and effects of the problem to be solved; management options are chosen that are feasible and that have benefits that are reasonably related to costs; prioritization of options preventing or reducing risks not just controlling them; alternatives to the command and control approach to regulation are important to consider; political, social, legal, and cultural considerations have to be carefully weighed; and incentives for innovation, evaluation, and research are long-term options.

(5)   Take actions to implement the decisions: decision implementation is carried out by one group of managers in one agency, but they are required to coordinate decision implementation with other agencies; and compliance will be better if all the stakeholders agree with the decision-making process and the actions taken.

(6)   Conduct an evaluation of the action's results: this should determine whether the decisions resulted in the intended actions and whether the cost–benefit estimates were accurate; what information gaps prevented full implementation of the actions; whether new information has emerged during the process that would be helpful to improve the decision-making process; whether the risk management framework and policy were effective or needed changing; how stakeholders reacted to the process and

how valuable was their input; and what overall lessons were learned to help other risk management decisions.

If new information becomes available, the framework is revisited and may change the need for or the nature of the risk management. Health Canada has based its framework on this Presidential/Congressional Commission approach, but it also uses some other steps, such as cost-benefit analysis, to assess risks along with possible benefits or negative effects.

It is not the purpose of this chapter to go into detail concerning the whole risk framework but to focus on the assessors and managers and to see how they address a problem, decide on what assessment step to take, and integrate this information into the management decision-making process. Such a framework involves interested and affected parties throughout the process. Each step involves a decision point as to whether to proceed to the next step, revisit a previous step, or end the process. The process is flexible in that one may move back and forth between steps or revisit steps based on available information.

Risk Evaluation

A risk evaluation is the formative stage of risk management. This involves several steps, such as identifying a food safety problem, establishing a risk profile, ranking the hazard for risk assessment and risk management priority, establishing the scope and risk assessment policy, commissioning the risk assessment, and considerng the risk assessment results. At the outset of a microbiological risk management activity, the managers and assessors should be clearly identified. However, the urgency with which the work should be done, and the resources available, are the risk manager's responsibility; however, they are often beyond the control of the individual manager.

Identification of a Food Safety Problem

Risk managers take the initiative to identify a food safety problem, although assessors and others may alert managers to new or emerging issues. Most assessments so far have been in response to pressure for some kind of public health action to reduce illness or in response to a change in policy if major stakeholders see it as being ineffective. An example is *Listeria monocytogenes*, where a total zero-tolerance policy has been questioned both by industry within the United States and by other countries that import ready-to-eat (RTE) foods into the United States. The FDA has finalized its relative risk ranking for *L. monocytogenes* in RTE foods with additional exposure data (Food and Drug Administration, Center for Food Safety and Applied Nutrition, U.S. Department of Agriculture, Food Safety and Inspection Service, Centers for Disease Control and Prevention 2003.). In addition, FAO/WHO has initiated a quantitative risk assessment process for *L. monocytogenes* in four RTE foods in response to requests by member countries of the Codex Alimentarius Commission (WHO 1999, see International Issues). A major reason for both types of risk assessments is to improve ways of controlling the pathogen in RTE products without overburdening industry for foods where there is little or no history of illness.

Establishing a more proactive approach with assessments to manage potential problems is new to microbial issues. It is done for plant and animal diseases and pests so as to restrict or minimize the risks of diseases being transported across borders. The Harvard risk assessment for risks of products containing protein from bovine spongiform encephalopathy (BSE)-infected cattle coming into the United States (Cohen et al. 2001) is an example of preparing for an event that has not occurred but that potentially could happen in the next few years, as it has in other countries. These examples are as much trade issues as they are national public health concerns.

The primary goal of any risk management strategy must be to choose the best measures or options that ensure an appropriate level of health protection. Because measurable public health goals relating to a food safety problem have rarely been attempted, managers usually have to make choices on a case-by-case basis as they relate to a general policy. Risk management goals may be risk-related (e.g., to reduce the incidence of adverse health effects), may involve public values (e.g., protecting the most sensitive subpopulation), and may consider economic effects (e.g., achieving an acceptable level of health protection without causing loss of jobs). Risk management goals may also be determined by legislative requirements, policy, or national or international obligations, or they may be influenced by priorities that have previously been established or by priorities dictated by limited resources. Goals may be revised as new information is obtained. One of the first steps is normally compiling a risk profile of the problem.

Establishment of a Risk Profile

A risk profile is developed to gather preliminary information to guide further action and to clarify risk management questions and goals. For this process, the risk manager needs to involve a team, including risk assessors and other scientific specialists who have detailed knowledge of the specific problem, as well as of food safety issues in general. Some governments keep participation in the whole risk analysis process confined to government employees or contractors. Others bring in outside groups to represent a broader stakeholder interest. A risk profile is more than an accumulation of scientific data on agents and foods. It looks at how the food in question is produced and distributed, prepared, consumed, and regulated. It considers distribution of risk to the commodity producers and consumers and considers those who would benefit and those who would not if new management steps are taken. It looks at human health and cultural, esthetic, economic, and other values. It anticipates risk perceptions of the public for both current and future management programs. It notes any contractual, national, or international agreements that may affect any management decisions. It summarizes the description of the issue and recommends the next steps to be taken.

Hazard Ranking for Priority of Risk Assessment and Management

A risk profile should be documented in such a way that risk managers can use it to prioritize a food safety issue in relation to other issues. Ranking of issues depends on a number of criteria including the human disease burden and its costs, severity of illnesses arising from consumption of the agent, political or international pressure, stakeholder concerns, time elapsed since the problem was first identified, and public perceptions of the issue. The profile may look at one specific hazard or at a range of likely hazards expected in a particular food product. These hazards may be ranked on the basis of frequency and severity as to which one should be studied first. For instance, although many pathogens can grow in shell eggs, the main organism of concern is *Salmonella* (Todd 1996).

Risk Assessment Scope and Risk Assessment Policy Development

A result of a risk profile may be a recommendation that no further control is required. Another choice is strengthening an existing program, such as through stricter hygienic practices, or recommending an enhanced educational program without requiring any more data or an assessment. If an assessment is to be done, the scope, range, and policy of the risk assessment need to be determined in collaboration with the risk assessors and, preferably, stakeholders. The scope of the risk assessment is dependent on the reason for doing the assessment and the risk management goals that were identified. Traditionally, risk assessment has often been considered simply as a means to provide an estimate of risk of the hazard, but it also can estimate the relative value of different mitigation strategies in reducing risks to consumers. The scope of the risk assessment will define the parts of the food chain considered and will provide the necessary focus for the scientific work involved in risk assessment. It is important to have a risk assessment policy so that assessors know how they must work with managers. This policy is a series of steps written down that should be followed to give guidance to the assessment team. It is often generated as an assessment is being done, and modified as more experience is obtained for subsequent teams.

Commissioning the Risk Assessment

Detailed discussion of conducting a risk assessment lies outside the scope of this chapter, but there are four generally recognized steps: hazard identification (what can go wrong), exposure assessment (information on the likelihood and level of the pathogen being in the food along a production, processing, transportation, home storage, and cooking pathway ending with meal servings), dose–response assessment (probability of illness resulting from different doses in the food), and risk characterization (combining exposure and dose response assessment data to estimate the level of risk in a meal serving or in a population). From this, the acceptability of risk can be ascertained (management decision) and modeling of mitigation strategies required to reduce the risk (assessor action). Commissioning of a risk assessment implies agreement that this is a project worth supporting with resources until its completion. Full quantitative risk assessments take time to implement and should not be started if there are limited data and if results are required within a few days or weeks. The establishment of a risk assessment team is an important part of the commissioning, and its composition will vary with each assessment. The team should clearly understand the risk management goals and, after the scope of the assessment, prepare a plan with a timeline for its completion. Collecting data can take a long time, and modeling may be difficult if there are no preexisting models. As microbial risk assessments develop and become more standardized, however, there may be ways to use existing assessments and to substitute new databases or do partial studies, such as an exposure pathway from retail to consumer rather than all the way from environment or preharvest to the consumer.

The risk manager is responsible for a review process using external independent reviewers to ensure risk assessment validation. The review process should involve experts who can assess the technical aspects of the work, as well as those who understand food production, storage, cooking, consumption practices, and disease implications.

Consideration of Risk Assessment Results

Risk assessors must communicate risk estimates effectively to be able to evaluate different risk management options. These options may indicate that the current level of risk is acceptable or indicate the need to consider one or more interventions. An extreme situation would be to ban the food product altogether because no intervention strategy would be acceptable. Risk assessors must attempt to communicate the results of the assessment appropriately, for both scientific and non-scientific audiences. If a quantitative risk assessment has been done, the assumptions, uncertainties, and outcomes have to be made clear enough that risk managers and stakeholders can comprehend them. In the written report, much of the detail can be put into appendices. Visual presentations with varying degrees of detail may be useful for helping different audiences understand the issue and the results. How well the results fit into existing knowledge of the issue and the peer review results will give a degree of confidence in the risk assessment process. However, any unexpected results, such as estimating a 10-fold greater increase in the number of cases arising from infection by a foodborne pathogen than was expected, needs to be explored thoroughly before the assessment is released to the public. It is recognized that risk assessments are never complete, and as new data and new mathematical models become available, they should be incorporated into the assessment. If the new additions result in substantial changes to the outputs, there may be necessary modifications to an existing management action such as a recommended change in a regulation.

Implementation of the Management Option

Once one or more options are considered to be scientifically valid to bring the risk down to an acceptable level, a plan has to be introduced to actually achieve this. This may be in the form of a specific regulation or other options. Non-regulatory options include published advisories, economic and technological incentives, educational programs, or general improvements to hygiene such as voluntary hazard analysis and critical control point system, and upholding the current management strategy. A process has then to be set up to implement the strategy, and this is where stakeholder input is critical. If it does not happen before implementation, there will likely be challenges to the process after it becomes obvious to those most affected; that is, the industry or the consumer. Therefore, any major change in strategy to reduce pathogen levels in foods needs to be accepted by all sectors, which will take much consultation and time. The plan will have to incorporate measures and resources to bring about change (new equipment and training by industry and increased inspection by government). It is important to determine the pathogen baseline in the food of interest before changing a monitoring method. Continuing to monitor the food following the introduced change will determine how effective the new strategy has been in reducing the risk to an acceptable level. An independent audit process to ensure that the desired safety standards are being achieved may also be required. Cost–benefit analyses need to be done ahead of any changes to determine whether the management option is viable from an economic point of view. Any costs to industry may eventually lead to an increase in the price of the product, or to some industries stopping production of the product because of lower profit margins, meaning that the consumer has less choice than before. The burden to government for implementing regulatory policy, political fallout, and consumer education may also be unacceptably high.

Evaluation of the Risk Management Policy

Once a management decision has been made and the process set up to carry it out, an evaluation should be built into the overall management framework. This should be planned for a specific period of time such as 1 or 2 years after the announcement of the policy. This will show whether the goals were achieved; e.g., a reduction of a foodborne disease or a lower contamination of targeted food items. There may be regional differences and certain areas identified in which the policy could be made more effective. Unexpected implementation costs, reallocation of resources, or resistance by some stakeholder groups may have reduced the policy's effectiveness. It may also generate new information from an industry or consumer reaction that may either point to having a new assessment done or reconsider the management option selected. Thus, the correct management option may result from an iterative process encompassing many years, with improvements to reflect better approaches and changing demands by stakeholders.

International Issues

FAO, WHO, Office international des épizooties, and CAC have shown an increasing interest and commitment to cooperating in international food safety issues, although each has its own mandate within the larger food safety area. The future is not clear, however, because these agencies or commissions have not been given enough resources to move forward in a major way to reduce foodborne and waterborne illness. They have only been able to hold meetings and consultations and organize small projects within some countries. A good example is the Joint FAO/WHO Expert Meetings on Microbiological Risk Assessment for putting together a number of pathogen-food risk assessments by world experts, with the summaries to be delivered to the CAC (WHO 1999). It is unclear how these minimal actions are going to help developing countries understand or use the assessments (which must be modified for each specific country) to improve management strategies. These assessments are complex enough that only a very few people would be able to grasp their complexities and tailor-make them for the needs in each developing country. International aid agencies, such as World Bank and USAID, could fund pilot projects to see how these strategies could be used effectively. The issue, however, is more than adapting existing risk assessments; it is putting together a comprehensive program that deals with both reducing domestic foodborne disease and producing fewer contaminated food products for overseas trade. These are rarely considered together within one governmental department and rarely have sufficient political input or have accessible resources because of perceived higher priorities.

A recent evaluation of the CAC program identified improvements in four key areas (FAO/WHO 2002b): greater speed in Codex and expert scientific advice especially for risk assessments; increased inclusiveness of developing-member countries in the Codex standard development process; Codex standards, which are of greater usefulness to Member Nations in terms of relevance to their needs and timeliness for consumer protection; and more effective capacity building for development of national food control systems. Recommendations in the area of risk are substantial:

(1) In Codex there should be a clearer distinction between risk assessment and risk management; Codex committees should concentrate on risk management and not confuse it with assessment; questions of assessment should be referred to scientific expert committees or *ad hoc* consultations.
(2) In light of the growing importance of microbiological hazards, Joint FAO/WHO Expert Meetings on Microbiological Risk Assessment should be ratified as a permanent committee and resources allocated to increase its output.
(3) The increased funding of risk assessment is a top priority.
(4) A high priority for WHO and FAO is to support data collection covering a much wider range of diets and production processes, including building essential capacity.

Future Directions at National and International Levels

Microbiological risk evaluation requires a more sophisticated approach than simply determining an acceptable level of risk. Many terms and concepts are currently used to describe the decision-making process of risk managers in the risk evaluation step (FAO/WHO 2002a, International Commission on Microbiological Specifications for Foods 2002, WHO 2000). These terms include: As Low As Reasonably Achievable (ALARA), Tolerable Level of Risk, Appropriate Level of (Sanitary or Phytosanitary) Protection (ALOP), and Food Safety Objectives (FSOs).

The ALARA approach, which balances risks and benefits, lies somewhere between precautionary principle and a program built entirely on the science of prevention and assumption of worst-case scenarios. ALARA divides risk into three categories—acceptable, acceptable only if a compensatory benefit is available, and unacceptable—in contrast to a "bright-line" approach (above which the risk is unacceptable and below which the risk is acceptable). Tolerable Level of Risk is similar to ALARA, and some managers prefer the word tolerable to acceptable, as no risk is really acceptable but, rather, one they are willing to live with until better ways of managing risks are found. The Proposed Draft Principles and Guidelines for the Conduct of Microbiological Risk Management under development in the Codex Committee on Food Hygiene refer to tolerable risk that is determined primarily by human health considerations (CAC 2001). This approach assumes a bright-line level of tolerable risk, unlike ALARA.

The agreement on the application of sanitary and phytosanitary measures (SPS) Agreement of the World Trade Organization refers to the ALOP, which is defined in the agreement as the level of protection deemed appropriate by the member establishing a sanitary or phytosanitary measure to protect human, animal, or plant life or health within its territory (SPS agreement; World Trade Organization 1994). This approach recognizes that there will be differences between countries but assumes there will be agreement within a country on the ALOP. It also assumes that there is a single level (a bright line) above which the risk is unacceptable. An ALOP could be reduction of an enteric disease by a certain percentage in a certain number of years; for example, 50% reduction of listeriosis cases per 100,000 people in a country from 2000 to 2010. This is measurable, but it is unclear what happens if the goal is not met. Presumably goals are adjusted along the timeframe to better reflect progress and respond to roadblocks.

FSO is a term used in Codex documents, although its final definition remains to be agreed on by members. The International Commission on Microbiological Specifications for Foods (2002) defines a FSO as a statement of the frequency or maximum concentration of a microbiological hazard in a food considered acceptable for consumer protection. An ALOP is an overall public health goal. FSOs are related to ALOPs by translating them into specific objectives that can be carried out by the food industry. FSOs are expressed as a maximum concentration or probability of occurrence of a pathogen, toxin, or metabolite in a food. The availability of quantitative information on risks (e.g., dose response data) allows the identification of a FSO. Once there is agreement what a FSO should be, either for a domestic market or for trade, criteria to achieve this can be set. Process criteria have been in place a long time; for example, for canning or pasteurization. Thus, a time/temperature process criterion could be set to achieve the desired kill step to satisfy a FSO. However, this does not give the industry any options to implement improved or less expensive approaches for control. Ideally, a FSO allows industries to use different technologies to achieve the same level of consumer protection because as long as the objective is achieved, the means of getting there is acceptable. This approach is consistent with the SPS agreement for harmonization of microbiological criteria for foods in international trade with equivalency of food standards. This is best achieved with a performance criterion, such as a reduction in the number of a pathogen per gram, or prevention of any increase in numbers. Because a FSO is set at the consumer stage, it has to allow for some consumer practices; for instance, use of an improper storage temperature or a heating step. Thus, the performance criterion will be more stringent than the FSO if growth in the stored product is expected, but less stringent if the product is to be cooked.

Conclusion: Future Areas for Improving Management Practices

Managing risk will continue to evolve as more agencies attempt new approaches to achieve safer food products. At the same time, there will be attempts to harmonize these. In principle this can occur at the international level, but specific management steps will still be independent in the immediate future, depending on the problem and expertise available. Some areas in which improvement can be made are listed here:

(1) Early detection and surveillance of foodborne disease to speed up responses to outbreaks and improving risk assessments; e.g., next generation of FoodNet and Pulsenet and with expansion to Europe, Canada, and Mexico; and understanding antibiotic and biocide resistance patterns to predict emerging problems in food animals and humans.

(2) Identification and characterization of emerging pathogens; e.g., parasites, viruses, and opportunistic bacterial pathogens, as well as prions, for susceptible populations, and changing profiles of antimicrobial resistance in bacteria.

(3) Characterization and support of the increasing numbers of those who are susceptible to foodborne disease, including the normal aging population as well as those who have underlying conditions.

(4) Encouragement of research to support food safety by using both cooperative agreements and competitive grants to address data gaps in microbial food safety.

(5) Development of quantitative and qualitative risk assessment models for different management purposes.

(6) Focus on better ways to integrate risk assessment, risk management, and risk communication to produce timely recommendations for the specific problem areas.

(7) Creation of a better compliance environment by more effectively coordinating the various agencies (federal, state/province/local) responsible for food safety.

(8) Improvement of food worker behaviors and food preparation practices that directly relate to foodborne illnesses in retail food/foodservice establishments.

(9) Improvement of public educational programs by building on existing campaigns, more effective labeling, coordination of different communication media with the same types of messages, tailor-making the kind of information to each audience, and understanding risk perception and ways of changing opinions.

(10) More monitoring of imported food by targeting foods of greatest concern both in the importing and exporting countries.

(11) The encouragement and monitoring of new food processing and packaging technologies to reduce risks.

(12) The understanding and simplification of food laws and regulations both within a country and between trading nations.

(13) Partnering with academia in multidisciplinary projects to understand and reduce risks of foodborne disease.

(14) More Member Country support of international harmonization for managing risk and improving trade.

(15) Adapting existing and future management and assessment approaches for food safety to respond to real and potential bioterrorism attacks using food as a vehicle.

References

Codex Alimentarius Commission. 1999. Principles and Guidelines for the Conduct of Microbiological Risk Assessment, Joint FAO/WHO Food Standards Programme (CAC/GL 30). Rome: Food and Agriculture Organization.

Codex Alimentarius Commission. 2001. Codex Committee on Food Hygiene. Proposed draft principles and guidelines for the conduct of microbiological risk management, CX/FH 01/7 (Step 3 of the Procedure). Rome: Food and Agriculture Organization.

Codex Alimentarius Commission. 2002. Joint FAO/WHO Food Standards Programme Committee on General Principles, 17th Session, Paris, France, April 17–19, 2002, Proposed Draft Working Principles for Risk Analysis, CX/GP 02/3. Rome: Food and Agriculture Organization.

Cohen, J. T., Duggar, K., Gray, G. M., Kreindel, S., Abdelrahman, H., HabteMariam, T., Oryang, D. and Tameru, B. 2001. Evaluation for the Potential for Bovine Spongiform Encephalopathy Harvard Center for Risk Analysis. Harvard School of Public Health and College of Veterinary Medicine, Tuskegee University, Tuskegee, Ala., November 26. Available at: http://www.hcra.harvard.edu/pdf/madcow_report.pdf.

Food and Agriculture Organization. 1997. Risk Management and Food Safety Report of a Joint FAO/WHO Expert Consultation. FAO Food and Nutrition Paper No. 65. Rome: Food and Agriculture Organization.

Food and Agriculture Organization/World Health Organization. 2002a. Federal Ministry for Consumer Protection, Food and Agriculture, Germany. Principles and Guidelines For Incorporating Microbiological Risk Assessment in the Development of Food Safety Standards, Guidelines and Related Texts. Report of a Joint FAO/WHO Consultation, Kiel, Germany, March 18–22, 2002. Rome: Food and Agriculture Organization.

Food and Agriculture Organization/World Health Organization. 2002b. Report of the evaluation of the Codex Alimentarius and other FAO and WHO food standards work. Available at: http://www.fao.org/docrep/meeting/005/y7871e/y7871e00.htm.

Food and Drug Administration, Center for Food Safety and Applied Nutrition, U.S. Department of Agriculture, Food Safety and Inspection Service, Centers for Disease Control and Prevention. 2003. Draft assessment of the relative risk to public health from foodborne *Listeria monocytogenes* among selected categories of ready-to-eat foods. Available at: http://www.foodsafety.gov/~dms/lmrz-toc.html.

International Commission on Microbiological Specifications for Foods. 2002. Microorganisms in Foods 7 Microbiological Testing in Food Safety Management. New York: Kluwer Academic/Plenum.

Krewski, D., Balbus, J., Butler-Jones, D., Haas, C., Isaac-Renton, J., Roberts, K. J. and Sinclair, M. 2002. Managing health risks from drinking water—a report to the Walkerton Inquiry. *Journal of Toxicology and Environmental Health Part A* **65**:1635–1823.

Presidential/Congressional Commission on Risk Assessment and Risk Management. 1997. The Framework for Environmental Health Risk Management, Vol. 1. Washington, D.C.: RiskWorld. Available at: http://www.riskworld.com/nreports/1997/risk-rpt/html/epajana.htm.

Todd, E. C. D. 1996. Risk assessment of cracked eggs in Canada. *International Journal of Food Microbiology* **30**:125–143.

World Health Organization. 1999. Risk Assessment of Microbiological Hazards in Foods, Report of a Joint FAO/WHO Expert Consultation, Geneva, Switzerland, March 15–19, 1999, WHO/SDE/PHE/FOS/99.5. Geneva: World Health Organization.

World Health Organization. 2000. Interaction between assessors and managers of microbiological hazards in foods, WHO/SDE/PHE/FOS/007 Report of Expert Consultation, Kiel, Germany, March 21–23. Geneva: World Health Organization. Available at: http://www.who.int/fsf/mbriskassess/ InteractionConsultationinKiel/Kiel_report.pdf.

World Trade Organization. 1994. The WTO Agreement on the Application of Sanitary and Phytosanitary Measures (SPS Agreement). Available at: http://www.wto.org/english/tratop_e/sps_e/spsagr_e.htm.

Afterword

Agricultural Biosecurity: An Important Component of Homeland Security

Neville P. Clarke

The Threat

Federal and state governments now clearly recognize and are reacting to the threat of deliberate acts of terrorism by rogue nations, terrorist organizations, or individuals. The agents of terrorism may be drawn from the technology that contributes to weapons of mass destruction—radiological, chemical, and biological. The capacity to use these materials is exemplified by the discovery of the major offensive biological weapons program that had been underway in the Soviet Union from 1969 onward, despite the Soviets' having signed the international treaty banning such development. Before and after the collapse of the former Soviet Union, rogue nations and terrorist groups were and still are actively involved in developing agents for bioterrorism. The provision of biological agents by rogue states to terrorist groups either intentionally or otherwise is one plausible mechanism for their gaining access to these materials.

In the former Soviet Union, the efforts of 10,000 of the 60,000 workers engaged in offensive biological weapons development were directed at agriculture. Multiple organisms affecting both livestock and crops were collected—their potency enhanced through biotechnology—and methods for delivery in various munitions were developed. Before and after the former Soviet Union program, there have been plans and attempts to use biological agents against food and agriculture.

The ability and will of terrorist groups to independently develop and use chemical or biological agents against civilian populations was well demonstrated by the nerve gas attack in Tokyo's subways by the Aum Shinriyko in 1995 and by the earlier use in 1984 of *Salmonella* organisms against local citizens in restaurants in Oregon by the Rajneeshees cult. In the latter instance, the purpose of the attack was political—to alter election results in a small community. The use of the postal service to distribute anthrax is illustrative of the overall vulnerability of the United States to such attacks, whether by individuals or by organized groups. It is clearly not necessary to postulate the use of missiles and bombs as delivery mechanisms. Simpler delivery methods and means are readily available, especially for attacks against the agricultural sector.

The catastrophic events of September 11, 2001, increased public awareness of the threat of and will to engage in acts of terrorism and set in motion a set of broad national and state initiatives aimed at homeland defense to prevent, protect, and mitigate such attacks. In the case of bioterrorism, the focus was initially preponderantly on people. Now, increasing concern and effort is being directed to food and agriculture.

Vulnerability

The Texas Agricultural and Natural Resources Summit Initiative conducted a summit meeting on "Biosecurity: Safeguarding Our Agriculture and Food Supply" in May 2002 in Austin, TX. Some 130 leaders from the food and agricultural sector, consumer groups, and state and federal agencies were involved. They discussed the threat, the current state–federal capabilities and responsibilities to protect the food and agriculture system, the related issues facing Texas agricultural and food industries, and recommendations on enhancing preparedness through improved policies, procedures, science, and communication. The report provides a sobering assessment of the current vulnerability of the food and agriculture system. It particularly highlights the vulnerability of Texas agriculture because of the more than 1,200 miles of international border, the high volume of agricultural goods flowing through its several major ports, the diversity and complexity of Texas agriculture, and the particular vulnerability related to intensive and concentrated poultry, cattle, and swine operations in the state. It included concerns about the substantial processing and manufacturing industries, all of which are currently highly vulnerable to intentional acts of bioterrorism or are even vulnerable to unintentional or accidental introductions of exotic disease into the state.

Both livestock and crop species are highly susceptible to exotic diseases that have been successfully excluded from the United States over the last 50 years. Breeding programs to achieve improved yield and quality, especially in crops, have narrowed the genetic base of these species and made them more vulnerable to exotic diseases. The presence of wildlife that are also susceptible to exotic diseases, intermingled with livestock in many parts of Texas, enhances the vulnerability and complicates the prevention and mitigation following a successful attack.

The ease of access by terrorists to multiple organisms capable of creating rapidly spreading disease in crops and livestock and the relative ease of dissemination in the U.S. food and agriculture system suggests that an attack would be simple and relatively inexpensive and that the consequences could be very substantial.

The existing systems for protecting against and reacting to accidental introduction and outbreaks of exotic disease have worked well in the past. New emergency management procedures and application of modern biology and information technology are bringing a new generation of capacity to deal more effectively with both the accidental introduction and deliberate use of biological agents that create foodborne illness and threaten animal and plant health. These systems, based on the concept of dual use for either situation, are coming on line. However, it is recognized that an intentional attack may involve multiple locations, rather than a single point, and organisms that may be more virulent than those occurring naturally and that may also spread more rapidly. Multiple organisms could be used to confound and complicate the detection, identification, and response. Existing response capacities could likely be overwhelmed in the event of an attack involving such multiple foci. This means that a "surge capacity" will be needed to apply the new capacity for emergency response in case of broader introductions.

Attacks on the food chain after harvest would tend to threaten public health, not crops or animals. The postharvest industry has already aggressively pursued steps to reduce their vulnerability to intentional introduction of biological agents or contaminants after harvest. Despite this, the overall vulnerabilities remain very substantial.

Consequences

Attacks on livestock and crops would primarily have economic consequences. Attacks involving the food chain after harvest would involve human pathogens and have their primary effect on public health. In the case of plant or animal diseases that also affect humans, preharvest attacks would also have human health implications. Attack on any part of the food and agriculture system would have immediate ramifications on all other parts. Economic destabilization would result from the chain of events, beginning with loss of product, followed by loss of markets, then hitting the industries involved in providing inputs and processing products. The price of food could increase very substantially, especially on a regional basis.

The cost of containment and destruction of crops and animals, using present strategies, would be staggering, as would the cost of compensation for destroyed property. The public reaction to the current draconian methods of containment would be very substantial. Disposal of a large number of carcasses of destroyed animals poses very serious environmental concerns and costs. The magnitude of the economic impact is illustrated by the recent outbreak of foot-and-mouth disease in the United Kingdom, where compensation paid to farmers for slaughter of their animals amounted to more than $14 billion. The multiplier effect of this on the total economy of the country was at least a factor of five. The total industry affected in the United Kingdom is small relative to even the size of the industry in just the state of Texas.

The threat to human health would have consequences far beyond the economic impact. In addition to the potential loss of life and sickness, loss of public confidence in the safety of the food supply—and even the government's ability to govern—would be a serious problem, as demonstrated by public reaction after the recent outbreaks of animal disease in Europe. In extreme situations, this could result in substantial social instability and panic if human lives are threatened or lost.

What's the likelihood that any of this will actually occur? Our national leadership obviously believes that terrorism is going to be around for a long time—hence the major emphasis on homeland security. Other forms of terrorism may be more likely. Explosions, chemical or nuclear materials or cyberterrorism are viable candidates. However, bioterrorism is regarded by experts as being a credible threat—some refer to it as "a poor man's nuke"—and most people believe that bioterrorism directed at the food, agriculture, and water systems of this country is achievable and attractive to terrorists because of the relatively high vulnerability of the sector and the consequences that substantially contribute to the achievement of the objectives of those who would create terror. If one balances the combination of the probability of attack and its consequences, if successful, agricultural bioterrorism seems to be a threat that cannot be disregarded. Because of its geographic position, the size of its agriculture and its position in international and national trade, Texas is a state that requires special attention.

Index